T0181861

Lecture Notes in Computer Science 13639

More information about this series at https://link.springer.com/bookseries/558

Utkarsh Porwal · Alicia Fornés ·
Faisal Shafait (Eds.)

Frontiers in
Handwriting Recognition

18th International Conference, ICFHR 2022
Hyderabad, India, December 4–7, 2022
Proceedings

 Springer

Editors
Utkarsh Porwal
Walmart Inc.
Hoboken, NJ, USA

Alicia Fornés ⓘ
Universitat Autònoma de Barcelona
Barcelona, Spain

Faisal Shafait ⓘ
National University of Sciences
and Technology (NUST)
Islamabad, Pakistan

ISSN 0302-9743 ISSN 1611-3349 (electronic)
Lecture Notes in Computer Science
ISBN 978-3-031-21647-3 ISBN 978-3-031-21648-0 (eBook)
https://doi.org/10.1007/978-3-031-21648-0

This Springer imprint is published by the registered company Springer Nature Switzerland AG
The registered company address is: Gewerbestrasse 11, 6330 Cham, Switzerland

Preface

It is an immense pleasure to present this volume of the proceedings for the 18th International Conference on Frontiers in Handwriting Recognition (ICFHR 2022) held on December 4–7, 2022, in Hyderabad, India. ICFHR is a flagship conference series that started with the first ICFHR being held in Montreal, Canada, in 2008. Subsequent ICFHRs were organized in Kolkata, India (2010), Bari, Italy (2012), Crete, Greece (2014), Shenzhen, China (2016), and Niagara Falls, USA (2018). ICFHR 2020 was planned to be held in the city of Dortmund, Germany; however, due to the COVID-19 pandemic, it was realized as a fully virtual conference. This year ICFHR was held in a physical mode in the city of Hyderabad, India. It was hosted by the International Institute of Information Technology, Hyderabad. Hyderabad is a modern city and a hub of numerous IT firms. This offered an excellent opportunity for the ICFHR community to get connected to industrial research activities in the broad area of handwriting and document analysis.

ICFHR is a major event endorsed by the International Association for Pattern Recognition (IAPR) and it's Technical Committee TC-11 (Reading Systems). The call for papers for ICFHR 2022 resulted in 61 high-quality paper submissions by researchers from across the globe. Each paper underwent a double-blind review process and was reviewed by at least three researchers in the respective field. Based on the decisions on the reviews, we accepted 37 papers for oral presentation. As a result, this proceedings covers a broad spectrum of topics related to online and offline handwriting analysis and recognition. Popular topics of interest included online signature verification, historical document analysis, and automatic handwriting generation. The majority of the papers focused on the evaluation of various attention-based methods for solving classic handwriting recognition and analysis problems. Furthermore, sophisticated end-to-end architectures for specific tasks were also proposed. There were some papers that presented handwriting datasets for regional languages like Gurmukhi, Urdu, and Chinese. Mathematical expressions, musical scores, and layouts of the handwritten documents were also addressed in some of the papers.

The success of a large event like ICFHR depends on the support of many institutions and individuals. Firstly, we would like to gratefully acknowledge the financial assistance of our sponsors. We are also grateful to our General Chairs C.V. Jawahar, Apostolos Antonacopoulos, and Venu Govindaraju for their valuable guidance and support. Our special thanks go to the Workshop Chairs Anand Mishra and Veronica Romera Gomez and the Competition Chairs Maroua Mehri and Rohit Saluja for their active support. We appreciate the efforts of our Publication Chairs Imran Siddiqi and Momina Moetesum. We also extend our cordial appreciation to all members of the local organizing committee for their untiring efforts in making this event a success.

Our sincere thanks go to the Program Committee members for their time and efforts in reviewing the papers and providing constructive feedback to the authors. We are also thankful to the keynote speakers, the authors, and the participants of the conference for their participation.

We hope that you enjoy reading this proceedings book and are looking forward to attending the next ICFHR.

December 2022

Alicia Fornés
Faisal Shafait
Utkarsh Porwal

Organization

General Chairs

Apostolos Antonacopoulos	University of Salford, UK
C. V. Jawahar	IIIT Hyderabad, India
Veno Govindaraju	University at Buffalo, USA

Program Chairs

Alicia Fornes	Universitat Autònoma de Barcelona, Spain
Faisal Shafait	National University of Sciences and Technology (NUST), Pakistan
Utkarsh Porwal	Walmart Inc., USA

Organizing Chairs

Anoop M. Namboodiri	IIIT Hyderabad, India
Manish Gangwar	ISB Hyderabad, India
Ravi Kiran Sarvadevabhatla	IIIT Hyderabad, India
Shruti Mantri	ISB Hyderabad, India

Publication Chairs

Imran Siddiqi	Bahria University, Pakistan
Momina Moetesum	Bahria University, Pakistan

Workshop Chairs

Anand Mishra	IIT Jodhpur, India
Veronica Romera Gomez	Universitat de València, Spain

Competition Chairs

Maroua Mehri	National Engineering School of Sousse, Tunisia
Rohit Saluja	IIIT Hyderabad, India

Program Committee

Alessandro Koerich	University of Québec, Montréal, Canada
Andreas Fischer	University of Applied Sciences and Arts, Switzerland
Angelo Marcelli	Università di Salerno, Italy
Anurag Bhardwaj	eBay Research Labs, USA
Aurélie Lemaitre	Université de Rennes, France
Basilis Gatos	Institute of Informatics and Telecommunications, Greece
Brian Kenji Iwana	Kyushu University, Japan
Carlos David Martinez Hinarejos	Universitat Politècnica de València, Spain
Chawki Djeddi	Larbi Tebessi University, France
Cheng-Lin Liu	Institute of Automation, Chinese Academy of Sciences, China
Christopher Kermorvant	TEKLIA, France
Christopher Tensmeyer	Adobe Research, USA
Cristina Carmona-Duarte	Universidad de Las Palmas de Gran Canaria, Spain
Daniel Lopresti	Lehigh University, USA
Ekta Vats	Uppsala University, Sweden
Eric Anquetil	IRISA, INSA Rennes, France
Fei Yin	Institute of Automation, Chinese Academy of Sciences, China
Florian Kleber	TU Wien, Austria
Foteini Simistira Liwicki	Luleå University of Technology, Sweden
Gernot A. Fink	TU Dortmund, Germany
Haikal El Abed	German Agency for International Cooperation (GIZ), Germany
Harold Mouchère	Universite de Nantes, France
Ioannis Pratikakis	Democritus University of Thrace, Greece
Jean-Christophe Burie	La Rochelle Université, France
Joan Andreu Sanchez	Universitat Politècnica de València, Spain
Josep Llados	Universitat Autònoma de Barcelona, Spain
Julian Fierrez	Universidad Autonoma de Madrid, Spain
Kenny Davila	Universidad Tecnológica Centroamericana, Honduras
Laurent Heutte	Université de Rouen, France
Lianwen Jin	South China University of Technology, China
Luiz Oliveira	Universidade Federal do Paraná, Brazil
Marc-Peter Schambach	Siemens Logistics GmbH, Germany
Maroua Mehri	Université de Sousse, Tunisia

Contents

Symbol and Graphics Recognition

Handwriting Recognition and Understanding

Handwriting Datasets and Synthetic Handwriting Generation

Document Analysis and Processing

Historical Document Processing

A Few Shot Multi-representation Approach for N-Gram Spotting in Historical Manuscripts

Giuseppe De Gregorio[1] , Sanket Biswas[2(✉)] , Mohamed Ali Souibgui[2] ,
Asma Bensalah[2] , Josep Lladós[2] , Alicia Fornés[2] , and Angelo Marcelli[1]

[1] DIEM - Department of Information and Electrical Engineering and Applied
Mathematics, University of Salerno, Fisciano, Italy
{gdegregorio,amarcelli}@cvc.uab.es

[2] Computer Vision Center and Computer Science Department, Universitat
Autònoma de Barcelona, Barcelona, Spain
{sbiswas,msouibgui,abensalah,josep,afornes}@cvc.uab.es

Abstract. Despite recent advances in automatic text recognition, the
performance remains moderate when it comes to historical manuscripts.
This is mainly because of the scarcity of available labelled data to train
the data-hungry Handwritten Text Recognition (HTR) models. The Keyword Spotting System (KWS) provides a valid alternative to HTR due to
the reduction in error rate, but it is usually limited to a closed reference
vocabulary. In this paper, we propose a few-shot learning paradigm for
spotting sequences of a few characters (N-gram) that requires a small
amount of labelled training data. We exhibit that recognition of important n-grams could reduce the system's dependency on vocabulary. In
this case, an out-of-vocabulary (OOV) word in an input handwritten
line image could be a sequence of n-grams that belong to the lexicon. An
extensive experimental evaluation of our proposed multi-representation
approach was carried out on a subset of Bentham's historical manuscript
collections to obtain some really promising results in this direction.

Keywords: N-gram spotting · Few-shot learning · Multimodal
understanding · Historical handwritten collections

1 Introduction

Historical document digitization is essential for preserving and maintaining the
integrity of these documents' records. Handwriting recognition is a central task
in facilitating different services related to historical manuscripts (e.g., searching, indexing, storing, etc.). In this context, handwritten historical documents
processing is considered a challenging problem because images often suffer from
degradation as a result of smears, artifacts, pen strokes, show-through, stains,
bleed-through effects, and uneven illumination [3]. A further obstacle is the writing style variability related to documents' writers living over different time periods [8]. Moreover, most of these records have unique and complex structured

© The Author(s), under exclusive license to Springer Nature Switzerland AG 2022
U. Porwal et al. (Eds.): ICFHR 2022, LNCS 13639, pp. 3–17, 2022.
https://doi.org/10.1007/978-3-031-21648-0_1

layout patterns, which make them difficult to handle [18]. This makes the usual recognition-based techniques for document processing (like OCR and HTR) not directly applicable to historical handwritten documents. The goal of an HTR system is to correctly map an input text image into a machine-encoded format at the word or character level. However, when writing is cursive, separation and recognition of characters is even more difficult [9]. Alternatively, the word-level prediction could be performed with greater accuracy in numerous cases. Hence, the KWS technique was introduced as a surrogate to recognition-based methods for word retrieval.

KWS can be defined as the task of finding all the instances of a keyword in a document image without explicitly recognizing it [31]. This approach makes the task more flexible and realizable, albeit the case when automatic transcription is impossible. Hence this technique is successful for documents of historical interest. The best performance of the systems mentioned above is achieved by lexicon-based systems [18]. In contrast, the major drawback is their inability to correctly recognize words that are not lexicon elements [21] (OOV words). This is a major constraint when building a scalable retrieval system, as obtaining labelled data is resource expensive and sometimes impossible in the case of historical documents. The handwriting process can be considered a complex combination of motor and cognitive skills. As any intricate motor skill, it is acquired through the two learning principles: repetition and memorization. A neural scheme for motor learning has been proposed in [19], where the authors state that sensory information is processed in the brain and appropriate motor commands are generated to execute the desired movement. During the learning phase of these movements, constant repetition enables the definition of motor automatisms that are subsequently activated when the same movement has to be performed again. Following the same approach to handwriting, we can presume that the handwriting learning process sets out an arrangement of automated movements corresponding to the best-learned writing motor primitives. The set of primitives depends on the writer's most familiar and best-learned sign sequence. This leads to the hypothesis that during the handwriting learning phase, an individual develops automated motor sequences to write the more frequent short sequences of characters, referred to as n-grams. Next, localizing these n-grams within a word could help with word-level recognition [22]. Thus, the ability to identify parts of a word could ease the recognition of the OOV words in a classical HTR system. Figure 1 shows the effect of N-gram spotting on an OOV word. It can be seen how it is possible to partially recognise the word by focusing on the detection of some of its n-grams. These insights led us to develop an N-gram spotting system capable of recognizing partial sequences of words within a handwritten line of text. On the other hand, current HTR models require large training sets. Nevertheless, finding historical documents together with manually labeled and aligned transcriptions is quite rare. Accordingly, a "few-shot" learning paradigm [36] can be introduced and explored to train a recognition model. This kind of scenario allows the model to adapt to the limited training data provided. Following the ideas proposed in [34], we can train a network to learn a similarity function between a handwritten text image query and a support N-gram image. Thus, this helps create a small training set composed of the most frequent n-grams extracted from available data collections. Later, we can detect n-grams inside handwritten documents of a test set in order to recognize text word images.

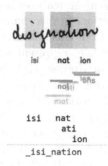

Fig. 1. **Illustration of the few shot N-gram spotting task:** Given an OOV word text image as input, our model gives an almost correct recognition.

The main contributions of this work are summarized as follows:

- We propose a few-shot N-gram spotting paradigm to deal with handwritten historical collections that have limited amount of transcription;
- We explore and investigate the potential of the proposed N-gram spotting solution in KWS scenario of predicting OOV word samples in handwritten text images;
- Correspondingly, a novel multi-representation fusion strategy of feature embedding attributes has been proposed for this task, to establish a tough-to-beat baseline.

The rest of the paper is organised as follows: Sect. 2 explains the related work, Sect. 3 presents the architecture of the proposed method, while Sect. 4 reports the experimental results. Finally, Sect. 5 presents the conclusions and future work.

2 Related Work

Keyword Spotting. KWS was introduced to bridge the gap between digital and historical documents since conventional visual recognition systems can hardly process them. In the classical approach, it is assumed that the text is already segmented into word candidates, and images of words are used as queries, referred to as Query-by-Example (QbE) in the literature. Thus the problem can be traced back to the definition of a distance measure between the different images of a single word [24]. Alternatively, segmentation-free systems do not provide any word-level segmentation, and the system searches within a line of text or on a whole page for regions similar to the example query [5,6,15]. If the search query is fed to the system in the form of a string of characters and not as an image, it has been defined as Query-by-String (QbS). In this scenario, the system must be able to compare strings with images of candidate regions. An interesting solution for this task is to use an embedding to project both the query string and the document image features into a shared space where it is possible to define a distance function. In this common space, it is, therefore, possible to execute queries starting from both a string and an image [2]. A powerful embedding function should be able to represent a word with its Pyramidal Histogram Of Characters (PHOC) attributes [1]. Using the PHOC representation, we could represent a generic string or a generic image word as a fixed-length vector by calculating the histogram of the characters of the input, concatenating the histogram of the first half and the second half, and so on until a certain depth is reached. Once the embedding space is defined, the search can be performed by learning a distance metric between the PHOC representations and then using the measure to spot keywords [32]. One set of KWS techniques that is particularly relevant in the case of historical documents is the lexicon-based approaches [23]. These systems are based on a list of predefined keywords that can be searched; hence any keyword not belonging to that list gives a null score. This stimulates the need for a robust KWS system that could provide a solution in the case of OOV keywords. In historical document analysis, the available data for training the recognition models is often very limited. In KWS, expanding the system lexicon and building large lexicons require large data sets. One method to mitigate the problem of finding OOV words is to define a similarity metric between generic words. In that way, it is possible to approximate the retrieval values of OOV words [23]. Another approach is to use language models at the n-gram level, where the obtained predictions from the model are entrusted with the task of recognizing proposed words that are not in the lexicon [16]. The performances of the proposed solutions do not exceed those of the lexicon-based systems, making the spotting of OOV words still an open problem.

Few Shot Learning for Handwriting Recognition. Few shot learning setting is emerging in machine learning as an efficient parading to supervise the process with a reduced number of samples. One of the first papers that attempted to address the KWS problem with a few-shot approach was proposed by Howe et al. [11]. Authors claimed their KWS system can perform the training phase with only a few labelled data. Since then, few-shot learning has shown a growing interest, and new solutions have been presented for handwriting analysis. Solutions ranged from generative systems capable of producing characters [37] to character recognition systems in different languages [28,35]. Also, synthetic generation has been further expanded to generation of text-line images [13] and layouts in whole-page documents [7]. One interesting solution was proposed by Souibgui et al. [30]. The authors proposed a few-shot approach to search for a set of symbols within a handwritten line of an encrypted document. All above mentioned and other works demonstrate that few-shot can be fruitfully applied for handwriting analysis.

3 Methodology

Inspired by the work in [30], we propose a few-shot learning-based model to tackle the N-gram spotting task in historical document collections. Given a handwritten input text image as a query and some of the most frequent N-gram examples as support, the model is trained to predict the position of the support N-grams within the query line image.

3.1 The Base Architecture

The proposed architecture is based on a faster R-CNN detector [25]. Then, it is adapted to a Siamese architecture by setting the goal of obtaining a similarity score between the images of the supporting N-grams and the boxes found within the line of handwritten text. Figure 2 shows the base architecture of the proposed system. Here, the supports are images of n-grams belonging to a given class c, and we form the model to detect all objects of class c in a query image of a handwritten line of text Q. The first stage is a feature extraction phase where the CNN network extracts features from the text images. It must be highlighted that the network used to extract the features of both the query and the supporting images share the same structure and the same weights, as required for a Siamese architecture. The choice of a feature extraction backbone is not fixed, and by changing the choice of the feature extraction network, a different search modality can be applied which allows performing the spotting in a search space each time different.

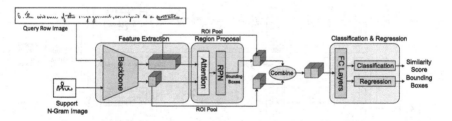

Fig. 2. Base architecture. A single backbone is used in feature extraction of the query and support images. The backbones for the query image and the support image share weights, as predicted by the Siamese model.

The region proposal phase follows the feature extraction stage. In this phase, the feature map of both the query image and the supporting n-gram image are fed to an attention module and, further, to a region proposal network (RPN). After this phase, ROI-pooling is introduced on top of the regions proposed by the RPN and the feature map of the supporting n-gram image. The feature maps are then combined and then fed to a classification and regression module. This module consists of a collection of fully connected layers divided into two heads. The first is a classification head, and the second is a regression head. The output of the classifier uses a sigmoid activation function. The sigmoid decides whether the proposed region belongs to the class of the supporting image (1) or not (0). In parallel, the regression model generates the coordinates of the bounding boxes within the handwritten line image with respect to the classified image parts.

3.2   The Multi-modal Architecture

The base architecture can be constructed in by using different types of backbones for the feature extraction phase, so that different search modalities can be combined with the aim to fuse the results obtained in different feature spaces to obtain an improved spotting performance. In this section, we propose an architecture that combines two different search modalities as shown in Fig. 3, to which we refer to from now on as *multi-modal architecture*. The two independent branches work concurrently, obtaining the two solutions Y_1 and Y_2 by using the backbones BB_1 and BB_2 respectively. At the end the two solutions are combined by a weighted concatenation as shown in Eq. 1.

$$Y = (w_1 \cdot Y_1)||(w_2 \cdot Y_2) \tag{1}$$

Through the selected weights of each backbone w_1 and w_2, it is possible to define the relative importance of solution Y_1 in comparison to Y_2.

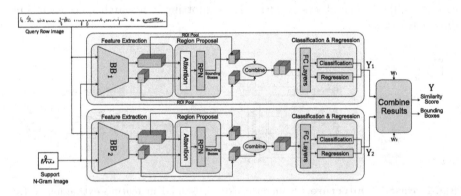

Fig. 3. The multi-modal architecture: Two independent solutions from different search modalities are fused.

In the final combined solution Y, a possibility would be that for the same region of the handwritten query line image, the multi-modal system could propose multiple interpretations of the same N-gram as illustrated in Fig. 4. In this case, the similar solutions are fused into a new unique solution whose score is recomputed by first computing the gain and then adding it to obtain the final score. When two overlapping solutions α_1 and α_2 belonging to the same N-gram class are detected, they are fused together, computing a new score s as in Eq. 2 defined by the maximum between the two scores s_1 and s_2 incremented by a gain γ.

$$s = max(s_1, s_2) + \gamma \tag{2}$$

where the additional gain γ is defined as:

$$\gamma = \delta \cdot \left(1 - \frac{|s_1 - s_2|}{max(s_1, s_2)} \right) \tag{3}$$

Here in Eq. 3, δ is the maximum increment step, and s_1 and s_2 are the scores computed in the two solutions α_1 and α_2. Thus, the new score s is directly proportional to the maximum score between s_1 and s_2 and to the difference between the two scores. Consequently, the cases where both branches propose the same interpretation with very high scores for the same text region are rewarded. If the new score exceeds the maximum allowable score value (in our case $s > 1$), the score of the recomputed solution is set equal to one, while the scores of all other interpretations that overlap for the same range of the query image are decreased by the excess value $1 - s$.

Fig. 4. An example of the combination of two similar solutions proposed for the same text area: a) shows two proposal of the 3-gram "the" with two different scores; b) shows the result of the fusion. It results with a new interpretation for the 3-gram "the" whose score is higher than that of the initial interpretations.

3.3 Multi-modal Architecture with Early Fusion

The previous multi-modal architecture performs the same region proposal, classification, and regression problems twice in both branches, using two different backbones before fusing the results at the end to obtain the final scores. Instead, an early fusion strategy that uses one branch for region proposal and the other branch for classification and regression could also be adopted. Thus, we present a modification of the previous multi-modal architecture as shown in Fig. 5. The key objective is to separate the feature space for the region proposal problem from the feature space for the classification and regression problem. Therefore, the first branch is used for the region proposal phase. It computes the features of the query line images and supporting n-grams used by the region proposal block. Then, it identifies the possible regions of the text row that are candidates for an instance of the supporting image N-gram. Afterward, the second branch calculates the features used in the classification and regression phase, using the regions proposed in the other branch.

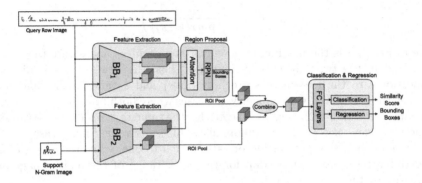

Fig. 5. The multi-modal architecture with early fusion: The architecture uses different modalities for region proposal and classification and regression.

4   Experiments

4.1   Experimental Setup

Among the premises of this work, the most restrictive is related to the amount of available training data. Indeed, our application scenario consists of very small handwritten document collections. Thereby, we considered a subset of the well-known Bentham Collection used in [26]. 20 pages were selected, 5 of which were used as a training set. The entire transcription and examples of frequent reference n-grams were extracted from this training set, obtaining a list of key n-grams consisting of 116 classes, each populated by a number of items ranging from five to seventy-six. The remaining 15 pages were used as a test set. All pages of the dataset were pre-processed by concatenating a binarization step [27], a text line segmentation, and lastly, a deslanting method [33].

Since limited training data is available, it may be useful to use different data sets for pre-training. In our case, we have augmented the pre-training dataset with two additional datasets. Additionally, we have created two synthetically generated datasets that are much larger than our real data. To create the first synthetic dataset, we have chosen the Omniglot dataset [17]. Omniglot consists of 1623 different characters handwritten by different scribes from 50 different alphabets. There are 20 examples of each character in this dataset. We generated

2000 lines of text with 964 different symbols by randomly spacing the symbols in the series with a high probability of symbol overlap. Next, for the second set of synthetic training data, lines of handwritten text were instead generated using a generative network capable of generating words from handwritten text [12,14]. The network was trained on the IAM dataset [20] and then used to generate random words (assembled into 2000 lines of text). In this case, it was possible to use these lines to highlight the contained n-grams and to use these n-grams as marked elements for the training set.

4.2 Evaluation Metrics

Fig. 6. Example of the result of the spotting on a query text line image: the image highlights all the boxes detected in correspondence with each n-gram reporting the scores associated with each box at the bottom.

As observed in the Fig. 6, the system offers different interpretation options for each text area of the text line, each with a different similarity score value between 0 and 1. However, the system obtains a single interpretation for each class of n-grams for each text line area. This is due to the options merging of the multi-modal architecture module described in Sect. 3.2. We address the task as a retrieval problem. Thus, the system is formulated as a multiple item recommendation system, with the peculiarity that the system cannot propose multiple options belonging to the same class. On the other hand, our interest lies in the fact that at least one of the top-k options proposed by the system is the correct one; thus, the transcription of the N-gram has been correctly identified. We then propose a modification of the Precision and Recall at k metrics as shown in Eq. 4 and Eq. 5 respectively.

$$p@k = \frac{true\_relevant\_ngrams@k}{retrieved\_ngrams} \qquad (4)$$

$$r@k = \frac{true\_relevant\_ngrams@k}{relevant\_ngrams} \qquad (5)$$

In the aforementioned formulations, $true\_relevant\_ngrams@k$ represents the number of N-grams correctly detected given the top k options for each area of the text line image, $retrieved\_ngrams$ denotes the number of all n-grams detected within the text line, while *relevant_ngrams* is the number of n-grams that make up the text line. It should be noted, however, that the system is not able to recognise n-grams outside the list of n-grams used for the search in the test phase but that it proposes an interpretation for all areas of the queried text line in any case, even if it has a low score. Therefore, it is interesting to limit the analysis to the n-grams that the system can actually recognise. We will refer to these n-grams by the term *in vocabulary*. In addition, we use a recall metric at k restricted to n-grams in the vocabulary, defined as in Eq. 6 where the *relevant_ngram_InVoc* term denotes the number of n-grams in the vocabulary that occur in the handwritten text line image.

$$r@k\_InVoc = \frac{true\_relevant\_ngrams@k}{relevant\_ngrams\_InVoc} \qquad (6)$$

4.3   Results and Discussion

Selecting Base Architecture Backbones: We selected several model backbones to test the base architecture and evaluated their performance. We have chosen to use two backbones related to two architectures that have performed well on different computer vision tasks: VGG16 [29] and Resnet18 [10]. We have also decided to test the PHOCnet [32] backbone as it is widely used in various handwritten word spotting tasks. We compare our approach with the solution presented in [30]. This work was designed to search for encrypted symbols in a line of handwritten text using a few-shot approach. Although there is a difference in domains, the commonality between these two problems is that both approaches have solutions based on a few-shot setting, and both solutions search for a symbol within a line of handwritten text image (whether it is an encrypted symbol or a Latin N-gram). Table 1 shows the results for a single branch architecture with different backbones compared to the results obtained with the model presented in [30]. The results are given for the cases $k = 1$ and $k = 5$ and for the number of shots of 1, 3, and 5. The different systems have been trained with the set of n-grams extracted from the Bentham collection training data, while for the model [30], the training conditions that showed better performance were chosen. We can infer from the table how the model performance changes depending on the backbone network used for the feature extraction phase. As outlined, the VGG16 network backbone in our base model achieves the best results.

The Significance of Pre-training: So far, the performance of the model seems unsatisfactory in most cases. To achieve a better performance, a pre-training phase is essential to be conducted using synthetic data. For this experimentation, we tested two backbones: the VGG16 backbone because it showed better performance compared to the architecture with the Resnet18 backbone, while the PHOCnet backbone was chosen due to its relevance to the application domain. Table 2 highlights the results and illustrates how the pre-training phase on synthetic data, followed by a fine-tuning phase on the real data, allows the system to significantly improve its performance in all possible metrics. It is worth mentioning that it was impossible to use the synthetic training data generated from

Table 1. Results for the base architecture evaluating different backbones (BB) for feature extraction:

	#shot	p@1	r@1	r@1_InVoc	p@5	r@5	r@5_InVoc
Souibgui et al. [30]	1	0.0325	0.0279	0.0557	0.0607	0.0513	0.1218
	3	0.0269	0.0256	0.0598	0.0757	0.0717	0.1864
	5	0.0122	0.0121	0.0348	0.0671	0.0641	0.1976
BB_{VGG16}	1	**0.0701**	0.0373	**0.0942**	0.1747	0.0854	0.2114
	3	0.0579	**0.0339**	0.0638	0.1947	0.1006	0.2302
	5	0.0465	0.0274	0.0513	**0.1980**	**0.1062**	**0.2732**
$BB_{RESNET18}$	1	0.0000	0.0000	0.0000	0.0533	0.0061	0.0150
	3	0.0333	0.0042	0.0167	0.0333	0.0042	0.0167
	5	0.0167	0.0026	0.0059	0.0333	0.0061	0.0150
$BB_{PHOCnet}$	1	0.0000	0.0000	0.0000	0.0167	0.0048	0.0083
	3	0.0310	0.0071	0.0392	0.0671	0.0153	0.0566
	5	0.0111	0.0048	0.0083	0.0954	0.0233	0.0780

the Omniglot dataset with the PHOCnet network since a PHOC representation of a word composed of symbols rather than letters is meaningless.

Comparison to Existing SOTA Methods: The model from Souibgui et al. [30] has been pre-trained with a dataset generated from the Omniglot dataset. For a fair comparison with our proposed approach, we have performed a fine-tuning phase on the real data. As denoted in Table 2, the results obtained with the model from Souibgui et al. [30] are comparable to the results of our model with the VGG16 backbone and the synthetic Omniglot dataset for pre-training. The model in [30] uses a VGG16 backbone for feature extraction, which could explain the comparable results. It is also interesting to note that the proposed system performs better in the case of $k = 5$ and when the number of shots is high.

Effectiveness of Multi-Modal Architecture: In this section we evaluate the model described in Sect. 3.2 using the two different backbones VGG16 and PHOCnet, respectively pre-trained with the dataset generated from the Omniglot dataset and with the synthetically generated dataset containing n-grams. The outcome of the model depends on the values assigned to the parameters w_1 and w_2 of the Eq. 3. Next, we evaluated the model by varying the two weights from the minimum value 0 to the maximum value 1 with a step size of 0.1 and testing all the weighs combination by performing a grid search [4]. Figure 7 shows the results of the search, and it is clear that the best results are obtained when combining the two different branches. Table 3 illustrates the best results for each metric in case of 5-shot scenario. As an overall trend, the weighting of the branch with VGG16 backbone is always greater than or equal to the weighting of the branch with PHOCnet backbone, except in the case of $r@5$. Based on the weights values, we can conclude that more discriminative feature spaces should have greater importance for combination purposes, complementing our intuition. Additionally, the performance of all indices is higher than the previous

Table 2. Comparative results with the Single-Branch pre-trained model, the Multi-modal model, and the Multi-modal model with Early Fusion:

		#shot	p@1	r@1	r@1_InVoc	p@5	r@5	r@5_InVoc
	Souibgui et al. [30] + finetuning	1	0.1613	0.1121	0.3408	0.3007	0.2042	0.6465
		3	0.2088	0.1427	0.4391	0.3436	0.2301	0.6948
		5	0.2241	0.1602	0.4703	0.3342	0.2364	0.7314
Single-branch	BB_{VGG16} + Omniglot	1	0.1823	0.1291	0.3376	0.3126	0.2171	0.6092
		3	0.2060	0.1405	0.4279	0.3462	0.2349	0.7170
		5	0.2216	0.1470	0.4436	0.3725	0.2450	0.7594
	BB_{VGG16} + Synth	1	0.1555	0.0778	0.2298	0.2940	0.1511	0.4908
		3	0.1823	0.1036	0.3357	0.3019	0.1644	0.5346
		5	0.1878	0.1117	0.3292	0.3336	0.1856	0.5832
	$BB_{PHOCnet}$ + Synth	1	0.1084	0.0612	0.1669	0.2682	0.1543	0.4156
		3	0.1369	0.0831	0.2105	0.2734	0.1638	0.4781
		5	0.1333	0.0791	0.1976	0.2861	0.1700	0.4809
Multi-modal	BB_{VGG16} + $BB_{PHOCnet}$	1	0.2140	0.1424	0.4505	0.3437	0.2392	0.7195
		3	0.2045	0.1495	0.4783	0.3536	0.2519	0.7498
		5	0.2303	**0.1808**	**0.5582**	0.3747	**0.2588**	**0.7975**
Early-fusion multi-modal	VGG16 ROI + PHOCnet class	1	0.0596	0.0553	0.1563	0.1504	0.1470	0.4027
		3	0.0793	0.0832	0.2192	0.1552	0.1619	0.4199
		5	0.0536	0.0582	0.1652	0.1715	0.1858	0.5105
	PHOCnet ROI + VGG16 class	1	0.2346	0.1179	0.3514	0.3374	0.1738	0.5762
		3	0.2234	0.1312	0.4258	0.3783	0.2185	0.6925
		5	**0.2419**	0.1392	0.4404	**0.4088**	0.2236	0.7043

performance with the single backbone model. This suggests that searching over multiple feature spaces may contribute to better performance.

Early Fusion Strategy: As discussed in Sect. 3.3, it may be interesting to investigate the early fusion strategy during the multi-modal combination of the two backbones. To this end, we have trained the model of Sect. 3.3 again using the two pretrained VGG16 and PHOCnet backbones, as in the case of the previous section. The architecture has been first implemented with the VGG16 backbone for the region proposal phase and the PHOCnet backbone for the classification phase, then with the PHOCnet backbone for the region proposal and the VGG16 for the classification. Table 2 presents information on the obtained

Table 3. Best recorded results with relative weights.

Measure	Max-Value	w_vgg16	w_phoc
Precision@1	0.2303	0.7	0.5
Recall@1	0.1808	0.5	0.5
Recall@1_InVoc	0.5582	0.4	0.4
Precision@5	0.3747	0.5	0
Recall@5	0.2588	0.8	1.0
Recall@5_InVoc	0.7975	0.6	0.3

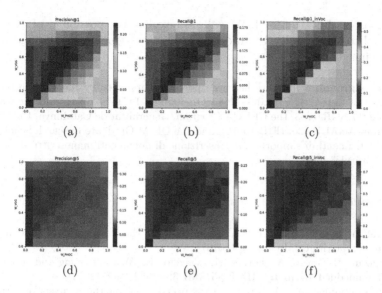

Fig. 7. Visualizing feature weights for multi-modal fusion. Variation of the (a) Precision@1 (b) Recall@1 (c) Recall@1_InVoc (d) Precision@5 (e) Recall@5 (f) Recall@5_InVoc according to the fusion weights for visual and phoc features in case of 5-shot scenario.

results. According to Table 2, the performance of the system decreases compared to the best shots with the previous solutions in terms of recall but increases in terms of precision. This entails that the two different strategies for the fusion are better suited to different scenarios, depending on whether the goal is to maximize recall rather than precision.

5   Conclusion

In this paper, we have presented a few-shot multi-representation model for N-gram spotting for historical manuscripts. Based on experimental results, we have demonstrated our approach's adequacy and that it is possible to detect n-grams of references within a line of handwritten text. We limited the analysis to some of the available feature spaces, but the results showed that using more than one representation can improve performance. In this direction, an exhaustive study on the different representations available could be performed by analyzing in detail different backbones architecture with different depths to understand the impact on the model. The method does not limit the search for n-grams only to the words that appear in the training pages, from which classical recognition systems build their reference lexicons, but offers the possibility to recognise n-grams also in the OOV words that appear only in the test pages. In the future, we will explore the possibility of extending the model with a "word reconstruction" module based on a language model that can reconstruct whole-word interpretations starting from the recognised n-grams to bring the focus of the problem back to the most useful word-level. We will also explore the possibility of extending the model to the page level in order to avoid the segmentation of handwritten text line images.

Acknowledgment. This work has been partially supported by the Spanish projects RTI2018-095645-B-C21, PID2021-126808OB-I00 and FCT-19-15244, and the Catalan projects 2017-SGR-1783, the CERCA Program/Generalitat de Catalunya, PhD Scholarship from AGAUR (2021FIB-10010), and the DIEM Graduate Research Scholarship entitled "Strumenti di supporto alla trascrizione di documenti manoscritti di interesse storico-culturale".

References

1. Almazan, J., Gordo, A., Fornés, A., Valveny, E.: Handwritten word spotting with corrected attributes. In: ICCV (2013)
2. Almazán, J., Gordo, A., Fornés, A., Valveny, E.: Word spotting and recognition with embedded attributes. IEEE TPAMI **36**, 2552–2566 (2014)
3. Antonacopoulos, A., Downton, A.C.: Special issue on the analysis of historical documents (2007)
4. Bergstra, J., Bengio, Y.: Random search for hyper-parameter optimization. JMLR (2012)
5. Biswas, S., Banerjee, A., Lladós, J., Pal, U.: DocSegTr: an instance-level end-to-end document image segmentation transformer. arXiv preprint arXiv:2201.11438 (2022)
6. Biswas, S., Riba, P., Lladós, J., Pal, U.: Beyond document object detection: instance-level segmentation of complex layouts. Int. J. Doc. Anal. Recogn. (IJDAR) **24**(3), 269–281 (2021)
7. Biswas, S., Riba, P., Lladós, J., Pal, U.: DocSynth: a layout guided approach for controllable document image synthesis. In: ICDAR (2021)
8. Bunke, H., Varga, T.: Off-line roman cursive handwriting recognition. In: Chaudhuri, B.B. (ed.) Digital Document Processing, pp. 165–183. Springer, London (2007). https://doi.org/10.1007/978-1-84628-726-8_8
9. Choudhary, A., Rishi, R., Ahlawat, S.: A new character segmentation approach for off-line cursive handwritten words. Proc. Comput. Sci. **17**, 88–95 (2013)
10. He, K., Zhang, X., Ren, S., Sun, J.: Deep residual learning for image recognition. In: CVPR (2016)
11. Howe, N.R.: Part-structured inkball models for one-shot handwritten word spotting. In: ICDAR (2013)
12. Kang, L., Riba, P., Rusinol, M., Fornés, A., Villegas, M.: Distilling content from style for handwritten word recognition. In: ICFHR (2020)
13. Kang, L., Riba, P., Rusinol, M., Fornés, A., Villegas, M.: Content and style aware generation of text-line images for handwriting recognition. IEEE TPAMI (2021)
14. Kang, L., Riba, P., Wang, Y., Rusiñol, M., Fornés, A., Villegas, M.: GANwriting: content-conditioned generation of styled handwritten word images. In: Vedaldi, A., Bischof, H., Brox, T., Frahm, J.-M. (eds.) ECCV 2020. LNCS, vol. 12368, pp. 273–289. Springer, Cham (2020). https://doi.org/10.1007/978-3-030-58592-1_17
15. Konidaris, T., Kesidis, A.L., Gatos, B.: A segmentation-free word spotting method for historical printed documents. Pattern Anal. Appl. **19**, 963–976 (2016)

16. Kozielski, M., Matysiak, M., Doetsch, P., Schlöter, R., Ney, H.: Open-lexicon language modeling combining word and character levels. In: ICFHR (2014)
17. Lake, B.M., Salakhutdinov, R., Tenenbaum, J.B.: Human-level concept learning through probabilistic program induction. Science **350**, 1332–1338 (2015)
18. Lombardi, F., Marinai, S.: Deep learning for historical document analysis and recognition-a survey. J. Imaging **6**, 110 (2020)
19. Marcelli, A., Parziale, A., Senatore, R.: Some observations on handwriting from a motor learning perspective. In: AFHA (2013)
20. Marti, U.V., Bunke, H.: The IAM-database: an English sentence database for offline handwriting recognition. IJDAR **5**, 39–46 (2002)
21. Parziale, A., Capriolo, G., Marcelli, A.: One step is not enough: a multi-step procedure for building the training set of a query by string keyword spotting system to assist the transcription of historical document. J. Imaging **6**, 109 (2020)
22. Poznanski, A., Wolf, L.: CNN-N-gram for handwriting word recognition. In: CVPR (2016)
23. Puigcerver, J., Toselli, A.H., Vidal, E.: Querying out-of-vocabulary words in lexicon-based keyword spotting. Neural Comput. Appl. **28**, 2373–2382 (2017)
24. Rath, T.M., Manmatha, R.: Word spotting for historical documents. IJDAR **9**, 139–152 (2007)
25. Ren, S., He, K., Girshick, R., Sun, J.: Faster R-CNN: towards real-time object detection with region proposal networks. In: NeurIPS (2015)
26. Sanchez, J.A., Toselli, A.H., Romero, V., Vidal, E.: ICDAR 2015 competition HTRtS: Handwritten text recognition on the transcriptorium dataset. In: ICDAR (2015)
27. Sauvola, J., Pietikäinen, M.: Adaptive document image binarization. Pattern Recogn. **33**, 225–236 (2000)
28. Shaffi, N., Hajamohideen, F.: Few-shot learning for Tamil handwritten character recognition using deep Siamese convolutional neural network. In: Mahmud, M., Kaiser, M.S., Kasabov, N., Iftekharuddin, K., Zhong, N. (eds.) AII 2021. CCIS, vol. 1435, pp. 204–215. Springer, Cham (2021). https://doi.org/10.1007/978-3-030-82269-9_16
29. Simonyan, K., Zisserman, A.: Very deep convolutional networks for large-scale image recognition. arXiv preprint arXiv:1409.1556 (2014)
30. Souibgui, M.A., Fornés, A., Kessentini, Y., Tudor, C.: A few-shot learning approach for historical ciphered manuscript recognition. In: ICPR (2021)
31. Stauffer, M., Fischer, A., Riesen, K.: Keyword spotting in historical handwritten documents based on graph matching. Pattern Recogn. **81**, 240–253 (2018)
32. Sudholt, S., Fink, G.A.: PHOCNet: a deep convolutional neural network for word spotting in handwritten documents. In: ICFHR (2016)
33. Vinciarelli, A., Luettin, J.: A new normalization technique for cursive handwritten words. Pattern Recogn. Lett. **22**, 1043–1050 (2001)
34. Vinyals, O., Blundell, C., Lillicrap, T., Wierstra, D., et al.: Matching networks for one shot learning. In: NeurIPS (2016)
35. Wang, T., Xie, Z., Li, Z., Jin, L., Chen, X.: Radical aggregation network for few-shot offline handwritten Chinese character recognition. Pattern Recogn. Lett. **125**, 821–827 (2019)
36. Wang, Y., Yao, Q., Kwok, J.T., Ni, L.M.: Generalizing from a few examples: a survey on few-shot learning. ACM Comput. Surv. (CSUR) **53**, 1–34 (2020)
37. Wong, A., Yuille, A.L.: One shot learning via compositions of meaningful patches. In: ICCV (2015)

Text Edges Guided Network
for Historical Document Super Resolution

Boraq Madi$^{(\boxtimes)}$ ⓘ, Reem Alaasam ⓘ, and Jihad El-Sana ⓘ

Ben-Gurion University of the Negev, Be'er Sheva, Israel
{borak,rym}@post.bgu.ac.il, el-sana@cs.bgu.ac.il

Abstract. Super-resolution aims to increase the resolution and the clarity of the details in low-resolution images, and document images are no exception. Although significant improvements have been achieved in super-resolution for different domains, historical document images have not been addressed well. Most of the current works in the text domain deal with modern fonts and rely on extracting prior semantic information from a recognizer to super-resolve images. The absence of a reliable handwritten recognizer for Arabic documents, where historical documents have a complex structure and overlapping parts, makes these text-domain works inapplicable. This paper presents a Text-Attention-ed Super Resolution GAN (TASR-GAN) to address this problem. The model deals with historical Arabic documents and does not rely on prior semantic information. Since our input domain documents, text edges are essential for quality and readability; thus, we introduce a new loss function called text edge loss. This loss function provides more attention and weight to text edge information and guides through optimization to super-resolve images with accurate small regions' details and fine edges to improve image quality. Experiments on six Arabic manuscripts show that the proposed TASR achieves state-of-the-art performance in terms of PSNR/SSIM metrics and significantly improves the visual image quality, mainly the edges of small regions details, and eliminates artifacts noises. Also, a grid search experiment has been conducted to tune the best hyperparameters values for our text edge loss function.

Keywords: Super-resolution · Historical handwritten documents · Generative adversarial networks

1 Introduction

Historical documents are an essential source of information about past societies, and this information can be extracted and interpreted for different purposes. However, historical document images often encounter quality degradation because of aging and lousy saving situations. This degradation leads to low resolution and blurry structures of these documents. This problem significantly impairs the performance of several tasks, such as optical character recognition

U. Porwal et al. (Eds.): ICFHR 2022, LNCS 13639, pp. 18–33, 2022.
https://doi.org/10.1007/978-3-031-21648-0_2

(OCR) [20,31,32] and text detection [22,41]. In order to achieve better performances in these tasks, we have to increase the resolution as well as enhance the visual quality of text in these document images.

Several super-resolution methods have been developed to improve image quality in the past years. Evident signs of progress have been obtained with the aid of deep learning-based approaches [4,6,34,35]. Existing super-resolution methods work primarily on supervised learning using realistic image datasets. These datasets include pairs of images, where each pair include low resolution(lr) and high resolution(hr) of the same image. The deep learning model is trained to super-resolve the lr image to obtain output similar to the hr image. With the strong capabilities of deep learning models, they can learn various prior knowledge and patterns from the trained data in order to provide good performance.

In recent years, vast super-resolution techniques have been developed in different domains, such as depth maps [10,13], which are used to increase the spatial resolution of these maps. Others were developed for face hallucination [3,42] tasks by incorporating prior facial knowledge to enhance face images. Despite the significant progress and usage of super-resolution in different domains, the text images domain was not addressed sufficiently well. The text image domain includes works such as TATT [23], TPGSR [21], which only deal with super-resolution for scene text images. These works use recognized semantics features of the text as prior information to produce better quality output. TPGSR and TATT super-resolve and correctly restore the text semantics of degraded images with good visual quality. Most of these works in the scene text images domain deal with Latin and modern fonts such as TextZoom [34] dataset.

However, using these works to super-resolve historical document images is not applicable because part of historical document languages, such as Arabic, do not have reliable recognizers to extract prior semantic information. Also, As far as we know, this is the first work that deals with a super-resolution for historical documents without relying on prior semantic information. Toward constructing experimental comparison, we compare our approach with state-of-the-art super-resolution approaches in the general domain, Single-image super-resolution (SISR). The models in the general domain [17,18,27,36,40] do not count on any prior information, such as the semantics of text or face features; they input lr images and output results similar to hr ones.

This paper presents novel architecture, Text-Attention-ed Super Resolution (TASR-GAN), for historical document text image super-resolution that does not depend on prior information. The framework of our method is training TASR-GAN on patches of pairs (lr,hr) extracted from various historical handwritten document datasets. We first build the pairs like the model in the general domain, such as ESRGAN [36]. Then we train the model on these pairs and compare the quality of the result between the output, hr^*, and the ground truth, hr, using two classical image quality assessments(IQA): PSNR and SSIM.

Overall, our contributions can be summarized as follows:

- We propose a novel architecture to super-resolve historical documents without considering prior information.

- We propose a novel loss function, *Text Edge Loss*, which guides the model to 'be-attention-ed' during super-resolving to the edges. It encourages the model to restore edges with high quality without corrupting them.
- Our proposed model is compared with state-of-the-art models in six different manuscripts that vary in writing styles and noise levels and achieve the highest results in most datasets.

In comparing with state-of-the-art methods, We show that the performance of our model is superior to them in the six manuscripts. In addition, it improves the resolution of patch image 16× while denoising background noise and constructing sharp edges to enhance the quality of the text in the images.

2 Related Work

Image super-resolution refers to the process of recovering high-resolution images from lower-resolution ones, which is a significant branch of image processing techniques in computer vision. It can be useful for wide range of different application such as surveillance and security [28,39], medical imaging [1,8,12]. Beside, helping to enhance image perceptual and visual quality, it help to improve various computer vision tasks [5,11].

Super resolution for images in general can be done using deep learning methods [37] or other various approaches [25]. We will focus only on deep learning methods for text document images. Some approached focus on the super resolution task for clean printed images. Dong *et al.* [6] used Super-Resolution Convolutional Neural Network (SRCNN), their approach takes a low-resolution printed text image and converts it to a super-resolution one. The result is then fed to an Optical Character Recognition (OCR) system for text recognition. Like the previous approach, Zhang *et al.* [38] also used a Convolutional neural network (CNN) but focused on a loss function that improve the OCR performance. The shortcoming of the two approaches is that they work only on noise-free printed text images, which thus do not work on handwritten images or historical images that include many noises and degradations. Other approaches focus on removing the noise from printed text images before applying super-resolution methods [2,9]. However, these approaches handle each task separately and hence they are time-consuming and work on specific type of noise (printer or scanner noise).

Recently, some approaches focused on achieving noise-free and super-restoration for text images from low-resolution noisy text images. Sharma *et al.* [30] use joint optimization of two CNNs, one handles the super-resolution task and the other handle the denoising to improve OCR performance and they tested their results with printed grayscale text images. In comparison, Lat and Jawahar [16] use Generative Adversarial Network (GAN) for both denoise and super-resolution at the same time and applied their method on color printed-text images. However, as mentioned before they focus on printer or scanner noise which is one type of noises only which is related to printed text, and would not be suitable for handwritten text images. Ray *et al.* [29] apply GAN framework for denoising followed by Deep Back Projection Network (DBPN) for super-resolution, they apply their method on printed and handwritten text. However, they used the handwritten IAM Database [24], which includes clean images of modern grayscale text and does not include many degradations.

Contrary to the methods mentioned above, our paper focuses on degraded historical handwritten images. These images include stains, bleed-through, faded ink and other type of degradation and may also include acquisition noise. We aim at improving the image quality via super-resolution and reducing noise at the same time with deep learning approach.

3 Dataset

This section overviews the datasets used in our experimental evaluation and explains their preparation procedure.

VML-HD [15] is a dataset that includes five Arabic manuscripts written between the years 1088-1451. Each manuscript has written by a unique writer and has its own style, as shown in Fig. 1. The dataset contains 668 pages with annotation at the sub-word level. Each sub-word is annotated with a bounding box and a transcription label. The pages in each manuscript are of high quality and contain varying writing styles, diacritics, and small touching components. We experimented with 20 pages from each one of the five manuscripts from the VML-HD data and we shall refer to them as VML_{Book1}, VML_{Book2}, VML_{Book3}, VML_{Book4}, and VML_{Book5}.

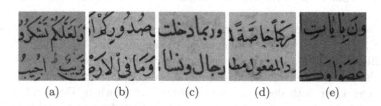

(a)   (b)   (c)   (d)   (e)

Fig. 1. The patch samples from left to right are from historical document pages of VML_{Book1}, VML_{Book2}, VML_{Book3}, VML_{Book4}, and VML_{Book5}, respectively.

AHTE is a historical Arabic handwritten dataset that includes pages selected from different manuscripts. It is available online for downloading. The characteristic of this dataset is complex layouts, crowded diacritics, and overlapping words and sentences, which make it a high-quality dataset with complex layouts, as seen in Fig. 2.

4 Method

We present **T**ext-**A**ttention-ed **S**uper **R**esolution(TASR-GAN), a novel framework for enhancing resolution and denoising manuscript on the image domain.

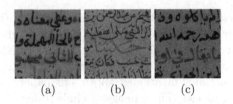

(a)   (b)   (c)

Fig. 2. A three patch samples of different writing styles from AHTE dataset are shown in (a), (b) and (c).

4.1   Model Framework

Generative Adversarial Networks(GAN) [7] is one of the machine learning classes that gained popularity in recent years. It is composed of two neural networks, generator(G) and Discriminator (D), which are trained against each other in the form of a zero-sum game. The generator produces an image, and the Discriminator needs to determine whether it is real or fake, i.e., the generator aims to produce images similar to the real ones.

The TASR-GAN framework consists of two modules: a generator, G, and a discriminator, D. Our goal is to train the model capable of increasing the resolution of manuscript patch image 16× times while focusing on constructing sharp edges and denoising background noise.

The Generator G inputs a low resolution image, lr, of size 64×64, and outputs a high resolution version, hr^* of size 256×256, as shown in Fig. 3. The output hr^* shares the same content lr while hr^* is a denoised version with sharp edges.

The architecture of G starts with two convolutional layers with a kernel of size 3×3. Then we stack a sequence of 16 sequence of Residual in Residual Dense Blocks(RRDBs) [36] layers and convolutional layer with a kernel of size 3×3. Using the skip-connection technique, we combine the features from the early stages layer with deep-semantic features, as shown in Fig. 3. Finally, we concatenate two up-sampling layers, each including three components: convolutional, leakyReLU, and PixelShuffle layers. The PixelShuffle [33] is used in super-resolution models to reduce channel dimension with r^2 while increasing the width and height of the result with r each. We set the reduction value, r, in our PixelShuffle layers to be 2. The last extension to our model is two convolutional layers with 3×3 kernel, which are separated by a leakyReLU layer, as shown in Fig. 3.

The Discriminator D is a relativistic discriminator [14], and its architecture includes four blocks, as shown in Fig. 4. Each block consists of these components in the following order: (1) convolutional layer with size 3×3, (2) batch norm, (3) LeakyReLU, (4) convolutional layer with size 3×3, (5) batch norm, (6) LeakyReLU. The first block in D includes one batch norm layer instead of two batch norm layers, as shown in Fig. 4.

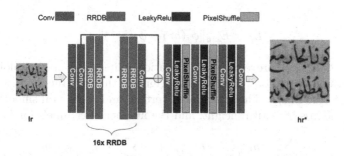

Fig. 3. The Generator G inputs a low resolution image, lr, of size 64×64, and outputs a high resolution version, hr^* of size 256×256.

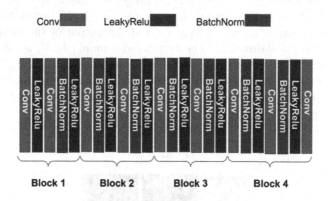

Fig. 4. The Discriminator D consists of four blocks, where each block consists of convolutional, batch norm, and LeakyReLU layers.

4.2 Objective Function

We present a new loss function that guides the model to give more attention to text edges, and we denote it by Text Edge loss. The objective function combines Pixel-Wise, Total Variation, Text-edge, Feature, and Adversarial losses. These loss functions are weighted by a hyperparameter, λ, that reflects the weight of each loss.

Pixel-Wise calculates the distance between the generated patch and the ground truth. Equation 1 presents the loss, where hr is the ground truth and generator input patch, lr, to generate hr^*.

$$L_{Pixel-Wise} = E_{x,y}||hr - hr^*||_1 \qquad (1)$$

Feature Loss compares the output of the generator G and the ground truth image hr on the feature spaces. We adopt a pre-trained VGG16 to extract high-level information and then apply the L_1 metric, as formulated Eq. 2), where ϕ_j is the activation function of the jth layer of the VGG16. We set $j = 4$ according to Liu [19] experimental study.

$$L_f = E_{x,y}\|\phi_j(hr) - \phi_j(hr^*)\|_1 \tag{2}$$

Total Variation Loss prevents color mutation, eliminates artifact noises, and encourages smooth regions. Equation 3 formulate this loss function, where hr^* is the transferred image. The loss applies vertical and horizontal smoothing. We normalize the loss by width, height, and the number of channels for hr^*.

$$L_{tv} = \sum_{(i,j)\in h^*} (hr^*_{i+1,j} - hr^*_{i,j})^2 + \sum_{(i,j)\in hr^*} (hr^*_{i,j+1} - hr^*_{i,j})^2 \tag{3}$$

Text Edge Loss avoids blur edges and encourages the model to construct sharp edges in the high-resolution output. Equation 4 is apply L_2 metric between the edge magnitude map of ground truth, hr, and the output of the model, hr^* in vertical and horizontal direction. The vertical edge map, $edgeV$, is calculated by subtract each pixel $p_{i,j}$ with $p_{i,j+1}$ (See Fig 5b) and horizontal one (See Fig 5c) is computed with $p_{i,j}$ with $p_{i+1,j}$. After, we apply absolute value to the two edge map to produce $edgeV_{hr}$, $edgeH_{hr}$, $edgeV_{hr^*}$, $edgeH_{hr^*}$ for hr and hr^*.

$$L_{te} = \|edgeV_{hr} - edgeV_{hr^*}\|_2 + \|edgeH_{hr} - edgeH_{hr^*}\|_2 \tag{4}$$

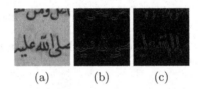

(a)   (b)   (c)

Fig. 5. The image (a) have vertical and horizontal edge map $edgeV$ and $edgeH$ as shown in (b) and (c) respectively.

Adversarial loss in Relativistic-GAN [14] depends on the output of D on both real, hr, and fake data, hr^*. We follow classical loss function choice [36], which is based on Binary Cross Entropy with Sigmoid[1].

$$L_{adv} = BCEWithLogitsLoss(D(hr) - D(hr^*), 1) \tag{5}$$
$$+ BCEWithLogitsLoss(D(hr^*) - D(hr), 0)$$

The loss computes the distance between the decision of Discriminator D over a real high-resolution image, hr, and fake one hr^*. The adversarial loss is expressed in Eq. 5 as adversarial loss for real and fake images.

The final objective function is the weighted sum of the previous loss functions, as shown in Eq. 6.

---

[1] https://pytorch.org/docs/stable/generated/torch.nn.BCEWithLogitsLoss.html.

$$L = \lambda_{Pixel-Wise} * L_{Pixel-Wise} + \lambda_f * L_f + \lambda_{tv} * L_{tv}$$
$$+ \lambda_{te} * L_{te} + \lambda_{adv} * L_{adv}. \qquad (6)$$

The components of our model G and D are optimized via the min-max criterion, $\min_G \max_D L$.

5 Experiment

We compare our framework, Text-Attention-ed Super Resolution(TASR) with state-of-the-art deep learning models, ESRGAN [36], SRGAN [17], SRFeat [27], RCAN [40], EDSR [18] and ERCA [26], over multiple datasets. We evaluate the performance of the approaches by measuring the quality of their generated images using Peak signal-to-noise ratio(PSNR) and Structural Similarity Index Measure(SSIM). For training, we used an Adam optimizer with a batch size of 4 over 20 epochs, and all the models were trained from scratch using a learning rate of 0.002.

For training TASR, we set the hyperparameters for the objective loss as follow: $\lambda_{Pixel-Wise} = 0.01$, $\lambda_f = 0.1$, and $\lambda_{adv} = 0.005$ as in [36]. We use the grid search method for setting the hyperparameter of Total Variance, λ_{tv}, and Text Edge, λ_{te}.

5.1 Data Preparation

Data preparation involves splitting $VML_{Book1},..., VML_{Book5}$ and $AHTE_{SET}$ pages to 60%, 10% and 30% for train, validation and test. After splitting, we sampled 40 patches of size 256×256 pixels from each of the split pages. For each patch with high resolution, hr, we generate a low resolution, lr, version of the same patch by resizing it with a factor of $1/4$ to get a patch of size 64×64, as shown in Fig. 6(a) and Fig. 6(b), respectively. The number of patches in train, validation, and test is 480, 40, and 120 for each data set. The preparation mentioned above is used in Sect. 5.3 and considered as standard preparation for super-resolution quality image evaluation.

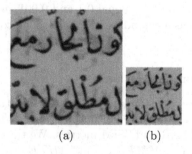

(a)   (b)

Fig. 6. (a) high resolution patch, hr, with size of 256×256. (b) low-resolution patch, lr, is obtained by resizing the hr with a factor of $1/4$ to get a patch of size 64×64.

5.2   Hyperparameters Tuning Using Grid Search

The model objective loss function 6 in Sect. 4.2 is composed of different loss functions and their hyperparameters(e.g., $\lambda_{Pixel-Wise}$, λ_{te}). Since Text Edge loss is a novel loss function and Total Variance did not integrated in many super resolution studies, there is no clue what values their hyperparameters, λ_{te},λ_{tv}, should take.

To get the best hyperparameter values, we use Grid Search to define a search space as a grid of hyperparameter values between λ_{te} and λ_{tv}, as shown in Fig. 7a. We choose $0.1, 0.01, 0.001, 0.0001$ values as candidate values for λ_{te} and λ_{tv}, in the grid search. Each configuration, include point in the grid, which set the value of λ_{te} and λ_{tv}, together with the values of $\lambda_{Pixel-Wise}$, λ_f, and λ_{adv} we mentioned above.

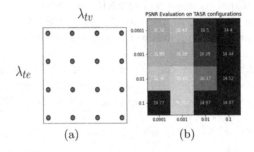

(a)   (b)

Fig. 7. (a) We use Grid Search to define a search space as a grid of hyperparameter values between λ_{te} and λ_{tv}. (b) Each cell is considered a configuration of TASR, where λ_{te} and λ_{tv} are the row and column values of that configuration.

We train TASR on 16 different configurations separately using the train sets, $VML_{Book1},..., VML_{Book5}$ and $AHTE_{SET}$. Next, we measure the performance of each TASR configuration on the test sets using PSNR. The PSNR value for each configuration is mentioned in Fig. 7b. According to the results, using a high value of λ_{tv} leads to excessive smoothness and construction of non-sharp edges. Thus, PSNR values are very low. The configuration with the highest PSNR value is obtained with λ_{te} and λ_{tv} equal 0.1 and 0.001, as shown in Fig. 7b. We adopt this configuration as a candidate in the following comparison experiments against other approaches.

5.3   Super-Resolution Evaluation

In this experiment, we use the standard preparation mentioned in Sect. 5.1. Each model in the experiment inputs lr and outputs hr^*, and then we evaluate the output's quality using PSNR and SSIM metrics. We train the approaches using the train sets, $VML_{Book1},..., VML_{Book5}$ and $AHTE_{SET}$. After, we measure the performance of each model on each test set separately using PSNR and SSIM metrics.

According to the results in both Table 1 and Table 2, our model outperforms other approaches in most test sets. As seen, the Nearest method generates distortions in images, especially in the text edges, as shown in Fig. 8a. Bilinear generates smoothed and blur image (See Fig. 8b) compared to the ground truth (See Fig. 8c). Also it corrupts image edges as exemplified in Fig. 8b.

Table 1. The PSNR values for the competing methods in the experimental comparison.

Method	VML_{Book1}	VML_{Book2}	VML_{Book3}	VML_{Book4}	VML_{Book5}	$AHTE_{SET}$
Nearset	27.51	26.02	24.16	27.58	24.71	27.17
Bilinear	28.05	27.26	26.72	29.02	26.89	29.91
SRGAN [17]	24.66	20.27	24.86	21.28	22.70	21.59
ERCA [26]	28.40	28.60	31.18	28.80	34.15	29.78
RCAN [40]	27.92	27.68	30.22	27.97	33.12	29.45
SRFeat [27]	25.74	25.62	26.74	25.83	28.87	26.55
ESDR [18]	28.61	28.72	31.41	28.92	34.41	30.049
ESRGAN [36]	28.52	29.69	30.05	30.48	34.81	28.97
TASR (Ours)	**30.90**	**32.63**	**34.83**	**33.39**	**38.15**	**32.09**

Table 2. The SSIM values for competing methods in the super-resolution experiment.

Method	VML_{Book1}	VML_{Book2}	VML_{Book3}	VML_{Book4}	VML_{Book5}	$AHTE_{SET}$
Nearset	0.75	0.793	0.834	0.80	0.871	0.749
Bilinear	0.786	0.835	0.872	0.86	0.92	0.781
SRGAN [17]	0.78	0.82	0.85	0.82	0.84	0.71
ERCA [26]	**0.8457**	0.8819	0.9185	0.8914	0.9439	0.8385
RCAN [40]	0.8141	0.8528	0.8855	0.8589	0.9247	0.8108
SRFeat [27]	0.7017	0.7175	0.7481	0.7138	0.7641	0.784
ESDR [18]	0.8394	0.8879	0.9226	0.8964	0.9472	**0.8416**
ESRGAN [36]	0.795	0.846	0.862	0.836	0.93	0.783
TASR (Ours)	0.844	**0.90**	**0.93**	**0.902**	**0.95**	0.833

The deep learning methods provide more visually sharp and better results than the interpolations ones. Visually, SRGAN model fails to generate results with a pattern similar to the ground truth images for all the test sets, as shown in Fig. 9a. In addition, the SRGAN results include color and noise artifacts, as shown in Fig. 10a and Fig. 10b, respectively.

In addition, ESRGAN model performance is better than SRGAN but not as TASR, and it fails to generate fine details and accurate edges for small details as shown in Fig. 9b. While TASR succeeds in preserving these small edges (See Fig. 9c) since it is optimized through the designed loss function, Text Edge loss.

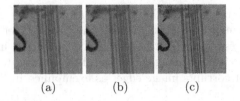

(a)   (b)   (c)

Fig. 8. (a) Nearest output for sample from VML_{Book1}. (b) The result of the super-resolution process using Bilinear. (c) The last column includes the ground truth.

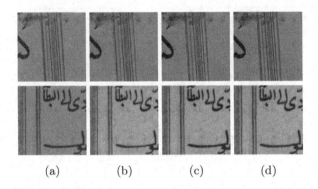

(a)   (b)   (c)   (d)

Fig. 9. (a) SRGAN network outputs for sample from VML_{Book3} and VML_{Book1}. (b) The result of the super-resolution process using ESRGAN. (c) The result of TASR on samples from VML_{Book3} and VML_{Book1} (d) The last column includes the ground truth.

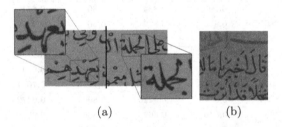

(a)   (b)

Fig. 10. (a) Shows the color artifacts in SRGAN output after zoom in, also it includes noisy edges. (b) The SRGAN adds texture artifact noise in the output, especially for $AHTE_{SET}$ images. (Color figure online)

Also, ESRGAN results suffer from texture noise, as shown in Fig. 11a, while TASR does not suffer from texture noise (See Fig. 11b). TASR is optimized using the total variance loss function, which aims to reduce noises by applying smoothness.

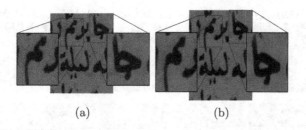

(a)   (b)

Fig. 11. (a) The ESRGAN adds texture artifact noise in the output, especially for $AHTE_{SET}$ images. (b) TASR outputs result without artifact results and reduce its background noises.

The last four models: SRFeat, ESDR, RCAN, and ERCA, have similar behavior to ESRGAN, where they fail to super-resolve edges for small regions, as shown in Fig. 12. Secondly, these four models also generate strong artifacts, particularly SRFeat, as shown in Fig. 13 and Fig. 12. Based on visual observations in Fig. 12, SRFeat generates noise blocks pattern above the image and also incoherently artificial colorful dots, as shown in Fig. 13. Moreover, these four models build a super-resolved version with non-sharp edges, as seen in Fig. 12 compared to ESRGAN, SRGAN, and ours.

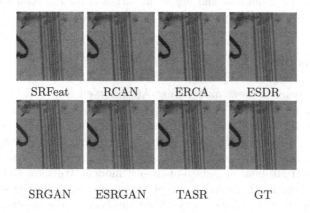

Fig. 12. Comparing models capability one preserving close edges and generating them well.

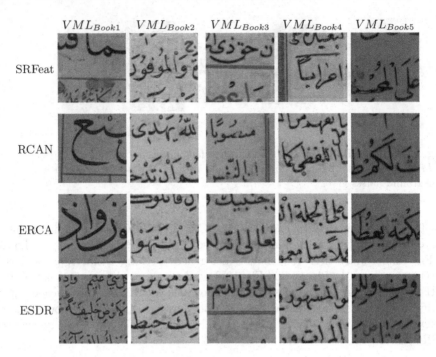

Fig. 13. Artificial colorful dots and regions in output of deep learning's models from multiple datasets. (Color figure online)

6   Conclusion

This work presents a new model Text-Attention-ed Super Resolution GAN (TASR-GAN), for historical documents super-resolution. In addition, we present a new loss function that guides the model to focus on the text edges on the documents and preserve small edges without corrupting them during the super-resolution process. To test our model, we use six historical datasets and compare our work with traditional and state-of-the-art models. We show that our model outperforms them in all datasets except two which we have the second highest result. This show the superiority of our method for the super-resolution of historical documents.

References

1. Super resolution techniques for medical image processing. In: 2015 International Conference on Technologies for Sustainable Development (ICTSD), pp. 1–6 (2015). https://doi.org/10.1109/ICTSD.2015.7095900
2. Banerjee, J., Namboodiri, A.M., Jawahar, C.: Contextual restoration of severely degraded document images. In: 2009 IEEE Conference on Computer Vision and Pattern Recognition, pp. 517–524. IEEE (2009)

3. Bulat, A., Tzimiropoulos, G.: Super-FAN: integrated facial landmark localization and super-resolution of real-world low resolution faces in arbitrary poses with GANs. CoRR abs/1712.02765 (2017). http://arxiv.org/1712.02765
4. Chen, J., Li, B., Xue, X.: Scene text telescope: text-focused scene image super-resolution. In: Proceedings of the IEEE/CVF Conference on Computer Vision and Pattern Recognition, pp. 12026–12035 (2021)
5. Dai, D., Wang, Y., Chen, Y., Gool, L.V.: How useful is image super-resolution to other vision tasks? CoRR abs/1509.07009 (2015). http://arxiv.org/1509.07009
6. Dong, C., Zhu, X., Deng, Y., Loy, C.C., Qiao, Y.: Boosting optical character recognition: a super-resolution approach. arXiv preprint arXiv:1506.02211 (2015)
7. Goodfellow, I., et al.: Generative adversarial nets. Adv. Neural. Inf. Process. Syst. **27**, 2672–2680 (2014)
8. Greenspan, H.: Super-resolution in medical imaging. Comput. J. **52**(1), 43–63 (2009)
9. Gupta, M.D., Rajaram, S., Petrovic, N., Huang, T.S.: Restoration and recognition in a loop. In: 2005 IEEE Computer Society Conference on Computer Vision and Pattern Recognition (CVPR 2005), vol. 1, pp. 638–644. IEEE (2005)
10. Haefner, B., Quéau, Y., Möllenhoff, T., Cremers, D.: Fight ill-posedness with ill-posedness: single-shot variational depth super-resolution from shading. In: Proceedings of the IEEE Conference on Computer Vision and Pattern Recognition (CVPR) (2018)
11. Haris, M., Shakhnarovich, G., Ukita, N.: Task-driven super resolution: object detection in low-resolution images. In: Mantoro, T., Lee, M., Ayu, M.A., Wong, K.W., Hidayanto, A.N. (eds.) ICONIP 2021. CCIS, vol. 1516, pp. 387–395. Springer, Cham (2021). https://doi.org/10.1007/978-3-030-92307-5_45
12. Huang, Y., Shao, L., Frangi, A.F.: Simultaneous super-resolution and cross-modality synthesis of 3D medical images using weakly-supervised joint convolutional sparse coding. In: Proceedings of the IEEE Conference on Computer Vision and Pattern Recognition, pp. 6070–6079 (2017)
13. Hui, T.-W., Loy, C.C., Tang, X.: Depth map super-resolution by deep multi-scale guidance. In: Leibe, B., Matas, J., Sebe, N., Welling, M. (eds.) ECCV 2016. LNCS, vol. 9907, pp. 353–369. Springer, Cham (2016). https://doi.org/10.1007/978-3-319-46487-9_22
14. Jolicoeur-Martineau, A.: The relativistic discriminator: a key element missing from standard GAN. arXiv preprint arXiv:1807.00734 (2018)
15. Kassis, M., Abdalhaleem, A., Droby, A., Alaasam, R., El-Sana, J.: VML-HD: the historical Arabic documents dataset for recognition systems. In: 1st International Workshop on Arabic Script Analysis and Recognition. IEEE (2017)
16. Lat, A., Jawahar, C.: Enhancing OCR accuracy with super resolution. In: 2018 24th International Conference on Pattern Recognition (ICPR), pp. 3162–3167. IEEE (2018)
17. Ledig, C., et al.: Photo-realistic single image super-resolution using a generative adversarial network. CoRR abs/1609.04802 (2016). http://arxiv.org/1609.04802
18. Lim, B., Son, S., Kim, H., Nah, S., Mu Lee, K.: Enhanced deep residual networks for single image super-resolution. In: Proceedings of the IEEE Conference on Computer Vision and Pattern Recognition (CVPR) Workshops (2017)
19. Liu, Y., Qin, Z., Luo, Z., Wang, H.: Auto-painter: cartoon image generation from sketch by using conditional generative adversarial networks. arXiv preprint arXiv:1705.01908 (2017)
20. Luo, C., Jin, L., Sun, Z.: Moran: a multi-object rectified attention network for scene text recognition. Pattern Recogn. **90**, 109–118 (2019)

21. Ma, J., Guo, S., Zhang, L.: Text prior guided scene text image super-resolution. arXiv preprint arXiv:2106.15368 (2021)
22. Ma, J., et al.: Arbitrary-oriented scene text detection via rotation proposals. IEEE Trans. Multimedia **20**(11), 3111–3122 (2018)
23. Ma, J., Zhetong, L., Zhang, L.: A text attention network for spatial deformation robust scene text image super-resolution (2022)
24. Marti, U.V., Bunke, H.: The IAM-database: an English sentence database for offline handwriting recognition. Int. J. Doc. Anal. Recogn. **5**(1), 39–46 (2002). https://doi.org/10.1007/s100320200071
25. Nasrollahi, K., Moeslund, T.B.: Super-resolution: a comprehensive survey. Mach. Vis. Appl. **25**(6), 1423–1468 (2014). https://doi.org/10.1007/s00138-014-0623-4
26. Nguyen, T.K., Hoang, H.T., Yoo, C.D.: GDCA: GAN-based single image super resolution with dual discriminators and channel attention. ArXiv abs/2111.05014 (2021)
27. Park, S.-J., Son, H., Cho, S., Hong, K.-S., Lee, S.: SRFeat: single image super-resolution with feature discrimination. In: Ferrari, V., Hebert, M., Sminchisescu, C., Weiss, Y. (eds.) ECCV 2018. LNCS, vol. 11220, pp. 455–471. Springer, Cham (2018). https://doi.org/10.1007/978-3-030-01270-0_27
28. Rasti, P., Uiboupin, T., Escalera, S., Anbarjafari, G.: Convolutional neural network super resolution for face recognition in surveillance monitoring. In: Perales, F.J.J., Kittler, J. (eds.) AMDO 2016. LNCS, vol. 9756, pp. 175–184. Springer, Cham (2016). https://doi.org/10.1007/978-3-319-41778-3_18
29. Ray, A., et al.: An end-to-end trainable framework for joint optimization of document enhancement and recognition. In: 2019 International Conference on Document Analysis and Recognition (ICDAR), pp. 59–64. IEEE (2019)
30. Sharma, M., Ray, A., Chaudhury, S., Lall, B.: A noise-resilient super-resolution framework to boost OCR performance. In: 2017 14th IAPR International Conference on Document Analysis and Recognition (ICDAR), vol. 1, pp. 466–471. IEEE (2017)
31. Shi, B., Bai, X., Yao, C.: An end-to-end trainable neural network for image-based sequence recognition and its application to scene text recognition. IEEE Trans. Pattern Anal. Mach. Intell. **39**(11), 2298–2304 (2016)
32. Shi, B., Yang, M., Wang, X., Lyu, P., Yao, C., Bai, X.: Aster: an attentional scene text recognizer with flexible rectification. IEEE Trans. Pattern Anal. Mach. Intell. **41**(9), 2035–2048 (2018)
33. Shi, W., et al.: Real-time single image and video super-resolution using an efficient sub-pixel convolutional neural network. In: 2016 IEEE Conference on Computer Vision and Pattern Recognition (CVPR), pp. 1874–1883. IEEE Computer Society, Los Alamitos, CA, USA (2016). https://doi.org/10.1109/CVPR.2016.207, https://doi.ieeecomputersociety.org/10.1109/CVPR.2016.207
34. Wang, W., et al.: Scene text image super-resolution in the wild. In: Vedaldi, A., Bischof, H., Brox, T., Frahm, J.-M. (eds.) ECCV 2020. LNCS, vol. 12355, pp. 650–666. Springer, Cham (2020). https://doi.org/10.1007/978-3-030-58607-2_38
35. Wang, W., et al.: TextSR: content-aware text super-resolution guided by recognition. CoRR abs/1909.07113 (2019). http://arxiv.org/1909.07113
36. Wang, X., et al.: ESRGAN: enhanced super-resolution generative adversarial networks. CoRR abs/1809.00219 (2018). http://arxiv.org/1809.00219
37. Wang, Z., Chen, J., Hoi, S.C.: Deep learning for image super-resolution: a survey. IEEE Trans. Pattern Anal. Mach. Intell. **43**, 3365–3387 (2020)

38. Zhang, H., Liu, D., Xiong, Z.: CNN-based text image super-resolution tailored for OCR. In: 2017 IEEE Visual Communications and Image Processing (VCIP), pp. 1–4. IEEE (2017)
39. Zhang, L., Zhang, H., Shen, H., Li, P.: A super-resolution reconstruction algorithm for surveillance images. Signal Process. **90**(3), 848–859 (2010)
40. Zhang, Y., Li, K., Li, K., Wang, L., Zhong, B., Fu, Y.: Image super-resolution using very deep residual channel attention networks. In: Ferrari, V., Hebert, M., Sminchisescu, C., Weiss, Y. (eds.) ECCV 2018. LNCS, vol. 11211, pp. 294–310. Springer, Cham (2018). https://doi.org/10.1007/978-3-030-01234-2_18
41. Zhou, X., et al.: EAST: an efficient and accurate scene text detector. In: Proceedings of the IEEE Conference on Computer Vision and Pattern Recognition (CVPR) (2017)
42. Zhu, S., Liu, S., Loy, C.C., Tang, X.: Deep cascaded bi-network for face hallucination. CoRR abs/1607.05046 (2016). http://arxiv.org/abs/1607.05046

Curt: End-to-End Text Line Detection in Historical Documents with Transformers

Benjamin Kiessling[1,2](✉) ⓘ

[1] École Pratique des Hautes Études, Paris, France
benjamin.kiessling@ephe.sorbonne.fr
[2] UMR 8546 CNRS-Université PSL (ENS-EPHE) - AOROC, Paris, France

Abstract. We present the curve transformer (CurT), a novel method of direct baseline detection that models document text line detection as set prediction of cubic Bézier curves, simplifying the layout analysis pipeline by removing the need for the laboriously hand-crafted postprocessing algorithms that are necessary with the current state of the art. CurT combines multiple appealing features: direct prediction enabling processing of material that is ill-suited for the prevailing methods adapting semantic segmentation backbones, a conceptually simple Transformer-based encoder-decoder architecture that can be extended to additional tasks beyond baseline detection, and increased computational efficiency in comparison to older approaches. In addition, we demonstrate that CurT achieves metrics that are competitive with methods based on semantic segmentation.

Training and inference code is available under Apache 2.0 license at https://github.com/mittagessen/curt.

Keywords: Document analysis · Machine learning · Text line detection · Object detection

1 Introduction

Document image analysis of historical material has seen a continued and rising interest over the last few decades, both from Computer Science researchers in search for more challenging material for their algorithms, scholars in the Human and Social Sciences aiming to apply computational analysis on ever larger corpora, and libraries and archives digitizing collections to ensure accessibility of humanity's collective cultural heritage. High quality document layout analysis is a keystone technology in any of those efforts reliant on retrodigitization on both the level of individual lines and higher order zones demarking textual and non-textual content.

While Automatic Text Recognition has achieved tremendous progress in the last decade, with typical character error rates well below 10% even for highly challenging handwritten material, these methods are uniquely dependent on accurate prior segmentation as both the best performing and most widespread systems rely on segmented text lines as inputs. While segmentation-less methods have been proposed from time to time, they uniformly suffer from incomparable requirements on training time and data, are less robust regarding the nature of the texts to be recognized, or are not competitive with pre-segmenting approaches. Thus text recognition workflows are almost exclusively constructed out of a preliminary text line extraction step followed by actual text recognition with any failures in the segmentation directly translating into text recognition errors.

As can be expected for its central role, a large number of methods and paradigms, with varying focuses over the years as text recognition methods have grown in capability, have been proposed to deal with various handwritten and machine-printed historical documents. Early algorithms to extract individual characters from a page utilizing conventional computer vision methods have largely been supplanted first by hand-crafted algorithms detecting whole lines such as [1] and, later, machine learning-based approaches. Through these advances many documents are now in the reach of digitization without close human supervision but significant obstacles remain: generalization on out-of-domain documents is generally poor, especially for degraded material or different writing supports, implicit assumptions on the nature of the text, e.g. writing direction, that often don't hold for non-Latin-script documents are widespread, and methodological limitations of many methods make detection of overlapping and rotated writing difficult.

Layout analysis methods based on object detection systems offer the promise to overcome many of the conceptual limitations of these earlier systems. While established *indirect* object detection algorithms employing surrogate regression and classification problems that are highly dependent on postprocessing steps such as non-maximum suppression to collapse near-duplicate predictions, designs of anchor sets, and heuristics assigning entities to anchors are difficult to adapt to non-box shaped objects required for text line detection in historical documents, a new class of direct object detectors based on vision Transformers are much more flexible in their output data models.

2 Related Work

2.1 Transformers for Computer Vision

Transformers are a class of artificial neural network architecture that is characterised by a self-attention mechanism that learns relationships between elements of sets. In contrast to conventional recurrent neural networks that process sequences recursively and in practice can model long-term relationships only in a limited manner, Transformers are able to attend to complete sequences. This particular attention mechanism computing attention tensors across multiple heads (multi-head self-attention) along with minimal inductive biases in comparison to recurrent (sequentiality, recursion) and convolutional (translation invariance, locality) neural networks through the exclusive use of fully connected layers are the Transformer's distinguishing features. While Transformer layers can be arranged in a number of different configurations depending on task, the original and most widely used one organizes them into a encoder-decoder configuration.

Originally proposed in [32] for Natural Language Processing, Transformers have demonstrated astounding improvement on the then current state of the art for language modelling tasks such as text classification, machine translation, or question answering. The ability of Transformer networks to be effectively scaled up and trained with very large parameter counts that consistently outperform prior more lightweight models, e.g. the 340 million parameter BERT, 175 billion parameter GPT-3, up to the latest Switch transformers with up to 1.6 trillion parameters, have achieved generalization and adaptability that makes the impact of these architectures difficult to overstate.

The breakthroughs in performance achieved with Transformers have caused great interest outside of the NLP domain and the computer vision community has started to adapt these models for vision and multi-modal learning tasks. The resulting systems can largely by divided into hybrid architectures combining CNN encoders and Transformer decoders and architectures replacing convolutions altogether. The Vision Transformer (ViT) [8] was one of the first showcases for a standard Transformer architecture operating on flattened image patches producing competitive results on a number of computer vision tasks, albeit requiring pre-training on the extremely large proprietary JFT dataset. DeiT [31] demonstrated training transformers on the more moderately sized ImageNet dataset with state-of-the-art results through a teacher-student distillation approach with a CNN teacher model. These fixed-scale methods perform well on sparse prediction tasks such as image classification but the quadratic complexity of self-attention limits their applicability to higher-resolution images. Multi-scale architectures that merge tokens reducing the sequence length along a cascade of hierarchical layers have been proposed as a better alternative for dense prediction, e.g. object detection or semantic segmentation. Examples of these are the Swin Transformer [19], Pyramid Vision Transformers [34], and Focal Transformers [36]. An extensive survey of self-attention and Transformer-like methods for a wide range of computer vision tasks can be found in [12].

2.2   DETR and Variants

DETR [2] is an object detector built upon a Transformer encoder-decoder architecture combined with a set-based loss that forces unique predictions for each ground-truth bounding box through bipartite matching.

The model operates on input feature maps extracted by a CNN backbone, in the originally proposed implementation ResNet-50 and ResNet-101, that are fed into a standard Transformer encoder-decoder architecture. The inputs of the decoder stage are the transformed image features from the encoder and N learned positional encodings called *object queries* that condition the decoder to produce N distinct output embeddings from the transformed image features. A simple linear projection and a 3-layer feed-forward network are used to decode the output embeddings into classes and regress the normalized bounding box coordinates $\mathbf{b} \in \{b_{cx}, b_{cy}, b_w, b_h\}$ respectively.

The architectural simplicity of DETR and lack of hand-crafted algorithms such as non-maximum suppression or anchors often required in non-direct methods such as Faster R-CNN [27], YOLOv3 [26], and SSD [18] among many others makes it an attractive design for object detection and derived tasks. Unfortunately the original design suffers from several major drawbacks. The first is the quadratic computational complexity of the attention weight computation in the Transformer encoder with regard to the input size, putting a low upper bound on the maximum input resolution which makes detection of small objects difficult.

The second is the slow convergence with the original approach requiring a very long training schedule of 500 epochs to converge on the COCO dataset, roughly 10 to 20 times slower than Faster R-CNN's typical 30 epochs. The primary reason identified in [37] for this slow convergence is the suboptimal initial initialization of the attention modules casting nearly uniform attention weights to all pixels in the input feature maps requiring many epochs to achieve sufficient sparsity for the decoder to detect object effectively. This slow convergence is exacerbated by the lack of pre-training of the Transformer.

Another contributing factor to these long training times is the instability of the bipartite matching during initial epochs, as the assignment is essentially random in the early phases of training [28].

An abundance of detection transformers variants are intended to resolve these problems. Deformable DETR [37] decouples the computational cost of the attention module from the input feature maps through a deformable attention mechanism that attends only to a small set of sampling points around a reference point while at the same time incorporating multi-scale features improving recall of small objects. Conditional DETR [22] narrows down the spatial range for localizing object regions via learning the decoder embedding conditioned on a spatial query. UP-DETR [6] pre-trains standard DETR using a multi-query patch pretext task. FP-DETR [33] proposes a way to pre-train an encoder-only Deformable DETR object detector on a classification task. [28] adapt an FCOS-like object detector [30] with an encoder-only Transformer block and a modified bounding-box specific matching scheme that improves stability in the early stages of training. DN-DETR [15] halves the training time of DETR-like methods by introducing an auxiliary denoising task that bypasses the bipartite matching to circumvent early instability while at the same time improving accuracy on baseline DETR.

2.3   Text Baseline Detection

As a representation of text lines, baselines have a long history in historical document layout analysis (see [16] for a survey of early methods) and has recently both enjoyed a resurgence in research interest and widespread use in a number of practical text recognition systems for handwritten and printed documents such as Transkribus [5] and eScriptorium [14]. While early methods employing this paradigm utilized conventional image processing methods, the challenging cBAD competitions in 2017 and 2019 have triggered the publication of a large number of deep learning-based methods.

Fig. 1. Example taken from the cBAD dataset visualizing the originally annotated baseline (purple), least squares fitted Bézier curve control points (blue circles), and Bézier curve interpolated at 20 equally spaced points.

Baselines are a term originating from typography referring to a virtual polyline on which most characters of a text line rest upon or hang from. Despite not being universal, or in fact sometimes not being located at the bottom of the line as is the case with Hebrew and various Indict scripts, analogues that serve the same purpose for text recognition purposes can often be devised for material lacking true baselines, e.g. approximate centerlines for Chinese characters. While not sufficient by themselves in most cases, baselines in combination with bounding polygons can be ingested by modern line-based text recognizers with minimal adaptation while at the same time requiring only modest effort for manual annotation.

The dominant approach to baseline detection with trainable methods employing artificial neural networks is pixel-wise classification on single or multi-scale feature maps to label baseline pixels or some derivation thereof such as corpus height lines, toplines, or centerlines. These are sometimes augmented by auxiliary labels to improve accuracy or enable computation of other line characteristics such as orientation. In a postprocessing step baseline instances are extracted through grouping of baseline pixels, typically through thresholding, local connectedness, skeletonization, or interline distance estimation. Examples of this class of systems are ARU-Net [11], dhSegment [23], and BLLA [13]. While sufficiently powerful for many applications, semantic segmentation-based text line extraction has limits: most systems lack a way to determine text line orientation, have a tendency to merge and/or split close lines, are conceptually unable to deal with intersecting lines, and often impose further limitations on possible line shapes in the postprocessing stage.

3   Contribution

The main contribution of this work is a document layout analysis system based on object detection paradigms that is:

1. largely **postprocessing free** having one parameter during inference, a simple threshold of the objectness score.

2. almost **unconstrained** with regard to shape, orientation, and overlapping of the **type of text lines** to be extracted.
3. **extensible** towards other tasks, e.g. text line boundary and region detection, document classification, or reading order determination.

4 The CurT Model

CurT is heavily inspired by the DETR system for object detection. As such it has the same fundamental components: a set prediction loss forcing unique matching between predicted and ground truth baseline curves, and an architecture that predicts in a single pass a set of objects and models their relation.

For the reasons described above the verbatim method proposed in the seminal paper is largely unsuitable as a layout analysis system for historical documents. The limits imposed by the computational and memory complexity on input image feature size cause suboptimal performance on the detection of small text lines and the long training times required for convergence make learning for new material, a frequent requirement when considering the variety of historical writing one might want to segment, impractical.

From the variants presented above we adapt Conditional DETR [22] and modify it to better reflect our data model.

4.1 Text Line Data Model

The principal difference between the output of an off-the-shelf object detector and our text line extractor is the modelisation of the detected objects instances as polylines that are placed typically on the bottom of the line corpus, the baseline. As DETR-style models regress object instances encoded with a fixed dimensionality a flexible, fixed-length line representation that is able to deal with arbitrarily shaped text is needed. While directly regressing the end points of line segments of a polyline is possible, the high output dimensionality required to accurately model complex text shapes and the smoothness of handwriting makes this approach unappealing. Bézier curves on the other hand are able to represent complex shapes with a low, fixed number of control points.

A Bézier curve represents a parametric curve $c(t)$ that uses the Bernstein Polynomial as its basis:

$$c(t) = \sum_{i=0}^{n} b_i B_{i,n}(t), 0 \leq t \leq 1 \tag{1}$$

with n being the degree of the curve, b_i the i-th control point, and $B_{i,n}(t)$ the Bernstein basis polynomial:

$$B_{i,n}(t) = \binom{n}{i} t^i (1-t)^{n-i}, i = 0, \ldots, n \tag{2}$$

To find the appropriate degree of the Bézier representation, a number of pages from the pre-training dataset (see Sect. 5.1) where sampled and the manually annotated polylines therein fit to curves of differing degrees. Cubic Bézier curves $n = 3$ were sufficient to model the lines in these document images to a sufficient degree (Fig. 1).

Converting the original polyline points to a cubic Bézier curve is done with a standard least squares fitting of the curve control points; the first and last points of the polyline are set as the first and last control points respectively. As polylines with a low number of line segments result in inaccurately parametrized curves, ground-truth lines are interpolated to contain at least 8 points.

4.2   Curve Detection Set Prediction Loss

As DETR and its derivations, CurT infers a set of N predictions of objects, in our case baseline curves encoded as cubic Bézier curves, and their associated class in a single decoder pass. Typically the number of possible text line predictions N is somewhat larger than the actual number of lines found in the image, so the possible classes are padded with a no object class $\varnothing$.

The loss operates in two stages: a matching phase where each predicted object (curve) is assigned to a ground truth object through an optimal bipartite matching computed with the Hungarian algorithm, followed by a loss optimizing the object-specific curve losses. The overall structure is similar to the set prediction loss proposed in [2] but modified to account for the prediction of baseline curves instead of bounding boxes.

Given the ground truth y and $\hat{y} = \{\hat{y}_i\}_{i=1}^N$, the set of N predictions with y padded with $\varnothing$ (no object) if smaller than N, we find an optimal bipartite matching between these two sets as a permutation of N elements $\sigma \in \mathfrak{S}_N$ with the lowest cost:

$$\hat{\sigma} = \arg\min_{\sigma \in \mathfrak{S}_N} \sum_i^N \mathcal{L}_{\text{match}}(y_i, \hat{y}_{\sigma(i)}) \tag{3}$$

where $\mathcal{L}_{\text{match}}(y_i, \hat{y}_{\sigma(i)})$ is a pair-wise matching cost between ground truth y_i and a prediction with index $\sigma(i)$.

The matching cost is a linear combination of the class prediction and the similarity between predicted and ground truth curves. Let the i-th ground truth element be $y_i = (t_i, c_i)$ where t_i is the target class label (which may be $\varnothing$) and $c_i \in [0, 1]^8$ is a vector defining the coordinates of the four control points relative to the image size.

For such an element y_i, we define the matching cost as $\mathcal{L}_{\text{match}}(y_i, \hat{y}_{\sigma(i)})$ as $-\mathbb{1}_{\{t_i \neq \varnothing\}} \alpha \mathcal{L}_{\text{focal}}(t_i, \hat{p}_{\sigma(i)}(t_i)) + \mathbb{1}_{\{t_i \neq \varnothing\}} \beta \ell_1(c_i, \hat{c}_{\sigma(i)})$ given for a prediction with index $\sigma(i)$ the probability of class t_i as $\hat{p}_{\sigma(i)}(t_i)$ and the curve prediction as $\hat{c}_{\sigma(i)}$. $\alpha = 1.0$ and $\beta = 5.0$ are free parameters defining the relative weight between class and curve score in the matching cost.

Once a optimal matching has been computed, the Hungarian loss is computed on the matched pairs. Similar to the matching cost it is a linear combination of class prediction focal loss and the curve loss:

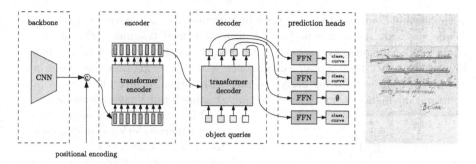

Fig. 2. CurT operates in a standard Transformer encoder-decoder configuration on feature maps computed with a conventional convolutional backbone. The feature maps from the backbone are flattened and positional encodings are concatenated to it before being passed into the encoder. The decoder then takes as input a fixed number of learned positional embeddings, called *object queries* that are mapped with a linear projection into reference points (see Fig. 3) and the encoder embeddings. Its output embeddings are then decoded into separate class scores and curve regressions with a shared feed-forward network (FFN).

$$\mathcal{L}_{\text{Hungarian}}(y, \hat{y}) = \sum_{i=1}^{N} \left[\gamma \mathcal{L}_{\text{focal}}(t_i, \hat{p}_{\hat{\sigma}(i)}(t_i)) + \mathbb{1}_{\{t_i \neq \varnothing\}} \epsilon \ell_1 (c_i, \hat{c}_{\hat{\sigma}}(i)) \right] \quad (4)$$

where $\hat{\sigma}$ is the optimal assignment computed in the first step (3) and $\gamma = 1.0$ and $\epsilon = 5.0$ are the relatives weights accorded to classification and regression losses respectively.

The component losses in both the matching and loss computation phase are focal loss with loss weight of four [17] for object classification and ℓ_1 distance for the curve regression.

4.3 CurT Architecture

The overall architecture of CurT depicted in Fig. 2 is based on a modification of the Conditional DETR [22] variant of the originally proposed Detection Transformer. It contains three main components: a convolutional backbone extracting an input feature representation, an encoder-decoder Transformer, and two feed forward networks predicting line class and the Bézier curve control points respectively.

Backbone. The backbone network is a conventional CNN backbone generating a lower-resolution feature map $\mathbf{f} \in \mathbb{R}^{C \times H \times W}$ from an input image $\mathbf{x} \in \mathbb{R}^{3 \times H_0 \times W_0}$. Resnet-50 (used by DETR and most variants), SegFormer [35] (a multi-head attention-based architecture originally devised for efficient semantic segmentation), and EfficientNetv2 [29] were evaluated informally as possible backbones. All 3 architectures produce output feature maps of size $\mathbf{f} \in \mathbb{R}^{C \times \frac{H}{32} \times \frac{W}{32}}$

with variable C and higher resolution feature maps available from earlier layers. While the tests were only performed on a shortened training cycle, backbone choice and configuration seems to not impact the training loss of the model drastically apart from slower convergence with the SegFormer backbone and steeper drop in loss during early training with larger segmentation maps (and concomitant higher memory consumption at same input image resolution).

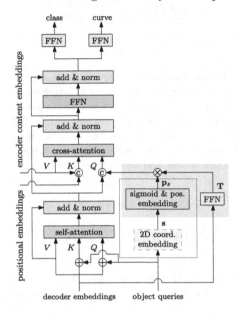

Fig. 3. A depiction of one decoder layer in CurT. The difference to the Conditional DETR decoder lies in the dimensionality of the reference points $s \in \mathbb{R}^8$ with $p_s \in [0,1]^8$ used in the conditional spatial query construction (red box) and the output regression feed-forward network. s represents the unnormalized reference points, p_s the normalized reference points including the fixed positional encoding. Raw object query dimensionality remains unchanged. Original figure from [22].

Thus, the backbone model chosen is an EfficientNetV2-L pretrained on Imagenet 21k with the last layer feature maps $\mathbf{f} \in \mathbb{R}^{640 \times \frac{H_0}{32} \times \frac{W_0}{32}}$ being used as the input embeddings of the Transformer part of the model. This choice optimizes convergence speed, memory consumption, and computational requirements.

Transformer Encoder. The CurT encoder follows the standard Transformer encoder layer construction of a multi-head attention module and a feed forward network (FFN).

Encoder inputs are reduced in dimensionality with a 1×1 convolution of the input feature map $\mathbf{f}$ with $C = 640$ to an embedding feature map $\mathbf{z_0} \in \mathbb{R}^{d \times H \times W}$ with $d = 256$. As the standard Transformer encoder layers expect a sequence the two-dimensional map $\mathbf{z_0}$ is collapsed into a $d \times HW$ tensor. As with most other applications of the Transformer architecture in vision a fixed sinusoidal positional encoding [24] is applied to the encoder inputs to account for the Transformer's permutation invariance.

Transformer Decoder. The CurT decoder cross-attention mechanism in the decoder layers is largely identical to the construction of the Conditional DETR decoder, which modifies the DETR decoder cross-attention by decoupling queries into a content and spatial part by decoding the object queries into explicit reference points which are then concatenated to the decoder embeddings with the aim to accelerate training.

The primary difference in the construction of our decoder is the use of multiple reference points in the construction of the conditional spatial query (see-gray-shaded box of Fig. 3) from object queries and decoder embeddings. Each object query is decoded not into a single center reference point $\mathbf{s_c} \in \mathbb{R}^2$ but four separate reference points $\mathbf{s} \in \mathbb{R}^8$. This is motivated in part by the formulation of the curve regression (see Sect. 4.3) but is primarily intended to aid the spatial attention mechanism to more easily deliminate the spatial extent of the baseline, similar to how singular reference points translate the attention to the extremities of the object box in the original architecture.

For a detailed description of the operation of the Transformer encoder and decoder we defer to [2,32] and [22] respectively.

Curve Regression. Following the regression scheme of Conditional DETR the control points of a candidate curve are predicted from each decoder layer as follows:

$$c = \text{sigmoid}(\text{FFN}(\mathbf{g}) + \mathbf{s}) \tag{5}$$

where $\mathbf{g}$ is the decoder embedding, $\mathbf{c} \in [0,1]^8$ an eight-dimensional vector of the normalized curve control points, and $\mathbf{s} \in \mathbb{R}^8$ the unnormalized coordinates of the reference points. This differs from the originally proposed:

$$\mathbf{b} = \text{sigmoid}(\text{FFN}(\mathbf{g}) + [\mathbf{s_c}^\top 0 \ 0]^\top) \tag{6}$$

with $\mathbf{b} = [b_{cx}b_{cy}b_wb_h] \in [0,1]^4$ where the reference point only impacts the regression of the bounding box center point and extremities of the bounding box are completely regressed from the decoder embeddings.

The FFN in the curve regressor is a three-layer multi-layer-perceptron with ReLU activation function, a hidden dimension of d (256 per default as per above) and output dimension of eight.

Line Class Prediction. The classification score for each candidate curve is directly predicted from the decoder embeddings through an FNN followed by a softmax activation from each decoder layer:

$$t = \text{softmax}(\text{FFN}(\mathbf{g})) \tag{7}$$

5   Experiments

5.1   Dataset and Evaluation Protocol

We perform experiments on the standard cBAD 2019 baseline detection dataset, containing 755 training, 778 validation, and 1511 test images. As this dataset is insufficient in size to train a CurT model from scratch, an auxiliary dataset of 38k annotated handwritten and machine-printed page images is assembled from the HTR-United repository [3], the NewsEye project [9], the Kuzushiji cursive Japanese dataset [4] with automatically annotated baselines, and an additional set of non-public data including highly challenging material from the Princeton

Geniza Project. The quality and annotation standards vary widely across this large dataset, often only containing annotations for parts of the text, a mixture of top-, center-, and baselines, and a variety of ontologies for text line and region classes. As there is little coherence across the chosen datasets and the standard cBAD evaluation scheme for text line detection disregards text line classification, line classes are merged into one default class and regions are suppressed.

5.2 Implementation Details

Strong augmentation is applied during training, with inputs being resized randomly to a longest edge size between 900 px and 1800 px, random rectangular crops followed by resizing, and random photometric distortion.

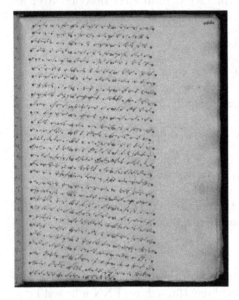

The number of object queries is increased to 1000 from the original 100 to account for the higher number of text lines on a typical page in comparison to objects annotated in the COCO 2017 images. As in other applications the number of queries is chosen to be in large excess of the possible number of objects in any input ($\mu = 7.19$ for objects per image for COCO 2017 resulting in 10–40 times the number of object queries for DETR and variants), following the same approach would increase computational requirements considerably for text line detection as the mean number of lines per page is 54.3 ($\sigma = 123.8$) with a small number of pages containing above 500 and even 1000 lines in comparison to the maximum 63 objects in a COCO image.

Fig. 4. Example output for a page taken from the cBAD dataset visualizing the Bézier curve control points (blue circles) and Bézier curve interpolated at 20 equally spaced points.

By default models are pretrained on the large general dataset of 38k page images for 100 epochs and then fine-tuned for an additional 50 epochs on the target dataset. The model is trained using the AdamW optimizer [20] with base learning rate of 10^{-4}, $\beta_1 = 0.9$, $\beta_2 = 0.999$, and weight decay of 10^{-4} with a lower learning rate of 10^{-5} for the convolutional backbone. Learning rate is scheduled according to a single cosine cycle with an initial warmup over 8000 training steps as the more widely used fixed schedule with a 10-fold decay after 80% of epochs results in convergence at very high losses for the text line detection task. The number of encoder and decoder layers is set to 3 respectively. Traditional dropout with $p = 0.1$ is applied to the transformer. Auxiliary losses are computed on the output embedding decoded with the prediction heads sharing weights at each decoder layer.

5.3   Overall Performance

We report precision, recall, and F-value averaged over the 1511 test set images of the cBAD 2019 dataset in Table 1. Baseline results are from the winning method of the complex track of the cBAD 2017 competition. The metrics were computed with the standard schema described in [10]. A sample from the output on the test set is shown in Fig. 4.

Table 1. CurT text baseline detection performance on cBAD 2019 dataset (values for other methods from [7])

Method	Precision	Recall	F-value
Baseline (DMRZ-17)	0.773	0.743	0.758
TJNU	0.852	0.885	0.868
UPVLC	0.911	0.902	0.907
DMRZ	0.925	0.905	0.915
Planet	0.937	0.926	0.931
CurT	0.909	0.908	0.908

As shown by the competitive results in comparison to semantic segmentation-based methods our approach is able to detect text baselines effectively under various challenging conditions such as faded ink, degraded writing surfaces, and variously oriented lines.

5.4   Ordered Prediction

In addition to models trained with the set loss described above, an alternative formulation without bipartite matching was also evaluated. The chief purpose is to determine the ability of the system to learn a basic reading order in addition to text line detection by enforcing that the prediction at $\hat{y}_i, i \leq N$ corresponds to the y_i in the original ground truth. The basic assumption underlying this experiment is that reading order can be determined using fixed geometric relationships, i.e. that the spatial attention conditioned on the object queries is sufficient to determine a basic reading order.

While such a basic system would evidently be insufficient for practical purposes without the introduction of additional semantic depth like the distinction of headings, notes, insertions, main text, etc. its capabilities would be in line with the current state of the art of heuristics, learned rule based systems [21], and recent neural approaches [25].

An obvious challenge for this approach to ordered prediction is that object query utilization is highly dependent on the spatial frequency of baselines in the source document, i.e. object queries need to attend to areas of the document occurring earlier in the reading order for documents with a high number of text lines and later areas for sparse documents. As somewhat expected CurT failed to converge for this considerably more challenging task.

5.5   Further Extensions

A straightforward next target for a text line detection system is the extension to region detection and text line boundary detection. In DETR these tasks were analogously modelled as panoptic segmentation, predicting pixel-wise maps for both stuff and thing classes in COCO with a multi-attentional (M) mask head that predicts $M \times N$ attention maps simultaneously from the decoder embeddings. These attention maps are then upsampled through a FPN-like architecture incorporating multi-scale feature maps from the convolutional backbone network, followed by a classification layer to produce the final output pixel maps.

The drawback of simultaneous prediction of all segmentation maps is the linear increase of memory consumption with the number of object queries, in addition to the high base memory requirements for high resolution inputs. A future extension to CurT is a mask head predicting regions and text line boundaries sequentially.

6 Conclusion

This work presents the first attempt to adapt a modern direct object detection system for the task of text baseline detection in historical documents. The capabilities of this approach are demonstrated on the widely used cBAD 2019 dataset where the proposed method was shown to perform well. While only rudimentarily explored at this time, the proposed framework offers the perspective to solve a number of ancillary tasks to document layout analysis such as region detection and text line boundary detection, or reading order computation.

References

1. Breuel, T.M.: Two geometric algorithms for layout analysis. In: Lopresti, D., Hu, J., Kashi, R. (eds.) DAS 2002. LNCS, vol. 2423, pp. 188–199. Springer, Heidelberg (2002). https://doi.org/10.1007/3-540-45869-7_23
2. Carion, N., Massa, F., Synnaeve, G., Usunier, N., Kirillov, A., Zagoruyko, S.: End-to-end object detection with transformers. In: Vedaldi, A., Bischof, H., Brox, T., Frahm, J.-M. (eds.) ECCV 2020. LNCS, vol. 12346, pp. 213–229. Springer, Cham (2020). https://doi.org/10.1007/978-3-030-58452-8_13
3. Chagué, A., Clérice, T., Romary, L.: HTR-united : mutualisons la vérité de terrain! In: DHNord2021 - Publier, partager, réutiliser les données de la recherche : les data papers et leurs enjeux. MESHS, Lille, France (2021)
4. Clanuwat, T., Bober-Irizar, M., Kitamoto, A., Lamb, A., Yamamoto, K., Ha, D.: Deep learning for classical Japanese literature. CoRR abs/1812.01718 (2018)
5. Colutto, S., Kahle, P., Hackl, G., Mühlberger, G.: Transkribus. a platform for automated text recognition and searching of historical documents. In: 15th International Conference on eScience, eScience 2019, San Diego, CA, USA, 24–27 September 2019, pp. 463–466. IEEE (2019)
6. Dai, Z., Cai, B., Lin, Y., Chen, J.: UP-DETR: unsupervised pre-training for object detection with transformers. CoRR abs/2011.09094 (2020)
7. Diem, M., Kleber, F., Sablatnig, R., Gatos, B.: cBAD: ICDAR 2019 competition on baseline detection. In: 2019 International Conference on Document Analysis and Recognition (ICDAR), pp. 1494–1498 (2019)

8. Dosovitskiy, A., et al.: An image is worth 16×16 words: transformers for image recognition at scale. In: 9th International Conference on Learning Representations, ICLR 2021, Virtual Event, Austria, 3–7 May 2021. OpenReview.net (2021)
9. Doucet, A., et al.: NewsEye: a digital investigator for historical newspapers. In: Estill, L., Guiliano, J. (eds.) 15th Annual International Conference of the Alliance of Digital Humanities Organizations, DH 2020, Ottawa, Canada, 20–25 July 2020, Conference Abstracts (2020)
10. Gruning, T., Labahn, R., Diem, M., Kleber, F., Fiel, S.: READ-BAD: a new dataset and evaluation scheme for baseline detection in archival documents. In: 13th IAPR International Workshop on Document Analysis Systems, DAS 2018, Vienna, Austria, 24–27 April 2018, pp. 351–356. IEEE Computer Society (2018)
11. Grüning, T., Leifert, G., Strauß, T., Michael, J., Labahn, R.: A two-stage method for text line detection in historical documents. Int. J. Doc. Anal. Recogn. **22**(3), 285–302 (2019). https://doi.org/10.1007/s10032-019-00332-1
12. Khan, S.H., Naseer, M., Hayat, M., Zamir, S.W., Khan, F.S., Shah, M.: Transformers in vision: a survey. CoRR abs/2101.01169 (2021)
13. Kiessling, B.: A modular region and text line layout analysis system. In: 17th International Conference on Frontiers in Handwriting Recognition, ICFHR 2020, Dortmund, Germany, 8–10 September 2020, pp. 313–318. IEEE (2020)
14. Kiessling, B., Tissot, R., Stokes, P.A., Ezra, D.S.B.: eScriptorium: an open source platform for historical document analysis. In: 2nd International Workshop on Open Services and Tools for Document Analysis, OST@ICDAR 2019, Sydney, Australia, 22–25 September 2019, pp. 19. IEEE (2019)
15. Li, F., Zhang, H., Liu, S., Guo, J., Ni, L.M., Zhang, L.: DN-DETR: accelerate DETR training by introducing query denoising. CoRR abs/2203.01305 (2022)
16. Likforman-Sulem, L., Zahour, A., Taconet, B.: Text line segmentation of historical documents: a survey. Int. J. Doc. Anal. Recogn. **9**(2–4), 123–138 (2007). https://doi.org/10.1007/s10032-006-0023-z
17. Lin, T., Goyal, P., Girshick, R.B., He, K., Dollár, P.: Focal loss for dense object detection. In: IEEE International Conference on Computer Vision, ICCV 2017, Venice, Italy, 22–29 October 2017, pp. 2999–3007. IEEE Computer Society (2017)
18. Liu, W., et al.: SSD: single shot multibox detector. In: Leibe, B., Matas, J., Sebe, N., Welling, M. (eds.) ECCV 2016. LNCS, vol. 9905, pp. 21–37. Springer, Cham (2016). https://doi.org/10.1007/978-3-319-46448-0_2
19. Liu, Z., et al.: Swin transformer: hierarchical vision transformer using shifted windows. CoRR abs/2103.14030 (2021)
20. Loshchilov, I., Hutter, F.: Decoupled weight decay regularization. In: 7th International Conference on Learning Representations, ICLR 2019, New Orleans, LA, USA, 6–9 May 2019. OpenReview.net (2019)
21. Malerba, D., Ceci, M., Berardi, M.: Machine learning for reading order detection in document image understanding. In: Marinai, S., Fujisawa, H. (eds.) Machine Learning in Document Analysis and Recognition, Studies in Computational Intelligence, vol. 90, pp. 45–69. Springer, Heidelberg (2008). https://doi.org/10.1007/978-3-540-76280-5_3
22. Meng, D., et al.: Conditional DETR for fast training convergence. In: 2021 IEEE/CVF International Conference on Computer Vision, ICCV 2021, Montreal, QC, Canada, 10–17 October 2021, pp. 3631–3640. IEEE (2021)
23. Oliveira, S.A., Seguin, B., Kaplan, F.: dhSegment: a generic deep-learning approach for document segmentation. In: 16th International Conference on Frontiers in Handwriting Recognition, ICFHR 2018, Niagara Falls, NY, USA, 5–8 August 2018, pp. 7–12. IEEE Computer Society (2018)

24. Parmar, N., et al.: Image transformer. In: Dy, J.G., Krause, A. (eds.) Proceedings of the 35th International Conference on Machine Learning, ICML 2018, Stockholmsmässan, Stockholm, Sweden, 10–15 July 2018. Proceedings of Machine Learning Research, vol. 80, pp. 4052–4061. PMLR (2018)
25. Quirós, L., Vidal, E.: Reading order detection on handwritten documents. Neural Comput. Appl. **34**(12), 9593–9611 (2022). https://doi.org/10.1007/s00521-022-06948-5
26. Redmon, J., Farhadi, A.: YOLOv3: an incremental improvement. CoRR abs/1804.02767 (2018)
27. Ren, S., He, K., Girshick, R.B., Sun, J.: Faster R-CNN: towards real-time object detection with region proposal networks. In: Cortes, C., Lawrence, N.D., Lee, D.D., Sugiyama, M., Garnett, R. (eds.) Advances in Neural Information Processing Systems 28: Annual Conference on Neural Information Processing Systems 2015, 7–12 December 2015, Montreal, Quebec, Canada, pp. 91–99 (2015)
28. Sun, Z., Cao, S., Yang, Y., Kitani, K.: Rethinking transformer-based set prediction for object detection. CoRR abs/2011.10881 (2020)
29. Tan, M., Le, Q.V.: EfficientNetV2: smaller models and faster training. In: Meila, M., Zhang, T. (eds.) Proceedings of the 38th International Conference on Machine Learning, ICML 2021, 18–24 July 2021, Virtual Event. Proceedings of Machine Learning Research, vol. 139, pp. 10096–10106. PMLR (2021)
30. Tian, Z., Shen, C., Chen, H., He, T.: FCOS: fully convolutional one-stage object detection. In: ICCV (2019)
31. Touvron, H., Cord, M., Douze, M., Massa, F., Sablayrolles, A., Jégou, H.: Training data-efficient image transformers & distillation through attention. In: Meila, M., Zhang, T. (eds.) Proceedings of the 38th International Conference on Machine Learning, ICML 2021, 18–24 July 2021, Virtual Event. Proceedings of Machine Learning Research, vol. 139, pp. 10347–10357. PMLR (2021)
32. Vaswani, A., et al.: Attention is all you need. In: NeurIPS (2017)
33. Wang, W., Cao, Y., Zhang, J., Tao, D.: FP-DETR: detection transformer advanced by fully pre-training. In: International Conference on Learning Representations (2022)
34. Wang, W., et al.: Pyramid vision transformer: a versatile backbone for dense prediction without convolutions. In: 2021 IEEE/CVF International Conference on Computer Vision, ICCV 2021, Montreal, QC, Canada, 10–17 October 2021, pp. 548–558. IEEE (2021)
35. Xie, E., Wang, W., Yu, Z., Anandkumar, A., Alvarez, J.M., Luo, P.: SegFormer: simple and efficient design for semantic segmentation with transformers. In: Ranzato, M., Beygelzimer, A., Dauphin, Y.N., Liang, P., Vaughan, J.W. (eds.) Advances in Neural Information Processing Systems 34: Annual Conference on Neural Information Processing Systems 2021, NeurIPS 2021, 6–14 December 2021, virtual, pp. 12077–12090 (2021)
36. Yang, J., et al.: Focal self-attention for local-global interactions in vision transformers. CoRR abs/2107.00641 (2021)
37. Zhu, X., Su, W., Lu, L., Li, B., Wang, X., Dai, J.: Deformable DETR: deformable transformers for end-to-end object detection. CoRR abs/2010.04159 (2020)

Date Recognition in Historical Parish Records

Laura Cabello Piqueras[1]([✉]) [ID], Constanza Fierro[1] [ID], Jonas F. Lotz[1,2] [ID],
Phillip Rust[1] [ID], Joen Rommedahl[3] [ID], Jeppe Klok Due[3] [ID], Christian Igel[1] [ID],
Desmond Elliott[1] [ID], Carsten B. Pedersen[4] [ID], Israfel Salazar[5] [ID],
and Anders Søgaard[1] [ID]

[1] University of Copenhagen, Copenhagen, Denmark
{lcp,c.fierro,jonasf.lotz,p.rust,igel,de,soegaard}@di.ku.dk
[2] ROCKWOOL Foundation, Copenhagen, Denmark
[3] The Danish National Archives, Copenhagen, Denmark
{jro,jkd}@sa.dk
[4] Centre for Integrated Register-based Research, Aarhus University,
Aarhus, Denmark
cbp@econ.au.dk
[5] Université Paris-Saclay, Gif-sur-Yvette, France
israfel.salazar@ens-paris-saclay.fr

Abstract. In Northern Europe, parish records provide centuries of lineage information, useful not only for settling inheritance disputes, but also for studying hereditary diseases, social mobility, etc. The key information to extract from scans of parish records to obtain lineage information is dates: birth dates (of children and their parents) and dates of baptisms. We present a new dataset of birth dates from Danish parish records and use it to benchmark different approaches to handwritten date recognition, some based on classification and some based on transduction. We evaluate these approaches across several experimental protocols and different segmentation strategies. A state-of-the-art transformer-based transduction model exhibits lower error rates than image classifiers in most scenarios. The image classifiers can nevertheless offer a compelling trade-off in terms of accuracy and computational resource requirements.

Keywords: Handwriting recognition · Parish records · Transfer learning · Robustness

1 Introduction

In 1968, the Danish state put a (digital) Civil Registration System (CRS) [36] into use. This system is a national register containing basic personal details on all individuals residing in Denmark, including ancestry information. To the

Supported by Novo Nordisk Foundation (grant NNF 20SA0066568).
L. C. Piqueras, C. Fierro, J. F. Lotz, and P. Rust—Equal Contribution.

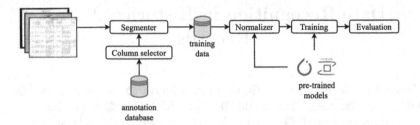

Fig. 1. Our end-to-end pipeline for handwritten date recognition. How the training data is generated depends on the segmentation strategy used in the segmentation module (Sect. 4.2) and the column of interest. Normalization of the input images depends on the chosen pre-trained model (Sect. 4.3). Evaluation (Sect. 4.4) is common across the different model architectures.

extent modern registers include civil registration numbers, the CRS enables us to study hereditary diseases, social mobility, etc., on the population that resides in Denmark after 1968. If we want to go further back in time, we will need to consult the parish records.

Parish records are registries handwritten in pre-printed books. Even though the layout of parish records is relatively consistent, it may vary across historical periods, regions, and countries that also established this tradition; but typically the books contain much of the same information, and there will be hundreds of thousands of records using the same format. However, since the records were handwritten by clergy, and the books were valuable, the pre-printed layouts were often used somewhat creatively, and many errors were introduced and corrected. The task of digitizing them is therefore both easy and hard at the same time. Examples of Danish parish records can be found on the Danish National Archives' website[1]. We are able to make some of our manual annotations publicly available for research purposes. See Sect. 2 for details.

The rationale for digitizing parish records is straight-forward: the parish records contain our family histories. This information is useful for studying a wide range of scientific topics; family histories can help get precision medicine off the ground [8] and are crucial to understand and prevent hereditary diseases [7]. Many of these diseases are of childhood onset [12], but there are also examples of hereditary diseases that have adult onset, e.g., breast cancer, diabetes, heart disease and blood clots, Alzheimer's disease and dementia, arthritis, depression, high blood pressure and high cholesterol [2].

2   Data

The original parish records are handwritten on pages with pre-printed layouts. The pre-printed books used and released as part of this project were digitized

---

[1] An example page can be found at: https://www.sa.dk/ao-soegesider/da/billedviser?
epid=17125564\#167405,28108453.

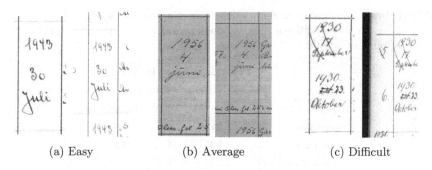

| (a) Easy | (b) Average | (c) Difficult |

Fig. 2. Examples of birth date cells in the different test splits. Each test split displays the same cell segmented with our two different approaches: U-Net (left) and fix-size (right) approaches. See Sect. 4.2 for further details.

and transcribed by paid Danish transcribers. These records comprise over 8100 files, including 26000 births and 7000 deaths approximately, from which nearly 7700 files are fully annotated. They cover information gathered from 10 parishes from 1920 to 1960. In total, the Danish National Archives (Rigsarkivet) stores 20.000 parish records from 1900–1980, comprising a total of approximately 12 million entries.

The books used for registering births and deaths use a pre-printed tabular layout, although with important variations in the information reported. Common aspects between both type of books are: (a) a sequential page number at the top-right corner of each page; (b) the title, indicating the type of registry (births [Fødte] or deceased [Døde]) and the gender of the persons on that page (men [Mandkøn] or women [Kvindekøn]); (c) the table header explaining the content of each column; and (d) the subsequent handwritten registries. Among the columns, the left-most one contains the row ID in both registries that, together with the page number, uniquely identifies each row. The IDs are sequential numbers with the exception of non-integers, indicating that the entry is duplicated because the actual birth took place in another parish. Birth dates are in the second column in the new-born registries and in the sixth column in the deceased registries (the second column contains date and place of death). The remaining columns are not of interest for the task of birth date recognition[2].

The data used is made publicly available[3], to enable further research on handwritten date recognition systems. Note that the dataset released includes only the columns containing the row ID and birth date.

3   Date Recognition

We distinguish between handwritten text recognition (HTR) and optical character recognition (OCR) as closely aligned but separate tasks: OCR is for machine

---

[2] Future work includes the integration of other columns containing dates in our training set.

[3] github.com/coastalcph/mgr-birthdates.

Table 1. Size of the different data splits depending on the segmentation strategy. #pages is the number of different images (book pages) in each split. #cells is the total number of birth date cells.

	Fix-size		U-Net	
	#Pages	#Cells	#Pages	#Cells
Train	6757	28308	5383	22978
Validation	746	3159	593	2565
Test easy	53	225	44	188
Test average	58	231	47	186
Test difficult	63	217	52	181
Total	7677	32140	6119	26098

printed text where the variation in style originates from the use of different fonts. Layout analysis is often integrated into the OCR engine [30,33,49] and the characters from a given document are processed individually. HTR, on the other hand, typically attempts to decode sequences from the data at word or sentence level and segmentation is a separate step in the HTR workflow.

Our approaches to segmentation (elaborated on in Sect. 4.2) attempt to identify the row and column structure of the parish records. Arriving at a grid structure is sufficient for transcription of birth dates and serves as a good starting point for the more advanced columns (future work) where text line detection can still be applied within the region of interest. With this approach, we do not have to address an additional problem of resolving what text lines refer to what row/column. After segmentation, the cells containing birth dates are passed, together with their row ID, to the models described in Sect. 4.3 which predict the content. Figure 1 depicts the end-to-end solution adopted for handwritten date recognition.

4   Experiments

4.1   Data Splits

The scanned images are split into separate sets for training, validation, and testing. The test set is further split into categories of easy, average, and difficult based on an assessment of the level of noise in the images. The easy and difficult test splits are constructed by manually inspecting individual rows across all parishes, selecting some that are clean and high quality for the former, and some that contain corrections or are in some other way hard to read for the latter (see examples in Fig. 2). For the average test split, we randomly select 58 pages. The remaining data are shuffled and 10% are kept for validation. Note that for constructing the easy and difficult test splits, we select individual *rows* rather than entire pages because the difficulty of cells can vary within pages. When selecting only a subset of the rows in a given page for testing, we add the remaining to the training set. Table 1 shows a summary of the final splits which, at the same time, depends on the deployed segmentation strategy (detailed below).

4.2 Segmentation

The first step in our pipeline is to segment every image into rows of the different individuals and columns with the different types of information. We utilize the tabular layout of the parish records and aim for a segmentation strategy that will return a grid of cells: each cell containing all the text lines from the corresponding record entry. The purpose of the segmentation is twofold: to identify the number of individuals registered on the given page and to locate the columns of interest. We implement two methods: a heuristic fix-size segmentation and a semantic segmentation based on a U-Net convolutional neural network architecture [45]. Figure 2 displays examples of birth date cells extracted with each of these methods. After obtaining the segmented columns for every individual, the cells containing the birth dates are fed to the text recognition models (described in Sect. 4.3) together with the annotated date as label.

Fix-Size Segmentation. The fix-size implementation takes advantage of the relatively consistent layout of the parish records. After reshaping the images to 1920 × 1080 pixels to align the sizes, the columns are detected based on a deterministic estimate of their horizontal offset. The horizontal image slices that split the rows are found using the number of annotations we have for each page and cropping the image into homogeneous segments. A small margin is added in each direction to include content that might be misaligned due to variation of the page position or writings close to the cell edges. See Fig. 2 (right side images) for an example.

It is important to note that this rather naïve method serves primarily as a baseline and that it does not account for any source of variation within the images. In addition, the provided annotations are not perfect and are an additional source of noise when deciding on the number of horizontal crops. To use this strategy in a real world application, we would need to train a separate model that predicts the number of rows used for cropping unlabeled images.

U-Net. We train a U-Net model to predict the boundaries of regions of interest, allowing for a more flexible segmentation that does not rely on the assumption that entries are uniformly distributed across the page. We manually annotate the cell boundaries for every row and column for 282 pages that are chosen at random from the training split. The implementation follows [45] but with ELU [9] activations instead of ReLU [31] and replacing the up-convolution with nearest neighbor up-sampling, as suggested in [37,38]. The model is trained from scratch for 1000 epochs with a batch size of 8 on 80% of the annotated pages. The remaining 20% are used for validation. During inference, we rely on the Harris corner detector [16] to obtain pixel coordinates of where the predicted row separators and column separators intersect. As a result, the segments retrieved are precisely aligned with the cell's margin, as we can see in the images (left side) in Fig. 2.

As seen in Table 1, this segmentation approach yields less data for the subsequent steps of our HTR pipeline than the fix-size approach. This is caused by a sanity check that discards the given page when the U-Net model and the number of annotations for that page do not agree. Out of the 1558 pages that are discarded, 1176 of them (75%) contain at least one segmentation error. For 208 of the discarded pages, the model incorrectly includes cells that only contained noise, e.g. squiggles or smudges, and should have been ignored. And in 174 of the cases, the U-Net gets the layout right but the provided annotations are wrong, typically because of too few registrations.

4.3   Models

We approach the problem of handwritten date recognition from the perspective of a multi-label image classification task and as a sequence-to-sequence text generation task. We benchmark results for two image classification models, one based on ResNet [17], the other on EfficientNet [48], and for a sequence-to-sequence model based on a pre-trained TrOCR model [27], which leverages the Transformer architecture [52].

Classification Models. We experiment with ResNet-18, the shallowest ResNet variant (11M trainable parameters) presented in [17] which is able to achieve competitive accuracy in our date recognition task. We also experiment with a larger classification model, EfficientNet-B4, which among the various models presented in [48] provides a good trade-off between accuracy and FLOPs (19M trainable parameters). We download the pre-trained ResNet-18 and EfficientNet-B4 models from Torchvision[4] and fine-tune them on our training datasets—obtained with either fix-size or U-Net segmentation—for 40 epochs, validating the performance on the validation split at the end of every epoch. We train with full precision in batches of 256 samples for the ResNet-18, and 128 samples for the EfficientNet-B4. Input images are normalized to 224×224 pixels. For both architectures, we perform a grid search over a set of specified learning rates, $\{5e-4, 1e-3, 3e-3, 5e-3, 1e-2\}$, and 5 different initializations, resulting in 5×5 runs over each training set. Model checkpoints are saved every 100 steps. After training, we select the checkpoint that achieved the highest validation accuracy to compare results. Evaluation on the test data is performed with the 5 versions of the model that reported the highest validation accuracy, i.e., we test the best-performing model along with the remaining 4 random initializations trained with the same learning rate.

In addition, we explore 2 different label encodings for computing the loss: (1) *datetime encoding* where every day since January 1 1800 up until December 31 1999 is encoded as a unique class; and (2) *digit encoding* where the year is

---

[4] https://pytorch.org/vision/stable/models.html.

encoded as digits from 0 to 9, the month as 1 digit from 1 to 12, and the date as 2 digits from 0 to 9.

We also investigated the effect of tailoring the input dimensions and different data augmentation techniques using the ResNet-18 architecture. We tested several standard augmentation transformations as well as the same augmentations used when training the TrOCR models as described below. Neither changing the input image size (including using the same size as used in the TrOCR experiments) nor the augmentations significantly altered the ResNet-18 results.

TrOCR Models. We initialize TrOCR with a BEiT-base [3] encoder and a RoBERTa-large [28] decoder from the pre-trained `trocr-base-handwritten` checkpoint provided by [27][5]. The full TrOCR model has 334M trainable parameters, making it significantly larger than our ResNet and EfficientNet models. We preprocess training examples by converting the labels' date format from *yyyymmdd* to one that matches the Danish handwriting, e.g., *"19500331"* to *"31 marts 1950"*. We then fine-tune for the text recognition task outlined in [27] on the training datasets obtained via fix-size or U-Net segmentation. Following [27], we perform data augmentation by selecting from a list of possible transformations at random with uniform probabilities: random rotation (-10 to $10°$C), Gaussian blur, image dilation, image erosion (all with kernel size 3), downscaling by a factor of 3, underlining with black pixels, and keeping the original. We train for up to 300 epochs and validate model performance based on the validation loss after every epoch. To reduce computational overhead, we use early stopping [40] with a patience of 20 validation steps, i.e., if the model does not improve its validation loss within 20 epochs, training is terminated preemptively. In practice, training was typically terminated after 30–80 epochs, depending on the learning rate. We train in batches of 256 input images of size 384×384 pixels, resulting in a sequence length of 576. We use the Adam optimizer [23] and perform a grid search over peak learning rates $\{6e - 6, 9e - 6, 2e - 5, 5e - 5, 8e - 5\}$. We warm up to these peak learning rates linearly from $1e - 8$ over the first 500 training steps and then decay with an inverse square root schedule. Weight decay is set to $1e - 4$. For each learning rate, we run 5 random initializations to account for randomness. We use automatic mixed precision (fp16) training with Nvidia Apex[6].

4.4 Evaluation Metrics

The different approaches are evaluated using accuracy. We compute both top-1 and top-5 accuracy for the full date, i.e., a classification is correct provided that day, month, and year are correct. Top-1 results are further decomposed, reporting values for the day, month and year individually. On each of the 3 test sets, we report the mean and standard deviation of the 5 random training initializations

---

[5] https://github.com/microsoft/unilm/tree/master/trocr#fine-tuning-and-evaluation.

[6] https://github.com/NVIDIA/apex.

Table 2. Top-1 validation set error rate of the best-performing models. TrOCR has the lowest validation error rate on the U-Net data whereas the EfficientNet models have the lowest on the (more noisy) data obtained via fix-size segmentation.

Model	Fix-size	U-Net
	Error (%)	Error (%)
ResNet-18 (1)	13.6	8.0
ResNet-18 (2)	13.8	8.0
EfficientNet (1)	10.1	7.8
EfficientNet (2)	10.2	6.2
TrOCR	12.7	6.0

that performed best on the validation set. Note that in this task, the Word Error Rate (WER) metric, commonly found in the OCR literature, would be complementary to the *full date* accuracy where any mismatch is counted as an error.

5   Results and Analysis

Looking at results from Table 3 and Table 4, we find that TrOCR achieves the highest *full date* recognition accuracy, outperforming the ResNet and Efficient-Net models across all three test splits. ResNet-18 has the lowest performance. The performance gap between TrOCR and ResNet-18 is around 5–7% of accuracy. EffienctNet-B4 is approximately 1–3% worse than TrOCR, depending on the test split and label encoding. This result is in line with [47] who have shown that TrOCR outperforms Transkribus [20] on historic handwritten records in Latin. The result is also expected considering the differences in model capacity: TrOCR has 334M trainable parameters whereas EfficientNet-B4 has 19M and ResNet-18 has 11M. With these differences in mind, the EfficientNet model may provide the most satisfactory trade-off in terms of computational requirements and performance.

For the classification models, there seems to be no clear winner among the two label encodings. The difference in mean top-1 accuracy between the two encodings is largest for our ResNet-18 models on the difficult test split (3.4% accuracy using fix-size segmentation and 1.4% with U-Net segmentation)—here in favor of the *datetime* encoding—although even in this case, both encodings are within one standard deviation for the U-Net data, and the top-5 accuracy in fact favors the *digit* encoding. The EfficientNet-B4 models are less sensitive to the choice of label encoding.

We now perform an error analysis of the best-performing models, i.e., models with the highest full date validation accuracy after hyperparameter tuning on the validation data. See Table 2 for an overview of the errors. We analyse errors across two dimensions: datasets and models (which can be further split into encodings for the two image classification models).

Table 3. Mean accuracy and standard deviation of 5 random initializations each using the crops generated by the fix-size segmentation heuristic. The numbers in parentheses refer to the label encoding used in each experiment ((1): datetime encoding, (2): digit encoding), as explained in Sect. 4.3. Highest mean accuracy is highlighted in **bold**.

Model	Test	Day	Month	Year	Full date	Top5
ResNet-18 (1)		$0.934 \pm 8e-3$	$0.977 \pm 5e-3$	$0.966 \pm 6e-3$	$0.896 \pm 6e-3$	$0.955 \pm 8e-3$
ResNet-18 (2)		$0.932 \pm 1e-2$	$0.968 \pm 6e-3$	$0.972 \pm 1e-2$	$0.892 \pm 1e-2$	$0.966 \pm 6e-3$
EfficientNet-B4 (1)	Easy	$0.948 \pm 2e-3$	$0.978 \pm 2e-3$	$\mathbf{0.988 \pm 4e-3}$	$0.933 \pm 2e-3$	$0.964 \pm 4e-3$
EfficientNet-B4 (2)		$0.948 \pm 1e-2$	$0.978 \pm 5e-3$	$0.984 \pm 5e-3$	$0.928 \pm 1e-2$	$0.971 \pm 3e-3$
TrOCR		$\mathbf{0.977 \pm 9e-3}$	$\mathbf{0.994 \pm 2e-3}$	$0.974 \pm 5e-3$	$\mathbf{0.947 \pm 7e-3}$	$\mathbf{0.996 \pm 3e-3}$
ResNet-18 (1)		$0.903 \pm 1e-2$	$0.970 \pm 1e-3$	$0.956 \pm 4e-3$	$0.867 \pm 1e-2$	$0.938 \pm 1e-3$
ResNet-18 (2)		$0.900 \pm 9e-3$	$0.968 \pm 5e-3$	$0.954 \pm 8e-3$	$0.859 \pm 5e-3$	$0.961 \pm 6e-3$
EfficientNet-B4 (1)	Average	$0.929 \pm 2e-3$	$0.970 \pm 2e-3$	$0.967 \pm 7e-3$	$0.897 \pm 5e-3$	$0.948 \pm 7e-3$
EfficientNet-B4 (2)		$0.925 \pm 6e-3$	$0.968 \pm 3e-3$	$0.965 \pm 6e-3$	$0.892 \pm 3e-3$	$0.965 \pm 5e-3$
TrOCR		$\mathbf{0.957 \pm 9e-3}$	$\mathbf{0.990 \pm 5e-3}$	$\mathbf{0.971 \pm 6e-3}$	$\mathbf{0.923 \pm 9e-3}$	$\mathbf{0.976 \pm 3e-3}$
ResNet-18 (1)		$0.811 \pm 1e-2$	$0.873 \pm 1e-2$	$0.892 \pm 1e-2$	$0.716 \pm 4e-3$	$0.865 \pm 1e-2$
ResNet-18 (2)		$0.785 \pm 1e-2$	$0.866 \pm 1e-2$	$0.882 \pm 7e-3$	$0.682 \pm 8e-3$	$0.905 \pm 6e-3$
EfficientNet-B4 (1)	Difficult	$0.841 \pm 6e-3$	$0.896 \pm 1e-2$	$\mathbf{0.925 \pm 5e-3}$	$0.760 \pm 1e-2$	$0.891 \pm 1e-2$
EfficientNet-B4 (2)		$0.849 \pm 8e-3$	$0.899 \pm 1e-2$	$0.919 \pm 4e-3$	$0.759 \pm 1e-2$	$\mathbf{0.924 \pm 1e-2}$
TrOCR		$\mathbf{0.910 \pm 4e-3}$	$\mathbf{0.904 \pm 5e-3}$	$0.863 \pm 8e-3$	$\mathbf{0.781 \pm 8e-3}$	$0.872 \pm 1e-2$

Table 4. Mean accuracy and standard deviation of 5 random initializations each using the crops generated by the U-Net model. The numbers in parentheses refer to the label encoding used in each experiment ((1): datetime encoding, (2): digit encoding), as explained in Sect. 4.3. Highest mean accuracy is highlighted in **bold**.

Model	Test	Day	Month	Year	Full date	Top5
ResNet-18 (1)		$0.950 \pm 1e-2$	$0.967 \pm 4e-3$	$0.985 \pm 1e-2$	$0.926 \pm 1e-2$	$0.971 \pm 2e-3$
ResNet-18 (2)		$0.951 \pm 4e-3$	$0.964 \pm 6e-3$	$0.994 \pm 7e-3$	$0.935 \pm 1e-2$	$0.972 \pm 4e-3$
EfficientNet-B4 (1)	Easy	$0.961 \pm 6e-3$	$0.971 \pm 6e-3$	$0.991 \pm 2e-3$	$0.950 \pm 2e-3$	0.973 ± 0
EfficientNet-B4 (2)		$0.960 \pm 1e-3$	$0.976 \pm 2e-3$	$\mathbf{0.996 \pm 5e-3}$	$0.954 \pm 4e-3$	$0.973 \pm 3e-3$
TrOCR		$\mathbf{0.994 \pm 6e-3}$	$\mathbf{0.998 \pm 3e-3}$	$0.990 \pm 4e-3$	$\mathbf{0.982 \pm 7e-3}$	$\mathbf{0.999 \pm 2e-3}$
ResNet-18 (1)		$0.933 \pm 8e-3$	$0.961 \pm 7e-3$	$0.980 \pm 4e-3$	$0.910 \pm 1e-2$	$0.952 \pm 5e-3$
ResNet-18 (2)		$0.933 \pm 1e-2$	$0.961 \pm 5e-3$	$0.967 \pm 6e-3$	$0.902 \pm 8e-3$	$0.969 \pm 6e-3$
EfficientNet-B4 (1)	Average	$0.952 \pm 9e-3$	$0.967 \pm 2e-3$	$0.976 \pm 4e-3$	$0.931 \pm 8e-3$	$0.954 \pm 2e-3$
EfficientNet-B4 (2)		$0.945 \pm 2e-3$	$0.963 \pm 4e-3$	$0.976 \pm 4e-3$	$0.923 \pm 4e-3$	$0.969 \pm 4e-3$
TrOCR		$\mathbf{0.982 \pm 3e-3}$	$\mathbf{0.995 \pm 5e-3}$	$\mathbf{0.981 \pm 4e-3}$	$\mathbf{0.960 \pm 4e-3}$	$\mathbf{0.985 \pm 5e-3}$
ResNet-18 (1)		$0.833 \pm 1e-2$	$0.898 \pm 1e-2$	$0.906 \pm 9e-3$	$0.743 \pm 2e-2$	$0.860 \pm 1e-2$
ResNet-18 (2)		$0.804 \pm 1e-2$	$0.901 \pm 7e-3$	$0.901 \pm 1e-2$	$0.729 \pm 1e-2$	$\mathbf{0.901 \pm 7e-3}$
EfficientNet-B4 (1)	Difficult	$0.836 \pm 7e-3$	$0.916 \pm 5e-3$	$0.913 \pm 9e-3$	$0.769 \pm 1e-2$	$0.864 \pm 1e-2$
EfficientNet-B4 (2)		$0.823 \pm 1e-2$	$\mathbf{0.917 \pm 7e-3}$	$\mathbf{0.927 \pm 4e-3}$	$0.770 \pm 9e-3$	$0.898 \pm 1e-2$
TrOCR		$\mathbf{0.928 \pm 8e-3}$	$0.905 \pm 1e-2$	$0.863 \pm 4e-3$	$\mathbf{0.797 \pm 1e-2}$	$0.842 \pm 7e-3$

Dataset Dimension. We observe that models trained on the U-Net segmentations make considerably fewer errors. This finding is expected considering that the fix-size segmenter makes a naïve assumption that all rows on a given page have the same height. Consequently, the examples obtained via fix-size segmentation are a lot noisier, in particular around the edges of a cell (see Fig. 2). This noise appears to not only cause more mispredictions overall, but also elicits a

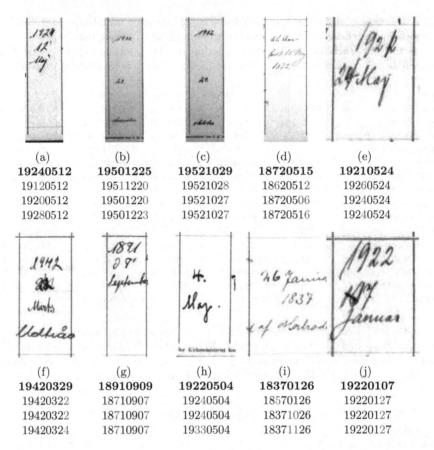

(a)	(b)	(c)	(d)	(e)
19240512	**19501225**	**19521029**	**18720515**	**19210524**
19120512	19511220	19521028	18620512	19260524
19200512	19501220	19521027	18720506	19240524
19280512	19501223	19521027	18720516	19240524

(f)	(g)	(h)	(i)	(j)
19420329	**18910909**	**19220504**	**18370126**	**19220107**
19420322	18710907	19240504	18570126	19220127
19420322	18710907	19240504	18371026	19220127
19420324	18710907	19330504	18371126	19220127

Fig. 3. A subset of errors shared among all models trained with data from U-Net segmentation. Beneath each image are displayed first the label in **bold**, then the predicted date from ResNet-18, the predicted date from EfficientNet-B4, and lastly the output from TrOCR. Both ResNet-18 and EfficientNet-B4 use the *datetime* encoding. Date format is "yyyyMMdd". The exact errors are highlighted in red. Common sources of errors that all models make are ambiguous handwriting, noisy cells with corrections or text belonging to other columns, and missing information.

more *consistent* misprediction behavior. In other words, the noise introduced in the fix-size data confounds all models in a similar manner; our best-performing ResNet and EfficientNet models share 58.1% of their prediction errors for the fix-size validation split, compared to only 49.8% using the U-Net validation split. Likewise, the benefit of employing the U-Net segmentation is evident across all of the *test* sets and models, in particular for TrOCR.

Model Dimension. Here, we only consider models trained on the U-Net data for which all models exhibited better performance. We find that approximately half of the *full date* errors (49.8% with *datetime* and 51.5% with digit encoding) made by the ResNet-18 and EfficientNet-B4 overlap. Overall, a total of 94 validation set examples were misclassified by all three models (ResNet, EfficientNet, and TrOCR). We visualize a subset of these in Fig. 3. Out of the 94 misclassified examples, we find that 37 were labeled incorrectly. Note that any incorrect labels were manually corrected in the dataset's test splits. Likewise, 38 of these examples were missing crucial information, usually the year component (see for instance example (h) in Fig. 3). Other common sources of mispredictions were unclear or ambiguous handwriting (for instance examples (c) and (g) in Fig. 3) and noisy cells containing corrections (examples (e), (f), (j)) or text belonging to adjacent columns (examples (d), (i)). We do not find clear error patterns related to the inherent visual ambiguity of some handwritten digits, e.g. between the visually similar digits 2 and 7 or 4 and 9 [25]. We believe that errors caused by ambiguous or unclear handwriting and incorrect labels most likely cannot be prevented by leveraging a different model architecture, segmentation strategy, or training on a larger dataset. They are, ultimately, results of the data creation and annotation processes that cannot easily be resolved post-hoc. However, the error patterns that can likely be attenuated are mispredictions caused by corrections in the handwriting and information leaking in from adjacent columns, in which case more robust training that learns to ignore irrelevant information could help. We will address these in future work.

Analyzing individual mispredictions, we observe a trend in the validation split that is also observable for the test splits (Table 3, Table 4): whereas the ResNet-18 and EfficientNet-B4 models consistently perform best on the year component (compared to days or months), the year component misleads TrOCR the most. The main source of errors within the year component is missing year information, as mentioned above. In such cases, the year is typically not written within each cell but once on top of the date of birth column, making the year effectively impossible to predict based on the segmented image alone. For TrOCR with U-Net segmentation, 51 out of the 79 incorrectly classified years fall into this category, which may also partly explain why TrOCR's performance is lowest on the year component. Note, however, that no examples without a year column are present in the manually curated test data. This source of error is merely an artifact of the data segmentation process but can be addressed with a post-processing step that could e.g. involve another model identifying and classifying years written at the top of a column. We will devise such a post-processing procedure in future work, which will improve overall performance even further.

6   Related Work

The work presented in [35] is similar to ours, using machine learning to transcribe 2.3 million handwritten occupation codes from the Norwegian 1950 population census. They combined convolutional and recurrent neural network (CNN-RNN) architectures to achieve an accuracy of 97%. [32] extracted handwritten data from the 1930 US census documents, targeting 10 different columns including names, gender, and age, also using a CNN-RNN model. [46] focused on developing a system for automatically extracting names and digits from a historical French population. [26] experimented with pre-trained OCR software Transkribus [30], Tesseract 4 [49], and OCRopy [34] when digitizing historical weather data in a tabular format from the 18th century. They showed that the OCR engines perform sub-optimally when applied out-of-the-box to detect and transcribe handwritten text in one step and that tuning the models to the specific data was required to obtain even modest results.

[21] showed that a HTR model based on the Transformer architecture with self-attention [53] can achieve competitive results using less training data while recognizing sentences at character level rather than word level. Integrating language models into digitization frameworks as a separate step have shown to boost overall accuracy when working with more complex manuscripts, guiding the search space of the optical recognition step to settle uncertainties about visually similar words or characters [22,44,51].

The problem of identifying the tabular structure in registry or census books, often referred to as document or layout analysis, is a common research problem [1,5,11,13,39]. Popular methods include relying on models from object detection such as Faster R-CNN [43] or YOLO [4,42], detection of horizontal and vertical ruling lines using Hough transform [19], and detection of individual text lines [6,15,24] and from there turning the problem into one of graph labeling to resolve which text lines comprise a cell [41]. [29] showed that in their setup errors during layout analysis have a greater negative impact on performance than errors during the recognition step.

The value of digitization stretches beyond the use cases of historical certificates and records [10,50] and can also be seen as a way of preserving culture [14].

7   Future Work

Reliable, automatic transcription of dates in the parish registries is the first step towards our visionary goal of creating a Multi Generation Registry (MGR) with familial relations for all Danes. The idea is to extend the Danish Civil Registration System (CRS) [36] backwards, going as far back as birth cohort 1920 by digitizing and linking historical Parish Registries with the CRS.

By effectively back-filling the CRS, which only contains individuals who were alive in 1968 or later, the final MGR will include all Danes from 1920 onwards. To accomplish this, we first need to have reliable, automatic transcription of names and birth dates from pairs of child and parents. Afterwards, the data transcribed from the parish registries needs to be linked with data in the CRS. Note that this will allow us to identify individuals who are present in *both* data sets. Experience can be drawn from [18] that describe the algorithms used to link the Norwegian Historical Population Register back to 1801. The MGR containing family relations for the majority of the Danish population will make it possible to analyze diseases and traits over a long period of time across several generations, as well as laterally between cousins in families with particular health histories.

8 Conclusion

We presented a new dataset of handwritten birth dates from Danish parish records which will help develop better handwriting recognition systems for use in future digitization efforts. We used the dataset to benchmark different models on handwriting recognition, including convolutional image classifiers and a sequence-to-sequence transformer architecture. We evaluated these approaches across three data splits of varying difficulty and two different strategies to segment scanned pages of the parish records. We found that the larger, transformer-based model, which was pre-trained for character recognition, performed better in most cases compared to the more resource-efficient standard image classifiers in our study.

Ethics Statement. The data published in this paper is in accordance with the applicable national law. Scanned books with personal information used in this project meet the European data protection laws. The original books are freely available to the public on the Danish National Archives' (Rigsarkivet) website. We release a curated subset of said data where individuals cannot be uniquely identified as only birth dates are provided without context, to favor academic research. We do not foresee any conflict of interest.

References

1. Andrés, J., Prieto, J.R., Granell, E., Romero, V., Sánchez, J.A., Vidal, E.: Information extraction from handwritten tables in historical documents. In: Uchida, S., Barney, E., Eglin, V. (eds) International Workshop on Document Analysis Systems, DAS 2022. LNCS, pp. 184–198. Springer, Cham (2022). https://doi.org/10.1007/978-3-031-06555-2_13
2. Bancroft, E.K.: Genetic testing for cancer predisposition and implications for nursing practice: narrative review. J. Adv. Nurs. **66**(4), 710–737 (2010). https://doi.org/10.1111/j.1365-2648.2010.05286.x
3. Bao, H., Dong, L., Piao, S., Wei, F.: BEit: BERT pre-training of image transformers. In: International Conference on Learning Representations (2022). https://openreview.net/forum?id=p-BhZSz59o4
4. Bochkovskiy, A., Wang, C.Y., Liao, H.Y.M.: Yolov4: optimal speed and accuracy of object detection. arXiv preprint. arXiv:2004.10934 (2020)
5. Boillet, M., Kermorvant, C., Paquet, T.: Multiple document datasets pre-training improves text line detection with deep neural networks. In: 2020 25th International Conference on Pattern Recognition (ICPR), pp. 2134–2141. IEEE (2021)
6. Boillet, M., Kermorvant, C., Paquet, T.: Robust text line detection in historical documents: learning and evaluation methods. Int. J. Doc. Anal. Recogn. (IJDAR) **95**, 1–20 (2022). https://doi.org/10.1007/s10032-022-00395-7
7. Boone, P.M.: Adolescents, family history, and inherited disease risk: an opportunity. Pediatrics **138**(2), e20160579 (2016). https://doi.org/10.1542/peds.2016-0579
8. Bylstra, Y.: Family history assessment significantly enhances delivery of precision medicine in the genomics era. bioRxiv (2020). https://doi.org/10.1101/2020.01.29.926139, www.biorxiv.org/content/early/2020/01/30/2020.01.29.926139

9. Clevert, D., Unterthiner, T., Hochreiter, S.: Fast and accurate deep network learning by exponential linear units (elus). In: Bengio, Y., LeCun, Y. (eds.) 4th International Conference on Learning Representations, ICLR 2016, San Juan, Puerto Rico, 2–4 May 2016, Conference Track Proceedings (2016). http://arxiv.org/abs/1511.07289
10. Dahl, C.M., Johansen, T.S., Sørensen, E.N., Westermann, C.E., Wittrock, S.F.: Applications of machine learning in document digitisation. arXiv preprint. arXiv:2102.03239 (2021)
11. Déjean, H., Meunier, J.L.: Table rows segmentation. In: 2019 International Conference on Document Analysis and Recognition (ICDAR), pp. 461–466. IEEE (2019)
12. Ross, L.F., Saal, H.M., David, K.L., Anderson, R.R.: Technical report: ethical and policy issues in genetic testing and screening of children. Genet. Med. 15(3), 234–245 (2013). https://doi.org/10.1038/gim.2012.176
13. Gao, L., et al.: ICDAR 2019 competition on table detection and recognition (cTDaR). In: 2019 International Conference on Document Analysis and Recognition (ICDAR), pp. 1510–1515. IEEE (2019)
14. Granell, E., Chammas, E., Likforman-Sulem, L., Martínez-Hinarejos, C.D., Mokbel, C., Cîrstea, B.I.: Transcription of spanish historical handwritten documents with deep neural networks. J. Imaging 4(1), 15 (2018)
15. Grüning, T., Leifert, G., Strauß, T., Michael, J., Labahn, R.: A two-stage method for text line detection in historical documents. Int. J. Doc. Anal. Recogn. (IJDAR) 22(3), 285–302 (2019). https://doi.org/10.1007/s10032-019-00332-1
16. Harris, C., Stephens, M., et al.: A combined corner and edge detector. In: Alvey vision conference, vol. 15, pp. 10–5244. Citeseer (1988)
17. He, K., Zhang, X., Ren, S., Sun, J.: Deep residual learning for image recognition. In: 2016 IEEE Conference on Computer Vision and Pattern Recognition (CVPR), pp. 770–778 (2016). https://doi.org/10.1109/CVPR.2016.90
18. Holden, L., Boudko, S., Thorvaldsen, G.: Lenking og kobling i historisk befolkningsregister. Heimen 57(3), 216–229 (2020)
19. Hough, P.V.: Method and means for recognizing complex patterns (1962). US Patent 3,069,654
20. Kahle, P., Colutto, S., Hackl, G., Mühlberger, G.: Transkribus-a service platform for transcription, recognition and retrieval of historical documents. In: 2017 14th IAPR International Conference on Document Analysis and Recognition (ICDAR), vol. 4, pp. 19–24. IEEE (2017)
21. Kang, L., Riba, P., Rusiñol, M., Fornés, A., Villegas, M.: Pay attention to what you read: non-recurrent handwritten text-line recognition. Pattern Recogn. 129, 108766 (2022)
22. Kang, L., Riba, P., Villegas, M., Fornés, A., Rusiñol, M.: Candidate fusion: integrating language modelling into a sequence-to-sequence handwritten word recognition architecture. Pattern Recogn. 112, 107790 (2021)
23. Kingma, D.P., Ba, J.: Adam: a method for stochastic optimization. In: Proceedings of the 3rd International Conference on Learning Representations (ICLR), San Diego, CA, USA (2015). http://arxiv.org/abs/1412.6980
24. Kodym, O., Hradiš, M.: Page layout analysis system for unconstrained historic documents. In: Lladós, J., Lopresti, D., Uchida, S. (eds.) ICDAR 2021. LNCS, vol. 12822, pp. 492–506. Springer, Cham (2021). https://doi.org/10.1007/978-3-030-86331-9_32
25. LeCun, Y., Bottou, L., Bengio, Y., Haffner, P.: Gradient-based learning applied to document recognition. Proc. IEEE 86(11), 2278–2324 (1998)

26. Lehenmeier, C., Burghardt, M., Mischka, B.: Layout detection and table recognition – recent challenges in digitizing historical documents and handwritten tabular data. In: Hall, M., Merčun, T., Risse, T., Duchateau, F. (eds.) TPDL 2020. LNCS, vol. 12246, pp. 229–242. Springer, Cham (2020). https://doi.org/10.1007/978-3-030-54956-5_17

27. Li, M., et al.: TrOCR: transformer-based optical character recognition with pre-trained models (2021). www.microsoft.com/en-us/research/publication/trocr-transformer-based-optical-character-recognition-with-pre-trained-models/

28. Liu, Y., et al.: RoBERTa: a robustly optimized BERT pretraining approach. arXiv preprint (2019)

29. Monroc, C.B., Miret, B., Bonhomme, M.L., Kermorvant, C.: A comprehensive study of open-source libraries for named entity recognition on handwritten historical documents. In: Uchida, S., Barney, E., Eglin, V. (eds.) DAS 2022. LNCS, vol. 13237, pp. 429–444. Springer, Cham (2022). https://doi.org/10.1007/978-3-031-06555-2_29

30. Muehlberger, G., et al.: Transforming scholarship in the archives through handwritten text recognition: transkribus as a case study. J. Doc. (2019)

31. Nair, V., Hinton, G.E.: Rectified linear units improve restricted boltzmann machines. In: Fürnkranz, J., Joachims, T. (eds.) Proceedings of the 27th International Conference on Machine Learning (ICML-10), 21–24 June 2010, Haifa, Israel, pp. 807–814. Omnipress (2010). https://icml.cc/Conferences/2010/papers/432.pdf

32. Nion, T., et al.: Handwritten information extraction from historical census documents. In: 2013 12th International Conference on Document Analysis and Recognition, pp. 822–826. IEEE (2013)

33. OCR, G.C.: https://cloud.google.com/vision/docs/ocr. Accessed 01 June 2022

34. OCRopy: https://github.com/ocropus/ocropy. Accessed 01 June 2022

35. Pedersen, B.R., Holsbø, E., Andersen, T., Shvetsov, N., Ravn, J., Sommerseth, H.L., Bongo, L.A.: Lessons learned developing and using a machine learning model to automatically transcribe 2.3 million handwritten occupation codes (2022)

36. Pedersen, C.B., Gøtzsche, H., Møller, J.O., Mortensen, P.B.: The danish civil registration system. a cohort of eight million persons. Dan. Med. Bull. **53**, 441–449 (2006)

37. Perslev, M., Dam, E.B., Pai, A., Igel, C.: One network to segment them all: a general, lightweight system for accurate 3d medical image segmentation. In: Shen, D., Liu, T., Peters, T.M., Staib, L.H., Essert, C., Zhou, S., Yap, P.-T., Khan, A. (eds.) MICCAI 2019. LNCS, vol. 11765, pp. 30–38. Springer, Cham (2019). https://doi.org/10.1007/978-3-030-32245-8_4

38. Perslev, M., Darkner, S., Kempfner, L., Nikolic, M., Jennum, P.J., Igel, C.: U-sleep: resilient high-frequency sleep staging. NPJ Digit. Med. **4**(1), 1–12 (2021)

39. Prasad, A., Déjean, H., Meunier, J.L.: Versatile layout understanding via conjugate graph. In: 2019 International Conference on Document Analysis and Recognition (ICDAR), pp. 287–294. IEEE (2019)

40. Prechelt, L.: Early stopping — but when? In: Montavon, G., Orr, G.B., Müller, K.-R. (eds.) Neural Networks: Tricks of the Trade. LNCS, vol. 7700, pp. 53–67. Springer, Heidelberg (2012). https://doi.org/10.1007/978-3-642-35289-8_5

41. Prieto, J.R., Vidal, E.: Improved graph methods for table layout understanding. In: Lladós, J., Lopresti, D., Uchida, S. (eds.) ICDAR 2021. LNCS, vol. 12822, pp. 507–522. Springer, Cham (2021). https://doi.org/10.1007/978-3-030-86331-9_33

42. Redmon, J., Farhadi, A.: Yolov3: an incremental improvement. arXiv preprint. arXiv:1804.02767 (2018)

43. Ren, S., He, K., Girshick, R., Sun, J.: Faster r-cnn: towards real-time object detection with region proposal networks. In: Advances in Neural Information Processing Systems, vol. 28 (2015)
44. Romero, V., Fornés, A., Granell, E., Vidal, E., Sánchez, J.A.: Information extraction in handwritten marriage licenses books. In: Proceedings of the 5th International Workshop on Historical Document Imaging and Processing, pp. 66–71 (2019)
45. Ronneberger, O., Fischer, P., Brox, T.: U-Net: convolutional networks for biomedical image segmentation. In: Navab, N., Hornegger, J., Wells, W.M., Frangi, A.F. (eds.) MICCAI 2015. LNCS, vol. 9351, pp. 234–241. Springer, Cham (2015). https://doi.org/10.1007/978-3-319-24574-4_28
46. Sibade, C., Retornaz, T., Nion, T., Lerallut, R., Kermorvant, C.: Automatic indexing of french handwritten census registers for probate geneaology. In: Proceedings of the 2011 Workshop on Historical Document Imaging and Processing, pp. 51–58 (2011)
47. Ströbel, P.B., Clematide, S., Volk, M., Hodel, T.: Transformer-based HTR for historical documents. arXiv preprint. arXiv:2203.11008 (2022)
48. Tan, M., Le, Q.: EfficientNet: rethinking model scaling for convolutional neural networks. In: Chaudhuri, K., Salakhutdinov, R. (eds.) Proceedings of the 36th International Conference on Machine Learning. Proceedings of Machine Learning Research, vol. 97, pp. 6105–6114. PMLR (2019). https://proceedings.mlr.press/v97/tan19a.html
49. Tesseract: https://github.com/tesseract-ocr/tesseract. Accessed 01 June 2022
50. Thorvaldsen, G.L., Sommerseth, H., Holden, L.: Anvendelser av norges historiske befolkningsregister. Heimen 57(3), 230–243 (2020)
51. Toledo, J.I., Carbonell, M., Fornés, A., Lladós, J.: Information extraction from historical handwritten document images with a context-aware neural model. Pattern Recogn. 86, 27–36 (2019)
52. Vaswani, A., et al.: Attention is all you need. In: Guyon, I., et al. (eds.) Advances in Neural Information Processing Systems. vol. 30. Curran Associates, Inc. (2017). https://proceedings.neurips.cc/paper/2017/file/3f5ee243547dee91fbd053c1c4a845aa-Paper.pdf
53. Vaswani, A., et al.: Attention is all you need. In: Advances in Neural Information Processing Systems, vol. 30 (2017)

Improving Isolated Glyph Classification Task for Palm Leaf Manuscripts

Nimol Thuon[1] , Jun Du[1(✉)] , and Jianshu Zhang[1,2]

[1] University of Science and Technology of China, Hefei, Anhui, China
{tnimol,xysszjs}@mail.ustc.edu.cn, jundu@ustc.edu.cn
[2] iFlytek Research, Hefei, Anhui, China

Abstract. Digitization of ancient palm leaf manuscripts is gaining momentum due to the limited datasets and complex features of text images of palm leaf manuscripts. Thus far, the previous studies did not deeply analyze the application of the trending techniques on the palm leaf manuscripts, considering how deep learning approaches require large datasets, while some isolated glyphs contain more than one character with complex grammatical components. Therefore, this paper explores the possibilities and practical methods for improving isolated glyph classification. In particular, we focus on both the front-end and the back-end processes involved in the image classification task. For the front-end analysis, we present multi-task preprocessing techniques, including data augmentation techniques, new datasets extraction, and image enhancement techniques to increase the quality and quantity of datasets. For the back-end side, we aim to study the visual backbones of deep learning techniques, especially CNNs (including VGG, ResNet, and EfficientNet) and attention-based models (including ViT, DeiT, and CvT). Furthermore, the analysis and evaluation examined how data augmentation techniques and preprocessing interact with the amount of data used in training. Evidently, we experimented on three palm leaf manuscripts, including Balinese, Sundanese, and Khmer scripts from the ICFHR contest 2018, SluekRith, AMDI LontarSet, and Sunda datasets. Regarding the quality of research, the experiment delivers an effective way of training palm leaf datasets for the document analysis community.

Keywords: Historical document analysis · Vision transformer · Palm leaf manuscript · Neural network

1 Introduction

Historically, thousands of scripts have been recorded on palm leaves to depict significant historical events, Buddhism practices, astrology, or even medical treatment [8]. Southeast Asian Palm Leaf Manuscripts Analysis (SEAPA) offers a known challenge for document analysis tasks. Among these, Balinese [8], Khmer [19], and Sundanese scripts [15] have not only received increasing attention but have also brought fresh challenges for researchers due to their unique writing

© The Author(s), under exclusive license to Springer Nature Switzerland AG 2022
U. Porwal et al. (Eds.): ICFHR 2022, LNCS 13639, pp. 65–79, 2022.
https://doi.org/10.1007/978-3-031-21648-0_5

formats. SEAPA datasets generally consist of text-line, isolated glyphs, and word/text. While SEAPA research is developing rapidly, most collections of documents are only used during the first step. Deep neural networks have the ability to automatically learn desirable representations of text images rather than designing the custom features manually, which is why this system is best known for its effectiveness when dealing with image classification problems. In addition to Deep Neural Networks, the Convolutional Neural Network (CNN), Very Deep Convolutional Networks (VGGNet) [13], Residual Neural Network (ResNet) [6], and EfficientNet [16] are also some of the well-known deep learning architectures. The recent developments of machine learning also center on ViTs [4,14,18], which is one of the trending architectures used to achieve rewarding results in the image classification task. Afterward, hybrid approaches like CNN-ViT [5,20] were proposed for solving deep learning tasks. However, the applications of those trending approaches on palm leaf datasets remain controversial. The previous studies, ICFHR 2016 [1] and experimental research on Southeast Asian palm leaf benchmarks, focused mainly on handcrafted feature extraction and traditional methods of Balinese scripts. Meanwhile, SEAPA classification tasks have only been analyzed sparingly in recent years. Over the last few years, numerous techniques have been developed and tested, particularly for Latin-based scripts. Regardless, low-resource languages like palm leaf manuscripts still require benchmarking on recent trending approaches for accuracy, data quantity, functional architecture, and training strategies.

Considering these challenges, this paper investigates the performance of front-end and back-end techniques for improving isolated glyph classification. Furthermore, due to the physical degradation of the manuscript datasets, different training ways, such as preprocessing, data augmentation, and visual backbones, are thoroughly investigated to determine their effectiveness and alternation on historical manuscripts.

In summary, the significant contributions of this work are as follows:

1. We present multi-task preprocessing techniques for boosting the accuracy rate. In particular, we extract new datasets from text-line and word datasets. Additionally, image enhancement and data augmentation techniques are presented.
2. We evaluate recent deep learning approaches with data augmentation on different benchmarks toward isolated glyph classification task.
3. We identify the problems with image quality. Based on our analysis, enhancing input images' color from RGB to greyscale to binary could improve their accuracy rates. As a result, binary datasets also showed better performance.
4. We investigate the performances of palm leaf manuscripts using the most effective deep learning approaches and our new datasets with preprocessing techniques.

The rest of this paper, we present relevant information on palm leaf manuscripts in Sect. 2. Section 3 describes the overall framework, including preprocessing and visual backbones of glyph classifications. Finally, Sect. 4 includes the experimental setup and results before we conclude our study in Sect. 5.

2 Palm Leaf Manuscripts from Southeast Asia

This section aims at providing descriptive information on the scales corpus of the palm leaf manuscript for our experimental studies. As shown in Fig. 1, the compilations consist of three palm leaf manuscripts, including Khmer script (Cambodia), Balinese, and Sundanese scripts (Indonesia).

2.1 Corpus and Languages

Balinese [9] is one of Indonesia's local and traditional languages, which is commonly found in many ancient manuscripts of Bali and is also the native language of Bali people. A total of 100 classes are presented in the Balinese language, including consonants, vowels, and special characters. In the previous contest [10], 133 classes of Balinese were obtained, 11,710 images of which were used for training and 7,673 for testing in the isolated glyph recognition task. Meanwhile, Khmer [19] is the official language of Cambodia, which is also commonly used to record Buddhist literature. The language consists of 111 classes, including consonants, vowels, numerals, special characters, and diacritics. The words can be formed in different ways by combining a few symbols, consonants, or vowels. Buddhist Institute or National Library collected most of the manuscripts, including 113,206 images for training and 90,669 for testing. Lastly, Sundanese [15] originated from Garut, West Java, and Situs Kabuyutan Ciburuy in Indonesia, containing 27 collections, each of which includes 15 to 39 pages. Similarly, Sundanese characters also include numbers, vowels, essential characters, and special characters. However, Sundanese contains only 60 classes, 4,555 images of which were used for training, while 2,816 others were for testing.

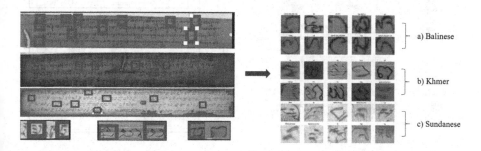

Fig. 1. Sample of data collections from the isolated character palm leaf datasets. (a) Balinese script from Indonesia; (b) Khmer script from Cambodia; (c) Sundanese script from Indonesia.

2.2 Challenges of Isolated Glyph Datasets

Palm leaf manuscripts are generally complicated considering the character classes, alphabets, and numerals, making historical document image analysis more complex to binarize than the scanned documents. In addition, the characters-based are similar, consisting of more than one way of layering up to form the other letters. Consequently, these ancient manuscripts require a sophisticated and practical system for recognizing and classifying these letters. Moreover, the publicly available datasets for each script of the isolated datasets are small and medium, the smallest of which is Sundanese scripts, in which the data for training and testing consists of approximately ∽4 K images and ∽2 K images, respectively. Meanwhile, Balinese and Khmer are considered as medium-sized datasets. From text-line and word datasets, we found that many potential isolated characters could be extracted based on palm leaf datasets [8,15,19]. Therefore, extracting more data to support deep learning methods is necessary to balance the training approaches.

3   Overall Frameworks

The overall architecture is shown in Fig. 2, including front-end and back-end. In the front-end, we present a simple and effective method for increasing datasets and improving image quality. Mainly, we extract isolated characters from text-line and word datasets, then perform image enhancement with data augmentation techniques. In the back-end, we present different deep learning techniques, especially on various CNNs and attention-based models.

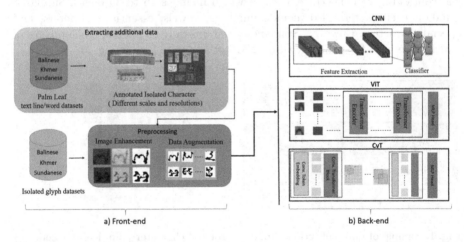

Fig. 2. An overview of our training strategy. As part of the first step of the front-end, we extract new collections from the text-line and word/text palm leaf datasets. Newly additional and the existing datasets are then enhanced using data augmentation and image enhancement techniques. Finally, we inspect the results based on the different data sizes by performing different visual backbones of CNNs and ViTs on the back-end side.

3.1   Data Pattern Generations

Since recent approaches like ViTs and Deep CNNs require a large amount of training data, we present a data pattern generation method to synthesize isolated glyph datasets from palm leaf manuscripts. We enhance and synthesize the isolated character images by taking 2 of the following steps:

1. Obtain additional data collections of text-line and text/word palm leaf datasets.
2. Deploy data augmentation techniques to increase the scale of the existing and the new data collections.

Extracting Additional Datasets. In this step, we aim to extract new collections based on the text-line and word datasets [10]. The publicly available datasets of the isolated glyph images contained limited images; therefore, we extract additional datasets from the previous works containing the three manuscripts. Specifically, 15 local university students from Southeast Asia were selected voluntarily as their comprehensive knowledge and understanding of the Cambodian and Indonesian languages would be more apprehensive in our experiment's classification and labeling process. As indicated in Fig. 1, we manually crop and label data using friendly interface annotated and label image tools[1,2]. As shown in some of the samples of the extracting characters, the 15 students are divided into two groups. The first group is the collectors, assigned to extract and identify the characters of the palm leaf images. Based on their general knowledge, they segment and label these palm leaf images upon various qualities of text and resolutions. The second group is responsible for validating the labels collected by the first group based on existing character classes and dictionaries from [10]. As a result, we collected ∽15 K images from palm leaf datasets. Mainly, we seek to increase 20–50% of the original datasets for training on small datasets such as Sundanese and Balinese manuscripts.

Data Augmentation Techniques. We aim to integrate two straightforward data augmentation techniques into the training of CNNs and ViTs. Firstly, we applied three basic image processing techniques in basic data augmentation using random cropping, random horizontal flipping, and random Gaussian noise. As we randomly resize the patch, crop it according to the scale, then resize it to the original size, we keep the aspect ratio fixed. A random Gaussian noise sample has a mean of 0 and a standard deviation of 0.01. Secondly, we selected regularization and data augmentation (AugReg) into training. In this case, regularization techniques are commonly adopted within the computer vision community [14]. According to [4], we then apply dropout to the intermediate activation of ViT and apply stochastic depth regularization technique [7] that drops layers with a linearly increasing probability. Apart from this, three data augmentation approaches are efficiently utilized to regularize training: RandAug [3], MixUp [22], and CutMix [21].

---

[1] https://github.com/donavaly/SleukRith-Set.
[2] https://markuphero.com.

3.2   Image Enhancement for Palm Leaf Manuscripts (IEPalm)

This section outlines the techniques to enhance the performance of the poor quality palm leaf manuscripts, by presenting a multi-task image processing to clean images called IEPalm in short. We begin by focusing on the contrast balance, and then we perform thresholding binarization to remove some noises from the background and separate the text to provide a better understanding.

Normalization. In the preprocessing step, we primarily focus on balancing the contrast of an image. However, due to the low contrast images, image enhancement is necessary to emphasize certain features or reduce ambiguity between regions of an image, thereby improving thresholding. Inspired by an updated version of contrast limited adaptive histogram equalization (CLAHE), [2] is selected for contrast enhancement in order to correct inconsistencies between text and background. Applying CLAHE can also reduce the noise levels and maintain an image's high spatial frequency content and medial filtering and edge sharpening. In addition to this, adaptive histogram clipping (AHC) can also be applied to the limited contrast technique. By using AHC, clipping levels are automatically adjusted, and over-enhancements are moderated.

With a uniform distribution, the CLAHE technique can be described by:

$$C = Pro(f) * [C_{max} - C_{min}] + C_{min} \tag{1}$$

In exponential distribution, gray level can be described as follows:

$$C = C_{min} - In\left[1 - Pro(f)\right]) * \left(\frac{1}{\delta}\right) \tag{2}$$

where C_{max} is the maximum pixel value, and C_{min} indicates the minimum pixel value. Therefore, C represents the computed pixel values, $Pro(f)$ is the cumulative probability distribution, and δ is the clip parameter.

Thresholding. WAN [12] was proposed to improve the Sauvola method for more reliability on low-quality images. However, Sauvola's approach struggles to segment text and background. For example, if the contrast between the foreground and the background is down, there may be less noises in the text image. Thus, we present an approach to calculate binarization thresholding for this stage, which is more likely to succeed for these types of degraded documents. The main benefit of WAN is that, it enhances binarization by shifting the threshold for detailed images. Furthermore, the investigation shows that *mean* value m significantly impacts the threshold values. Whenever non-text pixels and text pixels of gray values in an image are close to each other, this can enhance the appearance of the output image. However, if we increase the threshold value, the noise and artifact will remain in the image. Therefore, calculating the maximum threshold value as a replacement for the actual *mean* is necessary. For illustration, the following represents the maximum-mean equation:

$$i_{max} = \frac{\text{mean} + \max(a, b)}{2} \tag{3}$$

where $max(a, b)$ exemplifies the maximum contrast of the source image, whereas *mean*, the average contrast of the entire image; both of which allow us to calculate the average of the highest contrast that will consequently recondition the lost features and reduce noises and artifacts in the binarization results. In this case, we estimated the average of the highest contrast and *mean* of the image. Therefore, WAN is able to restore lost details and reduce noise and artifacts in the binarization results. Specifically, the following algorithm is presented:

$$T = (i_{\max}) \left[1 - k \left(1 - \frac{\sigma}{R} \right) \right] \tag{4}$$

where k and R values use a default value from the Sauvola method. k stands for gray level, m represents mean, σ is the standard deviation, and R stands for color (default value).

3.3   Training CNNs and ViTs

In this section, we investigate the performances of training with different backends. CNN and ViT variants have been used successfully in image classification for years. Hence, we explore the back-ends of CNN and ViT variants. The first back-end is CNN-based models. CNNs have also paved ways for convolutional networks, like translation equivalence, object classification, and recognition, which have gained attention in recent years [6,13,16]. The second back-end is attention-based models. In recent years, ViTs have performed well in ImageNet image classification tasks [4,18,20]. Despite this, implications of both back-ends on low-resource languages like palm leaf manuscripts remain in the studying step. Based on various layers and parameters, the study selects trending architectures as follows:

Very Deep Convolutional Networks (VGG). VGG is an architecture with 16 or 19 layers, introduced by [13] for large-scale image classification. In this case, VGG16 consists of ⌐138 million parameters and VGG19 consists of ⌐144 million parameters. As shown in Fig. 3(a), we fixed the convolution filter size to 3 × 3 for entire layers to reach deeper implementation while increasing nonlinearity functions for learning complex representations. Furthermore, after the convolution layers, the outputs were maxed before connecting to three fully-connected layers, leading to many different learnable parameters based on VGG16 and VGG19.

Residual Networks (ResNets). ResNet [6] is one of the most popular models in the history of CNN architectures. ResNet is a model that makes of the residual module involving shortcut connections. In this case, when stacking more convolution layers, gradients drop and vanish during back-propagation. Since adding layers without a plan reduces the accuracy and performance, residual learning adds the previous-layer output via "shortcut connections" to the stacked-layer output. These are proven to be less complex than the VGG networks due to the absence of complexity or networks parameters. Therefore, ResNet50 (⌐25 million parameters) and ResNet101 (⌐42 million parameters) are discussed in this study.

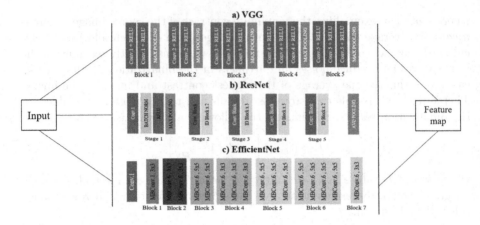

Fig. 3. CNN-based models, such as VGG, ResNet, and EfficientNet, are put into fair comparisons to evaluate their performances at various stages.

EfficientNet. EfficientNet [16] has been one of the trending approaches for improving the accuracy of CNNs. It proves we can achieve excellent results with reasonable parameters by carefully designing our architecture. Hereof, Efficient-Net showed excellent results with less parameters. In an effective but simple manner, EfficientNet scales up models using compound coefficients. In contrast to randomly scaling up width, depth, or resolution, compound scaling uniformly scales each dimension with a fixed set of scaling coefficients. Therefore, AutoML and the scaling method are being used. Seven models of various dimensions were developed, and they outperformed the state-of-the-art accuracy and efficiency of most convolutional neural networks. The architecture can be seen in detail in Fig. 3(c). In this study, we evaluate EfficientNetB0 to EfficientNetB3 models ranging from 4–10 million parameters.

Vision Transformers. ViT was proposed with comparable or higher accuracy rates than state-of-the-art approaches for image classification in ImageNet [4]. In most cases, an image is usually split into $K \times K$ overlapping patches. Each input embedding patch yields $K \times K$ input tokens. Figure 4(a) shows a ViT architecture based on transformer multi-attention layers that pair the model over-token intermediate representations. Then, the final grid embedding is used for discrimination. Several approaches use a "class token" to collect contextual information across the entire grid, while others use average global pooling to compact image representation. Lastly, an MLP head outputs a posterior distribution of target classes based on the whole image. Afterward, training on ViTs variants (DeiT) [18] was proposed using supervised pre-training but only with the more regularized and distillate ImageNet-1k datasets to seemingly demonstrate the progress of ViTs. Data-efficient image transformers are far more efficient in training the transformers for image classification tasks. Moreover, it requires far less data and computing resources than the original ViT models. In this case,

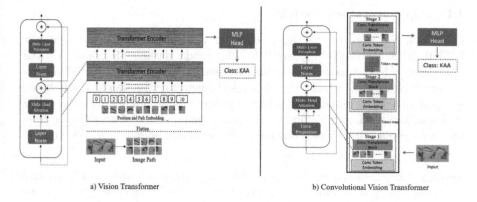

a) Vision Transformer   b) Convolutional Vision Transformer

Fig. 4. The overall diagram of ViTs. (a) The standard architecture of ViT (b) The overview of architecture that introduces convolutional to transformer (CvT).

ViT-S and DieT-S with 12 layers and ∽22 million parameters respectively, while ViT-B and DieT-B with ∽86 million parameters are selected for evaluations.

Convolutional Vision Transformer (CvT). CvT [20] was proposed to enhance the performance of standard ViT. As shown in Fig. 4(b), this work employs a multi-stage hierarchy design borrowed from CNNs, consisting of three stages in total. In summary, token embedding and convolutional projection are applied to transformer hierarchies and transformer blocks, respectively. As input, convolutional token embedding uses overlapping patches of tokens reshaped to the 2D spatial grid (the stride length can control the degree of the overlapping ones). Each step of CNN architecture lowers tokens (feature resolution) while increasing token width (feature dimension). As transformers stack on each level, the convolutional transformer block uses convolutional projection, whereby the MLP head predicts the output token's class. CvT13 with ∽20 million parameters and CvT21 with ∽32 million parameters are also included in our investigation.

4   Experimental Setups and Results

In this experiment, we aim to determine the front-end and back-end performance for improving isolated glyph classifications. In this case, different experiments are conducted thoroughly to investigate the performances. In addition to this, the experiments are categorized into three questions for quality assurance:

1. Which deep learning technique is the most effective back-end approach for each script in ancient palm leaf datasets?
2. How does the quality of the datasets affect accuracy?
3. Does our new datasets and IEPalm techniques effectiveness improve isolated glyph classification tasks?

4.1 Implementation Settings

Datasets. We evaluate all tasks based on the datasets extracted from the ICFHR 2018 contest [10], SleukRith [19], Sunda dataset [15], and AMADI LontarSet [8], which comprises palm leaf manuscripts for the isolated character/glyph classification task. The Balinese, Khmer, and Sundanese datasets are grouped into classes and divided accordingly for training and testing purposes, as shown in Table 1. The table further compares the number of data collected from the three manuscripts. Based on this table, the study presents the following:

Table 1. The palm leaf datasets contained the original datasets from the ICFHR 2018 contest and mixed them with newly extracted datasets, including classes, training, and testing images.

Script	Dataset	Classes	Training	Testing	Source
Balinese	Track 1	133	11,710	7,673	AMADI LontarSet [8]
	TrackMix 1		16,500		
Khmer	Track 2	111	113,206	90,669	SleukRith Set [19]
	TrackMix 2		120,206		
Sundanese	Track 3	60	4,555	2,816	Sunda Dataset [15]
	TrackMix 3		9,800		

Track 1. Dataset contained 133 classes for the Balinese characters, including 11,710 images of the training set and 7,673 images of the testing set.
TrackMix 1. Dataset contained 16,500 training images of the original Balinese isolated glyph dataset and our additional dataset.
Track 2. Dataset has 111 classes for the Khmer characters, in which 113,206 images were used for training and 90,669 images for testing.
TrackMix 2. Dataset contained 120,206 training images of the original Khmer isolated glyph dataset and our additional dataset.
Track 3. Dataset has 60 classes for the Sundanese characters, including 4,555 and 2,816 images for training and testing, respectively.
TrackMix 3. Dataset contained 9,800 training sets of the original Sundanese isolated glyph datasets and our additional dataset.

Training Settings. In this study, we train the models with NVIDIA 3090 GPUs 24 GB based on TensorFlow and PyTorch. In order to train on palm leaf datasets, we choose CNN architectures such as VGG16, VGG19, ResNet-50, ResNet101, and EffcientNetB0-B3 [6,13,16]. Due to limited data training, we perform and analyze data augmentation methods with large-scale pre-trained weights. Particularly, weights for the first CNN levels are frozen, and the remaining parameters are trained. For ViTs, we follow the training recipe of ViT-S, ViT-B, DeiT-S,

DeiT-B, CvT13, and CvT21 [4,17,18,20]. For pre-trained weights, ImageNet-1k (ILSVRC-2012) and ImageNet-21k (ImageNet-21k) are used as the image datasets. Firstly, data should be loaded and transformed into 224 × 224 in advance to regulate the model on an array of samples. Thus, the size of a loaded image must increase by one for each image with 224 × 224 pixels and three channels. Moreover, we optimize CNNs, and ViTs using Adam optimizer [11]. The initial learning rate is set to 5e-4, and the cosine learning rate scheduler is subsequently applied to decrease it. Note that both approaches have been trained with the same configurations of 100 epochs.

Evaluation Metrics. For all experiments, we evaluate the performances of the systems by using accuracy rates of the classification of isolated glyphs palm leaf manuscripts.

4.2   Results

Experiment 1. This study aims to compare results from different deep learning techniques with original datasets [10] with basic data augmentation techniques. As shown in Table 2, this section shows accuracy rates for CNNs and ViTs as well as data augmentation across different manuscripts. According to the results, EfficientNets and ResNets are generally the most reliable methods. Especially, EfficientNets presented impressive performance compared to ViTs methods. However, CvT has shown the best performance among ViT variants. In track 2, CvT also achieved comparable results to CNNs in terms of performances. Furthermore, we found that increasing the parameters of models did not significantly affect the accuracy rates.

Experiment 2. Data transformations were observed in this section. In most of the previous studies, RGB or greyscale images were used in the training step [1,10]. In this case, we train only the original datasets without any external datasets. Therefore, we compared the results of ResNet101, EfficientNetB0, DeiT-B, and CvT21 using color (RGB), greyscale (GS), and binary (BI) images. In particular, we transformed all training and testing sets into different types of formats. For binary images, we simply applied the WAN method [12] for thresholding. As shown in Fig. 5, isolated glyph classification can be improved by using data transformations. Interestingly, binary datasets outperform greyscale and RGB in Balinese and Khmer scripts, while Sundanese scripts produce equivalent performances. Consequently, preprocessing steps have proven necessary for palm leaf analysis based on the quality of datasets.

Experiment 3. As the central part of this study, we evaluated the effectiveness of our additional datasets and image enhancement methods (IEPalm). Particularly, we applied our strategy to the most effective CNNs and ViTs approaches, such as EfficientNetB0, DieT-B, and CvT-21. Our entire process is described in Sect. 3. To train these approaches, we used the datasets TrackMix 1, TrackMix 2, and TrackMix 3, whereby each manuscript used the same amount of testing datasets. As shown in Table 3, the results showed that our additional datasets

Table 2. The results of all the architectures trained for Track 1, Track 2, and Track 3 with CNNs, CNNs + data augmentation, ViTs, and ViTs + data augmentation.

Model	Track 1	Track 2	Track 3	Track 1	Track 2	Track 3
	CNNs			CNNs + data augmentation		
VGG16	87.45	88.98	84.50	88.98	89.08	87.20
VGG19	87.90	89.64	84.61	88.91	89.60	87.60
ResNet50	88.47	89.04	85.23	90.04	90.45	88.04
ResNet101	**88.78**	91.23	86.50	**91.65**	90.97	88.65
EfficientNetB0	88.45	90.56	**87.02**	89.49	**91.85**	**90.35**
EfficientNetB1	88.57	**91.67**	86.41	90.45	91.21	88.31
EfficientNetB2	87.44	91.05	86.04	90.77	91.55	89.45
EfficientNetB3	87.95	91.09	**87.65**	90.65	91.08	89.32
	ViTs			ViTs + data augmentation		
ViT-S_16	79.21	87.12	78.50	82.50	86.25	80.09
ViT-B_16	79.44	88.25	79.50	83.23	87.44	80.47
DeiT-S	80.75	88.04	79.20	85.04	88.45	80.78
DeiT-B	80.94	88.14	79.88	85.50	89.36	81.60
CvT-13	85.50	**90.28**	83.20	87.04	**90.45**	**84.54**
CvT-21	**85.80**	90.00	**83.59**	**87.50**	90.36	85.14

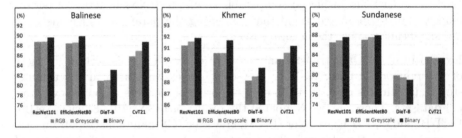

Fig. 5. The results of how data transformations can affect the accuracy achieves when performed with color, greyscale, and binary images.

mixed with the original ones had an improved accuracy rate for most cases in this experiment. Similarly, IEPalm techniques also showed their effectiveness in terms of accuracy. Sundanese and Balinese also present an interesting remark, which made significant improvements when used with larger datasets. Therefore, when it comes to boosting the accuracy rates, it is essential to take into account the importance of quality and quantity of the datasets.

Findings and Limitations. For all types of palm leaf manuscripts, the performance of ViTs falls slightly behind CNNs. The evaluation has showcased that both approaches still need significant support from datasets in addition to the original ones, mainly when trained on Sundanese and Balinese scripts. Specifically, the Sundanese would require more datasets in the training steps to achieve better results. Interestingly, EfficientNet showed impressive results despite using less hyper parameters than ResNet models. For attention based-models, ViTs are also constrained by limited memory and high computing requirements of the expensive quadratic attention in the encoder block. Accordingly, the first observation of CNNs and ViTs also indicates the possibility for improvement by adopting large-scale data such as additional datasets and data augmentation techniques. Lastly, the quality of datasets remains highly significant for boosting the accuracy rates. Thus, cleaning noises or restoring document approaches should be applied before post-processing.

Table 3. The evaluation results of our new collections using IEpalm techniques based on CNNs and ViTs approaches.

	Track 1	TrackMix 1	Track 2	TrackMix 2	Track 3	TrackMix 3
EfficientNetB0	88.45	91.47	90.56	91.47	87.02	91.54
EfficientNetB0 + **IEPalm**	**89.93**	**92.19**	**92.91**	**93.55**	**91.14**	**93.08**
DeiT-B	80.94	84.46	88.14	91.90	79.88	83.97
DeiT-B + **IEPalm**	**83.64**	**86.53**	**89.34**	**92.44**	**83.88**	**86.11**
CvT-21	85.8	87.44	90.00	91.57	83.54	88.65
CvT-21 + **IEPalm**	**87.89**	**89.64**	**91.55**	**93.88**	**85.48**	**90.23**

5   Conclusion

In conclusion, we presented a training way for improving isolated glyph classification tasks using additional datasets and preprocessing techniques. Additionally, we compared the results of CNNs and ViTs using different trending approaches. The results of various experiments indicate that ViTs remain behind CNNs on the isolated glyph classification task. Specifically, EfficientNets and ResNets perform well with or without external datasets and data augmentation techniques. Furthermore, Sundanese and Balinese require more datasets for training those deep learning approaches. In addition, it is essential to consider data augmentation and cleaning datasets to bring out better performance. Considering the results, ResNets and EfficientNets may be used to extract features for the palm leaf recognition system in future works.

Acknowledgements. This research is supported by Chinese Academic Science and World Academy of Science President's Fellowship (CAS-TWAS).

References

1. Burie, J.C., et al.: ICFHR 2016 competition on the analysis of handwritten text in images of balinese palm leaf manuscripts. In: 2016 15th International Conference on Frontiers in Handwriting Recognition (ICFHR), pp. 596–601. IEEE (2016)
2. Chang, Y., Jung, C., Ke, P., Song, H., Hwang, J.: Automatic contrast-limited adaptive histogram equalization with dual gamma correction. IEEE Access **6**, 11782–11792 (2018)
3. Cubuk, E.D., Zoph, B., Shlens, J., Le, Q.V.: Randaugment: practical automated data augmentation with a reduced search space. In: Proceedings of the IEEE/CVF Conference on Computer Vision and Pattern Recognition Workshops, pp. 702–703 (2020)
4. Dosovitskiy, A., et al.: An image is worth 16x16 words: transformers for image recognition at scale. arXiv preprint. arXiv:2010.11929 (2020)
5. Graham, B., et al.: Levit: a vision transformer in convnet's clothing for faster inference. In: Proceedings of the IEEE/CVF International Conference on Computer Vision, pp. 12259–12269 (2021)
6. He, K., Zhang, X., Ren, S., Sun, J.: Deep residual learning for image recognition. In: Proceedings of the IEEE Conference on Computer Vision and Pattern Recognition, pp. 770–778 (2016)
7. Huang, G., Sun, Yu., Liu, Z., Sedra, D., Weinberger, K.Q.: Deep networks with stochastic depth. In: Leibe, B., Matas, J., Sebe, N., Welling, M. (eds.) ECCV 2016. LNCS, vol. 9908, pp. 646–661. Springer, Cham (2016). https://doi.org/10.1007/978-3-319-46493-0_39
8. Kesiman, M.W.A., Burie, J.C., Wibawantara, G.N.M.A., Sunarya, I.M.G., Ogier, J.M.: Amadi_lontarset: the first handwritten balinese palm leaf manuscripts dataset. In: 2016 15th International Conference on Frontiers in Handwriting Recognition (ICFHR), pp. 168–173. IEEE (2016)
9. Kesiman, M.W.A., et al.: Benchmarking of document image analysis tasks for palm leaf manuscripts from southeast Asia. J. Imaging **4**(2), 43 (2018)
10. Kesiman, M.W.A., et al.: ICFHR 2018 competition on document image analysis tasks for southeast asian palm leaf manuscripts. In: 2018 16th International Conference on Frontiers in Handwriting Recognition (ICFHR), pp. 483–488. IEEE (2018)
11. Kingma, D.P., Ba, J.: Adam: a method for stochastic optimization. arXiv preprint. arXiv:1412.6980 (2014)
12. Mustafa, W.A., Yazid, H., Jaafar, M.: An improved sauvola approach on document images binarization. J. Telecommun. Electron. Comput. Eng. (JTEC) **10**(2), 43–50 (2018)
13. Simonyan, K., Zisserman, A.: Very deep convolutional networks for large-scale image recognition. arXiv preprint. arXiv:1409.1556 (2014)
14. Steiner, A., Kolesnikov, A., Zhai, X., Wightman, R., Uszkoreit, J., Beyer, L.: How to train your vit? data, augmentation, and regularization in vision transformers. arXiv preprint. arXiv:2106.10270 (2021)
15. Suryani, M., Paulus, E., Hadi, S., Darsa, U.A., Burie, J.C.: The handwritten sundanese palm leaf manuscript dataset from 15th century. In: 2017 14th IAPR International Conference on Document Analysis and Recognition (ICDAR), vol. 1, pp. 796–800. IEEE (2017)
16. Tan, M., Le, Q.: Efficientnet: rethinking model scaling for convolutional neural networks. In: International Conference on Machine Learning, pp. 6105–6114. PMLR (2019)

17. Tolstikhin, I.O., et al.: Mlp-mixer: an all-mlp architecture for vision. In: Advances in Neural Information Processing Systems, vol. 34 (2021)
18. Touvron, H., Cord, M., Douze, M., Massa, F., Sablayrolles, A., Jégou, H.: Training data-efficient image transformers & distillation through attention. In: International Conference on Machine Learning, pp. 10347–10357. PMLR (2021)
19. Valy, D., Verleysen, M., Chhun, S., Burie, J.C.: A new khmer palm leaf manuscript dataset for document analysis and recognition: Sleukrith set. In: Proceedings of the 4th International Workshop on Historical Document Imaging and Processing, pp. 1–6 (2017)
20. Wu, H., et al.: Cvt: introducing convolutions to vision transformers. In: Proceedings of the IEEE/CVF International Conference on Computer Vision, pp. 22–31 (2021)
21. Yun, S., Han, D., Oh, S.J., Chun, S., Choe, J., Yoo, Y.: Cutmix: regularization strategy to train strong classifiers with localizable features. In: Proceedings of the IEEE/CVF International Conference on Computer Vision, pp. 6023–6032 (2019)
22. Zhang, H., Cisse, M., Dauphin, Y.N., Lopez-Paz, D.: mixup: beyond empirical risk minimization. arXiv preprint. arXiv:1710.09412 (2017)

Signature Verification and Writer Identification

Impact of Type of Convolution Operation on Performance of Convolutional Neural Networks for Online Signature Verification

Chandra Sekhar Vorugunti[1]($\boxtimes$)(ID), Balasubramanian Subramanian[2](ID),
Avinash Gautam[3](ID), and Viswanath Pulabaigari[1](ID)

[1] Indian Institute of Information Technology, SriCity 517646, Andhra Pradesh, India
{chandrasekhar.v,viswanath.p}@iiits.in
[2] Sri Sathya Sai Institute of Higher Learning,
Puttaparthi 515134, Andhra Pradesh, India
sbalasubramanian@sssihl.edu.in
[3] Birla Institue of Technology and Science, Pilani 333031, Rajasthan, India
avinash@pilani.bits-pilani.ac.in

Abstract. An Online signature is a multivariate time series, a commonly used biometric source for user verification. Deep learning (DL) is increasingly becoming ubiquitous as a paradigm for solving problems that come with a wealth of data. Convolution has been its main workhorse. Recently, DL had marked its entry in online signature verification (OSV), a standard bio-metric method that has been mostly dealt with in traditional settings. However, embracing a DL solution to a problem requires certain issues to be tackled, viz. (i) type of convolution, (ii) order of convolution, and (iii) input representation. In this work, we experimentally analyse each of the issues mentioned above regarding OSV, and subsequently present a superior model that reports state-of-the-art (SOTA) performance on three widely used data-sets namely MCYT-100, SVC, and Mobisig. Specifically, the proposed model reports an equal error rate (EER) of 9.72% and 3.1% in Skilled_01 categories of MCYT-100 and SVC data-sets, with gains of around 4% and 3% over the next best performing methods, respectively. The experimental outcome confirms that the interrelationship between the type and order of convolution operation and the input signature representation plays a significant role in the performance of OSV frameworks.

Keywords: Online signature verification · Deep learning · Impact of convolution · Step size · One shot learning

1 Introduction

Deep learning (DL) is increasingly becoming ubiquitous as a paradigm for solving problems that come with a wealth of data [3,6,15,29]. Convolution has been its main workhorse. Recently, DL had marked its entry in online signature verification (OSV), a common biometric method that has been mostly dealt with in

U. Porwal et al. (Eds.): ICFHR 2022, LNCS 13639, pp. 83–97, 2022.
https://doi.org/10.1007/978-3-031-21648-0_6

traditional setting $[1, 13, 24, 25, 30, 37, 38, 40]$. However, embracing DL solution to a problem requires certain issues to be tackled viz. (i) type of convolution, (ii) order of convolution and (iii) input representation.

Concerning the type of convolution, earlier work on OSV [23] compared the standard convolution against separable convolution and showed that separable convolutions enhance performance. Recently, dilated convolutions [8] have become a popular option. We compare separable convolution against dilated convolutions. Regarding order of convolution, as depicted in Fig. 2, since an online signature can be viewed as a matrix, there is a possibility of 2-d convolutions to be performed on the input signature. Does this add value? We experimentally analyse this by comparing 2-d convolutions against 1-d convolutions. Finally, how do we represent the online signature as input? Each online signature is a set of n samples wherein each sample is a 5-tuple of numbers including x coordinate, y coordinate, pressure, altitude and azimuthal angle (see Fig. 2). So, should the input be represented as a $n \times 5$ matrix, or a $5 \times n$ matrix that corresponds to each sample being viewed as an input feature, or a $n \times 5$ matrix of differences between adjacent samples where adjacency is quantified by a step size? In the latter case, what is the step size? Further, should we consider all the samples of an online signature? We analyse these cases experimentally. The experimental analysis on all the aforementioned issues result in an efficient model that reports SOTA performance on three widely used OSV data-sets. In summary, our contributions are as follows:

1. Comparative analysis on the superiority of separable convolution over dilated convolutions for OSV
2. Comparative analysis on 1-d vs 2-d convolutions for OSV
3. Efficient input representation for OSV, including analysis of writer specific signature length
4. An efficient DL based OSV model that reports SOTA performance on three widely used OSV data-sets

2   Related Work

Until recently, classical techniques have been driving the research on OSV, the prominent one being dynamic time warping [26–28]. In fact, a detailed survey on research in OSV over the last decade is available in [18]. Since our focus is on DL based techniques, we will look at related work in that direction.

An early DL based work [21] used a sequence model (LSTM) to analyse the time varying samples of the online signature. Three scenarios viz user specific model, recognition model and verification model were proposed but none of them performed well. Subsequently, an improvised sequence model [39] augmented by siamese architecture was developed, which fared better than a traditional DTW based method, but only in skilled forgeries, and not on random forgeries. Other sequence model based OSV include [30] and [1]. The former uses gated autoregressive units while the latter uses LSTM autoencoder. The former performed well on random forgeries while the latter performed well on skilled forgeries.

Though sequence models lend themselves to be a natural fit for time varying samples of online signature, they did not exhibit exceedingly superior performance, in general. Only recently (in 2019), convolutional neural networks (CNNs) have been tested in OSV [13,16,23–25,40]. These offer better performance than sequence models for OSV. [40] combined a siamese CNN with DTW. The Siamese network is used to extract features from the input signature sequence. The temporal synchronisation of two input signatures is carried out by a DTW block. In [23], separable convolutions have been found to be efficient, even under difficult few shot learning conditions. The authors of [23] improvised in [24] by fusing deep features and statistical features wherein the deep features are extracted from a hybrid CNN-LSTM architecture. Similar idea of feature fusion is explored in [25]. The only difference is that the deep features are extracted using a convolutional autoencoder. Further, the training model does not have LSTM. [13] proposed a 1-d CNN for fixed-length representations of online signatures. Further, the work studied the potential of using synthesized forgeries to eliminate the need for skilled forgeries during training.

Most of the aforementioned CNN based works have their hyperparameters tuned manually. Also, the input to some of the methods includes handcrafted features fused with deep features. There are others where the representation obtained from deep architecture is fused with traditional features. Additionally, very rarely, influence of length of the input is studied. Further, dilated convolutions and 2-d convolutions have not been explored. In this work, we attempt to fill these gaps and build an efficient model for OSV.

3 Proposed OSV Framework

We now describe in detail about the model selection, type of convolution, order of convolution and input representation.

The point to be noted is that the type and order of convolution we had considered during the search is depth-wise separable 1-d convolution. The reason is that depth-wise separable convolution has been shown to be superior over standard convolution for OSV in [23].

3.1 Input Representation, Type of Convolution and Order of Convolution

Now that we have the OSV model ready, we focus on the ideal input representation to this model, the type, and the order of convolution to be used. Each online signature is a set of n samples of 5-tuple of numbers, as depicted in Fig. 2. Figure 1 plots the genuine and forged signatures of user1 from the MCYT dataset.

The online signature can be represented as a $n \times 5$ matrix. We call this representation1. The transpose of this representation is called as representation2. Figure 2 shows these two representations pictorially. In representation1, the signature is a collection of local features at each sample point or time unit. In

86   C. S. Vorugunti et al.

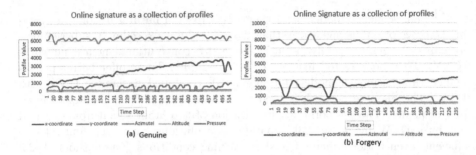

(a) Genuine

(b) Forgery

Fig. 1. Representation of online signature as a multivariate time series. (a) and (b) represents the profiles of genuine and forgery signatures of user 1 of MCYT data-set.

representation2, the signature is a collection of individual feature profiles across time. It can be viewed as a multivariate time series.

Given these two representations, it is clear that both the representations allow for both 1-d and 2-d convolutions. Figure 3 and Fig. 4 depict these convolutions in both representations. A 1-d convolution on representation1 would summarize each sample as weighted linear combinations based on its own 5-tuple feature representation, whereas the same on representation2 would summarize each feature across samples. A 2-d convolution, whether representation1 or representation2, can be viewed as feature re-computation using neighboring samples and neighboring features. Note that we emphasized on the phrase neighboring features because this is an ambiguous phrase that can change its meaning based on the order of components of the 5-tuple. Hence, we can predispose that 2-d convolution on either of the representations is not semantically strong and may not be useful for OSV. We experimentally verify this claim later in this section. Next, we look at the type of convolution - depth-wise separable or dilated. The reason we look at both these convolutions is because both of them result in lesser number of model parameters and computations, and also improve accuracy [33,42] in comparison to the standard convolution. We already know the potential of depth-wise separable convolution for OSV through [23,25] while we want to analyse the efficacy of dilated convolution for OSV.

In summary, we have 4 combinations (2 types of convolution and two orders of convolution) across representation1 and representation2 to be tested. We use MCYT-100 data-set for performing the analysis. Table 1 and 2 reports the analysis across type and order of convolutions for representation1 and representation2, respectively. The prefixes **S** and **R** in column headings denote skilled and random forgery settings, respectively. The number in the suffixes of each column heading denote the number of signatures per user used during training. It is to be noted that for 2-d convolution the filter size used is 6×6 (with required padding) while for 1-d it is 12 as estimated by BO based hyperparameter search. Further, the dilation factor used in dilated convolutions is 2. Larger dilation factors are not useful since in either of the representations, one of the dimensions is fixed to 5.

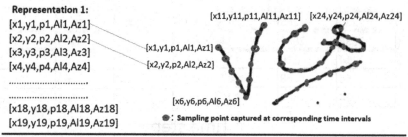

Representation 1:
[x1,y1,p1,Al1,Az1]
[x2,y2,p2,Al2,Az2]
[x3,y3,p3,Al3,Az3]
[x4,y4,p4,Al4,Az4]
....
....
[x18,y18,p18,Al18,Az18]
[x19,y19,p19,Al19,Az19]

[x11,y11,p11,Al11,Az11]   [x24,y24,p24,Al24,Az24]

[x1,y1,p1,Al1,Az1]
[x2,y2,p2,Al2,Az2]

[x6,y6,p6,Al6,Az6]

●: Sampling point captured at corresponding time intervals

Representation 2: (Transpose of Representation 1)
[x1,x2,x3,x4,x5,x6,x7,x8,.....,x17,x18,x19]
[y1,y2,y3,y4,y5,y6,y7,y8,····· ,v17.v18. v19]
[p1,p2,p3,p4,p5,p6,p7,p8,···· p17,p18,p19]
[Al1,Al2,Al3,Al4,Al5, Al6, ···· ,Al18, Al19]
[Az1,Az2,Az3,Az4,Az5,Az6,···· ,Az18,Az19]

Fig. 2. The two possible representations of online signature data. Representation1(R1): A signature as a collection of local features captured at each sample point. Representation2(R2): A signature as a collection of individual feature profiles (multivariate time series).

1D Convolution and 2D Convolution

	X	Y	P	Az	Al		X	Y	P	Az	Al
Time step	0.21693	0.56374	0.54962	0.78781	0.98939	Time step	0.71758	0.24838	-0.14549	-0.90672	-0.99375
	0.97619	0.82595	0.83542	0.61592	0.14529		-0.69648	-0.96866	-0.98936	-0.42173	-0.11166
	55.317	76.276	90.973	69.814	144.53		47.381	40.262	34.366	47.424	89.56
	32.28	14.765	33.377	90.708	47.854		24.739	15.811	40.497	47.074	34.438
	47	53	44	36	14		-14	-20	-27	-38	-21
	59	87	105	104	99		0	-38	-58	-134	-136
	0	0	0	0	0		1	2	0	1	0
	1	0	0	0	0		0	0	-2	1	0
	8	7	13	16	35		5	-1	3	6	2

[   ] :1D Convolution

(a)

[ ] :2D Convolution

(b)

Fig. 3. (a) Represents the 1D convolution operation on signature profiles, which covers profiles/ features of only one time step.(b) Represents the 2D convolution operation on signature, which covers features of multiple time steps.

It is clear that Table 1 reports much lower EER in comparison to Table 2. Therefore, representation1 is superior than representation2. Further, under representation1 setting, 2-d convolutions under-perform in comparison to 1-d convolutions, as expected. Between depth-wise separable and dilated convolutions, the former performs well on both skilled and random forgeries setting while the latter performs well on random forgeries setting.

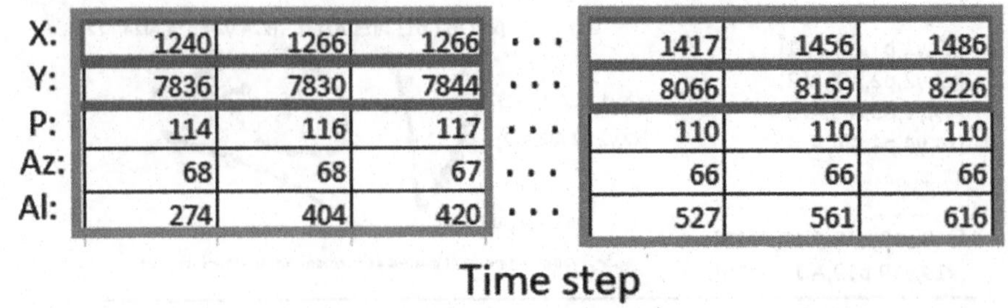

X:	1240	1266	1266	···	1417	1456	1486
Y:	7836	7830	7844	···	8066	8159	8226
P:	114	116	117	···	110	110	110
Az:	68	68	67	···	66	66	66
Al:	274	404	420	···	527	561	616

Time step

Fig. 4. Red colour represents the 1D convolution operation on signature profiles, which covers a single profile. Green colour represents the 2D convolution operation on signature profiles, which covers multiple single profiles. (Color figure online)

Table 1. EER outcome on MCYT-100 data-set under representation1 (S:Skilled, R:Random categories)

Conv Type	S_1	S_5	S_10	S_15	S_20	R_1	R_5	R_10	R_15	R_20
Separable Conv 1D	15.12	**5.97**	3.57	**1.40**	**0.90**	17.69	**6.19**	**2.96**	**2.14**	**1.53**
Dilated Conv 1D	**13.33**	6.00	**3.50**	1.70	0.98	19.99	16.79	12.76	9.28	5.76
Separable Conv 2D	16.94	8.62	5.17	2.9	1.6	**13.68**	12.49	8.9	3.25	2.93
Dilated Conv 2D	16.87	10.36	6.38	5.10	3.87	19.07	14.82	9.63	6.14	5.72

Table 2. EER outcome on MCYT-100 data-set under representation2 (S:Skilled, R:Random categories)

Conv Type	S_1	S_5	S_10	S_15	S_20	R_1	R_5	R_10	R_15	R_20
Separable Conv 1D	19.08	11.37	9.47	7.65	**5.60**	**15.64**	**11.91**	12.78	7.09	**4.10**
Dilated Conv 1D	**15.15**	**11.27**	**8.97**	**7.40**	5.70	19.91	16.79	12.76	9.28	5.76
Separable Conv 2D	39.94	27.62	16.87	11.17	9.00	19.07	12.94	**12.06**	**3.25**	4.24
Dilated Conv 2D	46.27	38.37	27.17	22.90	11.60	26.10	22.65	17.36	15.82	11.59

We also performed the analysis on other two data-sets namely SVC [34] and Mobisig [14,30] data-sets. Summary statistics of these data-sets are available in Table 6. Table 3 and 4 tabulate the performance. Again, as before, depth-wise separable 1-d convolution under representation1 performs the best. An important observation from these tables is that the 2-d convolution oscillates in its performance across representations, empirically confirming our earlier stress on ambiguity in the phrase neighboring features related to it. Hence, we infer that depthwise separable 1-d convolution under representation1 setting is the efficient setting for OSV.

Table 3. EER outcome on Mobisig data-set (S:Skilled, R:Random categories, R1: Representation 1 and R2: Representation 2)

Conv Type	S_1	S_05	S_10	S_15	R_1	R_5	R_10	R_15
Separable Conv 1D (R1)	**10.13**	**2.98**	**1.02**	1.00	**17.66**	**1.07**	**0.54**	**0.51**
Separable Conv 2D (R1)	11.27	4.33	2.47	1.63	23.96	3.33	1.86	1.45
Separable Conv 1D (R2)	13.86	8.31	5.24	4.17	23.7	9.46	5.15	3.61
Separable Conv 2D (R2)	10.29	3.36	1.27	**0.86**	18.35	13.46	1.29	1.24

Table 4. EER outcome on SVC data-set (S:Skilled, R:Random categories, R1: Representation 1 and R2: Representation 2)

Conv Type	S_1	S_5	S_10	S_15	R_1	R_5	R_10	R_15
Separable Conv 1D (R1)	**5.34**	2.73	2.1	**1.2**	12.12	**3.09**	**1.84**	**0.68**
Separable Conv 2D (R1)	8.42	5.7	4.35	3.9	12.48	8.04	4.18	2.43
Separable Conv 1D (R2)	5.87	2.97	2.67	1.9	12.76	7.89	4.04	2.39
Separable Conv 2D (R2)	5.97	**2.53**	**1.4**	2.3	**11.9**	5.41	3.37	1.5

3.2 Analyzing the Impact of Signature Length

Inter and intra writer variations is a challenging characteristic to deal with in OSV. Figure 5(a) and Fig 5(b) shows signatures of two different users with significant variance in length among their own signatures at different points in time. The question is that should we consider the entire length of the signature while processing. To analyse this, we recorded the best EER for each signature of each user in MCYT-100 data-set, with their lengths varying across the range of minimum to maximum. We found that only 40% to 50% of the entire length is required for each signature as depicted in Fig. 6. More specifically, Fig. 7(a) shows the optimal length required for each user (optimal in the sense of lowest EER) in MCYT-100 data-set and Fig. 7(b) shows the corresponding EER. Such significant reduction in length would be advantageous in terms of storage complexity and processing time.

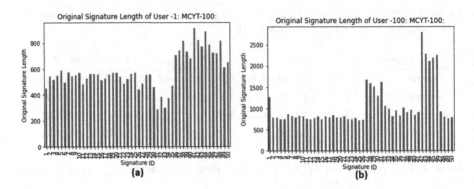

Fig. 5. (a) Represents the varying signature lengths of 25 genuine and 25 forgery signatures of user ID - 1 of MCYT-100 data-set. (b) Represents the varying signature lengths of 25 genuine and 25 forgery signatures of user ID - 100 of MCYT-100 data-set.

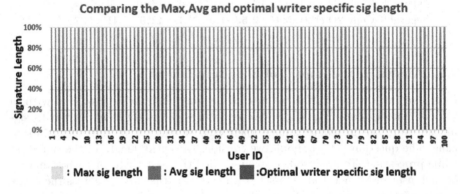

Fig. 6. The maximum, average and optimal length of signatures of each writer in case of MCYT-100 data-set.

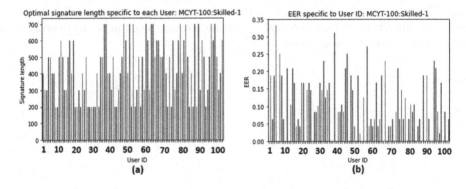

Fig. 7. (a) represents the optimal signature length for each user. (b) represents the EER outcome specific to each user considering the optimal signature length in case of Skilled_01 category of MCYT-100 data-set.

3.3   Further Improvement of Input Representation

We would like to make our model more robust to forgery detection. Input representation could play a significant role towards this end. Even though we found by our earlier analysis that representation1 is good, there is further scope for refinement. The reason is that a genuine and a forged signature may not differ much in the raw representation as shown in Fig. 8(a) and Fig. 8(d). Therefore, we converted the input representation1 to input of differences i.e. each row is a difference of two samples step size apart in the representation1. The step sizes included are $\{1, 2, 3, 4, 5, 6, 7, 8, 9, 10\}$. We found step size 6 to be robust to forged signature detection. A pictorial representation of this for a particular user is available in Fig. 8. Figure 8(b) and Fig. 8(e) show the difference representation using step size 1 for both genuine and forged signatures. Clearly, the representation is not very sensitive to the forged signature. Figure 8(c) and Fig. 8(f) show the difference representation using step size 6 for both genuine and forged signatures. As can be seen, towards the right one-third of the figures, the difference is clearly visible. Hence, instead of using raw representation1, we find difference based representation with step size 6 to be more efficient. This is further confirmed by the results reported in Table 5.

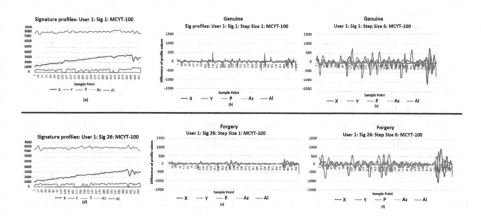

Fig. 8. Effect of difference of samples representation. a) and d) represents the genuine and forged signature profiles of user 1 in MCYT-100 data-set. b) and e) represents the profiles with one time step difference between original profiles. c) and f) represents the profiles with six time step difference between original profiles

Table 5. The step size and the corresponding EER outcome by the proposed model in case of Skilled_01 category of MCYT-100 data-set.

Step Size	1	2	3	4	5	6	7	8	9	10
Overall EER	10.73	10.17	10.13	10.09	10.05	**9.72**	10.54	11.04	11.75	11.98

4 Comparison with SOTA Methods

We have evaluated our proposed OSV framework comprehensively on three widely used data-sets viz. MCYT-100 [9], SVC [34], MobiSig [14,30]. Summary statistics of these data-sets are available in Table 6. Experiments were carried out on Nvidia Titan-X 12 GB GPU. We have considered five categories for evaluating performance, i.e. Skilled_01 (S_1), Skilled_5 (S_5), Skilled_10 (S_10), Skilled_15 (S_15) and Skilled_20 (S_20) in skilled forgery category; Random_01 (R_1), Random_5 (R_5), Random_10 (R_10), Random_15 (R_15) and Random_20 (R_20) in the random forgery category. The protocol for testing and training is as follows. Suppose the data-set consists of U number of users, each user with G genuine and F forged signatures. In the case of Skilled_C (S_C) category, the framework is trained with C genuine and C forged signatures per user. It is tested with $G - C$ genuine and $F - C$ forged signatures per user. In case of Random_C (R_C) category, the framework is trained with C genuine and $U - 1$ other signatures per user. These other signatures are considered as forged signatures with respect to the user in consideration. The way these $U - 1$ signatures are selected is by uniformly sampling one genuine signature from each of the other users (leaving aside the user in consideration). For testing, $G - C$ genuine signatures and $U - 1$ other signatures are considered per user. $U - 1$ other signatures are selected using the same strategy adopted for training but the only condition is that these signatures do not overlap with the $U - 1$ other signatures in the training set. It is to be noted that, in the comparative study results tabulated in various tables, the model with first best EER (the lesser the better) is marked as * and the second best is marked as **. We also emphasize that ours is the only study that has experimented on variety of categories, and as will be reported later, we perform well even on one shot learning setting (Skilled_01 category).

Table 7, 8 and 9 presents the comparative performance of our method against other methods from the literature on all the three data-sets. It is clear from the table that many of the other methods do not report results on all the categories. Limited evaluation of any model inhibits its real time applicability. Our framework has been evaluated on a large variety of categories and it is clear from the tables that our framework has done exceedingly well on both skilled and random categories across all the data-sets. In fact, in the most difficult one shot learning setting of Skilled_01 category, our framework reports over $4 - 6\%$ gain in MCYT-100 data-set and over $3 - 15\%$ gain in SVC-Task 2 data-set. In Mobisig data-set, our framework is almost on par with LSTM based method in the Skilled_01 category. It is to be noted that only two standard works are reported in the literature which evaluated their frameworks with Mobisig data-set.

We also compared our model with most recent non-CNN based model i.e. vision transformer with attention mechanism. It can be observed from the Tables 7, 8 and 9 that the vision transformer results in an EER slightly higher than the proposed framework in case of Skilled categories and performs quite badly in Random categories. For example, as illustrated in Table 7, in case of Skilled_01 category of MCYT-100 data-set, the proposed framework and the Transformer based model yields EER's of 9.72% and 9.88%, respectively. In case

Table 6. Details of data-sets used in experimental evaluation.

data-set	MCYT-100	SVC	MobiSig
Number of Users	100	40	83
Genuine signatures per user	25	20	40
Forgery signatures per user	25	20	25
Total genuine signatures	2500	800	3320
Total forgery signatures	2500	800	2075
Total signatures	5000	1600	5395

Table 7. Comparison of the proposed framework with SOTA models on MCYT-100 data-set. '*' represents the first best EER value. '**' indicates the second best EER value. '-' denotes that the corresponding model has not evaluated for this experimentation category.

Method	S_1	S_5	S_10	S_15	S_20
proposed: Impact of Convolution:	**9.72***	**0.38***	**0.10***	**0.05***	**0.00***
proposed: vision transformer with attention:	**9.88****	2.97	0.73	0.98	0.60
few shot learning [23]	13.42	7.03	5.7	3.95	2.2
feature fusion [25]	13.38	3.02	1.83	1.25	1.2
Clustering + Feature fusion [24]	**13.26****	2.66	2.58	3.01	1.2
LSTM+CNN [4]	15.57	1.88	**0.67****	**0.73****	**0.00***
Histogram + Manhattan [22]	–	4.02	–	–	–
Discriminative feature vector [22]	–	4.02	–	–	2.72
VQ+DTW [26]	–	1.55	–	–	–
Stroke-Wise [5]	13.72	–	–	–	–
Target-Wise [5]	13.56	–	–	–	–
Information Divergence-Based Matching [36]	–	3.16	–	–	–
Secure KNN-Regional features [7]	–	4.65	–	–	–
WP+BL DTW [28]	–	2.76	–	–	–
DTW (Skilled forgery) [12]	–	1.62	–	–	–
DTW (Random forgery) [12]	–	1.81	–	–	–
KNN-Global features [7]	–	5.15	–	–	–
DTW-Normalization(F13) [7]	–	8.36	–	–	–
Probabilistic-DTW(case 1) [2]	–	–	–	–	11.23
Stability Modulated DTW (F13) [7]	–	13.56	–	–	–
Probabilistic-DTW(case 2) [2]	–	–	–	–	–
Curvature feature [10]	–	10.22	8.25	6.38	–
Torsion Feature [10]	–	9.22	7.04	5.12	–
Curvature feature +Torsion Feature [10]	–	6.05			
VSA-DTW [17]	–	3.24	–	–	–
VSA-DTW [17]	–	2.68	–	–	–
Time-series averaging+Gradient Boosting [19]	–	1.28	–	–	–
Time-series averaging+DTW [20]	–	**0.72****	–	–	–
Gene-expression programming [35]	–	3.62	2.57	–	–

Table 8. Comparison of the proposed framework with SOTA models on SVC data-set.

Method	S_1	S_5	S_10	S_15	
proposed: Impact of Convolution:	**3.1***	**0.1***	**0.00***	**0.00***	
proposed: vision transformer with attention	5.45	2.47	0.55	0.40	
few shot learning [23]	5.83	**0.87****	**0.35****	0.2	
feature fusion [25]	17.96	5.17	2.07	–	
Clustering + Feature fusion [24]	15.84	4.95	1.68	–	
Target-Wise [5]	18.63	–	–	–	
LSTM+CNN [4]	6.71	1.05	**0.00***	**0.10****	
Classifier ensemble-1 [31]	–	2.62	–	–	
Classifier ensemble-2 [32]	–	6.00	–	–	
Stroke-Wise [5]	18.25	–	–	–	
DTW based (Common Threshold) [28]	–	7.80	–	–	
Stroke Point Warping [11]	–	1.00	–	–	
Probabilistic-DTW(case 1) [2]	–	–	–	–	
Curvature feature + Torsion Feature [10]	–	9.83	6.61	3.10	
SPW+mRMR+SVM(10-Samples) [11]	–	1.00	–	–	
Variance selection [41]	–	–	13.75	–	
Relief-2 [41]	–	–	5.31	–	
Probabilistic-DTW(case 2) [2]	–	–	–	–	
PCA [41]	–	–	7.05	–	
Classifier Set [31]	–	2.62	–	–	
Relief-1 (using the combined features set) [41]	–	–	8.1	–	
RNN+LNPS [30]	–	–	–	–	
Time-series averaging+Gradient Boosting [19]	–	4.98	4.26	–	
Time-series averaging+DTW [20]	–	2.08	1.53	–	
Gene-expression programming [35]	–	4.38	4.11	–	
Sig-2D [16]	**5.37****	–	–	–	–
Sig-2D(without CWL) [16]	6.55	–	–	–	–

of Random_01 category, the proposed and the transformer based model reports EER's of 0.58% and 17.32%, respectively. However, the Transformer based model exhibits similar trend of the proposed model by reporting decreasing EER values with increasing number of training samples. Similar analysis can be found in Tables 8 and 9. In summary, the proposed model performs better than the latest non-CNN based vision transformers.

Table 9. Comparison of the proposed framework with SOTA models on Mobisig dataset.

Method	S_1	S_5	S_10	S_15
proposed: NAS + Impact of Convolution:	**18.95****	**10.2***	**5.91***	**2.88***
proposed: vision transformer with attention:	24.61	19.32	17.65	12.56
GARU + DTW [30]	–	10.9**	–	–
LSTM [14]	16.1*	–	–	–

5 Conclusion and Future Work

In this work, we have proposed an end to end optimised OSV framework with four fold contribution towards satisfying the important requirements of a OSV system. Selection of BO based CNN hyperparameters, the efficient type and order of convolution, useful input representation and writer specific optimal signature length resulted in a light-weight framework with increased representation efficiency, propelling SOTA results on even difficult one-shot learning setting. As a part of future work, we would focus on selecting the writer specific optimal signature subset for OSV.

References

1. Ahrabian, K., Babaali, B.: On usage of autoencoders and siamese networks for online handwritten signature verification. Neural Comput. **31**(1), 1–14 (2018)
2. Al-Hmouz, R., Pedrycz, W., Daqrouq, K., Morfeq, A.: Quantifying dynamic time warping distance using probabilistic model in verification of dynamic signatures. Soft. Comput. **23**(11), 407–418 (2019). https://doi.org/10.1007/s00500-017-2782-5
3. Dutta, A., Verma, Y., Jawahar, C.V.: Recurrent image annotation with explicit inter-label dependencies. In: Vedaldi, A., Bischof, H., Brox, T., Frahm, J.-M. (eds.) ECCV 2020. LNCS, vol. 12374, pp. 191–207. Springer, Cham (2020). https://doi.org/10.1007/978-3-030-58526-6_12
4. Chandra Sekhar, V., Doctor, A., Viswanath, P.: A light weight and hybrid deep learning model based online signature verification. In: 2nd International Workshop on Machine Learning (ICDAR-WML), pp. 53–59 (2019)
5. Diaz, M., Fischer, A., Ferrer, M., Plamondon, R.: Dynamic signature verification system based on one real signature. IEEE Trans. Cybern. **48**(1), 228–239 (2018)
6. Dikshant, G., Aditya, A., Nehal, M., Vineeth, N.S., Jawahar, C.V.: A multi-space approach to zero-shot object detection. In: Winter Conference on Applications of Computer Vision (WACV), pp. 1209–1217 (2020)
7. Doroz, R., Kudlacik, P., Porwika, P.: Online signature verification modeled by stability oriented reference signatures. Inf. Sci. **460**(1), 151–171 (2018)
8. Fisher, Y., Vladlen, K.: Multi-scale context aggregation by dilated convolutions. ICLR (2016)
9. Garcia, O.J., Aguilar, J.F., Simon, D.: MCYT baseline corpus: a bimodal database. IEEE Proc. Vis. Image Sig. Process. **150**, 3113–3123 (2003)

10. He, L., Tan, H., Huang, Z.: Online handwritten signature verification based on association of curvature and torsion feature with Hausdorff distance. Multimed. Tools Appl. **78**(1), 253–278 (2019). https://doi.org/10.1007/s11042-019-7264-6
11. Kar, B., Mukherjee, A., Dutta, P.: Stroke point warping-based reference selection and verification of online signature. IEEE Trans. Instrum. Meas. **67**(1), 2–11 (2018)
12. Lai, S., Jin, L.: Recurrent adaptation networks for online signature verification. IEEE Trans. Inf. Forensics Secur. **14**(6), 1624–1637 (2018)
13. Lai, S., Jin, L., Lin, L., Zhu, Y., Mao, H.: SynSig2vec: learning representations from synthetic dynamic signatures for real-world verification. In: AAAI Conference on Artificial Intelligence (2020)
14. Li, C., Zhang, X., Lin, F.: A stroke-based RNN for writer-independent online signature verification. In: International Conference on Document Analysis and Recognition (ICDAR), pp. 526–532 (2019)
15. Li, W., Dong, L., Yousong, Z., Lu, T., Yi, S.: Dual super-resolution learning for semantic segmentation. In: 2020 Conference on Computer Vision and Pattern Recognition (CVPR), pp. 13774–13784 (2020)
16. Liyang, X., Zhongcheng, W., Xian, Z., Yong, L., Xinkuang, W.: Writer-independent online signature verification based on 2D representation of time series data using triplet supervised network. Measurement **80**, 1–28 (2022)
17. Moises, A., Miguel, F., Jose, J.: Anthropomorphic features for on-line signatures. IEEE Trans. Pattern Anal. Mach. Intell. (PAMI) **41**(12), 2807–2819 (2019)
18. Moises, D., Miguel, A.F., Donato, D.: A perspective analysis of handwritten signature technology. ACM Comput. Surv. **117**(1), 117–139 (2019)
19. Okawa, M.: Online signature verification using single-template matching with time-series averaging and gradient boosting. Pattern Recogn. **102**(1), 1–39 (2020)
20. Okawa, M.: Time-series averaging and local stability-weighted dynamic time warping for online signature verification. Pattern Recogn. **112**(1), 1–39 (2020)
21. Otte, S., Liwicki, M., Krechel, D.: Investigating long short-term memory networks for various pattern recognition problems. In: Perner, P. (ed.) MLDM 2014. LNCS (LNAI), vol. 8556, pp. 484–497. Springer, Cham (2018). https://doi.org/10.1007/978-3-319-08979-9_37
22. Sae-Bae, N., Memon, N.: Online signature verification on mobile devices. IEEE Trans. Inf. Forensics Secur. **9**(6), 933–947 (2014)
23. Sekhar, V., Gorthi, R.S., Viswanath, P.: Online signature verification by few-shot separable convolution based deep learning. In: 15th International Conference on Document Analysis and Recognition (ICDAR), pp. 1125–1129 (2019)
24. Sekhar, V.C., Viswanath, P., Prerana, M., Abhishek, S.: DeepFuseOSV: online signature verification using hybrid feature fusion and depthwise separable convolution neural network architecture. IET Biometrics **9**(6), 259–268 (2020)
25. Vorugunti, C.S., Pulabaigari, V., Gorthi, R.K.S.S., Mukherjee, P.: OSVFuseNet: online signature verification by feature fusion and depthwise separable convolution based deep learning. Neurocomputing **409**(7), 157–172 (2020)
26. Sharma, A., Sundaram, S.: An enhanced contextual DTW based system for online signature verification using vector quantization. Pattern Recogn. Lett. **84**(1), 22–28 (2016)
27. Sharma, A., Sundaram, S.: A novel online signature verification system based on GMM features in a DTW framework. IEEE Trans. Inf. Forensics Secur. **12**(3), 705–718 (2017)
28. Sharma, A., Sundaram, S.: On the exploration of information from the DTW cost matrix for online signature verification. IEEE Trans. Cybern. **48**(2), 611–624 (2017)

29. Sindhu, H., Prajwal, R., Rudrabha, M., Vinay, N., Jawahar, C.V.: Visual speech enhancement without a real visual stream. In: Workshop on Applications of Computer Vision (WACV), pp. 1–10 (2021)
30. Songxuan, L., Jin, L.: Recurrent adaptation networks for online signature verification. IEEE Trans. Inf. Forensics Secur. **14**(6), 1624–1637 (2019)
31. Subhash, C.: Verification of dynamic signature using machine learning approach. Neural Comput. Appl. **32**(5), 11875–11895 (2020). https://doi.org/10.1007/s00521-019-04669-w
32. Chandra, S., Singh, K.K., Kumar, S., Ganesh, K.V.K.S., Sravya, L., Kumar, B.P.: A novel approach to validate online signature using machine learning based on dynamic features. Neural Comput. Appl. **33**(19), 12347–12366 (2021). https://doi.org/10.1007/s00521-021-05838-6
33. Sun, W., Zhang, X., He, X.: Lightweight image classifier using dilated and depthwise separable convolutions. J. Cloud Comput. **9**(55) (2020)
34. SVC: Svc-2004 task 1 and task 2 dataset. https://cse.hkust.edu.hk/svc2004/download.html (2004)
35. Tan, H., He, L., Huang, Z.C., Zhan, H.: Online signature verification based on dynamic features from gene expression programming. Multimed. Tools Appl. **80**, 1–27 (2021). https://doi.org/10.1007/s11042-021-11063-z
36. Tang, L., Kang, W., Fang, Y.: Information divergence-based matching strategy for online signature verification. IEEE Trans. Inf. Forensics Secur. **13**(4), 861–873 (2018)
37. Tolosana, R., Vera-Rodriguez, R., Fierrez, J., Ortega-Garcia, J.: Biometric signature verification using recurrent neural networks. In: 14th International Conference on Document Analysis and Recognition (ICDAR), pp. 652–657 (2017)
38. Tolosana, R., Vera-Rodriguez, R., Fierrez, J., Ortega-Garcia, J.: DeepSign: deep on-line signature verification. Arxiv **20**(1), 1–10 (2017)
39. Tolosana, R., Vera-Rodriguez, R., Fierrez, J., Ortega-Garcia, J.: Exploring recurrent neural networks for on-line handwritten signature biometrics. IEEE Access **6**, 5128–5138 (2018)
40. Xiaomeng, W., Akisato, K., Brian, I.K., Seiichi, U., Kunio, K.: Deep dynamic time warping: end-to-end local representation learning for online signature verification. In: 14th International Conference on Document Analysis and Recognition (ICDAR), pp. 1103–1110 (2019)
41. Yang, L., Cheng, Y., Wang, X., Liu, Q.: Online handwritten signature verification using feature weighting algorithm relief. Soft. Comput. **22**(3), 7811–7823 (2018). https://doi.org/10.1007/s00500-018-3477-2
42. Zhengyang, W., Shuiwang, J.: Smoothed dilated convolutions for improved dense prediction. In: 24th International Conference on Knowledge Discovery and Data Mining, pp. 2486–2495 (2018)

COMPOSV++: Light Weight Online Signature Verification Framework Through Compound Feature Extraction and Few-Shot Learning

Chandra Sekhar Vorugunti[1]([✉]) [iD], Balasubramanian Subramanian[2] [iD],
Prerana Mukherjee[3] [iD], and Avinash Gautam[4] [iD]

[1] Indian Institute of Information Technology, SriCity 517646, Andhra Pradesh, India
chandrasekhar.v@iiits.in
[2] Sri Sathya Sai Institute of Higher Learning, Prasanthi Nilayam 515134, Andhra Pradesh, India
sbalasubramanian@sssihl.edu.in
[3] Jawaharlal Nehru University, New Delhi 110067, India
prerana@jnu.ac.in
[4] Birla Institute of Technology and Science, Pilani 333031, Rajasthan, India
avinash@pilani.bits-pilani.ac.in

Abstract. Online Signature Verification (OSV) is a systematically used biometric characteristic to endorse the genuineness of a user to access real time applications like healthcare, m-payment, etc. Because OSV frameworks are used in real-time applications and it is difficult to acquire a sufficient number of signature samples from users, they must meet a critical requirement: they must be able to detect skilled and random signature presentation attacks effectively with fewer training signature samples and a faster response time. To meet these needs, we developed a depth wise separable (DWS) convolution-based OSV framework that realizes one/few shot learning in inference phase. In addition to it, we have designed a compound feature extraction technique, which extracts maximum seven features from a set of 100 features in MCYT-100, and 3 features from a set of 47 in case of {SVC, SUSIG} datasets. The framework uses only three to seven features per signature to resist the signature presentation attacks. We have extensively evaluated our framework, by performing thorough experiments with three datasets i.e. MCYT-100, SVC and SUSIG. The model results state of the art EER in all skilled categories of SVC and SUSIG datasets.

Keywords: Online signature verification · Deep learning · Compound feature · One shot learning

1 Introduction

Amongst the extensive choice of biometric traits, online signatures are recognized as significant security means to endorse a user in diverse applications, e.g., banking, e-commerce, m-payments, etc. [1, 18, 32]. The advances in pen and touch-based human–machine interfaces allow capturing online signatures on a digital tablet, which samples

U. Porwal et al. (Eds.): ICFHR 2022, LNCS 13639, pp. 98–111, 2022.
https://doi.org/10.1007/978-3-031-21648-0_7

signatures at regular intervals resulting in a set of sampling points. Each sampling point is represented with a set of local features comprising pattern data (x, y coordinates) and the correlated temporal features (pressure, azimuth, tilt angle of device) etc. [1, 2, 7, 25].

In literature, usually, online signature verification frameworks broadly categorized into feature based [1, 4] and function based [6, 12, 17] approaches. In feature based techniques, sophisticated hand-crafted features are extracted that statistically describe the local or global attributes of an online signature. In function based techniques, the complete sequence of sample points of test signature is matched against reference signatures to classify the test signature. The function-based approaches adopt various processes like Gaussian Mixture Models [24], divergence based [10], stability based [12], feature fusion based [17], feature weighing based [19], feature fusion [20], DTW [14, 20–23], Hidden Markov models [6], sequence matching [10], neural network based [24], stroke based [31], deep learning based [16, 20, 25, 26, 35] etc.

2 Literature Survey

In this work, we present a deep learning based online signature verification, in which the focus is on developing a light weight OSV with least number of trainable parameters and least response time. Therefore, in the rest of this section, we summarize the specifics of recent works proposed based on the similar techniques in the literature of online signature verification.

Sekhar et al. [2], developed an OSV framework in which original '100' features are reduced to generate a set of '80' latent features based on a Class Covariance Score. The reduced set of 80 features is passed on to the depthwise separable convolution based CNN. Similarly, in subsequent contribution, Vorugunti et al. [3], designed a hybrid CNN and LSTM based OSV framework in which features are clustered using KNN clustering technique by setting K = 80. These 80 global features computer per signature are submitted to CNN and LSTM based hybrid architecture.

Songxuan et. al [20] presented an OSV model, in which LSTM is used to learn 'normalized path length (LNPS)' to categorize the signature as real or fake. The model yields an EER of 2.37% in Skilled-01 category of SVC dataset. Chandra et al. [35], proposed a compound feature based OSV framework in which the feature set is grouped into three categories namely strong, medium and weak and weighted average per category is computed. The proposed model resulted in an EER of 1.67% in skilled-01 category of MCYT-100, where signatures of 90% of users are used to train the framework and signatures of 10% users are used to test the framework. In [35], the authors focused on Zero Shot learning, in this work, as an improvement, we have focused on achieving Few Shot Learning with least number of features, as maximum as 3 per signature.

Deep learning-based models, have a high computational complexity, a huge parameter count, expensive to train and are prone to overfitting. To solve the aforementioned issues, only a few works have been suggested in the literature on the few shot learning based OSV.

Despite the fact that multiple OSV frameworks have been contributed based on sophisticated approaches, there is still a gap in satisfying the following requirements:

R1. A lightweight OSV framework that can efficiently learn to categorize an input test signature with one-shot learning i.e. one signature sample per user.

R2. A light weight OSV frameworks which can efficiently classify an input test signature with as minimum as 5 features extracted per signature or less.

Very minimal amount of contributions is done in the literature to address the first requirement R1. Javier et al. [34] designed an OSV structure in which Hidden Markov Models are used to spawn plausible fake signature samples by replicating single real signature. Diaz et al. [23] designed one shot learning based OSV framework, in which the artificial signature samples are created by computing kinematic theory based sigma-lognormal features. The model yielded an EER of 13.56%. Recently Subash et al. [37, 38] proposed a series of OSV frameworks in which a set of machine learning classifiers are trained and the classification outcome is increased by applying ensembling techniques. There is no contribution in OSV literature addressing R2. Motivated to address these downsides, we have proposed an OSV which address the requirements R1 and R2.

The following is a breakdown of how the manuscript is organized. The various components of our proposed OSV framework are described in Sect. 3. Segment 4 discusses the experimental analysis, results, and comparisons with contemporary models. In segment 5, the manuscript is concluded.

The main contributions of our work are structured as follows:

1. In this work, we present a feature dimensionality reduction technique that performs feature extraction using the Eigenvalues produced by the Principal Component Analysis of a feature set. The proposed technique decreases the feature set size from 100 to 7 (maximum) and in certain cases 100 to 3 (minimum) in case of MCYT-100 dataset. In case of SUSIG and SVC datasets, the feature set size is reduced from 47 to 3.
2. We use separable convolutions to create a Depthwise Separable Convolution Neural Network (DWSCNN) based online signature verification system, as a result, the number of parameters and computations required are significantly reduced.
3. Extensive experimentation and comprehensive set of assessment with the sate-of-the-art models based on three most widely used datasets.

3   Proposed Online Signature Verification Framework

3.1   Proposed Novel Dimensionality Reduction Algorithm

Our proposed dimensionality reduction algorithm supports simultaneous feature selection and feature extraction, is based on Principal Component Analysis (PCA) that provides best ordered linear approximation to a given high-dimensional data e.g. Feature set. PCA performs centering, rotating and scaling of input data and models the subspace with the maximum variance with descending order of eigenvalues, which outcomes the principal components in the order of significance. Top Eigenvectors (with higher Eigenvalues) comprise maximum variance and maximum discrimination information.

Algorithm 1: Computing writer specific Compound features based on NDL and three types of features discussed above.

Input: U: Set of Users, $\{u_1, u_2, u_3 \dots, u_n\}$

F_i: Original Feature Set of u_i, $F = \{F1, F2, \dots, Fd\}$, $'F1'$ represents a column vector.

T1: Predefined threshold values. We set T1 to 0.1.

N: Number of signature samples for each user.

Output: C – Writer specific compound feature set.

---

for each User $u_i \in U$ do

    Compute $\lambda = PCA(F_i)$ where $\lambda = \{ \lambda1, \lambda2, \dots, \lambda d\}$ be the resulting Eigen values, where $\lambda1 > \lambda2 \cdots > \lambda d$. // *Computational complexity:* $O(N)$.

    for each eigen vector Vp:

        Compute: $NDL^p = \frac{\lambda p}{\sum_{i=1}^{d} \lambda i}$ i.e. the normalized reduction loss when the Eigen vector set is reduced from size $'d'$ to $'p'$.

        // classify each feature vector $'Fp'$ as either weak (W), or moderate (M) or strong (S).

        If $(NDL^p \leq 0)$,

            WeakfeatureSet = WeakfeatureSet $\cup \{Fp\}$

        else if $(NDL^p > 0$ and $NDL^p \leq T)$

            ModeratefeatureSet = ModeratefeatureSet $\cup \{Fp\}$

        else

            StrongfeatureSet = StrongfeatureSet $\cup \{Fp\}$

        end

end // *Computational complexity:* $O(d)$.

for all the weak features $w \in W$,

    compute: $\sum_w Vw. \lambda w$ //results a column feature vector.

end // *Computational complexity:* $O(1)$.

for all the moderate features $m \in M$,

    compute: $\sum_m Vm. \lambda m$ //results a column feature vector.

end // *Computational complexity:* $O(1)$.

compute: Writer specific Compound feature vector $C = $ StrongfeatureSet $\cup$ $\{\sum_w Fw. \lambda w\} \cup \{\sum_m Fm. \lambda m\}$. // *Computational complexity:* $O(1)$.

Return C.

The loss incurred while dropping a dimension is proportional to the corresponding eigenvalues, which is termed as 'Dropping Loss'. Let $F = \{f1, f2, f3, \dots, fd\}$ represents the feature set. Let PCA be the dimensionality reduction technique i.e. $PCA(F) = \lambda = \{\lambda1, \lambda2, \lambda3, \lambda4, \dots, \lambda d\}$ be the resultant Eigen values, where $\lambda1 > \lambda2 \cdots > \lambda d$, λd represents the information alongside the dth component and the volume of loss occurred on dropping the dth component. Sreevani et al. [29] devised a metric named 'Normalized Dropping Loss' (NDL) of the feature set $F = \{f1, f2, f3, \dots, fd\}$ when dimension is condensed from 'd' to 'd – 1' i.e.,

$$NDL^{d,d-1} = \frac{\lambda d}{\sum_{i=1}^{d} \lambda i}. \tag{1}$$

Correspondingly, the loss occurred while dropping a dimension $p = NDL^p = \frac{\lambda p}{\sum_{i=1}^{d} \lambda i}$

Based on NDL, we are defining three types of features.

Weak: A feature 'w' is classified as a weak-feature, if it's 'Normalized dropping loss' is zero i.e. $NDl^{d,w} = 0$;

Moderate: A feature 'm' is classified as moderate-feature, if $NDl^{d,m} > 0$ and ≤ 0.1.

Strong: A feature 's' is classified as strong-feature, if $NDl^{d,s} > 0.1$.

In our proposed dimensionality reduction algorithm, we are applying PCA on the feature set of size 'd' corresponding to each user. PCA on the feature set results in an ordered set of Eigen values $\{\lambda 1 > \lambda 2 \cdots > \lambda d\}$. To select the features which results in greater classification accuracies, the features are grouped into three sets 1. Weak 2. Moderate 3. Strong features. Weak features have no influence (loss) on the classification result when dropped from the feature set, whereas moderate features have a small impact (loss) when removed.

Strong characteristics are those that have a significant impact on the classification outcome. The loss is computed based on the 'Normalized Dropping Loss' (NDL) discussed above. Strong features are taken into account without any operations (feature selection), while weak features are subjected to a scalar product between the weak features and the appropriate Eigen values. The summation of the resulted columns out comes a single column feature vector. Finally, the aggregation of strong features and extracted column vectors from weak and moderate features are considered as writer specific feature set.

3.2 Proposed Separable Convolution Operation Based OSV Framework:

In this paper, we examine and present an OSV framework based on depth wise separable convolution operation, inspired by recent contributions [27, 28]. We describe a CNN framework in which, instead of a vanila or standard convolution operation, each input signature is treated to a depth wise separable one-dimensional convolution operation. The input signature is represented in Fig. 1 as a $1 \times N$ dimensional feature vector (row vector).

As shown in Tables 2 and 3, the depth wise separable 1D convolution uses 12.55% reduced weights and biases to train the framework than traditional convolution. The combination of separable convolution layers and framework optimization processes leads to better input signature representation learning and increased input signature classification accuracy.

3.2.1 Separable Convolution Layer

The initial DWS convolution layer receives an online signature of dimension $(1 \times d)$, where 'd' indicates the feature dimension. 'd' has a minimum value of 3 and a maximum value of 7 in our proposed framework.

A collection of 36 filters, each with a dimension of 1×3, performs a depth wise separable convolution operation, as discussed in [6], to produce N feature maps, each with a size of 1×36. A 1×1 pointwise convolution operation is performed on intermediate feature maps, resulting in N feature maps, each of size 1×36. We used batch normalisation [9, 25] on the output of the first DWS convolution layer to regularize the inputs of each layer and make the model less sensitive to the original set of weights. The second DWS convolution layer, like the previous set of DWS convolutional and

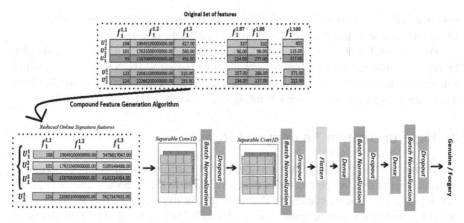

Fig. 1. Synopsis of the proposed compound feature generation and separable Conv1D based OSV framework.

Table 1. Parameter comparison for Conv1D and DWS Conv1D of the proposed framework

Convolution layer type	Trainable params:	Non-trainable params:	Total
Conv1D	20,342	400	20,742
SeparableConv1D	17,789	400	18,189
% of reduction	**12.55%**	–	**12.31%**

batch normalization layers, comprises of 36 filters, each of size 1×3, that produce 'd' feature maps, each of dimension 1×36. A 1×1 pointwise convolution operation is performed on these intermediate feature maps, yielding a feature vector of dimension 1×36.

Apart from batch normalization technique, to obtain better generalization and to prevent overfitting, we have introduced a dropout of 50% to both the DWS convolutional layers. Dropout is a technique in which a random group of nodes is removed from the framework's hidden layers [30]. The deep representational features from the second DWS convolutional layer is passed on to the fully connected layers for classification.

3.2.2 Fully Connected Layers

We employed a three-layered Fully Connected Multilayer Perceptron (FC-MLP) as a classifier to categorize the output from the depthwise separable convolution layers. The MLP has one input, one hidden layer, and one classification layer. In each dense layer, there are 64 neurons. The deep features resulted from the second DWS convolution layer of size $(5 \times 36) = 180$ is formed as an input to the classifier. The proposed framework's weights and bias are initialized with 'random_uniform' to decrease overfitting. The output layer is a sigmoid layer that produces two probability values that can be mapped to output classes 'Genuine', 'Forgery'. We've set 'binary crossentropy' as a loss function to compute the loss between the model output and the ground truth.

We employed 'ReLU' as an activation function in every DWS convolution and hidden layers of the framework to achieve faster learning, sparsity, and a lower likelihood of vanishing gradient. For both the hidden layers, there is a 40% dropout rate. We have used 'adam' as an optimizer with a learning rate of 0.004 and a batch size of 8.

4 Experimentation Analysis and Results

We comprehensively evaluate our framework, by carrying out thorough experiments on three commonly recognized datasets i.e. MCYT-100 [5, 7], SVC - Task 2 [10, 15], SUSIG [24, 25]. The outcomes are demonstrated in the tables below. We have carry out our complete set of experiments on Nvidia RTX 2080 GPU. Table 2 provides all of the information regarding the datasets that we used in our experiments, the number of signatures available in each dataset. In the skilled forgery category, we investigated five categories of experimentation, namely Skilled-01(S-01), Skilled-05(S-05), Skilled-10 (S-10), Skilled-15 (S-15), and Skilled-20 (S-20), as well as corresponding Random-categories, to thoroughly examine the proposed framework.

Table 2. Details of the datasets used in the assessment of the proposed framework

DataSet	MCYT-100	SUSIG	SVC
Total number of writers	100	94	40
Features computed per signature	100	47	47
Total genuine/real signatures per user	25	20	20
Total forgery/ fake signatures per user	25	10	20

Let the dataset contains of 'U' Writers/Users, each with 'R' genuine/real and 'F' forgery/fake signature samples, the Skilled-N category uses 'N' genuine and 'N' forgery signature samples to train the framework, with the remaining 'R-N' genuine, 'F-N' forgery signature samples for each user being considered for testing.

Tables 3, 4, 5 represents the comparison of Equal Error Rate (EER) with the latest state-of-the art OSV frameworks. We have evaluated our framework with Random categories also. Due to space limits, we are unable to present the values in the tables. As discussed in previous sections, subject to MCYT-100, all the verification frameworks using global features require more than 100 features. In case of local features, the computational complexity is $O(n)$, 'n' characterizes the amount of stroke points of a signature. In general, the stroke points are greater than 100. Hence the number of features required in both the cases is generalized to 100. The framework, capable to yield state of the art outcomes with only 3 to 7 features per signature.

Even though the EER difference between the first best and our proposed framework is 9.52 points, but the number of features required is reduced from 100 to 7. One optimistic observation is that, with increase of training signature samples, the EER reduced proportionately by achieving decent EER in S-20 category. Considering the experiments on SVC dataset, the proposed framework resulted in the state-of-the-art results in S-01, 05, 10 and 15 categories. It yielded an EER of 1.67% with 3 features compared to 5.83% with 40 features by our previous work. Similar is the case of with SUSIG, the framework outperformed the EER resulted by the recent state-of-the art [7] 6.67% with a good margin by yielding an EER of 0.98%.–

Table 3. Comparative analysis of the proposed framework with the recent MCYT (DB1)

Method	S-01	S-05	S-20	Number of Features for each signature
Compound feature selection	22.94	8.58	2.65	**(Minimum: 3, Maximum:7)**
Few shot learning[2]	**13.42***	7.03	2.2	80
CNN + LSTM [3]	15.57	1.88	**0.00***	80
Prob-DTW(case 2) [15]	–	–		100
Writer dependent parameters (IV) [5]	–	2.51	**0.03****	100
GMM + DTW [13]	–	3.05	–	100
Stroke-Wise [7]	13.72	–	–	100
DTW-Normalization(F13) [14]	–	8.36	–	100
Curvature feature [16]	–	6.05	–	100
Manhattan [11]	–	4.02	–	100
Writer specific features (conventional) [5]	–	6.79	**0.00***	100
RL(Random forgery) [16]		1.81		100
RL(Skilled forgery) [16]		**1.62****		100
Target-Wise [7]	13.56**	–	–	100
KNN-Regional features [14]	–	4.65	–	100
Common feature and threshold (IV) [5]	–	10.36	5.82	100
Cancelable templates [6]	–	13.30	–	100
Curvature feature [16]	–	10.22	–	100
Writer specific features (Symbolic) [8]	–	2.2	0.6	100
Divergence Matching [9]	–	3.16	–	100
VQ + DTW[12]	–	**1.55***	–	100

(*continued*)

Table 3. (*continued*)

Method	S-01	S-05	S-20	Number of Features for each signature
Common parameters (conventional) [5]	–	13.12	11.23	100
Torsion Feature [16]	–	9.22	–	100
KNN-Global features [14]	–	5.15	–	100
Cancelable templates [6]	–	10.29	–	100
WP + BL DTW [10]	–	2.76	–	100
Histograms [11]	–	4.02	2.72	100
Prob-DTW (case 1) [15]	–	–		100
Stability Modulated (F13) [33]	–	13.56	–	100
Two-tier ensemble [36]	–	2.84	–	100

Table 4. Comparative analysis of the proposed framework with the recent SOTA SVC

Method	S-01	S-05	S-10	S-15	# of Features per signature
Compound feature selection	**1.67***	**0.00***	**0.00***	**0.0***	**Maximum: 3**
Few shot learning[2]	**5.83****	**0.87****	**0.35****	0.2	40
Stroke Point Warping [18]	–	1.00	–	–	47
RNN + LNPS [20]	–	–	–	–	47
CNN + LSTM [3]	6.71	1.05	**0.00***	**0.10***	40
Relief-2 [19]	–	–	5.31	–	47
Target-Wise [7]	18.63	–	–	–	47
Variance selection [19]	–	–	13.75	–	47
Torsion Feature [16]	–	9.83	6.61	3.10	47
LCSS (User Threshold) [17]	–	–	5.33	–	47
mRMR (10-Samples) [18]	–	1.00	–	–	47
Probabilistic-DTW(case 1) [15]	–		–	–	47

(*continued*)

Table 4. (*continued*)

Method	S-01	S-05	S-10	S-15	# of Features per signature
DTW based (Common Threshold) [10]	–	7.80	–	–	47
Stroke-Wise [7]	18.25	–	–	–	47
Relief-1 [19]	–	–	8.1	–	47
Probabilistic-DTW(case 2) [15]	–	–	–	–	47
Multi-Scale Siameese [34]	11.74	2.33	–	–	47
Two-tier ensemble [36]	–	2.20	–	–	47
Classifier ensemble-2 [37]	–	2.62	–	–	47
Classifier ensemble-1 [38]	–	6.00	–	–	47

The proposed framework outcomes the best EER values in all skilled categories. As shown in Table 3, 4, 5, although the frameworks designed in [5, 7, 15, 19, 36–38] are yielding higher EER values equated to the proposed framework, the main drawback of these frameworks is limited experimental evaluation. These models are not evaluated end to end with all possible categories of experimentation to appreciate the framework performance. The framework in [5] is not evaluated in Skilled-01 category. The framework in [7] is not evaluated in Skilled-05 category. The model by [19] is evaluated only in S-10 category but not in S-01, 05, 15 categories. Yielding superior results in one or few categories, doesn't ratify for real time deployment. In our case, the proposed framework is evaluated with all possible categories of experiments and the performance is appraised.

Table 5. Comparative analysis of the proposed framework with the recent SOTA SUSIG

Method	S-01	S-05	S-10	S-15	Number of Features for each signature
Compound Feature Selection	**0.98***	**0.1***	**0.0***	**0.0***	**Maximum: 3**
Few shot learning [2]	10.41	**0.8****	0.63	–	40
writer specific classifiers [26]	–	–	1.92	–	47
CNN + LSTM [3]	13.09	1.95	**0.47****	–	47
Target-Wise [7]	**6.67****	–	–	–	40

(*continued*)

Table 5. (*continued*)

Method	S-01	S-05	S-10	S-15	Number of Features for each signature
stable domain [24]	–	–	2.13	–	47
Cosα + enhanced DTW [21]	–	–	3.06	–	47
Stroke-Wise [7]	7.74	–	–	–	47
Fractional Distance [26]	–	–	3.52	–	47
pole-zero models [22]	–	2.09	–	–	47
Collection of domain [24]	–	–	3.88	–	47
Hausdorff distance [19]	–	7.05	–	–	47
Divergence Matching [9]	–	1.6	2.13	–	47
Kinematic Theory [23]	7.87	–	–	–	47
DCT [25]	–	–	0.51	–	47

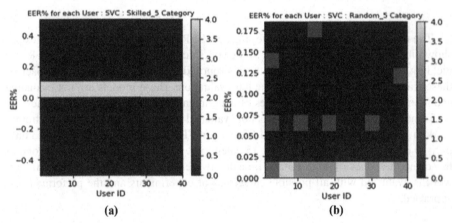

Fig. 2. a) The EERs of 40 users of SVC dataset for Skilled-05 b) The EERs of 40 users of SVC dataset for Random-05 category.

As depicted in 2D-Histogram of Fig. 2.a), the proposed framework results in zero EER for all the users with five training signature samples and as shown in Fig. 2.b) with slight deviations, the EER increases in seven cases as represented in blue square boxes. Figure 3 depicts the True Acceptance Rate (TAR) and False Acceptance Rate (FAR) for each user for MCYT-Skilled 20 and SUSIG Random-01 categories. Figure 3.b) reveals that with one genuine signature sample, the framework achieves zero FAR and a decent TAR.

To summarize the experimental analysis, we see that our proposed OSV framework based on a novel dimensionality reduction technique and depth wise separable convolution operation achieves state of the art EER values with as minimum as one

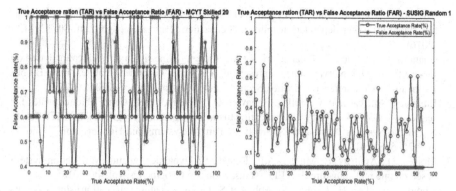

Fig. 3. The ROC curves: (a) The TAR and FAR under Skilled-20 category of MCYT-100 dataset. (b) The TAR and FAR under Random-01 category of SUSIG

trained signature samples and a maximum of seven features per signature. This confirms the accurate learning about the inter and intrapersonal variability of the samples. Although the framework outcome state-of-the art and decent EER in leading categories, the Tables 3, 4, 5 recapitulates that there is a scope of improvement in Random-01 and Random-05 categories.

5 Conclusion and Future Work

Our contribution in this work is twofold. We presented a new dimensionality reduction method that divides the feature set into weak, moderate, and strong features. We chose a maximum of seven features per user based on the MCYT-100 dataset and three features based on the SVC and SUSIG datasets. In addition, we have proposed an OSV framework based on DWSC that allows for a substantial decrease in the total number of parameters and computation complexity while still achieving higher classification accuracies by learning the inter and intrapersonal variability specific to each user, even from a single signature sample, and achieves higher classification accuracy in detecting signature presentation attacks. In comparison to SOTA models, the proposed model attains higher accuracy. In the case of random categories, the proposed OSV framework has the potential for advancement.

References

1. Devanur, G., Koushik, M., Manjunath, S., Somashekara, M.: Interval valued symbolic representation of writer dependent features for online signature verification. Elsevier J. Expert Syst. Appl. **80**, 232–243 (2017)
2. Sekhar, C., Sai, G., Viswanath, P.: Online signature verification by few-shot separable convolution based deep learning. In: 15th International Conference on Document Analysis and Recognition (ICDAR 2019) Sydney, Australia, pp. 1125–1129 (2019)
3. Vorugunti, S., Anoushka, D., Prerana, M., Viswanath, P.: A light weight and hybrid deep learning model based online signature verification. In: ICDAR WML 2019 2nd International Workshop on Machine Learning, 2019, pp. 53–59 (2019)

4. Koushik, M., Shantharamu, M., Devanur, G., Somashekara, M.T.: Online signature verification based on writer dependent features and classifiers. Pattern Recogn. Lett. **80**, 129–136 (2016)
5. Chandra, V., Devanur, G., Pulabaigari, V.: An efficient online signature verification based on feature fusion and interval valued representation of writer dependent features. In: IEEE 5th International Conference on Identity, Security and Behavior Analysis (ISBA) (2019)
6. Emanuele, M., Patrizio, C., Julian, F., Javier, O., Alessandro, N.: Cancelable templates for sequence-based biometrics with application to on-line signature recognition. In: IEEE Transactions on Systems, Man, and Cybernetics - Part A: Systems and Humans, vol. 40, no. 3, pp. 525–538 (2010)
7. Moises, D., Andreas, F., Miguel, A.F., Réjean, P.: Dynamic signature verification system based on one real signature. In: IEEE Transactions on Cybernetics, vol. 48 (2018)
8. Abdul, A., Madasu, H., Jaspreet, K., Abhineet, S.: Online signature verification using segment-level fuzzy modelling. IET Biometrics **3**(3), 113–127 (2014)
9. Lei, T., Wenxiong, K., Yuxun, F.: Information divergence-based matching strategy for online signature verification. In: IEEE Transactions on Information Forensics and Security, vol. 13 (2018)
10. Abhishek, S., Suresh, S.: On the exploration of information from the DTW cost matrix for online signature verification. In: IEEE Transactions on Cybernetics, vol 48 (2018)
11. Sae-Bae, N., Nasir, M.: Online signature verification on mobile devices. In. Transactions on Information Forensics and Security, vol. 9, no. 6, pp. 933–947 (2014)
12. Abhishek, S., Suresh, S.: An enhanced contextual DTW based system for online signature verification using vector quantization. Pattern Recogn. Lett. **84**, 22–28 (2016)
13. Chandra Sekhar, V., Devanur, G., Prerana, M., Viswanath, P.: OSVNet: convolutional siamese network for writer independent online signature verification. In: 15th ICDAR, Sydney, Australia, pp. 1470–1475 (2019)
14. Rafal, D., Przemyslaw, K., Piotr, P.: Online signature verification modeled by stability oriented reference signatures. Inf. Sci. **460–461**, 151–171 (2018)
15. Rami, A., Witold, P., Khaled, D., Ali, M., Ahmed, A.L.: Quantifying dynamic time warping distance using probabilistic model in verification of dynamic signatures. Elsevier-Soft Comput. vol. 23, pp. 407–418 (2019)
16. He, L., Tan, H., Huang, Z.-C.: Online handwritten signature verification based on association of curvature and torsion feature with Hausdorff distance. Multimedia Tools Appl. **78**(14), 19253–19278 (2019). https://doi.org/10.1007/s11042-019-7264-6
17. Devanur, G., Prakash, H.N.: Online signature verification and recognition: an approach based on symbolic representation. In: IEEE Transactions on Pattern Analysis and Machine Intelligence, vol. 31 (2009)
18. Biswajit, K., Anirban, M., Pranab, K.: Stroke point warping-based reference selection and verification of online signature. In: IEEE Transactions on Instrumentation and Measurement, vol. 67 (2018)
19. Yang, L., Cheng, Y., Wang, X., Liu, Q.: Online handwritten signature verification using feature weighting algorithm relief. Soft. Comput. **22**(23), 7811–7823 (2018). https://doi.org/10.1007/s00500-018-3477-2
20. Songxuan, L., Lianwen, J., Weixin, Y.: Online signature verification using recurrent neural network and length-normalized path signature descriptor. In:14th ICDAR (2017)
21. Mostafa, I., Mohamed, M., Hazem, M.: Enhanced DTW based on-line signature verification. In: Proceedings of the 16th IEEE International Conference on Image Processing (ICIP) (2009)
22. Saeid, R., Ali, F., Farzad, T.: Authentication based on pole-zero models of signature velocity. J. Med. Signals Sens. vol 3, pp.195–208 (2013)

23. Diaz, M., Andreas, F., Réjean, P., Miguel, F.: Towards an automatic on-line signature verifier using only one reference per signer. In: International Conference on Document Analysis and Recognition (ICDAR), Tunis, Tunisia, pp. 631–635 (2015)

24. Alireza, A., Srikanta, P., Umapada, P., Michael, B.: An efficient signature verification method based on an interval symbolic representation and a fuzzy similarity measure. In: IEEE Transactions on Information Forensics and Security, vol. 12 (2017)

25. Javier, G., Julian, F., Marcos, M., Javier, O.: Improving the enrollment in dynamic signature verfication with synthetic samples. In: ICDAR, pp. 1295–1299, Barcelona, Spain (2009)

26. Ruben, T., Ruben, V., Julian, F., Javier, O.: Biometric signature verification using recurrent neural networks. In: 14th ICDAR, Kyoto, Japan (2017)

27. Lukasz, K., Aidan, G., Francois, C.: Depthwise separable convolutions for neural machine translation. In: 6th International Conference on Learning Representations (ICLR) (2018)

28. Francois, C.: Xception: deep learning with depthwise separable convolutions. In: The IEEE Conference on Computer Vision and Pattern Recognition (CVPR), USA, pp:1251–1258 (2017)

29. Sreevani, Murthy, C.A.: Bridging feature selection and extraction compound feature generation. In: IEEE Transactions on Knowledge and Data Engineering, vol. 29, pp: 757–770 (2017)

30. Rohit, K., Richa, S., Mayank, V.: Guided dropout. In: Proceedings of the AAAI Conference on Artificial Intelligence, vol. 33, pp 4065–4072 (2019)

31. Chuang, L., Xing, Z., Feng, L.: A stroke-based RNN for writer-independent online signature verification. In: 15th ICDAR, pp. 526–532 (2019)

32. Chandra, S., Prerana, M., Devanur, G., Viswanath, P.: Online signature verification based on writer specific feature selection and fuzzy similarity measure. In: Workshop on Media Forensics, CVPR 2019, Long Beach, USA, pp. 88–95 (2019)

33. Antonio, P., Moises, D., Miguel, A., Angelo, M.: SM-DTW: Stability modulated dynamic time warping for signature verification. PRL, vol. 121, pp. 113–122. 15 April 2019

34. Javier, G., Julian. F., Marcos, M., Javier, O.: Improving the enrollment in dynamic signature verificationwith synthetic sample. In: ICDAR, Tunis, Tunisia, pp. 1295–1299 (2015)

35. Vorugunti, C.S., Pulabaigari, V., Mukherjee, P., et al.: COMPOSV: compound feature extraction and depthwise separable convolution-based online signature verification. Neural Comput. Appl. **34**, 10901–10928 (2022). https://doi.org/10.1007/s00521-022-07018-6

36. Bhowal, P., Banerjee, D., Malakar, S., et al.: A two-tier ensemble approach for writer dependent online signature verification. J. Ambient Intell. Hum. Comput. **13**, 21–40 (2022). https://doi.org/10.1007/s12652-020-02872-5

37. Chandra, S.: Verification of dynamic signature using machine learning approach. Neural Comput. Appl. **32**(15), 11875–11895 (2020). https://doi.org/10.1007/s00521-019-04669-w

38. Chandra, S., Singh, K.K., Kumar, S., Ganesh, K.V.K.S., Sravya, L., Kumar, B.P.: A novel approach to validate online signature using machine learning based on dynamic features. Neural Comput. Appl. **33**(19), 12347–12366 (2021). https://doi.org/10.1007/s00521-021-058 38-6

Finger-Touch Direction Feature Using a Frequency Distribution in the Writer Verification Base on Finger-Writing of a Simple Symbol

Isao Nakanishi$^{(\boxtimes)}$ ⓘ, Masaya Yamazaki, and Takahiro Horiuchi

Tottori University, 4-101 Koyama-minami, Tottori 680-8552, Japan
nakanishi@tottori-u.ac.jp

Abstract. In this study, individuals were asked to draw a symbol by using their fingertips on a digital device screen. This study focused on finger-touch direction information that can be extracted from a smartphone screen. To suppress rapid changes in the detected direction data, preprocessing was introduced, and its effectiveness was confirmed by evaluating the verification performance. Finally, representing the direction data as a frequency distribution was introduced as a new feature, which was demonstrated to improve the verification performance.

Keywords: Biometrics · Writer verification · Simple symbol · Finger-touching direction · Frequency distribution

1 Introduction

Recently, facial images, iris images, and fingerprints have been used for person authentication when using smartphones and tablet devices. These biometrics provide good usability and achieve a high authentication rate. However, they are constantly exposed, and thus, they can be acquired (stolen) easily. Furthermore, they cannot be changed like passwords, compounding their vulnerability.

Writer verification/identification, which authenticates individuals using information obtained from their writing actions [1], is attracting attention as it cannot be easily imitated. Signature verification has been employed for writer verification [2–4]. However, owing to the small displays of smartphones, smartphone users find it difficult to create a signature. In addition, signing is time consuming, and using a dedicated pen is inconvenient.

We have proposed an authentication method based on writing a simple symbol using a finger [5–7], as it is easy and fast for users to write simple symbols even on small displays of smartphones, and it does not require a dedicated pen. This is for achieving the most convenient way for writer verification. In addition, we have investigated the characteristics of finger-pressure and finger-touching areas in Ref. [8]. This information is completely private and not known to a

© The Author(s), under exclusive license to Springer Nature Switzerland AG 2022
U. Porwal et al. (Eds.): ICFHR 2022, LNCS 13639, pp. 112–121, 2022.
https://doi.org/10.1007/978-3-031-21648-0_8

third party or to the users themselves. Additionally, in Ref. [9], finger-touching direction has been used as a possible third authentication characteristic.

In this study, finger-touching direction was introduced as a feature in the writer verification method by writing a simple symbol with a finger and evaluating the verification performance. However, there were rapid changes in direction in the detected finger-touching direction data, which were caused by fluctuations in the touching area of the finger on the pad of a smartphone display. To suppress these rapid changes, preprocessing was introduced, and its effectiveness was confirmed by evaluating the verification performance. Furthermore, representing the finger direction as a frequency distribution was adopted as a method to improve the verification performance, and its effectiveness was also confirmed through an evaluation of the verification performance.

2   Writer Verification Based on Finger-Writing of a Simple Symbol

In this section, writer verification based on finger-writing of a simple symbol proposed in Refs. [5–7] is briefly introduced. As candidates for a simple symbol, ○, △, □ are considered. When writing a symbol, coordinate values of the fingertip position on a display, finger pressure, and finger touching area data are extracted from a device. Using the extracted coordinate values, the start and end points of writing and information on writing time, speed, and acceleration are calculated as individual features. The mean, maximum, and minimum values of finger pressure and finger-touching area are also used as individual features. During the enrollment stage, these features are extracted from regular users and their averaged values are stored as templates. During the test stage, features from a candidate who claims to be a regular user are extracted and compared with the user's templates. After that, Euclidean distance matching is used for verification. The applicant is regarded as genuine if the distance between the test and template data is smaller than a threshold.

Using 29 experimental subjects, Ref. [7] found that the equal error rate (EER) was approximately 30%. EER is defined as the value which makes the false rejection rate (FRR) equal to the false acceptance rate (FAR) and is used for evaluating the verification performance. A smaller EER indicates better performance. However, the obtained EER was not satisfactory. Thus, new features are necessary for improving the verification performance.

3   Introduction of Finger-Touching Direction Feature

In Refs. [6–8], pen-pressure and pen-touching area information were extracted and used as individual features. However, the finger-touching direction is also extractable [9] but has not been used as a feature in writer verification based on the finger-writing of a simple symbol. The finger-touching direction is the direction of the finger pad that touches the tablet screen and is detectable in some smartphones. In this study, we introduce the finger-touching direction feature into our writer verification method and examine the verification performance in finger-writing of a simple symbol.

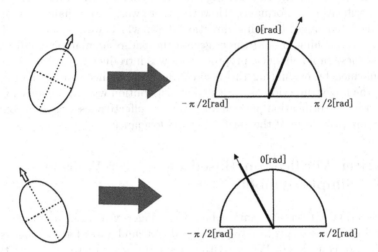

Fig. 1. Examples of detected finger-touching directions.

3.1 Finger-Touching Direction

In the development environment, Android Studio on a Fujitsu ARROWS NX F-04G smartphone was used to extract the finger-touching direction data. As illustrated in Fig. 1, an ellipse is fitted to the finger-touching area on a tablet screen and the long axis of the ellipse is detected as the direction of the finger-touching area using "getOrientation" in Android Studio [9]. This is called the finger-touching direction. When the touching areas are assumed to be on the left side of Fig. 1, their directions are detected as originating from the right side. When vertical, the direction of a smartphone screen is defined as 0, and the detection ranges from $-\pi/2$ rad to $\pi/2$ rad.

However, during actual writing, the finger-touching direction was never constant. Thus, the mean of the detected direction values was calculated as an individual feature. In addition, the directions at the start and end points of writing and the mean values of the directions of several points near the start and end points of writing are used as features. The reason for using them as features is that they are influenced by the user's way of holding a smartphone, which emphasizes individualities in finger writing. In addition, finger pressure and finger touching area data are also detectable; thus, the directions at the point of maximum and minimum pressure and area are also recorded. The features using the finger-touching direction are summarized in Table 1.

Table 1. Features using finger-touching direction.

Mean of directions (MD)
Direction at the start point (DS)
Direction at the end point (DE)
Mean of directions near the start point (MDS)
Mean of directions near the end point (MDE)
Direction at the maximum pressure (DmaxP)
Direction at the minimum pressure (DminP)
Direction at the maximum area (DmaxA)
Direction at the minimum area (DminA)

3.2 Evaluation of Verification Performance

Ten subjects were asked to write three simple symbols ten times, using their finger only. Among the 10 data obtained from each subject, five were used for making their template and the remaining five were used for testing. In addition, cross-validation was performed ten times. This reduced the influence of selecting data for the creation of a template and testing the verification performance. The averaged EERs in the 10 cross-validations performed for each direction feature and symbol are listed in Table 2. The smallest EER of 24.7% was obtained when drawing a circle and using the mean of directions. In other symbols, the mean of the direction feature achieved better performance. By averaging the direction data, differences in the finger-touching directions between individuals could be detected.

Table 2. EERs(%) of finger-touching direction features.

	○	△	□
MD	24.7	35.8	39.3
DS	31.2	42.6	45.2
DE	29.5	46.5	43.2
MDS	27.8	42.8	42.0
MDE	32.0	41.7	39.9
DmaxP	40.4	51.5	42.6
DminP	37.5	43.8	46.0
DmaxA	35.4	51.4	40.4
DminA	36.6	41.8	41.8

Compared with the EERs of drawing a circle, EERs when drawing a triangle and square were clearly larger. This was because when drawing a triangle and square, the drawing motion must change direction at the corners, while a circle can be drawn as a single stroke. When changing the drawing motion, such as at corners, the finger-touching direction is significantly changed, which might degrade the verification performance.

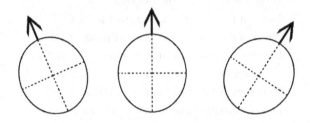

Fig. 2. Fluctuation of detected finger-touching direction.

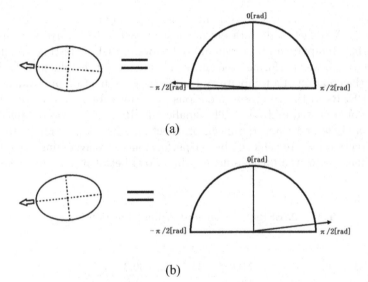

Fig. 3. Detected large change in detected finger-touching direction.

3.3   Considerations

Compared with the EERs obtained using the individual features in Ref. [6], the verification performance was not high. The reasons for this are as follows:

"getOrientation" used for detecting finger-touching direction, fits an ellipse into a touching area. If the touching area is close to a circle, it becomes difficult to determine the direction of the long axis. Therefore, the detected direction easily fluctuated by slightly changing the touching area, as indicated in Fig. 2 and may have resulted in degradation of the verification performance.

Furthermore, as shown in Fig. 3, the detection range of the direction was $-\pi/2$ to $\pi/2$ rad. If the long axis of an ellipse fitted is around $-\pi/2$ rad, as shown in (a), but it is slightly changed to (b) by fingertip movement, the direction is supposed to be around $\pi/2$ rad and significantly changed from $-\pi/2$ to $\pi/2$ rad. Figure 4 shows the time variation of the detected finger-touching direction. The detected direction changes rapidly. For instance, a negative value is changed to a positive value at two successive sampled points. However, it is unlikely that such a rapid change in the finger direction occurs on a smartphone screen. This may degrade the verification performance.

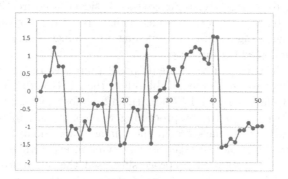

Fig. 4. An example of detected finger-touching direction.

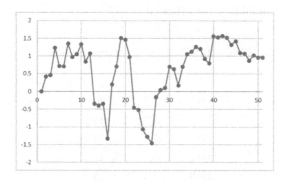

Fig. 5. An example of preprocessed direction data.

4   Introduction of Preprocessing

To prevent rapid changes in the detected direction, preprocessing was introduced. As explained using Fig. 3, slight changes in the touching surface around $|\pi/2|$ rad cause quite large fluctuations, for instance, from $-\pi/2$ to $\pi/2$ rad in detected direction, which unlikely in practical situations. Thus, if the absolute difference between two successive direction data is larger than $\pi/2$, the detected direction is inverted, that is, multiplied by -1.

Figure 5 shows an example of the preprocessed direction data, where the original data are shown in Fig. 4. It can be confirmed that several rapid changes that occurred, as shown in Fig. 4, were suppressed. However, a few significant changes remained. Here, it is noted that the finger direction is never the same as the finger-touching direction. As mentioned above, "getOrientation", which is used for detecting the finger-touching direction, fits an ellipse into a finger-touching area. Depending on the touching condition of the finger pad, the shape of the touching area changes regardless of the physical finger direction. Therefore, even when the finger direction does not change, the finger touching directions can change. Large changes caused by touching-area changes occur naturally when detecting a finger-touching direction.

To evaluate the effectiveness of the proposed preprocessing method, its verification performance was evaluated. Table 3 lists the EERs with pre-processing in three symbols. Compared with EERs in Table 2, almost all EERs were reduced. Thus, the effectiveness of the preprocessing was confirmed. In particular, the reduction of EER was remarkable in the triangle and square symbols. EERs using DS features were not changed in any symbol, even when using the pre-processing, since the proposed preprocessing used the difference between two successive sampled data and was never performed at the start point. In contrast, EERs using MDS and DminP for some symbols were slightly increased. The proposed preprocessing forcibly changed the detected directions. However, there is no proof that the changed directions were optimal. In the future, the optimal direction should be investigated.

Table 3. EERs(%) with preprocessing.

	○	△	□
MD	22.2	29.4	30.1
DS	31.2	42.6	45.2
DE	27.4	35.2	32.2
MDS	26.6	43.0	42.3
MDE	31.4	39.5	35.5
DmaxP	36.7	42.7	42.5
DminP	38.9	45.8	44.0
DmaxA	34.5	40.6	39.9
DminA	35.5	41.1	40.8

5   Frequency Distribution as a New Feature

The finger-touching direction feature evaluated in the previous sections is a one-dimensional feature. The smaller number of feature dimensions is not effective for pattern matching. Thus, we propose that the direction feature should be multi-dimensionalized. However, to directly use finger-touching direction data in pattern matching makes the direction feature dependent on the written shape (content). The aim of this study is to authenticate users using features independent of the written shape. Thus, we introduce a frequency distribution into the direction feature.

The frequency distribution of the finger-touching direction data is calculated and then the distribution is represented as a histogram, where the total frequency is normalized to one, since each histogram has a different total frequency. The number of bins and bin widths are determined empirically. Figure 6 shows an example of a histogram in which the bin width is 32.

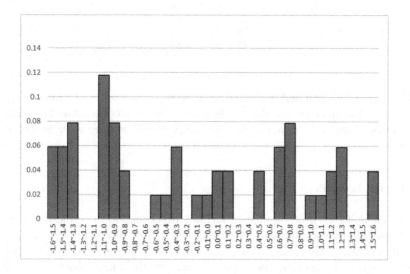

Fig. 6. A histogram of frequency distribution in direction data.

In the enrollment stage, a histogram is obtained by processing each template for the direction, and then the obtained histograms are ensemble-averaged, that is, the frequencies in each class are averaged, resulting in a template histogram. In the test stage, the histogram of an applicant is compared with the template histogram using Euclidean distance matching, which is defined in Eq. (1).

$$d = \sqrt{\sum_{i=1}^{n}(a_i - b_i)^2},$$

(1)

where a_i is the frequency of the template data and b_i is the frequency of the test data. n denotes the number of classes. Using the Euclidean distance, dissimilarity (%) is defined as follows:

$$\text{Dissimilarity} = \frac{d}{\sqrt{2}} \times 100. \tag{2}$$

If two histograms are the same, the dissimilarity is 0, and on the other hand, if they are completely different, the dissimilarity becomes 100. When dissimilarity is smaller than a threshold, the candidate is regarded as a regular user.

We evaluated the verification performance using the frequency-distribution feature. The results are presented in Table 4. Compared with EERs in Table 3, that is, the one-dimensional case, the smallest EERs were obtained for all symbols. Thus, the effectiveness of multi-dimensionalization was confirmed.

Table 4. EER(%) using the frequency distribution feature of direction.

○	△	□
22.0	25.3	24.7

However, the obtained EERs were 22% to 25% and it is difficult to use the proposed frequency distribution feature of the direction alone for verifying individuals. Thus, it is required to fuse the proposed feature with conventional features.

6   Conclusions

In this paper, we focused on finger-touching direction and introduced it into the writer verification method based on the finger-writing of a simple symbol. However, it was found that rapid changes occurred in the detected finger-touching direction data. To prevent the rapid changes in the finger-touching direction data, preprocessing was introduced, which disallowed extremely large changes in two successive sampled points. In addition, for multi-dimensionalizing the direction feature, the frequency distribution of the direction data was proposed to be used as a new feature. The effectiveness of the method was confirmed by the evaluation of the verification performance. However, the obtained verification performance was insufficient and could be improved by fusing the direction features with conventional ones.

References

1. Sreeraj, M., Idicula, S.M.: A survey on writer identification schemes. Int. J. Comput. Appl. **26**(2), 23–33 (2011)

2. Jain, A.K., Griess, F.D., Connell, S.D.: On-line signature verification. Pattern Recogn. **35**(12), 2963–2972 (2002)
3. Dimauro, G., Impedovo, S., Lucchese, M.G., Modugno, R., Pirlo, G.: Recent advancements in automatic signature verification. In: Proceedings of the 9th International Workshop on Frontiers in Handwriting Recognition, pp. 179–184 (2004)
4. Fierrez, J., Ortega-Garcia, J.: On-line signature verification. In: Jain, A.K., Flynn, P., Ross, A.A. (eds.) Handbook of Biometrics. Springer, New York (2007)
5. Nakanishi, I., Takahashi, A.: A study on writer verification based on finger-writing of a simple symbol on a tablet. In: Proceedings of 2018 IEEE R10 Conference (TENCON2018), pp. 2226–2230 (2018)
6. Takahashi, A., Nakanishi, I.: Authentication based on finger-writing of a simple symbol on a smartphone. In: Proceedings of International Symposium on Intelligent Signal Processing and Communication Systems, pp. 411–414 (2018)
7. Takahashi, A., Masegi, Y., Nakanishi, I.: A smartphone user verification method based on finger-writing of a simple symbol. In: Stephanidis, C., Antona, M., Ntoa, S. (eds.) HCII 2021. CCIS, vol. 1420, pp. 447–454. Springer, Cham (2021). https://doi.org/10.1007/978-3-030-78642-7_60
8. Masegi, Y., Takahashi, A., Nakanishi, I.: Investigation of detection characteristics of finger pressure and touch area and their application to pre-classifier in writer verification. J. Adv. Inf. Technol. **12**(1), 78–83 (2021)
9. Frank, M., Biedert, R., Ma, E., Martinovic, I., Song, D.: Touch-alytics: on the applicability of touchscreen input as a behavioral biometric for continuous authentication. IEEE Trans. Inf. Forensics Secur. **8**(1), 136–148 (2012)

Self-supervised Vision Transformers with Data Augmentation Strategies Using Morphological Operations for Writer Retrieval

Marco Peer$^{(\boxtimes)}$ ⓘ, Florian Kleber ⓘ, and Robert Sablatnig ⓘ

Computer Vision Lab, Institute of Visual Computing and Human-Centered Technology, TU Wien, Favoritenstraße 9/193-1, Vienna, Austria
{mpeer,kleber,sab}@cvl.tuwien.ac.at
http://www.cvl.tuwien.ac.at

Abstract. This paper introduces a self-supervised approach using vision transformers for writer retrieval based on knowledge distillation. We propose morphological operations as a general data augmentation method for handwriting images to learn discriminative features independent of the pen. Our method operates on binarized 224×224-sized patches extracted of the documents' writing region, and we generate two different views based on randomly sampled kernels for erosion and dilation to learn a representative embedding space invariant to different pens. Our evaluation shows that morphological operations outperform data augmentation generally used in retrieval tasks, e.g., flipping, rotation, and translation, by up to 8%. Additionally, we evaluate our data augmentation strategy to existing approaches such as networks trained with triplet loss. We achieve a mean average precision of 66.4% on the Historical-WI dataset, competing with methods using algorithms like SIFT for patch extraction or computationally expensive encodings, e.g., mVLAD, NetVLAD, or E-SVM. In the end, we show by visualizing the attention mechanism that the heads of the vision transformer focus on different parts of the handwriting, e.g., loops or specific characters, enhancing the explainability of our writer retrieval.

Keywords: Writer retrieval · Unsupervised learning · Morphological operations · Document analysis

1 Introduction

Writer retrieval describes the task of finding documents within a database written by the same author as a given query document. State-of-the-art methods [1–3] focus on deep learning-based approaches using algorithms like SIFT for selecting dominant characteristics contained in the handwriting. While achieving high performances on publicly available datasets, e.g., accuracies above 99%

U. Porwal et al. (Eds.): ICFHR 2022, LNCS 13639, pp. 122–136, 2022.
https://doi.org/10.1007/978-3-031-21648-0_9

[2,4] on CVL [5] or ICDAR2013 [6] dataset, approaches based on neural networks neither contribute to gaining insights into decision-making nor explaining embedding spaces formed during training [7]. Concerning writer retrieval, the lack of interpretability also causes issues regarding the characteristics of the learned embedding space: For example, does the use of different pens affect the retrieval performance of the network, or does the network overfit regarding specific pens? Such aspects are relevant for the use of deep learning e.g., in forensics, where experts desire to distinguish between writers based on features regarding pure handwriting, ignoring external characteristics like different styles of pens. With the rise of attention-based network architectures in vision [8], in particular Vision Transformers (ViTs) [9], one step towards explainability and whitening the black box of deep learning and neural networks was accomplished by investigating and visualizing the attention mechanism [10,11].

In this paper, we encounter those challenges with the following contributions: Firstly, we investigate an unsupervised approach for writer retrieval based on knowledge distillation presented by Caron et al. [12]. We train two networks, a student and a teacher, both of the same ViT architecture, on two augmented subparts of documents, as shown in Fig. 1. Secondly, we introduce Morphological Operations (MOs) as a data augmentation technique for writer retrieval tasks, learning discriminative features independent of the utensil used for writing and tackling the intraclass-variance with respect to the pen used for writing. In our experiments, these augmentations significantly enhance the performance. Additionally, we investigate the effect of MOs for a ResNet50 trained with triplet loss and show that MOs are a valid data augmentation strategy on binarized images of handwriting, easily pluggable into any architecture for writer retrieval. We evaluate our methods with Mean Average Precision (mAP) as well as the Top-1 accuracy by calculating the global page descriptor and applying a leave-one-image-out procedure, hence every document is once used as a query. In the end, a visualization of the attention mechanism of the ViT is presented to make a step towards interpretability of the handwriting features learned by the network to distinguish between different writers.

We start with related work in Sect. 2. Section 3 contains the methodology used in this paper. In Sect. 4, we describe our conducted experiments, followed by the conclusion in Sect. 5.

2   Related Work

The first approach for deep-learning based writer retrieval was established by Fiel and Sablatnig [13] using a convolutional neural network. They calculate the global page descriptor by mean pooling of the activations of the penultimate layer.

Christlein et al. [1] introduce a method based on clustered 32×32-sized patches extracted at SIFT keypoint locations. The cluster label is then used to train a residual network; they apply VLAD with multiple vocabularies to encode their embeddings. This approach achieves state-of-the-art results on the

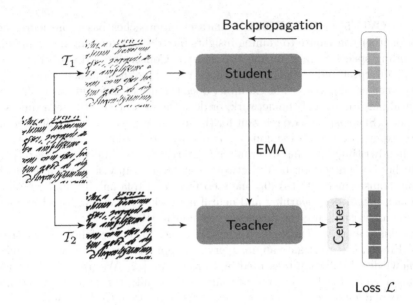

Fig. 1. Overview of our unsupervised approach used for writer retrieval. Two different views of the input patch defined by T_1 and T_2 are generated and forwarded to the student and the teacher, both two instances of the same architecture [12]. The student learns by backpropagating the cross-entropy loss whereas the teacher is updated by EMA.

Historical-WI [14] dataset with a mAP of 76.2% when adding an Exemplar-SVM (E-SVM).

While Keglevic et al. [3] and Rasoulzadeh et al. [2] investigate encoding schemes like VLAD or NetVLAD on networks trained with triplet losses and patches extracted at SIFT keypoint locations, Wang et al. [15] propose to subdivide the image, simultaneously train a U-Net for binarization and aggregate their embeddings by a residual encoding layer.

Our self-supervised approach is based on the work of Caron et al. [12], where they focus on image classification and segmentation using the attention mechanism of the trained ViT. Their embedding space relies on choosing appropriate data augmentation strategies T_1 and T_2 to train two networks to learn similar embeddings. Related self-supervised approaches such as *data2vec* [16] mainly differ in their strategies to avoid collapsing of training, e.g., they average the output of the last k layers of the teacher network for calculating their loss.

3   Methodology

In the following, each part of our writer retrieval approach is explained in detail. An overview is illustrated in Fig. 1. Firstly, we describe the preprocessing, followed by the ViT architecture and MOs as our data augmentation strategy. We conclude with the training strategy and the aggregation of the embeddings to obtain a global page descriptor.

3.1 Preprocessing

Our method is based on binarized images due to MOs. Although these operations are defined for grayscale images, we do not consider using them, in particular for historic datasets where background noise exists. Additionally, the datasets used are provided as binarized images. Note that image enhancing methods [17,18] for historic datasets can be applied for binarization.

For training, each document is split into patches of 256×256 with a stride of 128. The patch is randomly cropped to a size 224×224 online. Validation and test data are directly cropped to 224×224 patches with stride 224. Images with less than 1.5% black pixels are considered as noise and have been removed.

3.2 Vision Transformer

Self-attention-based neural networks achieve state-of-the-art results on several computer vision tasks like image classification [19], segmentation [20] and retrieval. Therefore, to learn discriminative embeddings in a self-supervised manner, we apply the ViT architecture, as proposed by Dosovitskiy et al. [9], with patch size 16 followed by an MLP head. We forward patches of our input images to the ViT where an attention-based encoder extracts features. The MLP head is only used during training; for inference and the retrieval task, we drop the MLP head and directly work on the L2-normalized embeddings of the transformer encoder. An overview of the architecture is shown in Fig. 2.

The CLS token is a learnable embedding extending the list of linear projected patches of the input image. Its output of the last encoder layer serves as an image representation forwarded to the MLP. Given an input sequence $\mathbf{z} \in \mathbb{R}^{n_k \times d_k}$ and an embedding dimension d_h, the encoder contains the self-attention defined by the matrices $\mathbf{Q}, \mathbf{K}, \mathbf{V} \in \mathbb{R}^{n_k \times d_h}$ which are learned through linear projections from $\mathbf{z}$. Thus, the self-attention $\mathbf{A} \in \mathbb{R}^{n_k \times d_h}$ yields

$$\mathbf{A}(\mathbf{Q}, \mathbf{K}, \mathbf{V}) = \mathrm{softmax}\left(\frac{\mathbf{Q}\mathbf{K}^{\mathrm{T}}}{\sqrt{d_h}}\right)\mathbf{V}, \tag{1}$$

where the softmax term is used for visualizing the attention maps [12,21]. By using a ViT and the self-attention mechanism, we can disregard algorithms like SIFT for preidentifying relevant areas of the handwriting. Instead, our transformer learns to select patches containing distinguishing characteristics.

3.3 Morphological Operations

Our self-supervised approach with a student and a teacher network relies on learning similar features from two different views of the input image defined by

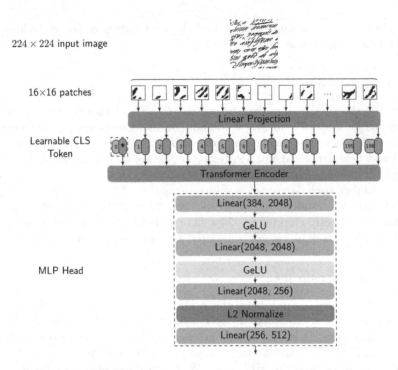

Fig. 2. ViT architecture with patch size 16 for student and teacher network. During inference, the MLP head is removed, and the L2-normalized output of the last encoder layer is considered for the retrieval task.

the transformations $\mathcal{T}_1$ and $\mathcal{T}_2$. Therefore, we propose MOs as a data augmentation strategy to improve the embedding space. More precisely, we want our network to learn features independent of the pen used for writing. While state-of-the-art approaches do not apply data augmentation [1,3,15] or use augmentation mainly designed for image classification or object detection, e.g., random rotation or translation [13], we randomly apply erosion and dilation with different kernels $\mathbf{K}_i \in \mathbb{Z}_2^{3\times3}$, where i is the number of ones of the kernel, during training. Other MOs like closing or opening are not considered since they focus on removing noise such as small objects or holes. Instead, we want to mimic different styles of pens. In Fig. 3, the effects of erosion and dilation on handwriting are shown.

For our approach, we focus on different sampling strategies of kernels applied for erosion or dilation since a kernel either enhances or decreases specific characteristics of the handwriting depending on its structure. We choose sampling kernels which contain 2×2 blocks and kernels containing either a row or a column filled with ones, as illustrated in Fig. 4. Additionally, we also evaluate using randomly generated kernels varying n_a, the number of times a MO is applied, and i in our experiments in Sect. 4.

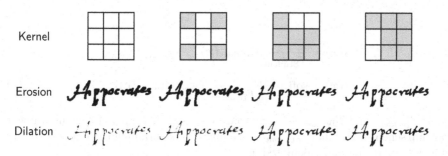

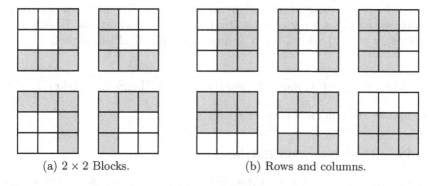

Fig. 3. Effect of erosion and dilation on a sample of handwriting images taken from the Historical-WI dataset. With MOs we can imitate different styles of pens and thicknesses depending on the location and the total number of ones (white) and zeros (black) of the kernel.

(a) 2 × 2 Blocks.   (b) Rows and columns.

Fig. 4. Kernel sampling strategies: a) Using 2 × 2 blocks and b) kernels where either a row or a column is filled with ones (white).

3.4   Self-supervised Training

For our unsupervised training, we follow an approach called DINO (**d**istillation with **n**o labels) introduced by Caron et al. [12], assuming two networks, a student f_{Student} with its parameters θ_s and a teacher f_{Teacher} with θ_t. In contrast to knowledge distillation approaches, the teacher network has the same architecture and is not pretrained on data of the same domain. While backpropagation is applied to train the student, the teacher's parameters are updated via EMA described by

$$\theta_t \leftarrow m\theta_t + (1 - m)\theta_s, \tag{2}$$

where m is called momentum. To introduce the cross-entropy loss, we first define the forward passes of the student and teacher as

$$s(\boldsymbol{x}) = \frac{1}{T_{\text{Student}}} f_{\text{Student}}(\boldsymbol{x}) \tag{3}$$

$$t(\boldsymbol{x}) = \text{softmax}(\frac{1}{T_{\text{Teacher}}}(f_{\text{Teacher}}(\boldsymbol{x}) - \boldsymbol{c})), \tag{4}$$

where T_{Student} and T_{Teacher} are hyperparameters with T_{Teacher} following a linear warmup schedule to equal T_{Student}. $\boldsymbol{c}$ is a bias term containing the batch statistics to avoid collapsing and calculated as

$$\boldsymbol{c} \leftarrow m_c \boldsymbol{c} + (1 - m_c) \frac{1}{B} \sum_{i=0}^{B-1} f_{\text{Teacher}}(\text{Concat}(\boldsymbol{x}_1, \boldsymbol{x}_2)) \tag{5}$$

with m_c the center momentum and B the batch size. The networks operate on two augmented views $\boldsymbol{x}_1, \boldsymbol{x}_2$ defined by the data augmentations $\mathcal{T}_1$ and $\mathcal{T}_2$ as $\boldsymbol{x}_i = \mathcal{T}_i(\boldsymbol{x})$, $i \in \{1, 2\}$ with a batch of images $\boldsymbol{x}$. The final loss is then

$$\mathcal{L}(\boldsymbol{x}_1, \boldsymbol{x}_2) = -t(\boldsymbol{x}_1)\text{LogSoftmax}(s(\boldsymbol{x}_2)) - t(\boldsymbol{x}_2)\text{LogSoftmax}(s(\boldsymbol{x}_1)) \tag{6}$$

to update the student's parameters.

3.5 Page Descriptor and Retrieval

Since we split each document in patches of size 224×224, we aggregate the L2-normalized embeddings $\boldsymbol{p}_i$ with $i = 0, \ldots, N - 1$ of a page consisting of N patches via sum-pooling described by

$$\mathbf{P} = \frac{1}{N} \sum_{i=0}^{N-1} \boldsymbol{p}_i. \tag{7}$$

To obtain a global page embedding, each descriptor $\mathbf{P}$ is transformed by a PCA with whitening fitted on the train set. We apply intra-normalization followed by a L2-normalization of each descriptor. We report the performance of our approach based on mAP and Top-1 accuracies.

4   Experiments

In this section, we describe the dataset used, our experiments, and their settings.

4.1   Historical-WI Dataset

The dataset used in this work was part of the ICDAR2017 Competition on Historical Document Writer Identification (Historical-WI) [14]. Its documents are available in color as well as binarized. The train set consists of 1182 pages from 394 writers, each contributing three pages. The test set provides 3600 pages from 720 writers resulting in five pages per writer. The dataset is chosen since intra-class variances regarding the used pen are observed, as shown in Fig. 5.

We split each page into patches as described in Subsect. 3.1. For validation, we randomly select 10% of the authors of the train set, ending up with 39 writers. The total patches used in this paper are 107k for training, 4.2k for validation, and 133k for testing.

As shown in Fig. 5, the dataset is selected due to intra-class variance with respect to the used pen. Other datasets for writer retrieval such as ICDAR2013 [6] or CVL [5] do not own this feature, hence, we concentrate on ICDAR2017.

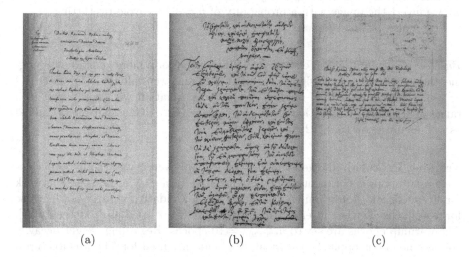

(a)   (b)   (c)

Fig. 5. Samples of one writer of the Historical-WI training set (ID 62). Variance in regards to the thickness and style of the pen is observed.

4.2 Evaluation

The following gives details about the network architectures, the data augmentations, and the training setup applied for our methodology.

Network Architecture. For our experiments based on DINO, we use ViT-Small/16 and ResNet50, both pretrained on ImageNet by Caron et al. [12]. ViT-Small/16 consists of 12 encoder layers, and the input is split into patches of size 16 × 16. The MLP head input dimension is 384 for ViT-Small/16 and 2048 for ResNet50 since we drop the last linear layer. Inputs are normalized to the ImageNet mean and standard deviation.

Data Augmentation. Our self-supervised approach relies on generating augmented views to learn meaningful features. In the following, we describe the different augmentation strategies used for evaluation. T_1 and T_2 consist of the same augmentations if not stated otherwise. During training, we randomly crop patches of size 256×256 to size 224 and scale them with a factor between 0.75 and 1.

MOs . As described in Sect. 3.3, we choose erosion or dilation and our default setting is random sampling of kernels with up to $i \leq 4$ ones. We also vary the number of iterations, denoted as n_a. Afterward, we apply a Gaussian blur. For T_2, we invert the image with a probability of $p = 0.2$ to avoid overfitting.

Baseline Augmentations. To show that MOs outperform more general data augmentations, specifically designed for other domains, we evaluate them on DINO as well. They consist of a random affine transformation with a rotation in a range of $\pm 10°$ and translation from -10% to $+10\%$. A Gaussian blur is added as well as inversion ($p = 0.2$) for T_2.

Training Protocol. In the following, we give an overview of the hyperparameters used in our experiments. They are listed in Table 1. We train each network for 40 epochs and stop training if the mAP on the validation set does not increase for five epochs. The last layer is frozen for the first five epochs as proposed in [12]. The warmup epochs are set to $n_{\text{warmup}} = 10$ epochs. Regarding weight decay, a cosine schedule is applied to start at 10% and only used for ViT architectures; the learning rate has a linear warmup schedule and is then cosine scheduled to zero until the end of training.

Table 1. Overview of the training settings for our experiments if not stated otherwise.

Hyperparameter	Value
Batch size B	32
Learning Rate l_r	$5 \cdot 10^{-4}$
Weight decay w_d	0.4 (10^{-4} for ResNet)
Momentum m	0.9995
Center momentum m_c	0.9
Temperatures $T_{\text{Teacher}}, T_{\text{Student}}$	0.04
Teacher warmup temperature T^0_{Teacher}	0.02
MLP output dimension D	512

Testing. For evaluating our approach, each document of the corresponding set is once used as a query (leave-one-image-out procedure); the ranked list is created by calculating the cosine distance to all page descriptors. We report the performance in terms of mAP and Top-1 accuracies as common metrics in the field of writer retrieval.

4.3 Results

Firstly, we report the performance of applying MOs as data augmentation for our approach and a metric learning-based method. Additionally, we evaluate the influence of the output dimension N of the MLP and visualize the attention maps of the ViT architecture. In the end, we give a comparison to the state-of-the-art.

Morphological Operations. Firstly, we evaluate different kernel sampling strategies as described in Sect. 3. The results are shown in Table 2. Random sampling kernels outperform applying specific patterns of kernel structures in each of our experiments. We assume that a wider variety of different kernels improves the generalization since 2×2 blocks (four kernels in total) perform the worst with a mAP of 57.4%. Our best result is achieved by the setting with $n_a = 1, i \leq 4$ with a mAP of 66.2%. Augmentations with a higher n_a or i are performing worse, e.g., we report a drop of 3.5% when increasing n_a; applying MOs more than once may affect the quality of the handwriting. We conclude that using randomly sampled kernels work best as a data augmentation strategy since it introduces the most variation to the training process.

Table 2. Results for different kernel sampling strategies for ViT-Small/16 pretrained on ImageNet by Caron et al. [12]. (Reported in %)

Sampling		mAP	Top-1
2×2 blocks	$n_a = 1$	57.4	76.9
Rows/Columns	$n_a = 1$	62.3	80.5
Random	$n_a = 1, i = 1$	64.6	82.3
Random	$n_a = 1, i \leq 4$	**66.2**	**83.6**
Random	$n_a = 1, i \leq 9$	63.7	81.4
Random	$n_a = 2, i \leq 4$	62.7	80.8

In Table 3, the performances in terms of mAP and Top-1 accuracies are given. Using no data augmentation, we obtain a mAP of 55.7% for ViT-Small/16 and 44.2% for ResNet50. In comparison, applying MOs, the mAP improves up to 66.2% (+10.5%). The baseline augmentations perform significantly worse than MOs with only 52.5% mAP and 41.8% for ResNet50. For experiments using ViTs where MOs are applied, we experience an increase in the retrieval performance and the effect for residual architectures is decreased. We conclude that proper augmentation techniques significantly impact the effect or our self-supervised approach, with MOs being the most effective, while the baseline augmentations even decrease the performance.

Table 3. Results for different augmentation strategies for ViT-Small/16 and ResNet50, both pretrained on ImageNet by Caron et al. [12]. MOs with $n_a = 1$ achieve the highest performance for both architectures. (Reported in %)

Data Augmentation	ViT-Small/16		ResNet50	
	mAP	Top-1	mAP	Top-1
No Augmentation	55.7	75.2	44.2	63.9
Baseline	52.5	72.8	41.8	62.2
MOs, $n_a = 1, i \leq 4$	**66.2**	**83.6**	**50.1**	**70.0**

Supervised Training. We train a ResNet50 with triplet loss (margin $m = 0.1$), a weakly supervised metric learning loss, e.g., used by [2,3]. The results are shown in Table 4, where the baseline is trained with data augmentation as described in Sect. 4.2. When applying MOs as data augmentation (we apply them with a probability of 50%), we observe an improvement of +5.8% for $n_a = 2$ in terms of mAP. Comparing the performance of both ResNet50 in Table 3 and Table 4, the network trained with triplet loss outperforms our proposed approach by 0.9%. Therefore, we consider MOs as a proper data augmentation strategy in other training strategies as well.

Table 4. ResNet50 trained with triplet loss. (Reported in %)

Data Augmentation	ResNet50	
	mAP	Top-1
Baseline	45.4	65.6
MOs, $n_a = 1, i \leq 4$	50.5	70.6
MOs, $n_a = 2, i \leq 4$	**51.2**	**71.2**
MOs, $n_a = 3, i \leq 4$	49.1	68.7

MLP Output Dimension. In the end, we evaluate the influence of the output dimension D of the MLP. As shown in Fig. 6, we observe the highest mAPs for low dimensions such as 256 (66.4%) and 512 (66.2%), although Caron et al. [12] propose values like 65536 for large datasets.

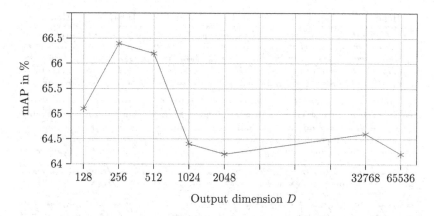

Fig. 6. Influence of the output dimension D of the MLP on the retrieval results of ViT-Small/16. The MLP head is only used during training.

Attention Maps. In the following, we provide the attention maps of our network trained with MO, $n_a = 1$ (see Table 3). We visualize the attention map $\mathbf{M}$ of the CLS token of the last layer defined by

$$\mathbf{M} = \mathrm{softmax}\left(\frac{\mathbf{QK}^{\mathrm{T}}}{\sqrt{d_h}}\right). \tag{8}$$

In Fig. 7, we highlight the top 10% of the attention values for each head on three different documents of the same writer. Note that our network retrieves each of the documents shown on top of the ranked list when using a page of this author as a query. Firstly, we observe that each head focuses on different zones, e.g., Head 1 has close attention on letters on the baseline, whereas the other heads rely on characteristics in the lower and upper zone. We follow that our network learns features for distinguishing between writers that are more frequently located at loops or curves, e.g., 'f' or 'l' (Head 2), 'g' or 'b' (Head 6) and smaller parts like dots or curves of special characters of a language (Head 5). However, there are also locations where highlighted areas do not contain relevant information (e.g. Patch 2, Head 1). These issues could be tackled by training on word-level, assuming no discriminative information is included in the spacing between words.

Comparison to State-of-the-Art. In Table 5, we compare our approach with state-of-the-art methods on the Historical-WI dataset. We achieve a mAP of 66.4% in our best experiment. Our method beats the winner of the competition where the dataset was introduced [14] and the work of Wang et al. [15] by more than 10%. While we are not able to beat the work of Christlein et al. [1] using an unsupervised approach with SIFT keypoint detection and mVLAD with an E-SVM, our approach does not need a sampling algorithm like SIFT for data generation and additionally, only considering Sum-Pooling as encoding, we

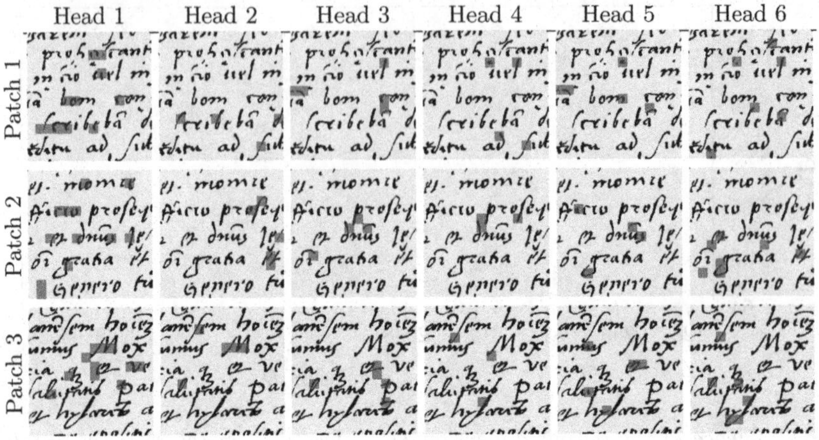

Head 1   Head 2   Head 3   Head 4   Head 5   Head 6

Fig. 7. Visualization of the CLS token attention of one 224×224 patch of three different documents with the top 10% highlighted for each head. (Writer ID 729)

are able to outperform their approach by 23.8%. Surprisingly, experiments with superior encodings like mVLAD or Generalized Max Pooling did not improve our results which we believe is caused by a smaller amount of samples used for learning a vocabulary since we are using 224 × 224 patches instead of 32 × 32 patches extracted with SIFT as [1] (107k vs. 480k regarding training set).

Table 5. Results compared to state-of-the-art methods on the Historical-WI dataset. (Reported in %)

	mAP	Top-1
Ours (ViT-Small/16, Sum-Pooling)	66.4	83.6
Wang et al. [15]	53.2	72.4
Gattal and Djeddi [14]	55.6	76.4
Christlein et al. (Sum-Pooling) [1]	42.6	$n\backslash a$
Christlein et al. (mVLAD + E-SVM) [1]	**76.2**	**88.9**

5   Conclusion

This paper introduced an unsupervised approach based on different views of an input image. Morphological operations were used to mimic different styles of pens. We showed that they significantly improved the results compared to traditional data augmentation strategies and can be applied in other network architectures and training setups. Additionally, the vision transformer enhanced the retrieval performance compared to ResNet50 as well as providing an attention mechanism to gain insights into similarities detected by the network without adding any complexity.

Future work includes further extending the use of MOs to existing approaches to combine the advantages of the methods. We also want to investigate generative adversarial networks and diffusion models to augment the available data from one writer while preserving the handwriting characteristics. Furthermore, transparency and explainability methods can be evaluated to gain deeper understanding of the similarities learned by the neural network.

Acknowledgment. The project has been funded by the Austrian security research programme KIRAS of the Federal Ministry of Agriculture, Regions and Tourism (BMLRT) under the Grant Agreement 879687.

References

1. Christlein, V., Gropp, M., Fiel, S., Maier, A.: Unsupervised feature learning for writer identification and writer retrieval. In: 2017 14th IAPR International Conference on Document Analysis and Recognition (ICDAR), pp. 991–997 (2017)
2. Rasoulzadeh, S., BabaAli, B.: Writer identification and writer retrieval based on NetVLAD with Re-ranking. IET Biometrics **11**(1), 10–22 (2022)
3. Keglevic, M., Fiel, S., Sablatnig, R.: Learning features for writer retrieval and identification using triplet CNNs. In: 2018 16th International Conference on Frontiers in Handwriting Recognition (ICFHR), pp. 211–216 (2018)
4. Christlein, V., Bernecker, D., Angelopoulou, E.: Writer identification using VLAD encoded Contour-Zernike moments. In: 2015 13th International Conference on Document Analysis and Recognition (ICDAR), pp. 906–910 (2015)
5. Kleber, F., Fiel, S., Diem, M., Sablatnig, R.: CVL-DataBase: an o-line database for writer retrieval, writer identification and word spotting. In: 2013 12th International Conference on Document Analysis and Recognition, pp. 560–564 (2013)
6. Louloudis, G., Gatos, B., Stamatopoulos, N., Papandreou, A.: ICDAR 2013 competition on writer identification. In: 2013 12th International Conference on Document Analysis and Recognition, pp. 1397–1401 (2013)
7. Buhrmester, V., Münch, D., Arens, M.: Analysis of explainers of black box deep neural networks for computer vision: a survey. Mach. Learn. Knowl. Extr. **3**(4), 966–989 (2021)
8. Khan, S., Naseer, M., Hayat, M., Zamir, S.W., Khan, F.S., Shah, M.: Transformers in vision: a survey. ACM Comput. Surv. **54**(10s), 1–41 (2021)
9. Dosovitskiy, A., et al.: An image is Worth 16x16Words: transformers for image recognition at scale. ICLR (2021)
10. Abnar, S., Zuidema, W.: Quantifying attention flow in transformers. In: Proceedings of the 58th Annual Meeting of the Association for Compu- tational Linguistics, Online: Association for Computational Linguistics, pp. 4190–4197 (2020)
11. Chefer, H., Gur, S., Wolf, L.: Transformer interpretability beyond attention visualization. In: IEEE/CVF Conference on Computer Vision and Pattern Recognition (CVPR), pp. 782–791 (2021)
12. Caron, M., et al.: Emerging properties in self-supervised vision transformers. In: Proceedings of the International Conference on Computer Vision (ICCV) (2021)

13. Fiel, S., Sablatnig, R.: Writer identification and retrieval using a convolutional neural network. In: CAIP (2015)
14. Fiel, S., et al.: ICDAR2017 competition on historical document writer identification (Historical-WI). In: 2017 14th IAPR International Conference on Document Analysis and Recognition (ICDAR), vol. 01, pp. 1377–1382 (2017)
15. Wang, Z., Maier, A., Christlein, V.: Towards end-to-end deep learning-based writer identification. In: INFORMATIK,: Gesellschaft für Informatik. Bonn, pp. 1345–1354 (2020)
16. Baevski, A., Hsu, W.N., Xu, Q., Babu, A., Gu, J., Auli, M.: Data2vec: a general framework for self-supervised learning in speech, vision and language (2022)
17. Souibgui, M.A., et al.: DocEnTr: an end-to-end document image enhancement transformer. arXiv preprint arXiv:-2201.10252 (2022)
18. Khamekhem Jemni, S., Souibgui, M.A., Kessentini, Y., Fornés, A.: Enhance to read better: a multi-task adversarial network for handwritten document image enhancement. Pattern Recognit. **123**, 108 370 (2022)
19. Liu, Z., et al.: Swin transformer V2: scaling up capacity and resolution. In: International Conference on Computer Vision and Pattern Recognition (CVPR) (2022)
20. Yan, H., Zhang, C., Wu, M.: Lawin transformer: improving semantic segmentation transformer with multi-scale representations via large window attention. ArXiv, vol. abs/2201.01615 (2022)
21. Vaswani, A., et al.: Attention is all you need. In: Advances in Neural Information Processing Systems, vol. 30 (2017)

EAU-Net: A New Edge-Attention Based U-Net for Nationality Identification

Aritro Pal Choudhury[1] (ID), Palaiahnakote Shivakumara[2](✉) (ID), Umapada Pal[1] (ID), and Cheng-Lin Liu[3] (ID)

[1] Computer Vision and Pattern Recognition Unit, Indian Statistical Institute, Kolkata, India
umapada@isical.ac.in
[2] Faculty of Computer Science and Information Technology, University of Malaya, Kuala Lumpur, Malaysia
shiva@um.edu.my
[3] Institute of Automation of Chinese Academy of Sciences, Beijing, China
liucl@nlpr.ia.ac.cn

Abstract. Identifying crime or individuals is one of the key tasks toward smart and safe city development when different nationals are involved. In this regard, identifying Nationality/Ethnicity through handwriting has received special attention. But due to freestyle and unconstrained writing, identifying nationality is challenging. This work considers words written by people of 10 nationals namely, India, Malaysia, Myanmar, Bangladesh, Iran, Pakistan, Sri Lanka, Cambodia, Palestine, and China, for identification. To extract invariant features, such as the distribution of edge patterns despite of the adverse effect of different writing styles, paper, pen, and ink, we explore a new Edge-Attention based U-Net (EAU-Net), which generates edge points for each input word image written by different nationals. Inspired by the success of the Convolutional Neural Network for classification, we explore CNN for the classification of 10 classes by considering candidate points given by EAU-Net as input. The proposed method is tested on our newly developed dataset of 10 classes, a standard dataset of 5 classes to demonstrate the effectiveness in classifying different nationalities. Furthermore, the efficacy of the proposed method is shown by testing on IAM dataset for gender identification. The results of the proposed and existing methods show that the proposed method outperforms the existing methods for both nationality and gender identification.

Keywords: Handwriting analysis · U-Net · CNN · Edge points detection · Nationality identification

1 Introduction

Assisting the forensic team to identify the crime plays a vital role in the development of a smart, safe city, especially when the crime involves multiple nationals. There are models and approaches for identifying people who commit crimes based on biometric features [1, 2]. However, those methods may not be robust to adverse situations, such as causes of open environment, illumination, distortion etc. This is because biometric features are

U. Porwal et al. (Eds.): ICFHR 2022, LNCS 13639, pp. 137–152, 2022.
https://doi.org/10.1007/978-3-031-21648-0_10

not invariant to the above-mentioned adverse situations. To alleviate this limitation and to assist forensic teams, nationality or ethnicity identification using handwriting analysis has been introduced. The reason is that processing of handwriting analysis is simple, robust, and accurate compared to biometric images. It is evident from the methods of the writer, gender identification [3–7], age estimation [8] and personality traits identification [9, 10], where the handwriting of different persons is studied for successful classification. However, in the case of national identification, there are no constraints of age, gender, pen, paper, or writing style, unlike other works which are usually confined to a particular category or range of ages. Therefore, nationality identification using handwriting analysis is complex and it is an open challenge. It can be seen from the sample multiple images of each nation shown in Fig. 1, where it is noted that English text written by different nationals exhibits arbitrary variations for the images of intra-and-inter-classes.

Despite the complex problem as mentioned above, one can expect some regular patterns for handwritten English texts of each country. The rationale behind this hypothesis is that writers of Chinese have practiced writing horizontal and vertical strokes because Chinese text involves writing more like boxes in a pyramid structure. When the same Chinese writer writes English text, it is expected more prominent horizontal and vertical strokes in him/her writing. This observation remains the same for any Chinese writer. In the same way, when an Indian who has practiced writing cursive characters (Indian script), writes English text, it is expected more prominent cursive nature in the writings. The same observation can be seen for English text written by any Indian. The same rationale is true for English handwritten by other nationals.

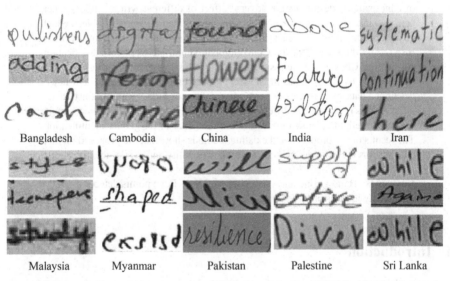

Bangladesh   Cambodia   China   India   Iran

Malaysia   Myanmar   Pakistan   Palestine   Sri Lanka

Fig. 1. Examples of English handwriting samples of ten different countries

To extract the above observation, we propose to detect edges (strokes of text) because the distribution of edge points is invariant to the number of characters, words, and style of writing as it represents the dominant information in the handwritten text. Therefore, for detecting prominent edges in the input handwritten image, inspired by the special property of U-Net which can separate dominant information in the image irrespective of contrast variation, and distortion [11], we explore the same U-Net in a novel way for detecting edges in the input images. The output of this step represents the dominant information in the image. For the purpose of detecting prominent edges, we explore the combination of Edge-Attention Network and U-Net, which is called Edge-Attention based U-Net (EAU-Net). There are methods for edge detection using deep learning approaches, such as Hussain et al. [12], which uses U-Net for blood vessel segmentation, Wang et al. [11], which uses Edge Attention for medical image segmentation and Chen et al. [13], which uses Edge attention for image instance segmentation. However, none of the methods [11–13] use the combination of U-Net and Edge Attention for edge detection. In addition, the methods are applied on general and medical images but not handwritten text images. Therefore, to the best of our knowledge, this is the first work for detecting edge points in the handwritten text images application to nationality identification. For classification, motivated by the success of the Convolutional Neural Network (CNN) for classification, we propose CNN for the classification of 10 classes.

The key contributions are as follows. (i) Exploring the combination of Edge Attention and U-Net for edge detection in handwritten text images. (ii) The way the proposed work combines EAU-Net and CNN for the classification of the handwritten text of ten different nationals. (iii) The results on nationality and gender classification show that the proposed method is independent of the number of classes, size of the datasets and applications.

The structure of the rest of the paper is organized as follows. The review of existing methods of different cases of handwritten text analysis is presented in Sect. 2. The new architecture of the proposed work for edge detection and classification of 10 different nationals is described in Sect. 3. Section 4 presents experimental analysis to validate the proposed method for the classification of different nationals. The concluding remark is summarized in Sect. 5.

2 Related Work

The related work for nationality or ethnicity identification is gender, writer identification, age estimation and personality traits identification. This section reviews the methods of the above-mentioned categories.

The methods developed for gender identification consider the handwritten text of male and female writers as input [3, 4]. For example, Dargan et al. [4] presented different methods for both gender and writer identification, where we can see a discussion on the performance of different classifiers and the methods. The methods work based on the assumption that the text written by males exhibits an inconsistent style of writing, while the text written by females exhibits a consistent writing style. To achieve two-class classification, the models extract geometrical features of characters or words or text lines to study regular and irregular patterns of writing. Since the focus of the method is to classify males and females, the methods may not be suitable for the classification of ten different nationals. This is because each class of different nationals can have written text of both males and females.

Similarly, models were developed for addressing the challenges of writer identification. For instance, Mridha et al. [5] used the combination of Gabor transform and convolutional neural networks for writer identification by targeting Indian scripts. Punjabi et al. [6] extracted patches for writer identification using deep neural networks. Purohit et al. [7] presented state-of-the-art methods for writer identification in offline mode. The intuition behind proposing methods for writer identification is that the style of each writer has unique properties in terms of strokes, shape, and style of characters. Therefore, the models extract features like strokes to differentiate text written by different people. Since the methods consider the classes of handwritten text written by the same person for writer identification, these methods may not be effective for nationality identification because each class of nationals can have the text of different persons. Therefore, these approaches are not effective for nationality identification.

Recently, the work in [8] proposed an age estimation using handwriting analysis, which uses disconnected features for age estimation. The method assumes that as age changes, disconnectedness increases due to the psychological effect of the human mind on writing. Therefore, the method extracts disconnected features using Hua moments and then K-means clustering has been used for classification. Since the method considers text written by fixed aged persons, it may not work for ten-classes of nationality identification.

There are methods for personality and emotions identification using handwriting analysis [9, 10]. For example, Gahmouse et al. [10] explored texture features for personality identification using handwriting analysis. The method assumes that the personality of a person reflects in handwriting. As personality changes, the writing style changes. The method extracts different handcrafted features to study the effect of personality in writing. In this case, there is no constraint that the personality class should contain the text of the same person. The class of text can contain any person, but the person should be in the same category of personality traits, such as extraversion, neuroticism, agreeableness, conciseness, Openness etc. However, the extracted features are sensitive to shape, strokes, and character appearance. These constraints may not be valid for nationality identification because the features must be invariant to shape, strokes, and character appearance for successful classification. Therefore, the methods may not be capable of nationality identification of ten classes.

However, a few methods have been developed in the past for nationality identification [14, 15]. Nag et al. [15] proposed nationality identification using handwriting analysis. The method explores Cloud of Line Distribution (COLD) for extracting features and then the features are fed to the SVM classifier for classification. The scope of the method is limited to five classes, namely, Bangladesh, Iran, India, China, and Malaysia. Since the method follows a conventional approach to solving the problem of nationality identification, the method does not have the ability to handle ten class classification problems. The reason for deficient performance is that the features do not have generalization ability.

In summary, it is noted from the review of the methods on various categories that only a few methods are available for classification of different nationalities using handwriting analysis. However, the scope of the methods is limited to two classes and five classes but not for the ten classes as proposed in this work. The methods developed for other categories are not effective for nationality identification because the scope and objective of the methods are confined to a particular target. Therefore, we can infer that nationality identification is still considered as an open problem for the researchers.

3 Proposed Model

The aim of the proposed work is to identify the different nationalities using handwriting analysis. This work considers words written by ten-nationals without imposing any constraints on writing, paper, ink, age and gender as input for classification. The ten-nationals are India, Bangladesh, China, Pakistan, Iran, Palestine, Myanmar, Sir Lanka, Malaysia, and Cambodia. The reason to consider these countries for nationality identification is that these countries share a common culture and hence the problem becomes complex. As discussed in the previous section, the challenge is to develop a model that should be invariant to number of characters, number of words, different writing style etc. We believe that the distribution of edge patterns (strokes) is invariant to the above threats. This is because the detected edge pattern represents prominent information in the images, which represents the global pattern of the input image. For example, in the case of text written by the Chinese, horizontal and vertical information is prominent compared to cursive. On the other hand, in the case of texts written by Indians, cursive is prominent compared to straightness. The same is true for all other classes where one can expect different prominent properties for each nationality.

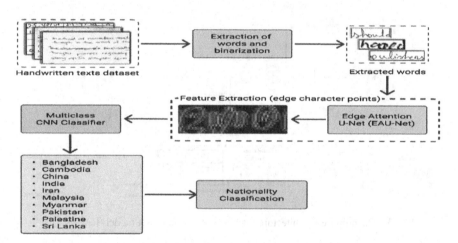

Fig. 2. Block diagram of the proposed work for nationality classification.

To extract the above observations, we explore the combination of Edge-Attention and U-Net for edge detection for the input images. The motivation to propose this combination is the excellent performance of U-Net for extracting dominant information, irrespective of scales, distortion, contrast variations in the images, which results in regions of interest in the images. For detecting edges of text information, we believe that attention networks serve our purpose. Thus, the combination for candidate points detection in the input images. For feature extraction and classification, we propose CNN for classification of ten-national classes. The block diagram of the proposed model is shown in Fig. 2.

3.1   Edge-Attention Based U-Net for Edge Detection

The proposed Edge-Attention based U-Net (EAU-Net) architecture is shown in Fig. 3. The EAU-Net utilizes the encoder and decoder modules from the original U-Net architecture. The network encoder provides a downsampling path for the input image. The encoder consists of four components to contract the image, increasing the number of channels by a factor of 2 in each of them. A single encoder component consists of two 3 × 3 convolutions, each followed by a rectified linear unit (ReLU). The output from one component is passed through a 2 × 2 Max pooling layer, with a stride of 2. The downsampled feature map from the lowest encoder unit is passed to the decoder.

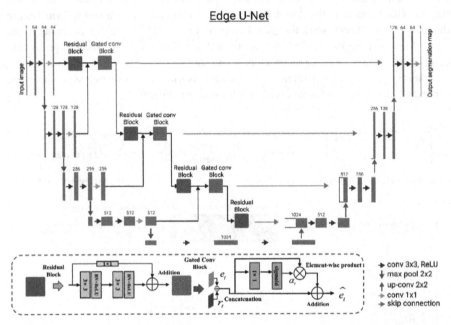

Fig. 3.   The proposed architecture for Edge-Attention U-Net model (EAU-Net).

The decoder upsamples the feature maps using repetitive 2 × 2 up-convolutions, which divides the number of feature channels. A skip connection from each encoder unit crops the encoder output and adds to the decoder upsampled feature map in each step in the original U-Net model. The skip connections are modified to extract special edge features using Edge Attention Block in our approach. The edge attention block extracts edge and contour features effectively. This facilitates the segmentation of sharp edges in images. This block includes a Residual unit, along with a Gated Convolution unit. The feature map from the encoder is processed by the edge attention block. The gated convolution limits the model to only process edge information.

The Residual Block uses skip connection with a single 1 × 1 convolution. The output from the residual block is aggregated with the output from the next encoder unit, which is used to extract the edge features using the edge feature module. The aggregated feature map is used for the edge stream attention. We get the attention map, $\alpha_t \in R^{H \times W}$ by concatenating the edge feature map, e_t and residual map, r_t followed by a 1×1 convolution with normalized $C(.)$, which, in turn, is followed by a sigmoid function as defined in Eq. (1).

$$\alpha_t = \sigma(C(e_t \oplus r_t)) \tag{1}$$

$\oplus$ represents concatenation of the feature maps. Gated convolution is applied on e_t as an element-wise product $\otimes$ with an attention map. This is followed by a channel-wise weight with kernel w_t and residual connection. Gated convolution at each layer is calculated as defined in Eq. (2).

$$e_t^{\widehat{(i,j)}} = \left(\left(e_{t_{(i,j)}} \otimes \alpha_{t_{(i,j)}} \right) + e_{t_{(i,j)}} \right)^T w_t \tag{2}$$

The feature map is passed to the decoder as a skip connection. This feature map is cropped and added to the upsampled feature map in the decoder. At the top layer, the decoder provides a single channel output, with the final edge feature image. This edge image finds the dominant character points in the dataset images. For training the model, mean squared error (MSE) loss criterion is used, along with Root Mean Squared Propagation (RMSProp) as optimizer. The learning rate of $1e^{-5}$, weight decay of $1e^{-8}$ and momentum of 0.9 was used. The EAU-Net model is used as a feature extractor in the classification model discussed in the next section.

The MSE Loss given by $\mathscr{L}$ is calculated as defined in Eq. (3).

$$\mathscr{L}(x, y) = L = \left\{ l_1, \ldots, l_N \right\}^T, l_n = (x_n - y_n)^2 \tag{3}$$

x_n denotes the output from the model for the $n^{th.}$ input image, and y_n denotes the ground truth for the $n^{th.}$ image. The architecture is trained with the samples of DRIVE benchmark dataset of vessel segmentation in Retina images [16]. Since the ground truth of edge information is available for Retina images, we use the same ground truth to train the proposed mode. The reason is that the objective (detecting prominent edges) of edges in Retina images and text images is the same.

The effectiveness of the proposed EAU-Net is illustrated in Fig. 4(a) and (b), where for each input of 10 different nationals, edges detected by the original U-Net, intermediate results of the proposed model and the final edge detected by the proposed are shown. It is observed from Fig. 4(a) and (b) that the results of the intermediate step of the proposed method and the results of the proposed model are better than the results of original U-Net. Since the original U-Net does not have an attention mechanism, it reports poor edge detection results compared to the results of the proposed model. The result of the intermediate step indicates that the edge attention contributes to detect prominent edges in the images. Therefore, the proposed EAU-Net detects most of the prominent edges with clear structure of text information.

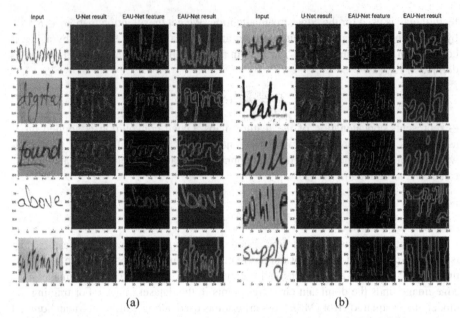

(a)   (b)

Fig. 4. EAU-Net for edge detection: Sample images of column 1 shown in (a) and (b), are given as input to the model. Columns 2 of (a) and (b) show the results of an original U-Net model, which is trained on the same DRIVE dataset [16]. Columns 3 of (a) and (b) are the results of intermediate steps of the EAU-Net encoder. Columns 4 of (a) and (b) show the output of the proposed model.

3.2 Nationality/Ethnicity Identification

It is noted from the edge detection results discussed in the previous section that the edges capture styles of writing of different nationals irrespective of age, gender, pen, paper, ink etc. For classification, we propose a simple Convolutional Neural Network (CNN) as shown in Fig. 5, where the architecture is designed as a classifier for classification of 10 classes. The classifier model uses 3 convolutional blocks, which consists of 3 × 3 convolutions, Batch Normalization, Rectified Linear Unit (ReLU) and dropout (with probability of 0.1) layers. The convolution layers are followed by 2 fully connected layers on the flattened feature map. LogSoftMax activation function is used to get the probability distributions of the predicted class labels. The classifier model is trained with Cross Entropy Loss, with Adam optimizer with a learning rate of 0.005 for 30 epochs. The experimentation details along with results are discussed in the following section.

To visualize the contribution of the proposed work for classification of 10 classes, activation maps of images of inter-classes are shown in Fig. 5. It is observed from the activation maps shown in Fig. 5 that the prominent edge in activation maps exhibits a distinct pattern for the images of inter-classes. Therefore, one can infer that the proposed model is capable of handling challenges of nationality identification of 10 nationals.

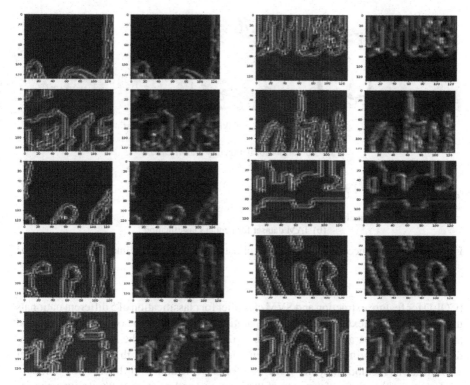

Fig. 5. Activation maps of the last 2D convolutional layer for the edge images given by EAU-Net.

4 Experimental Results

Since the problem considered in this work is new, we create our own dataset by collecting handwritten samples of 10 nationals, namely, India, Malaysia, Myanmar, Bangladesh, Iran, Pakistan, Sri Lanka, Cambodia, Palestine, China. We segment the words from the text manually for experimentation. Our dataset contains 800 images in total, with class imbalances ranging from 30 to 100 images in a class. Sample images of each class are shown in Fig. 6, where we can see arbitrary handwritten patterns for each class. To avoid adverse effects of imbalance class size and size of the images, the images are resized to standard size of 256 × 256 pixels and used augmentation techniques, such as flipping horizontally, randomly. Further, to test fairness of the performance of the proposed method, we use a standard dataset of 5 classes [15], which comprises India, Malaysia, Iran, China, Bangladesh. Each class provides 100 images, so in total 500 images are considered for experimentation. Overall, 1300 samples are used for evaluating the proposed and existing methods.

To show that the proposed method is generic, we consider the standard dataset of IAM for gender classification [17], which provides 16542 handwritten words written by male and 4946 handwritten word images written by females.

To show effectiveness of the proposed method, we implemented two state-of-the-art existing methods, namely, Nag et al. [15] which proposes Cloud of Line Distribution

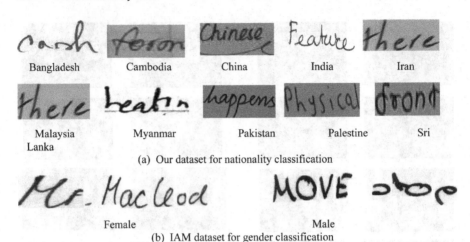

(a) Our dataset for nationality classification

Female   Male
(b) IAM dataset for gender classification

Fig. 6. Samples of successful classification of nationality and gender images of our and IAM datasets, respectively.

(COLD for feature extraction and SVM classifier for classification of 5 different nationals using handwritten text analysis for comparative study. This is the only relevant method available for comparative study. Similarly, one more state-of-the-art method for gender and writer identification [4] has been implemented to show that gender and writer classifications methods may not work well for classification of different nationalities.

For evaluating the performance of the proposed and existing methods, we generate a confusion matrix and calculate the Average Classification Rate (ACR), which is the mean of diagonal elements of the confusion matrix.

Implementation Details: Our dataset does not provide ground truth for edge detection, we train our model with the ground truth of edge detection provided by Digital Retinal Images for Vessel Extraction (DRIVE), which is an open-source dataset for retinal vessel segmentation. This dataset provides ground truth (edges) of retinal images for segmentation. Since the edges that represent dominant information are common for any images, we use the same ground truth to train the proposed model in this work. The classifier is trained on our curated 10-class nationality handwritten dataset. An 80%-20% training-validation ratio is considered for all the experiments in this work. The same experimental set up has been used for calculating measures using existing methods.

4.1 Ablation Study

In the proposed work, exploring U-Net for edge detection is the key step and therefore to validate the effectiveness of the key step, we conducted the following experiments on our dataset. The average classification rate is calculated and reported for each experiment. (i) For classification of 10 nationals, the proposed work uses the baseline U-Net for edge detection and then CNN for classification as reported in Table 1. This is to show the baseline U-Net model is not capable of handling challenges of nationality identification.

(ii) The average classification rate is calculated by feeding input images to CNN directly to assess the effectiveness of the proposed edge detection. (iii) Harris corner detection is popular for extracting dominant edge information. Therefore, this experiment uses Harris corners as input for classification using CNN instead of edge detection using EAU-Net. The average classification rate is reported in Table 1. This is to show that Harris corners are not sufficient to achieve the best results for classification of 10 nationals. (iv) This experiment includes the proposed EAU-Net and CNN for classification and the average classification rate is reported in Table 1. It is observed from Table 1 that the baseline U-Net model for edge detection, use of Harris corners for extracting dominant information and CNN without edge detection reports poor results compared to the proposed model. This shows that the proposed combination of edge attention and U-Net (EAU-Net) for edge detection is effective in achieving the best results.

Table 1. Ablation study to assess the effectiveness of the key steps of the proposed model in terms of Average Classification Rate (ACR)

Exp.	(i)	(ii)	(iii)	(iv)
Key steps	Original U-Net (baseline architecture)	Feeding input image to CNN directly	EAU-Net with Haris corner detection	Proposed Model
ACR	85%	84%	87%	96%

4.2 Experiments on Edge Detection

Qualitative results for samples of each class of the proposed EAU-Net are shown in Fig. 7, where it is noted that the prominent edges are detected for all the sample images. It is also noted that the detected edges preserve the structure of text. This indicates that the edge detection step facilitates subsequent classification to achieve the best results. This is the key contribution of the EAU-Net for edge detection in the images.

4.3 Experiments on Classification of Nationality

Qualitative results of the proposed method for successful classification can be seen in Fig. 6, where all the sample images are classified correctly. Similarly, quantitative results of the proposed and existing methods [4, 15] on our dataset of 10 classes and standard dataset of 5 classes are reported in Tables 2, 3 and 4. When we look at the confusion matrix and average classification rates of the proposed and existing methods on both the datasets, the proposed model is better than existing methods for both the datasets. This shows that the proposed model is effective, and it is independent of the number of classes. The reason for achieving the best results by the proposed model is the contribution of EAU-Net for edge detection. On the other hand, although the method [15] was developed for classification of five nationalities, it reports poor results for 10 classes as well five classes. The main reason for reporting the poor results is that extracted hand-crafted features are good for specific dataset and hence features are not generic unlike the proposed method, which does not use handcrafted features for classification. In the same way, the existing method [4] reports poor results for both the datasets because the features extracted for gender and writer classification but not nationality identification. This shows that the features used for gender and writer identification are not effective for classification of nationality identification.

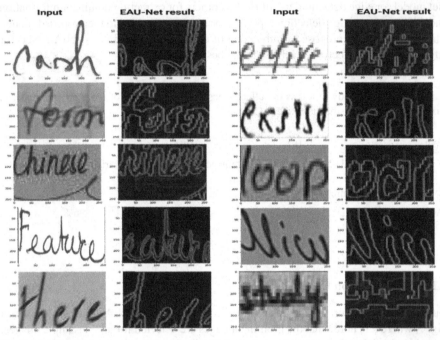

Fig. 7. Qualitative result of the proposed EAU-Net for edge detection in the images of 10 class nationality dataset.

However, it is noted that the proposed method obtains a better average classification rate for 10 classes compared to five classes. This is because when the proposed model is trained with considerable number of samples of 10 classes, there are chances of introducing an overfitting problem and hence it reports slightly poor results for five classes compared to 10 classes.

Table 2. Confusion matrix and Average Classification Rate (ACR) of the proposed method on our dataset in (%)

Classes	Bangladesh	Cambodia	China	India	Iran	Malaysia	Myanmar	Pakistan	Sri Lanka	Palestine
Bangladesh	92	0	4	0	0	0	0	0	0	4
Cambodia	0	100	0	0	0	0	0	0	0	0
China	0	0	100	0	0	0	0	0	0	0
India	0	0	0	100	0	0	0	0	0	0
Iran	0	0	0	0	100	0	0	0	0	0
Malaysia	0	0	0	0	0	100	0	0	0	0
Myanmar	4	0	0	0	0	3	93	0	0	0
Pakistan	0	0	0	0	0	0	0	100	0	0
Sri Lanka	0	0	0	0	0	0	0	0	94	6
Palestine	0	0	0	0	0	0	0	0	11	89
ACR	96.8%									

Table 3. Confusion matrix and average classification rate of the proposed method on the standard dataset of five classes in (%)

Classes	Bangladesh	China	India	Iran	Malaysia
Bangladesh	90	2	8	0	0
China	5	85	10	0	0
India	5	5	90	0	0
Iran	0	0	0	100	0
Malaysia	0	0	0	0	100
Average classification rate	93				

Table 4. Average classification rate of the proposed and existing methods for nationality identification in (%).

Datasets	Proposed	Nag et al. [15]	Dargan et al. [4]
Our dataset of 10 classes	96.8	74.2	68.7
Standard dataset of 5 classes	93	75	71.2

4.4 Gender Classification

To show versatility of the proposed model on classification, we test our method on the IAM dataset, which is the standard dataset available and provides ground truth for gender identification. We retrain the proposed method and existing methods [4, 15] on handwritten word images written by male and female and the results of the proposed and existing models are reported in Table 5 and Table 6. The results in Table 5 and Table 6 show that the proposed model is the best at Average Classification Rate (ACR) for gender classification. Compared to the existing methods, the method [4] reports better results than the method [15]. This makes sense because the method [4] was developed for writer and gender identification while the method [15] was developed for nationality identification. It is noted from the results of the proposed model on nationality and gender identification that our method reports almost consistent results. This shows that the proposed EAU-Net for edge detection is robust and generic for classification of different datasets and different applications.

Table 5. Confusion matrix and average classification rate of the proposed method on IAM dataset in (%)

Class	Male	Female
Male	93.1	6.9
Female	8.3	91.6
Average classification rate	92.3	

Table 6. Average classification rates of gender classification on IAM Dataset in (%)

Datasets	Proposed	Nag et al. [15]	Dargan et al. [4]
IAM dataset of male and female classes	92.3	81.9	89.5

4.5 Error Analysis

However, sometimes, due to severe degradations, writing style, changing color of foreground and background, there are chances of misclassification as shown in a few samples in Fig. 8. The reason for such misclassification is that when the images are affected by the above-mentioned adverse factors, images lose edge quality. In this situation, the proposed EAU-Net model may miss prominent edges in the input images and hence misclassifications occur. To overcome this limitation, there is a scope for developing robust edge detection and then classification. Another way is to develop a single robust deep learning model for classification instead of relying on edge detection. These ideas will be implemented in the near future.

| China-Cambodia | Sri Lanka-Iran | Sri Lanka-Malaysia | Maynmar-Malaysia | Pakistan-India |

Fig. 8. Unsuccessful classification of the proposed method on our dataset

5 Conclusion and Future Work

We have proposed a novel method for nationality identification through handwritten text analysis. The main contribution of the proposed work is to introduce Edge-Attention-U-Net (EAU-Net) for edge detection to extract prominent information in the input images. The features are extracted from edge images given by EAU-Net and classification is perfumed using simple Convolutional Neural Networks (CNN). Experimental results on our dataset of 10-mationality classes, standard dataset of five classes and male-female classes of IAM dataset show that the proposed model outperforms the existing methods in terms of the average classification rate. However, sometimes, due to severe degradations, writing style and variations in foreground and background colors, the proposed model misclassifies the images. To alleviate this limitation, we will plan to develop a robust edge detection step and new deep learning models for classification in the future.

References

1. Singh, M., Nagpal, S., Vatsa, M., Singh, R., Noore, A., Majumdar, A.: Gender and ethnicity classification of iris images using deep class-encoder. In: 2017 IEEE International Joint Conference on Biometrics (IJCB), pp. 666–673. IEEE (2017)
2. Uddin, M.A., Chowdhury, S.A.: An integrated approach to classify gender and ethnicity. In: 2016 International Conference on Innovations in Science, Engineering and Technology (ICISET), pp. 1–4. IEEE (2016)
3. Bouadjenek, N., Nemmour, H., Chibani, Y.: Robust soft-biometrics prediction from off-line handwriting analysis. Appl. Soft Comput. **46**, 980–990 (2016)
4. Dargan, S., Kumar, M.: Gender classification and writer identification system based on handwriting in Gurumukhi script. In: 2021 International Conference on Computing, Communication, and Intelligent Systems (ICCCIS), pp. 388–393. IEEE (2021)
5. Mridha, M.F., Ohi, A.Q., Shin, J., Kabir, M.M., Monowar, M.M., Hamid, M.A.: A thresholded Gabor-CNN based writer identification system for Indic scripts. IEEE Access **9**, 132329–132341 (2021)
6. Punjabi, A., Prieto, J.R., Vidal, E.: Writer identification using deep neural networks: impact of patch size and number of patches. In: 2020 25th International Conference on Pattern Recognition (ICPR), pp. 9764–9771. IEEE (2021)
7. Purohit, N., Panwar, S.: State-of-the-art: offline writer identification methodologies. In: 2021 International Conference on Computer Communication and Informatics (ICCCI), pp. 1–8. IEEE (2021)
8. Basavaraja, V., Shivakumara, P., Guru, D.S., Pal, U., Lu, T., Blumenstein, M.: Age estimation using disconnectedness features in handwriting. In: 2019 International Conference on Document Analysis and Recognition (ICDAR), pp. 1131–1136. IEEE (2019)
9. Fairhurst, M., Erbilek, M., Li, C.: Study of automatic prediction of emotion from handwriting samples. IET Biom. **4**(2), 90–97 (2015)
10. Gahmousse, A., Gattal, A., Djeddi, C., Siddiqi, I.: Handwriting based personality identification using textural features. In: 2020 International Conference on Data Analytics for Business and Industry: Way Towards a Sustainable Economy (ICDABI), pp. 1–6. IEEE (2020)
11. Wang, K., Zhang, X., Zhang, X., Lu, Y., Huang, S., Yang, D.: EANet: Iterative edge attention network for medical image segmentation. Pattern Recognit. **127**, 108636 (2022). https://doi.org/10.1016/j.patcog.2022.108636

12. Hussain, S., Guo, F., Li, W., Shen, Z.: DilUnet: a U-net based architecture for blood vessels segmentation. Comput. Methods Programs Biomed. **218**, 106732 (2022)
13. Chen, X., Lian, Y., Jiao, L., Wang, H., Gao, Y., Lingling, S.: Supervised edge attention network for accurate image instance segmentation. In: Vedaldi, A., Bischof, H., Brox, T., Frahm, J.-M. (eds.) ECCV 2020. LNCS, vol. 12372, pp. 617–631. Springer, Cham (2020). https://doi.org/ 10.1007/978-3-030-58583-9_37
14. Al Maadeed, S., Hassaine, A.: Automatic prediction of age, gender, and nationality in offline handwriting. EURASIP J. Image Video Process. **2014**(1), 1–10 (2014). https://doi.org/10. 1186/1687-5281-2014-10
15. Nag, S., Shivakumara, P., Wu, Y., Pal, U., Lu, T.: New cold feature based handwriting analysis for enthnicity/nationality identification. In: 2018 16th International Conference on Frontiers in Handwriting Recognition (ICFHR), pp. 523–527. IEEE (2018)
16. Staal, J., Abràmoff, M.D., Niemeijer, M., Viergever, M.A., Van Ginneken, B.: Ridge-based vessel segmentation in color images of the retina. IEEE Trans. Med. Imaging **23**(4), 501–509 (2004)
17. Marti, U.V., Bunke, H.: The IAM-database: an English sentence database for offline handwriting recognition. Int. J. Doc. Anal. Recognit. **5**, 39–46 (2002). https://doi.org/10.1007/s10 0320200071

Progressive Multitask Learning Network for Online Chinese Signature Segmentation and Recognition

Xunhui Qin[1,2], Hanyue Zhang[2,3], Xiao Ke[2], Zhonghao Shen[1,2(✉)], Songmao Qi[2], and Ke Liu[1,2]

[1] Chongqing Handwriting Big Data Research Institute, Chongqing, China
{qinxunhui,liuke}@aosign.cn
[2] Chongqing AOS Technology Co., Ltd., Chongqing, China
{kexiao,shenzhonghao,qisongmao}@aosign.cn
[3] Southwest University, Chongqing, China
zhanghanyue@swu.edu.cn

Abstract. Recently, the booming of electronic devices has revolutionized the way we sign in our daily life. The sudden surge of online Chinese signatures calls for need online Chinese signature segmentation as prerequisite for downstream tasks such as building up database of Chinese characters and verifying signatures based upon individual characters. However, common approaches deriving from over-segmentation do not apply well to Chinese signatures, which have little linguistic meanings but instead have flourish, artistic styles, composite structures or even word overlaps. To cope with those difficulties, this paper exploits the benefits of signature recognition to boost segmentation performance and proposes a progressive multitask learning network (PMLNet) for online Chinese signature segmentation and recognition. PMLNet consists of a dual channel stroke feature extraction block (DSF-Block), a stacked transformer encoder block (STE-Block) and a progressive multitask learning block (PML-Block). DSF-block is used to extract stroke-wise spatial features and semantic features through dual channels; STE-block is used to model long-range dependencies among different strokes and enhance their feature representations; PML-block is used to branch interactively and progressively the signature's final segmentation and recognition. Specifically, we introduce a progressive learning strategy in PML-block to fine-tune segmentation and recognition results by fully leveraging their reciprocal relationship. The experiment results on our private database show that PMLNet achieves 8.77% higher accurate rate (AR), 9.15% higher correct rate (CR) and 9.93% higher sample-level segmentation accurate rate (ACC_{seg}) than SOTAs.

Keywords: Online Chinese signature segmentation and recognition · Multitask interaction · Progressive learning · Transformer · PMLNet

U. Porwal et al. (Eds.): ICFHR 2022, LNCS 13639, pp. 153–167, 2022.
https://doi.org/10.1007/978-3-031-21648-0_11

1   Introduction

Commonly practiced in banking commercials and judicial procedures, signature is inscribed with one's valuable personal information and is therefore regarded as personal identification of the highest standard. Unlike handwritten texts, Chinese signatures seldom have linguistic meanings, yet have other difficulties for segmentation because of their signature flourish, artistically writing styles, composite structures and word overlaps, etc. In recent years, booming of electronic devices has witnessed the rise of online Chinese signatures, whose temporal property provides far richer information than offline signatures [31,33,34]. To enable downstream tasks such as building up database of Chinese characters and verifying signatures based upon individual characters, online Chinese signature segmentation is important but unfortunately rarely studied.

Over-segmentation-based methods deriving from text recognition in early years incorporate character over-segmentation [28], character classification and linguistic context modeling [16]. Wu et al. [30] introduced CNN models to traditional over-segmentation-based methods and ranked first in ICDAR 2013 competition [32]. However, over-segmentation-based methods have difficulty solving the problem of overlapping and intertwining among Chinese characters, whose composite structure further deteriorates the model efficiency. By contrast, most of the text recognition methods in recent years are based on segmentation-free methods [24,29], which has gradually become the mainstream of handwritten text recognition. Specifically, Su et al. [24] proposed a segmentation-free method based on Hidden Markov Model (HMM) for offline recognition of realistic Chinese handwriting. Wang et al. [29] proposed a writer-aware CNN based on HMM to improve the recognition accuracy. However, segmentation-free methods do not provide segmentation results, while online Chinese signatures need to be recognized and segmented to obtain individual characters to support subsequent research such as text removal and editing [22]. Therefore, segmentation-free methods are not suitable for Chinese signature segmentation and recognition task.

In this paper, inspired by the ideology of multitask learning in numerous deep learning applications [6,12,13], we propose a progressive multitask learning network (PMLNet) that predicts the segmentation and recognition results of Chinese signature through an end-to-end network. Specifically, PMLNet consists of a dual channel stroke feature extraction block (DSF-Block), a stacked transformer encoder block (STE-Block) and a progressive multitask learning block (PML-Block). DSF-Block first extracts hand-crafted features, including the 8-Directional features [2], and feeds those features to the following dual channels to obtain stroke-wise abstract features representative of spatial meanings and semantic meanings. STE-Block, composed of several stacked transformers to model long range dependencies among different strokes, further exploits the overall structure and layout of the whole signature. PML-Block has two parallel branches to predict segmentation results and recognition results of Chinese signature simultaneously. Furthermore, we note that the two tasks share common knowledge to a great extent; therefore, a progressive learning strategy [25] is introduced to fine-tune segmentation and recognition results by fully leveraging their reciprocal relationship [3]. Since there is no public Chinese dataset, our experiments are conducted on a Chinese signature database collected on our own. The ablation study proves the effectiveness of each block. Compared with other existing methods, PALNet achieves state-of-the-art performance with accurate rate (AR) of 86.49%, correct rate (CR) of 87.52% and segmentation accurate rate (ACC_{seg}) of 96.93%.

2   Methodology

2.1   Overview

Figure 1 shows the architecture of our proposed PMLNet. PMLNet consists of a DSF-Blcok, a STE-Block and a PMI-Block. Inspired by [23], DSF-Blcok is designed to combine traditional knowledge [2] and deep learning model [9] to extract feature representation of each stroke. STE-Block takes all stroke-wise feature as inputs and outputs the sequential feature of the signature as a whole. PML-Block is designed to predict segmentation results and recognition results simultaneously. In contrast with most conventional methods that only perform a single task (i.e. segmentation or recognition), PMLNet is a multi-task network capable of improving signature segmentation performance by exploiting the benefits of signature recognition.

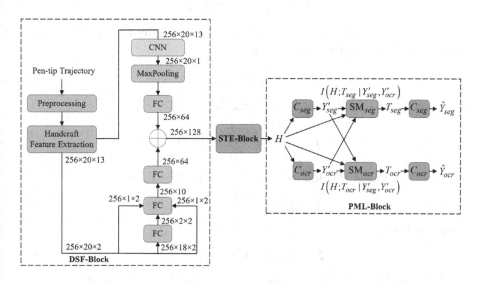

Fig. 1. The architecture of progressive multitask learning network (PMLNet).

To be more specific, DSF-Block consists of preprocessing, hand-craft feature extraction and a dual-channel sub-network. Preprocessing is used to normalize the original signature trajectory and to split the trajectory into strokes. Moreover, stroke-wise feature extraction is adopted to facilitate the subsequent feature extraction. Two channels of sub-network subsequently extract both spatial features and semantic features. STE-Block consists of several stacked transformer encoders to model long-range dependencies among different strokes and enhance their feature representations. PML-block consists of two branches where the recognition branch is added to boost the segmentation ability by the progressive learning strategy.

2.2   Dual Channel Stroke Feature Extraction Block (DSF-Block)

An electronic device can capture sequences of X, Y, S as a signature trajectory at a fixed time sampling rate. X, Y represent the signature's 2D coordinates in the canvas. S represents the pen-up/down info, where $S = 0$ denotes pen-down; $S = 1$ denotes pen movement on canvas; $S = 2$ denotes pen-up.

2.2.1   Preprocessing
Due to the fact that any Chinese signature is written horizontally with arbitrary length of words, we normalize X, Y by the span of Y while maintaining the height-width ratio.

$$\hat{Y} = \frac{Y - min(Y)}{max(Y) - min(Y)} \tag{1}$$

$$\hat{X} = \frac{X - min(X)}{max(Y) - min(Y)} \tag{2}$$

The angle of trajectory's tangent line θ can be easily obtained by taking derivatives of $\hat{X}$, $\hat{Y}$ with respect to time,

$$\theta = \arctan(\dot{\hat{Y}}/\dot{\hat{X}}) \tag{3}$$

where derivative is calculated by the formula below.

$$\dot{f}_i = \frac{\sum_{\epsilon=1}^{2} \epsilon \cdot (f_{i+\epsilon} - f_{i-\epsilon})}{2 \cdot \sum_{\epsilon=1}^{2} \epsilon^2} \tag{4}$$

We then segment the original signature T into M strokes,

$$T = \{T_i | i = 0, 1, ..., M - 1\} \tag{5}$$

where each stroke T_i consisting of N_i points

$$T_i = \{t_{ij} | j = 0, 1, ..., N_i - 1\} \tag{6}$$

must have its starting point t_{i0} satisfies either one of the two conditions:

(a) point t_{i0} is pen-down

$$S_{i0} = 0 \tag{7}$$

(b) point t_{i0} has a sudden swift change of angle

$$|\dot{\theta}_{i0}| > \frac{\pi}{8} \tag{8}$$

2.2.2 Hand-Craft Feature Extraction Consider the 8-uniformly-separated-direction set, which can be divided into two subsets $\mathcal{D}$ and $\mathcal{D}'$.

$$\mathcal{D} = \{\frac{\pi}{4}, \frac{3\pi}{4}, \frac{5\pi}{4}, \frac{7\pi}{4}\} \tag{9}$$

$$\mathcal{D}' = \{0, \frac{\pi}{2}, \pi, \frac{3\pi}{2}\} \tag{10}$$

The angle θ_{ij} of each point t_{ij} can be decomposed into its two nearby directions d_{ij}, d'_{ij} from $\mathcal{D}$ and $\mathcal{D}'$ respectively,

$$d_{ij} = \arg\min_{d \in \mathcal{D}} min\{\theta - d, 2\pi - (\theta - d)\} \tag{11}$$

$$d'_{ij} = \arg\min_{d' \in \mathcal{D}'} min\{\theta - d', 2\pi - (\theta - d')\} \tag{12}$$

with their corresponding values a_{ij}, a'_{ij} using the method 1 proposed in [2],

$$a_{ij} = |cos(\theta) - sin(\theta)| \tag{13}$$

$$a'_{ij} = \sqrt{2} \cdot min\{cos(\theta), sin(\theta)\} \tag{14}$$

Each point t_{ij} is then mapped to two directional feature vectors $v_d \in \mathbb{R}^4$, $v_{d'} \in \mathbb{R}^4$, a state feature vector $v_s \in \mathbb{R}^3$, and a coordinate feature vector $v_c \in \mathbb{R}^2$.

$$v_d(t_{ij}) = \sum_{d \in \mathcal{D}} a_{ij} \cdot \mathbf{1}(d_{ij} = d) \cdot e_d \tag{15}$$

$$v_{d'}(t_{ij}) = \sum_{d' \in \mathcal{D}'} a'_{ij} \cdot \mathbf{1}(d'_{ij} = d') \cdot e_{d'} \tag{16}$$

$$v_s(t_{ij}) = [\mathbf{1}(s_{ij} = 0), \mathbf{1}(s_{ij} = 1), \mathbf{1}(s_{ij} = 2)]^T \tag{17}$$

$$v_c(t_{ij}) = [\hat{x}_{ij}, \hat{y}_{ij}]^T \tag{18}$$

where $\mathbf{1}(\cdot)$ is indicator function; $\{e_d | d \in \mathcal{D}\}$ and $\{e_{d'} | d' \in \mathcal{D}'\}$ are standard bases vectors.

The directional feature vectors v_d and $v_{d'}$ are one-hot encodings of the decomposed directions; the state feature vector v_s is one-hot encoding of pen-up/down info S; the coordinate feature vector v_c is simply the normalized x-y coordinate.

Each stroke T_i can be therefore represented as a sequence of feature vectors. In order to keep subsequent network stable, we set the fixed stroke length as 20. Strokes with fewer than 20 points will be zero padded at the end whereas strokes with more than 20 points will be linearly resampled to 20 points. The resulting sequence of feature vectors now become $V_d(T_i) \in \mathbb{R}^{20 \times 4}$, $V_{d'}(T_i) \in \mathbb{R}^{20 \times 4}$, $V_s(T_i) \in \mathbb{R}^{20 \times 3}$ and $V_c(T_i) \in \mathbb{R}^{20 \times 2}$. Note that linear interpolation still maintains the trajectory profile, which is crucial in our signature segmentation and recognition tasks, in spite of losing its dynamic profile, which might be crucial in other tasks such as signature verification. These feature vectors can be concatenated as $V(T_i) \in \mathbb{R}^{20 \times 13}$.

$$V(T_i) = V_d(T_i) \oplus V_{d'}(T_i) \oplus V_s(T_i) \oplus V_c(T_i) \tag{19}$$

Almost all Chinese signature has characters less than four and any common Chinese character has no more than dozens of strokes, so we set the max number of strokes to be 256 and any signature with fewer strokes will be zero padded. Hence, a signature T can be represented as $V(T) \in \mathbb{R}^{256 \times 20 \times 13}$. Subsequent neural network will take these feature vectors as inputs accordingly.

2.2.3 Dual-Channel Subnetwork The subsequent dual-channel subnetwork consists of two different channels, one taking care of stroke's trajectory profile and the other responsible for its location.

The former channel, taking as inputs all types of features $V(T) \in \mathbb{R}^{256 \times 20 \times 13}$, has a 1-D convolution to capture the spatial information of strokes, followed by a max-pooling layer for dimension reduction and a fully connected layer to unite point-wise features into stroke-wise features.

The latter channel, fed with only coordinate features $V_c(T) \in \mathbb{R}^{256 \times 20 \times 2}$, has three fully connected layers in tandem. Note that during the first FC, the features of the starting and ending point are directly forwarded to subsequent network whereas the remaining middle points are learnt to represent the body position. Fully receptive of strokes' locations, this channel is able to capture the strokes' semantic representation. Specifically, inspired by the bottleneck structure in designing feature extraction block [11], we carefully set the inner dimension of FCs as 10 and the output dimension of FCs as 64.

At the end, both spatial and semantic stroke-wise features are concatenated for later block to extract features within strokes such as structure and layout of the whole signature.

2.3   Stacked Transformer Encoder Block (STE-Block)

Equipped with self-attention mechanism [5], transformer is capable of modeling long-range dependencies of sequence in parrellel, an edge over recurrent neural networks (RNN) and long short-term memory(LSTM) that require sequential computation. It has become the fundamental architecture in many research areas [10]. Therefore, instead of using conventional RNN and LSTM to extract features, we use several stacked transformer encoders to model long-range dependencies among different strokes and enhance their feature representation.

Figure 2 shows the architecture of two successive stacked transformer encoders. Transformer encoders basically consist of multi-head attention and point-wise, fully connected layers. A residual connection (denotes as "ADD" in Fig. 2) is employed around each of the two sub-layers, followed with layer normalization (denotes as "Norm" in Fig. 2). Positional embedding is added to the original input to make use of the order of the sequence. More details about transformer encoder can be found in [27].

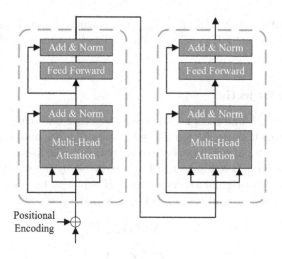

Fig. 2. Two successive stacked transformer encoders.

2.4 Progressive Multitask Interaction Block (PMI-Block)

Our previous efforts on signature segmentation heavily fail on Chinese characters with complicated left-and-right structures. The importance of learning what are those characters compels us to switch to multitask learning, where signature recognition functions as an auxiliary task that promotes signature segmentation. Further inspired by the discovery that Chinese signature's segmentation and recognition are two mutually dependent tasks, PMI-Block segments and recognizes the signature progressively and interactively.

The PMI-Block is designed similarly as in [25]. It first makes intermediate predictions on segmentation Y'_{seg} and recognition Y'_{ocr} independently, both of which may contain additional information that may facilitate the other task. The mixed information flow of the original representation H and intermediate predictions Y'_{seg}, Y'_{ocr} is regulated by stochastic maps SM_{seg} and SM_{ocr} respectively, to form better representations T_{seg} and T_{ocr}, from which the final segmentation and recognition results $\hat{Y}_{seg}$ and $\hat{Y}_{ocr}$ are learnt.

Recognition classifier C_{ocr}, consisting of a fully connected layer followed by a soft-max layer, predicts character to which each stroke belongs. Segmentation classifier C_{seg}, consisting of a fully connected layer followed by a conditional random field (CRF) layer, tags each stroke with beginning-inside-outside (BIO) format, where "B" denotes the beginning stroke of a character, "I" denotes stokes inside a character, and "O" denotes outliers outside any characters.

The module SM_{seg} balances the amount of information to represent the sufficient statistics T_{seg} from H and from Y'_{seg}, Y'_{ocr} by simulating two Gaussian distributions $r(t_{seg}|y'_{seg}, y'_{ocr})$ and $p(t_{seg}|h, y'_{seg}, y'_{ocr})$, as in variable information bottleneck (VIB) proposed by [1], where the former is the variational approximation of the latter. Their KL-divergence $KL(p(t_{seg}|h, y'_{seg}, y'_{ocr})\|r(t_{seg}|y'_{seg}, y'_{ocr}))$ sets the upper bound for the conditional mutual information $I(H; T_{seg}|Y'_{seg}, Y'_{ocr})$, where lesser divergence implies lesser injection of H into T_{seg}. The similar rationale and approach is applied to SM_{ocr}.

2.5 Training Objective

Segmentation branch and recognition branch together have three pairs of target losses. The intermediate prediction losses L'_{seg} and L'_{ocr} are minimized to improve the predictive powers of two independent early tasks. The former is the negative log likelihood of the labeled sequence Y_{seg} predicted by the conditional random field conditioned on original representation H whereas the latter is the cross entropy between the ground truth label Y_{ocr} and the prediction Y'_{ocr}.

$$L'_{seg} = NLL(Y_{seg}|H) \tag{20}$$

$$L'_{ocr} = CE(Y_{ocr}, Y'_{ocr}) \tag{21}$$

The KL divergences KL_{seg} and KL_{ocr} are minimized to improve the representations of T_{seg} and T_{ocr} as discussed in Sect. 2.4.

$$KL_{seg} = KL(p(t_{seg}|h, y'_{seg}, y'_{ocr})||r(t_{seg}|y'_{seg}, y'_{ocr})) \tag{22}$$

$$KL_{ocr} = KL(p(t_{ocr}|h, y'_{seg}, y'_{ocr})||r(t_{ocr}|y'_{seg}, y'_{ocr})) \tag{23}$$

The final prediction losses $\hat{L}_{seg}$ and $\hat{L}_{ocr}$ are minimized to improve the performance of our final predictions. The former is the negative log likelihood of the labeled sequence Y_{seg} predicted by the conditional random field conditioned on improved representation T_{seg} whereas the latter is the cross entropy between the ground truth label Y_{ocr} and the prediction $\hat{Y}_{ocr}$.

$$\hat{L}_{seg} = NLL(Y_{seg}|T_{seg}) \tag{24}$$

$$\hat{L}_{ocr} = CE(Y_{ocr}, \hat{Y}_{ocr}) \tag{25}$$

The total loss of our network is summed by all of the losses with regularization proposed by [18].

$$L_{total} = \sum_{L,c \in \mathcal{L}, C} \frac{L}{2c^2} + \ln(1 + c^2) \tag{26}$$

where $\mathcal{L}$ and C are respectively the set of all losses and the set of their corresponding regularization factors.

$$\mathcal{L} = \{L'_{seg}, L'_{ocr}, KL_{seg}, KL_{ocr}, \hat{L}_{seg}, \hat{L}_{ocr}\} \tag{27}$$

It is worth noting that the OCR result is stroke-wise. During training, however, only the last strokes of each character are trained for L'_{ocr} and $\hat{L}_{ocr}$ minimization; during testing, only the last strokes of each character predicted by the segmentation branch will be returned as the final OCR result. This way, the results of segmentation and recognition are assured to be consistent.

3   Experiments

3.1   Database

Since there is no open database for online Chinese signature segmentation and recognition, our experiments are based on a private database. The database is comprised of 98451 signatures comprising 1542 different Chinese characters signed by 1700 different users with stylus/finger on a variety of smartphones, pads and signing boards commonly seen in the market of China. Its source mainly origins from the government and banks, so most users sign their genuine signatures as they actually do. For security concerns, the database is not open to the public. For more details about the database, please refer to Table 1.

Table 1. Database details.

Items	Training set	Evaluation set	Test set	Total
#signatures	83758	13782	911	98451
#users	836	775	89	1700
#characters	1537	1335	178	1542
#avg chars/signatures	2.69	2.68	2.71	2.69

3.2   Evaluation Metrics

We use the same evaluation metrics i.e., accurate rate (AR) and correct rate (CR) [32] to evaluate the recognition performance:

$$AR = (N_t - D_e - S_e - I_e)/N_t \qquad (28)$$

$$CR = (N_t - D_e - S_e)/N_t \qquad (29)$$

where N_t is the total number of characters; D_e is the deletion errors; S_e is the substitution errors; I_e is the insertion errors of the testing samples.

We use sample-level segmentation accurate rate (ACC_{seg}) to evaluate segmentation performance:

$$ACC_{seg} = C/N \qquad (30)$$

where C is the number of completely correctly segmented signatures and N is the number of signatures in total.

3.3   Implementation Details

Implemented using PyTorch, PMLNet is trained with 100 epochs. We use AdamW [21] optimizer and set batch size to be 256. The initial learning rate is $1e-3$ and is multiplied by 0.1 after 80 epochs. There are 8 transformer encoders and 8 multi-heads.

Figure 3 illustrates the labeling of a Chinese signature in our database. For segmentation task, we adopt BIO format to denote stroke-wise belongings to characters. For recognition task, only last stroke of each character is labeled with its corresponding character index, while the remaining strokes are always labeled as 0.

We have also experimented with other methods. In particular, segmentation-free models deriving from text recognition such as Sun2016 [26], Peng2021 [23], DisGRU [19], VGG-DBLSTM [8] and CharNet-DBLSTM [4] are retrained on our database. Keysers2016 [14] is trained on CASIA database with 4022 individual characters. BiLSTM-CRF model [15] is fed with only x, y coordinates and pen-up/down information downsampled 20 Hz, and has two stacks of Bi-LSTMs with 128 dimensional hidden layers. BERT [7] has the similar set-up except that it has 8 stacks of transformers. K-Means [20] clusters each point solely based on x coordinates and timestamps after re-scaling the space both within and between strokes. DB-Net [17] is trained to visually detect individual characters without any post-process.

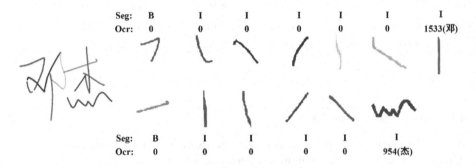

Fig. 3. Label illustration of Chinese signature

3.4   Qualitative Results

To demonstrate the effectiveness of our proposed PMLNet, we present qualitative results for better understanding. Actually, we observe diverse levels of difficulties in segmenting Chinese signatures. The simplest scenario (L1) is where all characters are unambiguously recognizable and completely horizontally separable. At the second level (L2), some characters have some portions overlapped or intertwined with each other and hence are not spatially separable, but each still maintains a compacted structure of its own. Things get interesting when moving up to the third level (L3), where some characters are loosely structured, awkwardly positioned relative to each other, or not semantically recognizable such that even the number of words are hard to foresee. The difficulty is ultimately raised to the forth level (L4) when the trajectory of nearby characters are connected with each other.

Figure 4 illustrates the segmentation results where PMLNet and other five models are compared at various difficulty levels. All models successfully segment "陈儒鑫" and pass the simplest difficulty level. However, BiLSTM-CRF [15] and Keysers2016 [14] are unable to correctly segment between "刘" and "梃" at the second difficulty level. Moving up to the third level, we observe that only PMLNet gives correct segmentation results for "伍成成" and "林夕" ; other models normally fail to split "成" and "成" as well as falsely split "林" into "木" and "木" . At the hardest level, almost all models struggle to segment at the inter-connection between characters "赵" and "鸣" as well as "王" and "州" . PMLNet passes the former case but fails the latter because of its stroke-wise feature design, where two characters are not allowed to share a common stroke as they are the case of "王州平" . On the contrary, models such as KMeans [20] and BiLSTM-CRF [15] based on point-wise segmentation might give a more flexible result. Nevertheless, we can undoubtedly conclude from these cases that teaching a model to recognize Chinese does help it to correctly segment Chinese signature that recognition-free models might otherwise find great difficulties.

Fig. 4. Signature segmentation comparison.

Table 2 shows the recognition results where PMLNet and other five models are compared at various difficulty levels. Substitution errors are marked as color red, insertion errors as color yellow, and deletion errors as color blue. All models except VGG-DBLSTM [8] succeed at the simplest level. However, only PLMNet, Sun2016 [26] and Peng2021 [23] are correct at the middle levels. Interestingly, all models fail at the hardest level and give recognition results of diverse kinds. Nevertheless, only PMLNet has the least substitution errors, insertion errors and deletion errors among all.

Table 2. Signature recognition comparison.

Difficulty level	Ground truth	PMLNet (ours)	Sun2016 [26]	Peng2021 [23]	DisGRU [19]	VGG-DBLSTM [8]	CharNet-DBLSTM [4]
L1	陈儒鑫	陈儒鑫	陈儒鑫	陈儒鑫	陈儒鑫	陈德鑫	陈儒鑫
L2	刘梃材	刘梃材	刘梃材	刘梃材	刘廷材	刘**财	刘梃材
L3	伍成成	伍成成	伍成成	伍成成	伍成成	伍成成	**成成
L3	林夕	林夕	林夕	林夕	林夕	林夕	林夕
L4	赵鹏	赵珊	赵高	赵亚	赵鹏	赵雄	赵谊
L4	王州平	王小平	王川宇	王川华	王川平	王**平	王**平

3.5 Quantitative Results

Table 2 shows the quantitative results. Some methods deriving from text recognition only have the recognition part, while some only have segmentation part. DB-Net [17], whose visual segmentation result is hard to be processed to online segmentation, does not show any result on the table. The nearly 10% boost of performance on all metrics demonstrates that our model outperforms other existing models to a great extent.

Table 3. Quantitative results.

Model	AR (%)	CR (%)	ACC_{seg} (%)
Sun2016 [26]	74.83	75.40	/
Peng2021 [23]	76.14	76.73	/
DisGRU [19]	75.66	77.03	/
VGG-DBLSTM [8]	70.46	71.38	/
CharNet-DBLSTM [4]	70.22	70.94	/
Keysers2016 [14]	77.72	78.37	87.00
BiLSTM-CRF [15]	/	/	89.39
BERT [7]	/	/	84.5
KMeans [20]	/	/	87.25
DB-NET [17]	/	/	/
PMLNet (ours)	**86.49**	**87.52**	**96.93**

3.6 Ablation Studies

Furthermore, we conduct ablation studies to demonstrate the strength of each block, as shown in Table 3. Observe that the underlying stroke-wise features are the most fundamental part in our network. Also, the transformer indeed outperforms LSTM by an obvious margin as expected. Lastly but not the least, The progressive and interactive learning strategy just makes the final result even better.

Table 4. Ablation study.

	AR (%)	CR (%)	ACC_{seg} (%)
w/o spatial representation in DSF	56.72	57.46	92.00
w/o semantic representation in DSF	81.64	82.89	96.27
w/o transformer (LSTM) in STE	83.28	83.91	96.48
w/o progressive in PML	82.91	83.83	95.39
w/o interaction in PML	85.27	86.43	95.61
w/o recognition branch in PML	/	/	95.59
PMLNet	**86.49**	**87.52**	**96.93**

4   Conclusion

In this paper, PMLNet is proposed for Chinese signature segmentation and recognition. The novel network is designed to extract stroke-wise features and subsequently learn the interplay among them. Furthermore, the network adopts a progressively and interactively learning strategy to build up better representations for signature segmentation and recognition. The results demonstrate that our PMLNet outperforms other 10 competitive models, achieving 8.77% higher AR, 9.15% higher CR and 9.93% higher ACC_{seg}. In the future, we expect to see more research and open database on Chinese signature segmentation and recogintion.

References

1. Alemi, A.A., Fischer, I., Dillon, J.V., Murphy, K.: Deep variational information bottleneck. arXiv preprint arXiv:1612.00410 (2016)
2. Bai, Z.L., Huo, Q.: A study on the use of 8-directional features for online handwritten Chinese character recognition. In: Eighth International Conference on Document Analysis and Recognition (ICDAR 2005), pp. 262–266. IEEE (2005)
3. Chen, K., et al.: Hybrid task cascade for instance segmentation. In: Proceedings of the IEEE/CVF Conference on Computer Vision and Pattern Recognition, pp. 4974–4983 (2019)
4. Chen, K., et al.: A compact CNN-DBLSTM based character model for online handwritten Chinese text recognition. In: 2017 14th IAPR International Conference on Document Analysis and Recognition (ICDAR), vol. 1, pp. 1068–1073. IEEE (2017)
5. Cheng, J., Dong, L., Lapata, M.: Long short-term memory-networks for machine reading. arXiv preprint arXiv:1601.06733 (2016)
6. Collobert, R., Weston, J.: A unified architecture for natural language processing: deep neural networks with multitask learning. In: Proceedings of the 25th International Conference on Machine Learning, pp. 160–167 (2008)
7. Devlin, J., Chang, M.W., Lee, K., Toutanova, K.: BERT: pre-training of deep bidirectional transformers for language understanding. arXiv preprint arXiv:1810.04805 (2018)

8. Ding, H., Chen, K., Hu, W., Cai, M., Huo, Q.: Building compact CNN-DBLSTM based character models for handwriting recognition and OCR by teacher-student learning. In: 2018 16th International Conference on Frontiers in Handwriting Recognition (ICFHR), pp. 139–144. IEEE (2018)
9. Graves, A., Fernández, S., Gomez, F., Schmidhuber, J.: Connectionist temporal classification: labelling unsegmented sequence data with recurrent neural networks. In: Proceedings of the 23rd International Conference on Machine Learning, pp. 369–376 (2006)
10. Han, K., et al.: A survey on vision transformer. IEEE Trans. Pattern Anal. Mach. Intell. (2022)
11. He, K., Zhang, X., Ren, S., Sun, J.: Deep residual learning for image recognition. In: Proceedings of the IEEE Conference on Computer Vision and Pattern Recognition, pp. 770–778 (2016)
12. Huang, J.T., Li, J., Yu, D., Deng, L., Gong, Y.: Cross-language knowledge transfer using multilingual deep neural network with shared hidden layers. In: 2013 IEEE International Conference on Acoustics, Speech and Signal Processing, pp. 7304–7308. IEEE (2013)
13. Ke, X., Zhang, X., Zhang, T.: GCBANET: a global context boundary-aware network for SAR ship instance segmentation. Remote Sens. **14**(9), 2165 (2022)
14. Keysers, D., Deselaers, T., Rowley, H.A., Wang, L.L., Carbune, V.: Multi-language online handwriting recognition. IEEE Trans. Pattern Anal. Mach. Intell. **39**(6), 1180–1194 (2016)
15. Lample, G., Ballesteros, M., Subramanian, S., Kawakami, K., Dyer, C.: Neural architectures for named entity recognition. In: Proceedings of the 2016 Conference of the North American Chapter of the Association for Computational Linguistics: Human Language Technologies (2016)
16. Li, N., Jin, L.: A Bayesian-based probabilistic model for unconstrained handwritten offline Chinese text line recognition. In: 2010 IEEE International Conference on Systems, Man and Cybernetics, pp. 3664–3668. IEEE (2010)
17. Liao, M., Wan, Z., Yao, C., Chen, K., Bai, X.: Real-time scene text detection with differentiable binarization. In: Proceedings of the AAAI Conference on Artificial Intelligence, vol. 34, pp. 11474–11481 (2020)
18. Liebel, L., Körner, M.: Auxiliary tasks in multi-task learning. CoRR abs/1805.06334 (2018). http://arxiv.org/abs/1805.06334
19. Liu, M., Xie, Z., Huang, Y., Jin, L., Zhou, W.: Distilling GRU with data augmentation for unconstrained handwritten text recognition. In: 2018 16th International Conference on Frontiers in Handwriting Recognition (ICFHR), pp. 56–61. IEEE (2018)
20. Lloyd, S.: Least squares quantization in PCM. IEEE Trans. Inf. Theory **28**(2), 129–137 (1982)
21. Loshchilov, I., Hutter, F.: Decoupled weight decay regularization. arXiv preprint arXiv:1711.05101 (2017)
22. Peng, D., et al.: Recognition of handwritten Chinese text by segmentation: a segment-annotation-free approach. IEEE Trans. Multimed. (2022)
23. Peng, D., et al.: Towards fast, accurate and compact online handwritten Chinese text recognition. In: Lladós, Josep, Lopresti, Daniel, Uchida, Seiichi (eds.) ICDAR 2021. LNCS, vol. 12823, pp. 157–171. Springer, Cham (2021). https://doi.org/10.1007/978-3-030-86334-0_11
24. Su, T.H., Zhang, T.W., Guan, D.J., Huang, H.J.: Off-line recognition of realistic Chinese handwriting using segmentation-free strategy. Pattern Recogn. **42**(1), 167–182 (2009)

25. Sun, K., Zhang, R., Mensah, S., Mao, Y., Liu, X.: Progressive multi-task learning with controlled information flow for joint entity and relation extraction. In: Proceedings of the AAAI Conference on Artificial Intelligence, vol. 35, pp. 13851–13859 (2021)
26. Sun, L., Su, T., Liu, C., Wang, R.: Deep LSTM networks for online Chinese handwriting recognition. In: 2016 15th International Conference on Frontiers in Handwriting Recognition (ICFHR), pp. 271–276. IEEE (2016)
27. Vaswani, A., et al.: Attention is all you need. Adv. Neural Inf. Process. Syst. 30 (2017)
28. Wang, Z.X., Wang, Q.F., Yin, F., Liu, C.L.: Weakly supervised learning for over-segmentation based handwritten Chinese text recognition. In: 2020 17th International Conference on Frontiers in Handwriting Recognition (ICFHR), pp. 157–162. IEEE (2020)
29. Wang, Z.R., Du, J., Wang, J.M.: Writer-aware CNN for parsimonious HMM-based offline handwritten Chinese text recognition. Pattern Recogn. 100, 107102 (2020)
30. Wu, Y.C., Yin, F., Liu, C.L.: Improving handwritten Chinese text recognition using neural network language models and convolutional neural network shape models. Pattern Recogn. 65, 251–264 (2017)
31. Xie, Z., Sun, Z., Jin, L., Ni, H., Lyons, T.: Learning spatial-semantic context with fully convolutional recurrent network for online handwritten Chinese text recognition. IEEE Trans. Pattern Anal. Mach. Intell. 40(8), 1903–1917 (2017)
32. Yin, F., Wang, Q.F., Zhang, X.Y., Liu, C.L.: ICDAR 2013 Chinese handwriting recognition competition. In: 2013 12th International Conference on Document Analysis and Recognition, pp. 1464–1470. IEEE (2013)
33. Zhang, X.Y., Yin, F., Zhang, Y.M., Liu, C.L., Bengio, Y.: Drawing and recognizing Chinese characters with recurrent neural network. IEEE Trans. Pattern Anal. Mach. Intell. 40(4), 849–862 (2017)
34. Zhou, X.D., Wang, D.H., Tian, F., Liu, C.L., Nakagawa, M.: Handwritten Chinese/Japanese text recognition using semi-Markov conditional random fields. IEEE Trans. Pattern Anal. Mach. Intell. 35(10), 2413–2426 (2013)

Symbol and Graphics Recognition

Musigraph: Optical Music Recognition Through Object Detection and Graph Neural Network

Arnau Baró[1]([✉])[iD], Pau Riba[2][iD], and Alicia Fornés[1][iD]

[1] Computer Vision Center and Computer Science Department,
Universitat Autònoma de Barcelona, Bellaterra, Catalonia
{abaro,afornes}@cvc.uab.cat
[2] Helsing AI, Berlin, Germany
pau.riba@helsing.ai

Abstract. During the last decades, the performance of optical music recognition has been increasingly improving. However, and despite the 2-dimensional nature of music notation (e.g. notes have rhythm and pitch), most works treat musical scores as a sequence of symbols in one dimension, which make their recognition still a challenge. Thus, in this work we explore the use of graph neural networks for musical score recognition. First, because graphs are suited for n-dimensional representations, and second, because the combination of graphs with deep learning has shown a great performance in similar applications. Our methodology consists of: First, we will detect each isolated/atomic symbols (those that can not be decomposed in more graphical primitives) and the primitives that form a musical symbol. Then, we will build the graph taking as root node the notehead and as leaves those primitives or symbols that modify the note's rhythm (stem, beam, flag) or pitch (flat, sharp, natural). Finally, the graph is translated into a human-readable character sequence for a final transcription and evaluation. Our method has been tested on more than five thousand measures, showing promising results.

Keywords: Object detection · Optical music recognition · Graph neural network

1 Introduction

The task of transcribing music scores into a human-readable format, such as MEI, MIDI, or MusicXML, is called Optical Music Recognition (OMR). During the last decades, the research community has shown an increasing interest, especially thanks to the emergence of deep learning. Several works have shown very good results using Convolutional Neural Networks (CNN), Recurrent Neural Networks (RNN), or both [3,7]. This task was tackled as a one-dimensional problem, similar to text recognition, in which text is seen as a sequence of characters.

© The Author(s), under exclusive license to Springer Nature Switzerland AG 2022
U. Porwal et al. (Eds.): ICFHR 2022, LNCS 13639, pp. 171–184, 2022.
https://doi.org/10.1007/978-3-031-21648-0_12

Although there are some similarities between music and text recognition, there are many differences. Music, in contrast to text, has to deal with a two-dimensional notation. It has to be read from left to right, but symbols have a duration (the rhythm) and a position in the staff, the pitch (melody). In addition, one can find groups of notes (e.g. chords) or symbols to provide musicality (e.g. ties, slurs, dynamic or tempo markings). Also, articulations or ornaments might appear below or above notes. Moreover, different notes (8th note, 16th note, etc) can be represented differently. For example, the stem might look up or down and these symbols could be found isolated with a flag, or together by a beam.

Since a music symbol can be drawn in different position and with different shapes, it is necessary to decompose those musical symbols which are composed of other symbols. In other words, not all the symbols are atomic, some of them can be divided into smaller symbols called primitives. For example, a quarter note is composed of a full notehead and a stem, an eighth note is composed of a full notehead, a stem and a beam or a flag (see Fig. 1).

Fig. 1. Graphic primitives. Isolated/atomic symbols shown in black color, whereas music primitives are shown in colors. (Color figure online)

In order to make the OMR system to understand which primitives form a symbol, a structure that relates them is needed. Music has a large variability of symbols, combining rhythm and melody, which is even increased in polyphonic scores. But, despite this variability, it shares a common set of primitives (e.g. stem, notehead, beam, flag, etc). Graphs are suitable for music description because of its flexibility and n-dimensional representation power. Music does not have a fixed representation (e.g. the measures do not have to be equal to each other), but each primitive will be related to others following the rules of music theory.

For the above reasons, in this paper we present *Musigraph*, a graph-based OMR system for music scores. Inspired by the success of Graph Neural Network models, we explore their adaptation to optical music recognition. Our method is based on Convolutional Neural Networks and Graph Neural Networks. As far as we know, this is the first OMR method based on GNNs. Besides, given the few labelled available music scores, we have created a labeled synthetic dataset to train such deep learning systems. We provide the groundtruth for music object detection and graph recognition.

The contributions are: 1) A model based on Convolutional Neural Networks and Graph Neural Networks and adapted to recognize music scores, 2) The creation of a new synthetic dataset, publicly available[1].

The rest of this paper is organized as follows. First, the problem statement is described in Sect. 2. Section 3 is devoted to describing the architecture. The new dataset is detailed in Sect. 4. Section 5 discusses the experimental results. Finally, conclusions and future work are drawn in Sect. 6.

2 Related Work

In this section we describe the most relevant approaches related to our work: Optical Music Recognition and Graph Neural Networks.

2.1 Optical Music Recognition (OMR)

During decades OMR used traditional techniques like Hidden Markov models [22,23] or methods for detection and recognition [10,24]. Besides, OMR works such as [2,8] proposed the use of rules to minimize errors and ambiguities in the recognition.

In the last decade the number of works in OMR has increased exponentially due to technological improvements such as the rise of deep learning. Hidden Markov models have been replaced by Recurrent Neural Networks (RNN), with a significant performance improvement. For example, in [3,7] the authors use long short-term memory recurrent neural networks (BLSTMs). In the same direction, in [1,30] the application of sequence-to-sequence models shows very good results, which are even improved by the incorporation of language models by Torras *et al.* [27]. Whereas some applications try to deal with all the OMR pipeline, others only focus on a part of it. For example, [19,28] uses Convolutional Neural Netwoks (CNN) for detecting symbols or primitives, whereas Calvo-Zaragoza *et al.* uses CNNs for staff line detection [6].

2.2 Graph Neural Network (GNN)

A graph is a symbolic data structure describing relations (*edges*) between a finite set of objects (*nodes*). Let L_V and L_E be a finite or infinite label sets for nodes and edges, respectively. A *graph* g is a $4 - $ tuple $g = (V, E, \mu, \nu)$ where, V is the finite set of *nodes*; $E \subseteq V \times V$ is the set of *edges*; $\mu \colon V \to L_V$ is the node labelling function; and, $\nu \colon E \to L_E$ is the edge labelling function.

GNNs introduced by Gori and Scarselli [12,26] were the first attempt to generalize neural networks to graphs. Then, Bruna *et al.* [5] proposed a new formulation based on the spectral graph theory. Later, the works of Henaff *et al.* [14], Defferrard *et al.* [9] and Kipf *et al.* [17] addressed these computational drawbacks. In its simplest form, a GNN layer $G_c(\cdot)$ is defined as:

---

[1] http://www.cvc.uab.es/people/abaro/ in Datasets.

$$h^{(k+1)} = G_c(h^{(k)}) = \rho \left(\sum_{B \in \mathcal{A}^{(k)}} B h^{(k)} \Theta_B^{(k)} \right) \tag{1}$$

where $h^{(k)}$ is the node hidden state at the k-th layer, ρ is a non-liniarity such as ReLU$(\cdot)$, $\mathcal{A}$ is a set of graph intrinsic linear operators that act locally on the graph signal and Θ are learnable parameters. In most cases, $\mathcal{A}$ it only contains the adjacency matrix [25].

More recently, Gilmer *et al.* [11] proposed the *Message Passing Neural Networks* (MPNNs) as a general supervised learning framework for graphs. MPNNs define its layers in terms of a message and update functions. Similarly, Hamilton *et al.* [13], proposed a graphSAGE layer which SAmples and aggreGatEs node features.

3   The Musigraph Model

As explained before, music recognition can be seen as a two-dimensional problem. The music score has to be read from left to right, but taking into account the rhythm and the pitch of each symbol. Thus, our approach will be composed of two steps. First, we will detect each atomic symbol or primitive by an object detector. Note that each of these primitives are just compounding parts of the symbols we consider on a music score. Second, we will relate the primitives detected by a graph neural network. Figure 2 shows the full pipeline.

We use the same set of music measures to train the object detector and the graph neural Network. During training, the graph neural network uses the bounding boxes obtained by the groundtruth. Concretely, to train the graph neural network we need to initialize the graph by specifying the edges (*i.e.* connecting nodes between each others), and later, the GNN will decide if the edge should exists or not. At test time, we use the bounding boxes provided by the object detector. So, once the object detector is trained, at test time, we obtain the candidate locations of each primitive and symbol, which will be postprocessed for improving the bounding-box prediction accuracy. Finally, these detections will be the input of the graph neural network during the test time.

3.1   Object Detector

The object detector will be used in the test phase. Note that during the training, we use the bounding boxes provided in the groundtruth to ensure that the graph neural network learns properly. Specifically, for object detection we have used the well-known library *Detectron 2* by Facebook AI Research [31]. This library provides the current state-of-the-art detection algorithms. Particularly, we have used the Faster R-CNN, which consists of a CNN with a final ROI pooling to extract a fixed-length feature vector from each region proposal. Finally, the fully connected layer is divided in two branches; a category softmax and a bounding box regression.

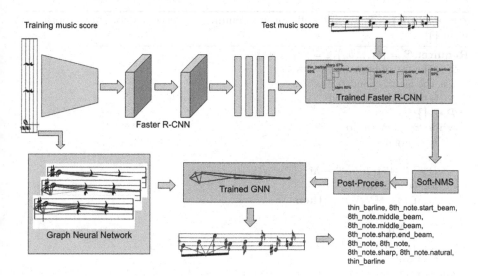

Fig. 2. Architecture Pipeline. In blue the training process, in red, testing. (Color figure online)

Once all the bounding boxes have been detected by the Faster R-CNN, we have used the soft non-maximum suppression (soft-NMS) algorithm [4]. The main difference with the classic NMS is that it removes a detection if the confidence is lower than a threshold. Moreover, in the soft-NMS when the degree of overlap reaches a threshold and the confidence is lower, it is not directly suppressed, instead, its confidence only is reduced. Algorithm 1 shows this difference (in blue the soft-NMS and in red the classic NMS).

Moreover, we have improved the beam and stem detections using morphological operations. Stems have been detected through the noteheads detections. By applying a vertical projection in the closest area of the notehead's bounding box, we find peaks that allow us to detect stems, a very thin and elongated primitive. For beam detection, we have eroded and dilated the image using as a kernel a vertical and horizontal line respectively; and then, we find those regions with an aspect ratio in which the width is three or more times the height.

Finally, having the noteheads' bounding boxes, we perform a horizontal projection to detect the staff lines, which allows us to recognize the pitch of each note.

3.2   Graph Neural Network

Given a graph g, our proposed graph neural network architecture $\varphi(\cdot)$ has N GNN layers and is defined using graphSAGE [13] layers, with mean aggregation function and ReLU activations.

In order to initialise the graph we have tried different input graphs by changing the heuristics about how to relate the nodes to each other. In all cases, the

Algorithm 1. Classic and soft non-maximum suppression algorithm. Reprinted from [4]

Require: $\mathcal{B} = \{b_1, .., b_n\}, \mathcal{S} = \{s_1, .., s_n\}$
  $\mathcal{B}$ is the list of bounding boxes
  $\mathcal{S}$ is the list of confidence scores
  $\mathcal{N}_t$ is the NMS threshold
  $\mathcal{D} \leftarrow \{\}$
  while $\mathcal{B} \neq empty$ **do**
    $m \leftarrow \arg\max \mathcal{S}$
    $\mathcal{M} \leftarrow b_m$
    $\mathcal{D} \leftarrow \mathcal{D} \cup \mathcal{M}; \mathcal{B} \leftarrow \mathcal{B} - \mathcal{M}$
    for b_i in $\mathcal{B}$ **do**
      if $iou(\mathcal{M}, b_i) \geq \mathcal{N}_t$ **then**
        $\mathcal{B} \leftarrow \mathcal{B} - b_i; \mathcal{S} \leftarrow \mathcal{S} - s_i$
      end if
      $s_i \leftarrow s_i f(iou(\mathcal{M}, b_i))$
    end for
  end while
  return $\mathcal{D}, \mathcal{S}$

---

nodes are the symbols or the primitives. During training, these correspond to the groundtruth, whereas in testing, these are the ones detected by the Faster R-CNN. The edges connect the nodes between them according to the heuristic selected, which is detailed next.

1. K-Nearest Neighbours (KNN) graph: The graph is initialized connecting each node with the K nodes closer to itself. We have used the Euclidean distance expressed in Eq. 2.

$$distance = \sqrt{(X_2 - X_1)^2 + (Y_2 - Y_1)^2} \tag{2}$$

2. Music Heuristic: This heuristic consists on analysing the music scores (frequent relations and distance) to avoid meaningless relationships. For each primitive we consider which other primitives are usually related to it, and how far they are. Thus, when the graph is constructed all this information is taken into account to avoid relations between primitives that are too far apart. Furthermore, we avoid meaningless primitive relations, in other words, we avoid the relation of atomic symbols (they are already a unit by themselves, for example a rest) with other elements.

Once the graph neural network is built, each node is embedded in a learned feature vector according to its detected class and bounding box using a fully-connected layer. Then, after each GNN layer (all but the last), the previous hidden state is concatenated. Afterwards, each node feature is normalized using the L_2-norm. Finally, an edge score is computed by performing an element-wise dot product between features of u (source node) and v (target node).

Our GNN has been trained using the binary cross entropy objective with the edge scores and its corresponding ground-truth. Our model has been optimized with the Adam [16] optimizer.

The end of the full pipeline finishes once the model is trained and the primitives are transformed into symbols through the detected graph for an evaluation at symbol level.

4 Dataset

As stated before, to recognize music scores one must deal with the two dimensional nature of music. It is read from left to right, but also from top to bottom. So, to properly train deep learning models, we need a considerable amount of labelled music scores. Indeed, the problem of current datasets is that they are not suited for training GNNs. First, because the few existing datasets labelled at graph level are very small (such as MUSCIMA++ [15]), and second, because most of them (e.g. DeepScores [28,29]) do not have an accurate labelling of symbols and primitives in the score and the relationship between them.

For this reason, we have created a new dataset in order to train our GNN model. Concretely, we have created a new printed-synthetic dataset using Lilypond[2]. This dataset contains more than 18K measures. We have splitted the dataset between train, validation and test in the following way: 60% Train, 20% Validations and 20% Test. Table 1 shows the number of primitves per each symbol that appears in the dataset. And Table 2 shows the different relations that the dataset has and how many we can find in the dataset. Finally, we provide the images and the groundtruth in the following formats:

1. At primitive level in COCO format [18] for object detection task.
2. At primitive level including the position in the score in case the symbol has pitch (e.g. notehead line 4).
3. At primitive level in XML format including the position in the score in case the symbol has pitch and the relations between primitives, so that one can build the music symbols (e.g. compound notes). Figure 3 shows an example of a music measure groundtruth. The arrows state the relationship, taking the notehead as the centre of the union of a musical symbol.

5 Experimental Validation

This section describes the experiments performed to validate our approach. On the one hand, we will show and discuss the object detection results, and on the other hand, we will show and analyze the graph neural network results.

---

[2] https://lilypond.org.

Fig. 3. Groundtruth of a music measure.

Table 1. Atomic symbols and number of appearances in the dataset.

Symbol	Number of appearances
f-clef	2462
thin_barline	37505
timeSig_2-2	2837
stem	98417
notehead-full	91338
8th_flag	12560
sharp	23962
natural	10957
quarter_rest	4790
half_rest	4484
flat	24272
16th_flag	20084
c-clef	2970
beam	16068
notehead-empty	7309
timeSig_cut	2595
8th_rest	3358
16th_rest	4436

5.1 Object Detection Results

To evaluate the object detection method we have used the mean average precision(mAP) [20,21]. The mAP is defined by

$$mAP = \frac{1}{n} \sum_{k=1}^{k=n} AP_k \qquad (3)$$

where n is the number of classes and AP_k is the average precision of class k.

Quantitative Results: Table 3 shows the comparison between the object detection method baseline and the post processing, including the non-maximum suppression and the beam and stem detection improvement. The first column

Table 2. Number of relations between primitives.

Source	Target	Number of relations
notehead-full	stem	91338
notehead-full	8th_flag	12560
notehead-full	sharp	22155
notehead-full	natural	10083
notehead-full	flat	22472
notehead-full	16th_flag	20110
notehead-full	beam	64218
notehead-empty	flat	1800
notehead-empty	stem	7079
notehead-empty	sharp	1807
notehead-empty	natural	1

shows the list of primitives and symbols. The second column shows the average precision using the Faster R-CNN from Detectron 2 per each primitive or symbol. And finally, the third column indicates the average precision after the non-maximum suppression and the beam and stem betterment. The last row of the table shows the mean average precision per each method. From the Table 3 we can observe that the Faster R-CNN performs very well in all symbols and primitives except those primitive in which one side is much larger than the other. For example, stems are very large and thin with lots of close primitives, whereas beams have the same particularity but horizontally. From the last column, we observe that the morphological operations have improved more than 15 points in the case of beams and almost 10 points in the case of stems. This betterment will help the creation of the graph because, when one node is not detected by the Faster R-CNN, the graph will not be able to recognize the complete symbol (*e. g.* if the network detects a notehead and stem and a beam is missed, the graph will recognize the note as a quarter-note instead of an 8th-note).

Qualitative Results: Figure 4 shows some qualitative results from the Faster R-CNN before applying the post processing. In Fig. 4a we can observe one of the main problems of the Faster R-CNN: the missdetections of beams (note that only one is detected in each group of 16th notes). In Fig. 4b we can observe that the third notehead in each 16th group of notes has missed the stem. Apart from this, in the second group of 16th notes multiples beams are detected and the non-maximum suppression is needed.

5.2   Graph Neural Network Results

Quantitative Results: Table 4 demonstrates the results of the Graph Neural Network using the different initialization options. The results are provided in terms of Music Error Rate (MER) at symbol level, which means that we first

Table 3. Detectron results using the Faster R-CNN. We show the results of the baseline and improving the stem and beam detection.

	AP %	AP % after processing
16th_flag	99.73	99.73
16th_rest	99.48	99.48
8th_flag	99.76	99.76
8th_rest	99.85	99.85
beam	84.48	100
c-clef	100	100
f-clef	100	100
flat	99.26	99.26
half_rest	99.56	99.56
natural	97.11	97.11
notehead-empty	99.93	99.93
notehead-full	99.48	99.48
quarter_rest	99.90	99.90
sharp	99.43	99.43
stem	79.71	88.85
thin_barline	99.17	99.17
timeSig_2-2	100	100
timeSig_cut	100	100
mAP	97.60	98.97

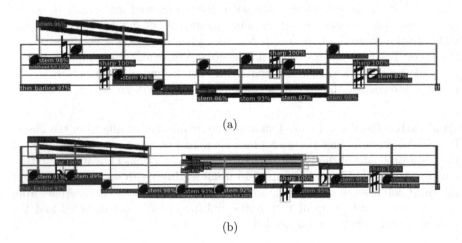

(a)

(b)

Fig. 4. Object detection results.

convert the different primitives into a symbol (*e.g.* flag+stem+notehead-full is converted into a 8th note). The Music Error Rate is defined as

$$MER = \frac{I + R + S}{T} \qquad (4)$$

where I, R and S are the number of insertions, deletions and substitutions to obtain the groundtruth sequence. T is the length of the groundtruth. Lower values mean better results. The first columns shows the different graph initialization techniques explained in Sect. 3. The second column shows the mean percentage of Music Error Rate and the standard deviation after five iterations.

From Table 4 we can observe that using the Music Heuristic as graph constructor, we obtain very good results (around 5% of MER). However, it has to be taken into account that KNN does not use any musical information, so having one error for every ten is quite understandable, even if it is almost twice the previous one.

Table 4. Results using our Graph Neural Network. We show which graph initialization has been used and the Music Error Rate. Note that the lower the better. The first number is the mean of the five executions, whereas the standard deviation is shown between parenthesis. All results use the Faster R-CNN and the post-processing to detect the nodes at test time.

Graph initialization	MER %
KNN	10.01 ($\pm$0.72)
Music Heuristic	**5.32** ($\pm$0.43)

Qualitative Results: Figures 5 and 6 show some qualitative results from all the pipeline. The first row of each image shows the groundtruth, the blue points are the centers of the nodes and the arrows, the edges. The second row shows the edges detected by the Graph Neural Network. The blue arrows mean that this edge was created by the graph initialiser, but the GNN has decided that the edge is not relevant. Oppositely, the red arrows are the relevant edges. Finally the third row is the graph translated into text as a final result. Figure 5a and 6a shows the results using the music heuristic in the graph constructor. Figure 5b and 6b shows results using the K-Nearest Neighbours. In Fig. 5a we can observe that the graph creator has not related the last flag with the stem, so the network is not able to predict this relation and finally it is detecting a quarter note. In contrast, in Fig. 5b, the KNN initialisation has many more relations, so this measure is perfectly recognized. In addition, in Fig. 6a the recognition is perfect but in Fig. 6b, the last notehead has an edge with the beam but the network has decided that edge is not important.

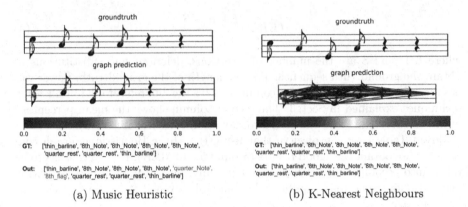

(a) Music Heuristic

(b) K-Nearest Neighbours

Fig. 5. Final qualitative results of a measure comparing the graph initialiser.

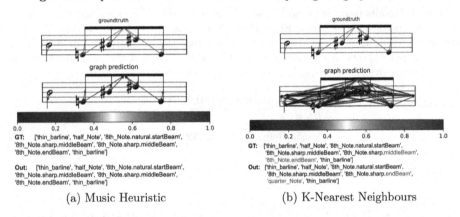

(a) Music Heuristic

(b) K-Nearest Neighbours

Fig. 6. Final qualitative results of another measure comparing the graph initialiser.

6   Conclusions and Future Work

In this work we have proposed Musigraph, an Object Detection and Graph Neural Network architecture for recognizing music scores. We have demonstrated that our model obtains very good results. With more than 98% of mAP in the object detection task and with almost 5% of Music Error Rate as a final result, our model proves to be a promising technique for recognizing music scores, or to help musicians when transcribing them.

As a future work, and given the two dimensionalities of music and the good results that the method has performed, the next step is to tackle the polyphonic music scores. The nature of graphs, which can easily create relations between primitives demonstrates that it is suitable for addressing music score recognition. Another point to investigate is the incorporation of language models to help correcting the misdetections of the network. Also, we believe that incorporating some music knowledge in the model could improve the results, for example the time measure is used to check if the number of beats in a bar unit is correct. Finally, we plan to adapt the method to the particularities of handwritten music scores.

Acknowledgment. This work has been partially supported by the Spanish projects RTI2018-095645-B-C21 and PID2021-126808OB-I00, and the CERCA Program/Generalitat de Catalunya. The FI fellowship AGAUR 2020 FI_B2 00149 (with the support of the Secretaria d'Universitats i Recerca of the Generalitat de Catalunya and the Fons Social Europeu). We gratefully acknowledge the support of NVIDIA Corporation with the donation of the Titan Xp GPU used for this research.

References

1. Baró, A., Badal, C., Fornés, A.: Handwritten historical music recognition by sequence-to-sequence with attention mechanism. In: International Conference on Frontiers in Handwriting Recognition, pp. 205–210 (2020)
2. Baró, A., Riba, P., Fornés, A.: Towards the recognition of compound music notes in handwritten music scores. In: International Conference on Frontiers in Handwriting Recognition, pp. 465–470 (2016)
3. Baró, A., Riba, P., Calvo-Zaragoza, J., Fornés, A.: From optical music recognition to handwritten music recognition: a baseline. Pattern Recogn. Lett. **123**, 1–8 (2019)
4. Bodla, N., Singh, B., Chellappa, R., Davis, L.S.: Soft-NMS – improving object detection with one line of code. In: Proceedings of the IEEE International Conference on Computer Vision, pp. 5561–5569 (2017)
5. Bruna, J., Zaremba, W., Szlam, A., LeCun, Y.: Spectral networks and locally connected networks on graphs. arXiv preprint arXiv:1312.6203 (2013)
6. Calvo-Zaragoza, J., Pertusa, A., Oncina, J.: Staff-line detection and removal using a convolutional neural network. Mach. Vis. Appl. **28**(5–6), 665–674 (2017)
7. Calvo-Zaragoza, J., Rizo, D.: End-to-end neural optical music recognition of monophonic scores. Appl. Sci. **8**, 1–23 (2018)
8. Coüasnon, B., Rétif, B.: Using a grammar for a reliable full score recognition system (1995)
9. Defferrard, M., Bresson, X., Vandergheynst, P.: Convolutional neural networks on graphs with fast localized spectral filtering. In: Advances in Neural Information Processing Systems, pp. 3844–3852 (2016)
10. Escalera, S., Fornés, A., Pujol, O., Radeva, P., Sánchez, G., Lladós, J.: Blurred shape model for binary and grey-level symbol recognition. Pattern Recogn. Lett. **30**(15), 1424–1433 (2009)
11. Gilmer, J., Schoenholz, S.S., Riley, P.F., Vinyals, O., Dahl, G.E.: Neural message passing for quantum chemistry. In: Proceedings of the International Conference on Machine Learning, pp. 1263–1272 (2017)
12. Gori, M., Monfardini, G., Scarselli, F.: A new model for learning in graph domains. In: IEEE International Joint Conference on Neural Networks, vol. 2, pp. 729–734 (2005)
13. Hamilton, W., Ying, Z., Leskovec, J.: Inductive representation learning on large graphs. In: Advances in Neural Information Processing Systems, vol. 30, pp. 1024–1034 (2017)
14. Henaff, M., Bruna, J., LeCun, Y.: Deep convolutional networks on graph-structured data. arXiv preprint arXiv:1506.05163 (2015)

15. Hajič, J., Pecina, P.: The MUSCIMA++ dataset for handwritten optical music recognition. In: Proceedings of the International Conference on Document Analysis and Recognition, pp. 39–46 (2017)
16. Kingma, D.P., Ba, J.: Adam: a method for stochastic optimization. In: Proceedings of the International Conference on Learning Representations (2015)
17. Kipf, T.N., Welling, M.: Semi-supervised classification with graph convolutional networks. In: Proceedings of the International Conference on Learning Representations (2017)
18. Lin, T.-Y., et al.: Microsoft COCO: common objects in context. In: Fleet, D., Pajdla, T., Schiele, B., Tuytelaars, T. (eds.) ECCV 2014. LNCS, vol. 8693, pp. 740–755. Springer, Cham (2014). https://doi.org/10.1007/978-3-319-10602-1_48
19. Pacha, A., et al.: Handwritten music object detection: open issues and baseline results. In: International Workshop on Document Analysis Systems, pp. 163–168 (2018)
20. Padilla, R., Netto, S.L., da Silva, E.A.B.: A survey on performance metrics for object-detection algorithms. In: International Conference on Systems, Signals and Image Processing, pp. 237–242 (2020)
21. Padilla, R., Passos, W.L., Dias, T.L.B., Netto, S.L., da Silva, E.A.B.: A comparative analysis of object detection metrics with a companion open-source toolkit. Electronics 10(3), 279 (2021)
22. Pugin, L.: Optical music recognition of early typographic prints using hidden Markov models. In: International Society for Music Information Retrieval, pp. 53–56 (2006)
23. Pugin, L., Burgoyne, J.A., Fujinaga, I.: Map adaptation to improve optical music recognition of early music documents using hidden Markov models. In: International Society for Music Information Retrieval, pp. 513–516 (2007)
24. Rebelo, A., Capela, G., Cardoso, J.S.: Optical recognition of music symbols: a comparative study. Int. J. Doc. Anal. Recogn. 13(1), 19–31 (2010)
25. Satorras, V.G., Estrach, J.B.: Few-shot learning with graph neural networks. In: International Conference on Learning Representations (2018)
26. Scarselli, F., Gori, M., Tsoi, A.C., Hagenbuchner, M., Monfardini, G.: The graph neural network model. IEEE Trans. Neural Netw. 20(1), 61–80 (2009)
27. Torras, P., Baró, A., Kang, L., Fornés, A.: On the integration of language models into sequence to sequence architectures for handwritten music recognition. In: International Society for Music Information Retrieval, pp. 690–696 (2021)
28. Tuggener, L., Elezi, I., Schmidhuber, J., Stadelmann, T.: Deep watershed detector for music object recognition. In: International Society for Music Information Retrieval, pp. 271–278 (2018)
29. Tuggener, L., Satyawan, Y.P., Pacha, A., Schmidhuber, J., Stadelmann, T.: The DeepScoresV2 dataset and benchmark for music object detection. In: International Conference on Pattern Recognition, pp. 9188–9195 (2021)
30. van der Wel, E., Ullrich, K.: Optical music recognition with convolutional sequence-to-sequence models. In: International Society for Music Information Retrieval, pp. 731–737 (2017)
31. Wu, Y., Kirillov, A., Massa, F., Lo, W.Y., Girshick, R.: Detectron2 (2019). https://github.com/facebookresearch/detectron2

Combining CNN and Transformer as Encoder to Improve End-to-End Handwritten Mathematical Expression Recognition Accuracy

Zhang Zhang[1]([⊠]) and Yibo Zhang[2]([⊠])

[1] Beijing National Day School, No. 66 Yuquan Road, Haidian District, Beijing, People's Republic of China
lillian_chandet@163.com
[2] Beijing Jiaotong University, No. 3 Shangyuancun, Haidian District, Beijing, People's Republic of China
98940299@bjtu.edu.cn

Abstract. The attention-based encoder-decoder (AED) models are increasingly used in handwritten mathematical expression recognition (HMER) tasks. Given the recent success of Transformer in computer vision and a variety of attempts to combine Transformer with convolutional neural network (CNN), in this paper, we study 3 ways of leveraging Transformer and CNN designs to improve AED-based HMER models: 1) *Tandem* way, which feeds CNN-extracted features to a Transformer encoder to capture global dependencies; 2) *Parallel* way, which adds a Transformer encoder branch taking raw image patches as input and concatenates its output with CNN's as final feature; 3) *Mixing* way, which replaces convolution layers of CNN's last stage with multi-head self-attention (MHSA). We compared these 3 methods on the CROHME benchmark. On CROHME 2016 and 2019, *Tandem* way attained the ExpRate of 54.85% and 58.56%, respectively; *Parallel* way attained the ExpRate of 55.63% and 57.39%; and *Mixing* way achieved the ExpRate of 53.93% and 55.64%. This result indicates that *Parallel* and *Tandem* ways perform better than *Mixing* way, and have little difference between each other.

Keywords: Handwritten mathematical expression recognition ·
Transformer · Encoder-decoder model · Convolutional neural network

1 Introduction

Handwritten Mathematical Expression Recognition (HMER) plays an important role in its wide applications including the intelligent office and intelligent education. It can greatly improve efficiency by simplifying the complex process of manually inputting formulas by converting handwritten text into lateX text that can be compiled by computers. On account of the high requirements for the recognition accuracy of symbol's size and position, and the need to identify the

U. Porwal et al. (Eds.): ICFHR 2022, LNCS 13639, pp. 185–197, 2022.
https://doi.org/10.1007/978-3-031-21648-0_13

spatial position relationship between symbols [1], the actual design of HMER is more difficult and complex than traditional Optical Character Recognition (OCR).

With the development of deep learning, the attention-based encoder-decoder (AED) model is gradually aroused in the solutions of HMER [5,31]. For example, in "Watch, attend and parse" (WAP) [31], it proposed an end-to-end neural network-based model, using CNN as the encoder and RNN as the decoder to output recognition results in lateX form, which achieves excellent recognition accuracy. With the success of WAP, the encoder-decoder model has gradually become the mainstream solution of HMER, and a lot of innovative attempts have been made to improve the performance of the AED framework from different perspectives.

[29] proposed a multi-scale attention mechanism based on WAP to recognize symbols of different scales and restore fine-grained details dropped by pooling layers. Paired adversarial learning was proposed by [25,26], which helped the model to focus on semantic-invariant features of patterns to solve the difficulties caused by writing-style variation and small sample size. [14] proposed pattern generation strategies to improve HMER. A transition probability matrix was introduced into decoder in [12] to capture long-range dependencies, and help the model to learn syntactical rules hopefully. [30] tried to replace string decoder of WAP with a tree-structured decoder to explicitly model mathematical expression trees. Scale augmentation and drop attention were used in [15], to improve HMER accuracy. An auxiliary symbol recognition loss is used in [22], which can provide weak supervision to improve HMER accuracy. In [19], a global context block was integrated into CNN encoder and Transformer [24] encoder was introduced to encode positional information. [27] replaced RNN based decoder with a CNN based one in a printed math expression recognition system to improve efficiency. [7] introduced multi-head attention and stacked decoder design into WAP design to improve accuracy. Graph neural network was combined with CNN and RNN encoder in [20] to model 2D spatial structure of math expression images better. A Bidirectionally Trained Transformer (BTTR) model was proposed in [32] to make the attention map more concise and improve accuracy. And [28] incorporated syntax information into AED framework to improve HMER system.

Recently, Transformer [24] has been on the rise with regards to its unique framework and has shown surprisingly good results in natural language processing (NLP)(e.g. [6]) and computer vision (CV) (e.g. [2,8]). Therefore, some work has been done to combine CNN and Transformer, trying to further improve the performance of the model by combining the advantages of both. Here are some examples: BoTNet [21], CoAtNet [4], and Conformer [9]. Based on ResNet [10], BoTNet replaces the 3×3 convolution kernel as Multi-Head Self-Attention (MHSA) [24] in the last stage of CNN bottleneck. This modification significantly enhanced Baselines' ability on tasks including instance segmentation and object detection on while also reducing the parameter size. CoAtNet improved the generalization and efficiency of the model by naturally unifying the convolutional and self-attention modules by relative attention, and vertically stacking the convolutional and attention layers. These hybrid models allow CoAtNet to largely compensate for Transformer's inductive bias and lack of training capabilities under larger resource constraints. Conformer is a combination of CNN and Transformer in the field of speech processing. Based on Transformer, this model changes the structure of the encoder part, which consists of 6 blocks. In each block, there is a Feed-Forward module before and after, and the middle layers are the MHSA module and Convolution module. Conformer combines the local feature extraction and global information processing capabilities of CNN and Transformer respectively.

In many attempts to combine CNN and Transformer, a better way to combine and maximize the advantages of the two models is expecting. In this paper, we used a Transformer decoder, and focus on the impact of structural changes of the encoder on model performance. The following three combinations are proposed: 1) *Tandem* approach: The CNN encoder (here we use DenseNet [13] to extract features from HME images) is used to extract local features, and the output is fed into Transformer encoder to capture global dependencies. So that it can effectively use local features, global dependencies, and interaction information at the same time, thus improving performance. 2) *Parallel* approach: use CNN and Transformer dual-branch encoders to process the raw image patches simultaneously, and finally concatenates the output of the two branches to get the final output of the encoder. 3) *Mixing* approach: replace the convolutional layer of CNN encoder in the last stage with MHSA. This captures global dependence while reducing the number of parameters, and give play to the advantages of Transformer.

In the CROHME2014, 2016, and 2019 data sets, the results of *Tandem* and *Parallel* have little difference, but both are better than the *Mixing* approach, of which the *Parallel* approach is slightly better.

2   Methodology

2.1   Baseline System

Our baseline system is based on BTTR [32] as shown in Fig. 1. In BTTR, a bidirectionally trained Transformer decoder is applied, which replaces the traditional RNN decoder in AED framework and improves the conciseness. In the CNN encoder part, DenseNet is applied to extract features. As a fully linked network, DenseNet adds direct connections to each layer to increase information flow, reduce parameters and optimize the gradient transfer process. After CNN feature is extracted by DenseNet, a 2D absolute positional encoding is added to help Transformer decoder better identifies the image feature vector locations.

In the decoder part, Transformer decoder is used, and it is characterized by bidirectional training, through which the input lateX text sequences are arranged in reverse order and marked with beginning and end identifiers (<SOS>,<EOS>) Since Transformer processes the input as a whole and the sequence order has no effect on the result, the model sends the forward and reverse sequence into the decoder for decoding training without affecting the accuracy of the result.

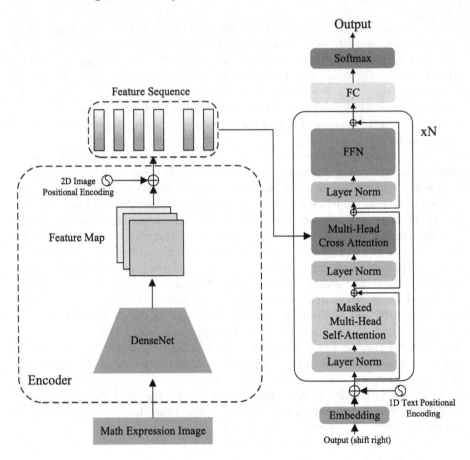

Fig. 1. Overview of our baseline model.

In this paper, we use the rough framework of The BTTR model as the baseline, and make two modifications: 1) Change the ReLU activation function in FFN (Feed Forward Net) to GELU [11]. GELU performs better on Transformer models, which not only improves learning speed and avoids the problem of gradient disappearance, but also makes GELU data more statistically significant compared to the sparsity of ReLU data. 2) Change the post-norm in decoder to the pre-norm, that is, put the layer-norm in front of the MHSA and FFN in decoder. Since post-norm is very sensitive to initial parameter values, it takes a lot of time to adjust learning parameters, and the structure of pre-LN omits warm-up initial training and further saves time. As a reference, the spirit of the combined methods in this paper is similar to [3] which is for recognizing textual documents. In [3], FCN encoder and Transformer decoder are used to solve the problem of image segmentation at the semantic level to improve the accuracy of document recognition.

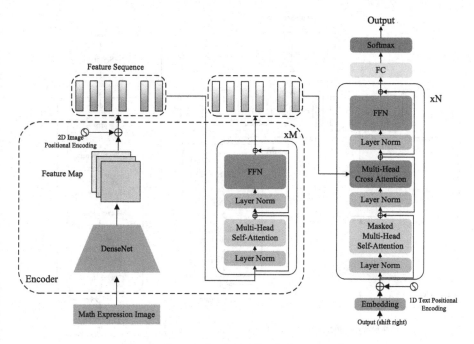

Fig. 2. Overview of our *Tandem* approach.

2.2 Tandem Approach

Inspired by the successful result of adding Transformer decoder in the BTTR model, we further change the structure of encoder part of the baseline model by adding Transformer encoder, and make it connect with the original CNN encoder as shown in Fig. 2. Since CNN and Transformer encoder are connected in tandem, we call it the *Tandem* approach. The output of CNN encoder is taken as the input of Transformer encoder. After passing through the Transformer encoder layers, the final output can capture more global contextual information, which we hope can improve accuracy.

2.3 Parallel Approach

Based on the thinking of the *Tandem* approach, we also try to combine CNN encoder and Transformer encoder in parallel. First, the features extracted from the original image data are operated by two encoders at the same time, and then the two feature sequences are concatenated to generate the final feature sequence by a linear mapping layer, as shown in Fig. 3. For the Transformer branch, we first patchify the input image into a sequence of 16 × 16 patches as ViT [8] does, and then feed this sequence into a linear embedding layer to generate the input sequences of Transformer encoder.

We hope that CNN and Transformer encoders can capture local and global contextual information independently to improve final recognition accuracy.

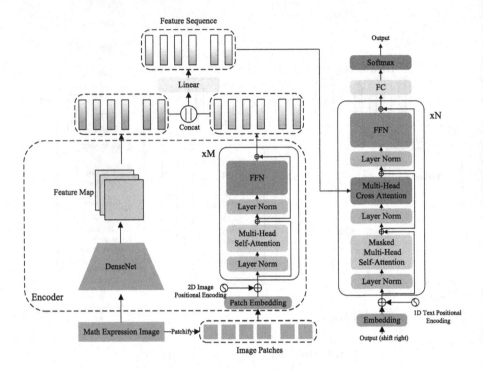

Fig. 3. Overview of our *Parallel* approach.

2.4 Mixing Approach

Inspired by BoTNet [21], we also try to replace the 3 × 3 convolution layers of DenseNet's last stage with MHSA layers. We call it *mixing* approach, and the resulting model as DenseSANet. The difference between DenseNet and DenseSANet is the addition of MHSA in the final stage.

The detailed architecture comparison has been listed in Table 1. Besides replacing conv 3×3 with MHSA, we also replace Batch Normalization (BN) with Instance Normalization (IN) [23] and ReLU activation function with GELU.

Compared with *Tandem* and *Parallel* approaches, we expect that *Mixing* approach can capture local and global context in a more parameter-efficient way.

3   Experimental Result

In this section we will describe experiments conducted according to the methodology in Sect. 3. Section 3.1 describes the setting of the comparative experiment; Sect. 3.2 comprehensively describes the comparison between the experimental results of the method in this paper and those in other work; In 4.3–4.5, the influence of some parameters of the three methods in this paper on the results is listed and analyzed to test the sensitivity of the method.

Table 1. Architecture comparison of DenseNet and DenseSANet. Assume input image size as 256×256

Layers	Output size	DenseNet	DenseSANet
Convolution	128×128	7×7 conv+BN+ReLU, stride 2	
Pooling	64×64	2×2 max pool, stride 2	
Dense (1)	64×64	$\begin{bmatrix} 1 \times 1 \text{ conv+BN+ReLU} \\ 3 \times 3 \text{ conv+BN+ReLU} \end{bmatrix} \times 16$	
Transition (1)	64×64	1×1 conv+BN+ReLU	
	32×32	2×2 average pool, stride 2	
Dense (2)	32×32	$\begin{bmatrix} 1 \times 1 \text{ conv+BN+ReLU} \\ 3 \times 3 \text{ conv+BN+ReLU} \end{bmatrix} \times 16$	
Transition (2)	16×16	1×1 conv+BN+ReLU	
	32×32	2×2 average pool, stride 2	
Dense (3)	16×16	$\begin{bmatrix} 1 \times 1 \text{ conv+BN+ReLU} \\ 3 \times 3 \text{ conv+BN+ReLU} \end{bmatrix} \times 16$	$\begin{bmatrix} 1 \times 1 \text{ conv+IN+GELU} \\ \text{MHSA} \\ 1 \times 1 \text{ conv+IN+GELU} \end{bmatrix} \times 16$
Post norm	16×16	BN	

3.1 Experimental Setup

In the experiment, we used the Competition on Recognition of Online Handwritten Mathematical Expressions (CROHME) datasets for training and testing. We used CROHME 2014 data set [17] as the training set, CROHME 2014 test sets as validation set, and CROHME2016 [18] and 2019 [16] as the test sets. CROHME 2014 dataset contains a total of 8,836 handwritten mathematical expressions, while CROHME 2014, 2016, and 2019 test sets contain 986, 147 and 1,199 expressions respectively.

Our implementation is based on open-sourced BTTR [32]. During training, no data augmentation is used, and we use dropout to avoid overfitting. For *Tandem* and *Parallel* approaches, the output dimension of MHSA in Transformer encoder layer is 256 and hidden dimension of FFN is 1024. For *Mixing* approaches, the output dimension of MHSA layers is 192. Beam search is used for decoding, and the beam size is set as 10.

The InkML data in the dataset contained handwritten brush tracing information and the ground truth in lateX format, which we converted into bitmap image format for subsequent training and testing. We apply size normalization on all images by zero-padding to make their widths and heights divisible by 16. After training and testing are completed, the recognized lateX sequence results and ground truth lateX texts in the test sets are converted to symLG format for evaluation through the official evaluation tool.

Experiment recognition rate (ExpRate) was used to evaluate and compare the results of each model; Also, the ExpRate with different recognition error tolerance of one, two, and three denoted as "≤ 1", "≤ 2", and "≤ 3" is included. With the official evaluation tool, we also evaluate the structure rates.

3.2   Overall Results

Table 2. Performance comparison of different systems on CROHME 2014/2016/2019 test sets. * denotes that data augmentation is used during training.

Dataset	System	ExpRate (%)	≤ 1 (%)	≤ 2 (%)	≤ 3 (%)	Structure rate (%)
CROHME 2014	WYGIWYS [5]	28.7	–	–	–	–
	WAP [31]	46.55	61.16	65.21	66.13	–
	DenseWAP [30]	43	57.8	61.9	–	–
	Pal-v2 [25]	48.88	64.5	69.78	73.83	–
	BTTR [32]	53.96	66.02	70.28	–	–
	SAN [28]	56.20	72.60	79.20	–	–
	Li et al.* [15]	56.59	69.07	75.25	78.60	–
	Ding et al.* [7]	58.70	–	–	–	–
	SAN* [28]	63.10	75.80	82.00	–	–
	Baseline	55.78	68.56	73.53	75.76	74.04
	Tandem	**57.20**	68.15	73.83	76.37	74.75
	Parallel	56.49	69.27	72.92	74.95	74.24
	Mixing	55.88	68.36	72.92	75.76	74.54
CROHME 2016	WAP [31]	44.55	57.1	61.55	62.34	–
	DenseWAP [30]	40.1	54.3	57.8	–	59.2
	PAL-v2 [25]	49.61	64.08	70.27	73.5	–
	BTTR [32]	52.31	63.9	68.61	–	69.4
	SAN [28]	53.60	69.60	76.80	–	–
	Li et al.* [15]	54.58	69.31	73.76	76.02	–
	Ding et al.* [7]	57.72	70.01	76.37	78.90	-
	SAN* [28]	61.50	73.30	81.40	–	–
	Baseline	53.8	65.5	72.66	75.37	72.4
	Tandem	54.85	67.86	74.50	76.86	74.06
	Parallel	**55.63**	68.82	74.24	76.86	74.76
	Mixing	52.93	66.29	71.88	74.85	73.01
CROHME 2019	DenseWAP [30]	41.7	55.5	59.3	–	60.7
	DenseWAP-TD [30]	51.4	66.1	69.1	–	69.8
	BTTR [32]	52.96	65.97	69.14	–	70.06
	SAN [28]	53.50	69.30	70.10	–	–
	Ding et al.* [7]	61.38	75.15	80.23	82.65	–
	SAN* [28]	62.10	74.50	81.00	–	–
	Baseline	54.14	68.92	74.19	76.36	74.27
	Tandem	**58.56**	72.43	75.69	77.86	76.52
	Parallel	57.39	71.18	75.86	77.94	76.02
	Mixing	55.64	68.76	73.02	75.36	73.52

In Table 2, we list the results of our proposed baseline, *Tandem*, *Parallel*, and *Mixing* approaches compared with other AED based systems, including WYGI-WYS, WAP, DenseWAP(-TD), PAL-v2, BTTR, SAN, Li et al. and Ding et al.. Our baseline performs slightly better than BTTR due to replacing ReLU with GELU and PostNorm with PreNorm in all Transformer layers. We vary the number of Transformer encoder layers in *Tandem* and *Parallel* models, and number of attention heads of MHSA layers in *Mixing* models, to get the bes-performing models of proposed approaches according to validation set (CROHME2014) ExpRates. Experimental results show that best-performing *Tandem* model contains 12 Transformer encoder layers, and the best *Parallel* model contains 8. The number of attention heads in each MHSA layer of best *Mixing* model is 12.

As can be seen, *Tandem* method performed best on CROHME 2014, which works as validation set. In CROHME2016 test set, *Parallel* method ExpRate achieves best results, reaching 55.63%. On the CROHME2019 test set, there was little difference in accuracy between the three methods, with *Tandem* way slightly higher. On the 3 test sets, the performances of the three methods are better than other AED based methods without using data augmentation, and *Tandem* method achieves the best overall.

3.3 Effects of Number of Transformer Encoder Layers to Tandem Approach

In this section, we investigate the effects of number of Transformer encoder layers on *Tandem* model performances. We tried 4/8/12/16 layers separately, each layer is a MHSA with 256 output neurons including a feed forward with 1024 neurons. As shown in Table 3, CROHME2014 and 2019 data sets, the number of 12 number layers reached the highest 57.20% and 58.56%, which increased by 1.93% and 0.92% respectively compared with that of 4 layers. Increasing the number of layers in *Tandem* to around 12 increases the model's performance, but too many layers burdened the model to be too complex so the efficiency falls back.

Table 3. Performance comparison of *Tandem* approach with different number Transformer layers on CROHME 2014/2016/2019 test sets

Dataset	# Layers	ExpRate (%)	≤ 1 (%)	≤ 2 (%)	≤ 3 (%)	Structure rate (%)
CROHME 2014	4	55.27	69.47	73.53	76.37	74.54
	8	55.38	67.14	72.82	75.66	74.14
	12	**57.20**	**68.15**	**73.83**	**76.37**	**74.75**
	16	56.49	68.15	72.72	75.56	74.04
CROHME 2016	4	**55.55**	**67.86**	**73.54**	**76.16**	**73.97**
	8	54.41	68.47	73.89	76.94	74.67
	12	54.85	67.86	74.50	76.86	74.06
	16	55.28	68.12	73.80	76.42	74.85
CROHME 2019	4	57.64	71.68	76.94	78.53	77.36
	8	57.48	71.18	75.52	77.11	75.86
	12	**58.56**	**72.43**	**75.69**	**77.86**	**76.52**
	16	57.98	71.93	76.69	78.45	76.86

Table 4. Performance comparison of *Parallel* approach with different number of Transformer encoder layers on CROHME 2014/2016/2019 test sets

Dataset	# layers	ExpRate (%)	$\leq$ 1 (%)	$\leq$ 2 (%)	$\leq$ 3 (%)	Structure rate (%)
CROHME 2014	4	56.39	69.07	74.65	77.08	75.46
	8	**56.49**	**69.27**	**72.92**	**74.95**	**74.24**
	12	55.07	67.95	72.41	74.95	73.23
	16	54.87	68.05	73.02	75.35	74.24
CROHME 2016	4	54.93	68.03	73.97	76.94	75.02
	8	**55.63**	**68.82**	**74.24**	**76.86**	**74.76**
	12	53.89	67.07	72.31	74.67	73.19
	16	54.93	67.42	73.01	75.72	73.89
CROHME 2019	4	56.98	71.26	75.61	77.44	76.19
	8	**57.39**	**71.18**	**75.86**	**77.94**	**76.02**
	12	55.56	68.92	73.35	76.02	74.27
	16	56.73	72.60	76.94	78.70	77.86

Table 5. Performance comparison of *Mixing* approach with different number of attention heads on CROHME 2014/2016/2019 test sets

Dataset	#heads	ExpRate (%)	$\leq$ 1 (%)	$\leq$ 2 (%)	$\leq$ 3 (%)	Structure rate (%)
CROHME 2014	1	54.67	67.14	72.11	74.04	72.52
	2	54.46	66.53	71.40	73.94	72.11
	4	54.77	67.75	72.62	75.15	73.33
	6	54.16	66.02	71.10	73.63	71.20
	8	55.38	68.36	74.04	76.17	74.44
	12	**55.88**	**68.36**	**72.92**	**75.76**	**74.54**
CROHME 2016	1	53.28	67.07	72.49	75.46	73.89
	2	54.85	68.03	74.24	76.68	74.93
	4	55.28	67.51	72.31	75.28	73.80
	6	53.54	66.64	72.40	75.20	73.01
	8	**55.46**	**69.17**	**74.59**	**77.55**	**75.37**
	12	52.93	66.29	71.88	74.85	73.01
CROHME 2019	1	56.81	70.68	75.52	77.19	76.19
	2	56.06	69.42	72.85	74.94	73.52
	4	54.05	68.09	72.18	74.44	73.27
	6	56.47	68.92	73.77	76.19	73.93
	8	**57.81**	**71.60**	**75.94**	**77.94**	**77.03**
	12	55.64	68.76	73.02	75.36	73.52

3.4 Effects of Number of Transformer Encoder Layers to Parallel Approach

Similarly, we also studied the effects of number of Transformer encoder layers to *Parallel* approach. We tried to set encoder layer number as 4/8/12/16, while MHSA output dimension and FFN hidden dimension in each layer are same with *Tandem* approach. As shown in Table 4, in all three test sets, the *parallel* model with 8 Transformer encoder layers achieved the highest ExpRate.

3.5   Effects of Number of Attention Heads to Mixing Approach

In *Mixing* approach, 3×3 conv layers of last DenseNet stage are replaced by MHSA. In order to study the effects of attention head number on final ExpRate, we vary it as $1/2/4/6/8/12$, while keep the final output dimension fixed as 192. As shown in Table 5, although best validation ExpRate was achieved with 8 attention heads, on CROHME 2016 and 2019 test sets, the 8-head *mixing* model was stable and reached the peak value of ExpRate (55.46% and 57.81%). Compared with single head model, 8-head model improved ExpRate by 2.18% and 1% on CROHME2016 and 2019, respectively.

4   Conclusion

In this paper, we studied 3 CNN Transformer combined encoder solutions to AED based HMER task. *Tandem* approach feeds CNN encoder output into several Transformer encoder layers to capture global context. *Parallel* approach concatenates CNN outputs with an independent ViT-like encoder to combine local and global context. *Mixing* approach replaces 3×3 conv layers in last stage of densenet with MHSA to capture global context in a more compact way. According to experimental evaluation on CROHME dataset, *Tandem* approach achieved best performance in terms of ExpRate when no data augmentation is used. However it is known that Transformer models require a large amount of training data to work well, we still need further investigation on large scale datasets, which we leave as our future work.

References

1. Anderson, R.H.: Syntax-directed recognition of hand-printed two-dimensional mathematics. In: Symposium on Interactive Systems for Experimental Applied Mathematics: Proceedings of the Association for Computing Machinery Inc., Symposium, pp. 436–459 (1967)
2. Carion, N., Massa, F., Synnaeve, G., Usunier, N., Kirillov, A., Zagoruyko, S.: End-to-end object detection with transformers. In: Vedaldi, A., Bischof, H., Brox, T., Frahm, J.-M. (eds.) ECCV 2020. LNCS, vol. 12346, pp. 213–229. Springer, Cham (2020). https://doi.org/10.1007/978-3-030-58452-8_13
3. Coquenet, D., Chatelain, C., Paquet, T.: DAN: a segmentation-free document attention network for handwritten document recognition. arXiv preprint arXiv:2203.12273 (2022)
4. Dai, Z., et al.: CoAtNet: marrying convolution and attention for all data sizes. Adv. Neural. Inf. Process. Syst. **34**, 3965–3977 (2021)
5. Deng, Y., Kanervisto, A., Ling, J., Rush, A.M.: Image-to-markup generation with coarse-to-fine attention. In: International Conference on Machine Learning, pp. 980–989. PMLR (2017)
6. Devlin, J., Chang, M.W., Lee, K., Toutanova, K.: BERT: pre-training of deep bidirectional transformers for language understanding. arXiv preprint arXiv:1810.04805 (2018)

7. Ding, H., Chen, K., Huo, Q.: An encoder-decoder approach to handwritten mathematical expression recognition with multi-head attention and stacked decoder. In: Lladós, J., Lopresti, D., Uchida, S. (eds.) ICDAR 2021. LNCS, vol. 12822, pp. 602–616. Springer, Cham (2021). https://doi.org/10.1007/978-3-030-86331-9_39

8. Dosovitskiy, A., et al.: An image is worth 16x16 words: transformers for image recognition at scale. arXiv preprint arXiv:2010.11929 (2020)

9. Gulati, A., et al.: Conformer: convolution-augmented transformer for speech recognition. arXiv preprint arXiv:2005.08100 (2020)

10. He, K., Zhang, X., Ren, S., Sun, J.: Deep residual learning for image recognition. In: Proceedings of the IEEE Conference on Computer Vision and Pattern Recognition, pp. 770–778 (2016)

11. Hendrycks, D., Gimpel, K.: Gaussian error linear units (GELUs). arXiv preprint arXiv:1606.08415 (2016)

12. Hong, Z., You, N., Tan, J., Bi, N.: Residual BiRNN based Seq2Seq model with transition probability matrix for online handwritten mathematical expression recognition. In: 2019 International Conference on Document Analysis and Recognition (ICDAR), pp. 635–640. IEEE (2019)

13. Huang, G., Liu, Z., Van Der Maaten, L., Weinberger, K.Q.: Densely connected convolutional networks. In: Proceedings of the IEEE Conference on Computer Vision and Pattern Recognition, pp. 4700–4708 (2017)

14. Le, A.D., Indurkhya, B., Nakagawa, M.: Pattern generation strategies for improving recognition of handwritten mathematical expressions. Pattern Recognit. Lett. **128**, 255–262 (2019)

15. Li, Z., Jin, L., Lai, S., Zhu, Y.: Improving attention-based handwritten mathematical expression recognition with scale augmentation and drop attention. In: 2020 17th International Conference on Frontiers in Handwriting Recognition (ICFHR), pp. 175–180. IEEE (2020)

16. Mahdavi, M., Zanibbi, R., Mouchere, H., Viard-Gaudin, C., Garain, U.: ICDAR 2019 CROHME+ TFD: competition on recognition of handwritten mathematical expressions and typeset formula detection. In: 2019 International Conference on Document Analysis and Recognition (ICDAR), pp. 1533–1538. IEEE (2019)

17. Mouchere, H., Viard-Gaudin, C., Zanibbi, R., Garain, U.: ICFHR 2014 competition on recognition of on-line handwritten mathematical expressions (CROHME 2014). In: 2014 14th International Conference on Frontiers in Handwriting Recognition, pp. 791–796. IEEE (2014)

18. Mouchère, H., Viard-Gaudin, C., Zanibbi, R., Garain, U.: ICFHR2016 CROHME: competition on recognition of online handwritten mathematical expressions. In: 2016 15th International Conference on Frontiers in Handwriting Recognition (ICFHR), pp. 607–612. IEEE (2016)

19. Pang, N., Yang, C., Zhu, X., Li, J., Yin, X.C.: Global context-based network with transformer for image2latex. In: 2020 25th International Conference on Pattern Recognition (ICPR), pp. 4650–4656. IEEE (2021)

20. Peng, S., Gao, L., Yuan, K., Tang, Z.: Image to LaTeX with graph neural network for mathematical formula recognition. In: Lladós, J., Lopresti, D., Uchida, S. (eds.) ICDAR 2021. LNCS, vol. 12822, pp. 648–663. Springer, Cham (2021). https://doi.org/10.1007/978-3-030-86331-9_42

21. Srinivas, A., Lin, T.Y., Parmar, N., Shlens, J., Abbeel, P., Vaswani, A.: Bottleneck transformers for visual recognition. In: Proceedings of the IEEE/CVF Conference on Computer Vision and Pattern Recognition, pp. 16519–16529 (2021)

22. Truong, T.N., Nguyen, C.T., Phan, K.M., Nakagawa, M.: Improvement of end-to-end offline handwritten mathematical expression recognition by weakly supervised learning. In: 2020 17th International Conference on Frontiers in Handwriting Recognition (ICFHR), pp. 181–186. IEEE (2020)
23. Ulyanov, D., Vedaldi, A., Lempitsky, V.: Instance normalization: the missing ingredient for fast stylization. arXiv preprint arXiv:1607.08022 (2016)
24. Vaswani, A., et al.: Attention is all you need. Adv. Neural Inf. Process. Syst. **30** (2017)
25. Wu, J.-W., Yin, F., Zhang, Y.-M., Zhang, X.-Y., Liu, C.-L.: Handwritten mathematical expression recognition via paired adversarial learning. Int. J. Comput. Vis. **128**(10), 2386–2401 (2020). https://doi.org/10.1007/s11263-020-01291-5
26. Wu, J.-W., Yin, F., Zhang, Y.-M., Zhang, X.-Y., Liu, C.-L.: Image-to-markup generation via paired adversarial learning. In: Berlingerio, M., Bonchi, F., Gärtner, T., Hurley, N., Ifrim, G. (eds.) ECML PKDD 2018. LNCS (LNAI), vol. 11051, pp. 18–34. Springer, Cham (2019). https://doi.org/10.1007/978-3-030-10925-7_2
27. Yan, Z., Zhang, X., Gao, L., Yuan, K., Tang, Z.: ConvMath: a convolutional sequence network for mathematical expression recognition. In: 2020 25th International Conference on Pattern Recognition (ICPR), pp. 4566–4572. IEEE (2021)
28. Yuan, Y., et al.: Syntax-aware network for handwritten mathematical expression recognition. In: Proceedings of the IEEE/CVF Conference on Computer Vision and Pattern Recognition, pp. 4553–4562 (2022)
29. Zhang, J., Du, J., Dai, L.: Multi-scale attention with dense encoder for handwritten mathematical expression recognition. In: 2018 24th International Conference on Pattern Recognition (ICPR), pp. 2245–2250. IEEE (2018)
30. Zhang, J., Du, J., Yang, Y., Song, Y.Z., Wei, S., Dai, L.: A tree-structured decoder for image-to-markup generation. In: International Conference on Machine Learning, pp. 11076–11085. PMLR (2020)
31. Zhang, J., et al.: Watch, attend and parse: an end-to-end neural network based approach to handwritten mathematical expression recognition. Pattern Recogn. **71**, 196–206 (2017)
32. Zhao, W., Gao, L., Yan, Z., Peng, S., Du, L., Zhang, Z.: Handwritten mathematical expression recognition with bidirectionally trained transformer. In: Lladós, J., Lopresti, D., Uchida, S. (eds.) ICDAR 2021. LNCS, vol. 12822, pp. 570–584. Springer, Cham (2021). https://doi.org/10.1007/978-3-030-86331-9_37

A Vision Transformer Based Scene Text Recognizer with Multi-grained Encoding and Decoding

Zhi Qiao⑩, Zhilong Ji(✉)⑩, Ye Yuan⑩, and Jinfeng Bai⑩

Tomorrow Advancing Life, Beijing, China
{qiaozhi1,jizhilong,yuanye8,baijinfeng1}@tal.com

Abstract. Recently, vision Transformer (ViT) has attracted more and more attention, many works introduce the ViT into concrete vision tasks and achieve impressive performance. However, there are only a few works focused on the applications of the ViT for scene text recognition. This paper takes a further step and proposes a strong scene text recognizer with a fully ViT-based architecture. Specifically, we introduce multi-grained features into both the encoder and decoder. For the encoder, we adopt a two-stage ViT with different grained patches, where the first stage extracts extent visual features with 2D fine-grained patches and the second stage aims at the sequence of contextual features with 1D coarse-grained patches. The decoder integrates Connectionist Temporal Classification (CTC)-based and attention-based decoding, where the two decoding schemes introduce different grained features into the decoder and benefit from each other with a deep interaction. To improve the extraction of fine-grained features, we additionally explore self-supervised learning for text recognition with masked autoencoders. Furthermore, a focusing mechanism is proposed to let the model target the pixel reconstruction of the text area. Our proposed method achieves state-of-the-art or comparable accuracies on benchmarks of scene text recognition with a faster inference speed and nearly 50% reduction of parameters compared with other recent works.

Keywords: Scene text recognition · Vision transformer · Self-supervised learning

1 Introduction

Scene text recognition has been studied for many years, the mainstream methods always use CNNs as the main feature extractor. Despite the excellent performance, CNNs have some major drawbacks. First, the inductive bias of it is proved to harm the performance of the model with the increasing amount of training data [8]. Second, the CNNs suffer from the limited receptive field due to the locality, to relieve this problem, some works [22,38,39,55] adopt RNN or Transformer [43] units to further extract the contextual features. Since CNN is

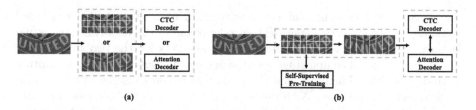

(a)   (b)

Fig. 1. Illustrations of ViT-based scene text recognition pipelines. (a) Existing pipelines with single-grained encoder and decoder (b) Our proposed method with multi-grained encoder and decoder, where the CTC-based and attention-based decoding are involved into a single decoder with deep interactions. The self-supervised pre-training is applied to the fine-grained encoder additionally in our method.

still the main part of the model, the problem of modeling long-range context dependency still exists, which is essential for text recognition.

Nowadays, Transformer [43] shows the effectiveness of long sequence processing in natural language processing (NLP). Thanks to the Vision Transformer [9] (ViT), the Transformer can also replace CNNs as a feature extractor for vision tasks. Specifically, ViT splits the images into several patches and uses the attention mechanism to model dependencies between each patch. We explain that the ViT also adapts to text recognition for three major reasons: 1) text recognition needs long-range dependencies and context information. VisionLAN [48] shows that the model can make use of linguistic information among visual levels, which further indicates the importance of the long-range dependencies for text recognition. 2) The need for large-scale training data for ViT is not a major problem for text recognition, since the text images can be synthesized easily. In other words, for text recognition, there can be enough data to boost the performance of ViT constantly. 3) Text recognition is a cross-modal task between visual images and linguistic texts, Transformer can be a union framework to bridge the gap between visual images and language.

Recently, there have been some works introducing the ViT into text recognition. ViTSTR [2] adopts a simple framework, which predicts all characters in parallel with the patches of text images as inputs. TRIG [41] and TrOCR [23] replace the CNN-based encoder with ViT, and adopt an autoregressive decoder to transcribe the target text. To improve the efficiency of ViT, TRIG uses 1D rectangle patches instead of the original 2D square patches, but an additional CNN-based rectification module is needed, which will introduce computing resources accordingly. As shown in Fig. 1(a), existing ViT-based methods usually capture the features with single-size patches and transcribe the target texts with a single decoding technique. In this paper, we improve the ViT-based text recognizer with multi-grained encoding and decoding. As shown in Fig. 1(b), a two-stage ViT encoder and a joint decoder are proposed:

Two-Stage ViT Encoder. Fine-grained patches help the model distinguish each character in the image but also affect the computational efficiency due to the attention operations with the number increment of the patches. Inspired by the scheme of visual-contextual feature extraction adopted by most CNNs-based methods, we propose a two-stage ViT encoder to target the visual and contextual features respectively, where the major difference between the two stages is the size of the patches. The first stage uses the fine-grained square patches to capture more detailed visual features, and the second stage considers the coarse-grained vertical rectangle patches. With the larger and fewer patches, the model ignores some details and focuses on the 1D contextual features. Since the self-attention in ViT is conducted between each patch, the efficiency of the model will also increase with the number of patches decreasing in the second stage.

Joint Decoder. Connectionist Temporal Classification (CTC) and attention mechanism are two major decoding techniques for text recognition with a local and global perspective respectively. Specifically, CTC depends on fine-grained local predictions and transcribes the target text with split-and-merge post-processing. Each patch corresponds to a specific character prediction or the additional blank token introduced by CTC. Instead, the attention mechanism predicts each character with a corresponding global attention weight, which aggregates some of the related patches to achieve the target character. Note that, the attention mechanism adopts a coarse-grained prediction perspective since each patch is no longer a single prediction unit. In this paper, we integrate the two decoding techniques into a single joint decoder. Different from the union training adopted by [17,25], our proposed joint decoder takes a deep interaction of the two-grained features belonging to the corresponding decoding techniques and lets them benefit from each other.

To improve the extraction of fine-grained features, we explore the pre-training with pixels reconstruction adopted in Masked Autoencoders [13] (MAE). For text recognition, the MAE can help the model be sensitive to the details of characters and improve the model in modeling long dependencies. Furthermore, we improve the original MAE for text recognition with a proposed focusing mechanism. The motivation is that the background in text recognition seems meaningless, so we want the model to focus on the pixels inner or near texts. To get the location of characters in a self-supervised manner, we propose to adopt Maximally Stable Extremal Regions [30,54] (MSER) to obtain the character candidate regions. The MSERs are converted into a segment map, which further works as a mask to let the model reconstruct the target regions.

In summary, the contributions of this paper are as follows:

1. We improve ViT for text recognition with a multi-grained encoder and decoder. Specifically, a two-stage ViT is adopted with different patchy strategies to capture both fine-grained visual features and coarse-grained sequence contextual features. A joint decoder integrated CTC-based and attention-based decoding is proposed, where a deep interaction is conducted between two-grained features belonging to two decoding processes respectively.

2. Pixel reconstruction-based self-supervised learning is explored for text recognition. To further help the model focus on the text regions, a focusing mechanism is proposed based on MSER.

3. As a result, we propose a fully ViT-based text recognizer. Compared with other works, our method achieves state-of-the-art or comparable results with a faster inference speed and nearly 50% reduction of parameters.

2 Related Works

2.1 Scene Text Recognition

Scene text recognition is a hot research topic with many challenges, recent methods usually adopt a deep-learning based framework. According to the decoding techniques, existing methods can be roughly separated into three major categories: CTC-based, attention-based, and segmentation-based.

CTC-based methods [15,17,38,40] usually extract the features with a combination of CNN and RNN, then transcribe the target text with CTC. The CTC-based methods can achieve fast inference speed but also suffer from some misalignment problems.

Inspired by machine translation, recent works [6,7,22,25,28,39,46,56] try to decode the text with attention mechanism, where the decoder focuses different image regions with the predicted attention weights and transcribes the corresponding characters. Among them, the arbitrary text recognition [7,22,29,39, 53,57], strong encoder [25,52,60] and semantic information [10,34,35,48,55] has been focused in recent years.

CA-FCN [24] first adopts FCN [27] to convert the text recognition to a semantic segmentation task. Since the segmentation results lack the order information of the text sequence, TextScanner [44] proposes to improve the segmentation-based decoding with additional order and localization maps. S-GTR [16] captures the 2D spatial context of visual semantics with GCN, and the proposed textual reasoning module can also benefit the CTC- and attention-based methods. Apart from these three major decoding techniques, some works try other prediction schemes for text recognition, such as the classification [18], aggregation cross-entropy [50] and visual matching [59].

2.2 Vision Transformer

Vision Transformer [9] (ViT) proposes a fully Transformer-based network to solve the image classification task, which splits the image into several patches and adopts self-attention to capture the dependencies between each patch. With the impressive performance of ViT, many works [26,42,47,49] improve the ViT in various aspects. For text recognition, some works also take a step to explore the effectiveness of ViT. ViTSTR [2] adopts ViT to directly predict all characters in order and conducts extensive analysis to verify the efficiency. TrOCR [23] replaces the CNNs with ViT for the encoder and adopts a Transformer-based decoder to transcribe the texts. Although significant accuracies are achieved in document and handwriting recognition, the efficiency is limited due to the number of patches and autoregressive decoding. TRIG [41] uses vertical rectangle patches based on the characteristics of text recognition, but the additional rectification module and the autoregressive GRU-based decoder burden the efficiency of the model. Different from previous methods, we propose a two-stage ViT encoder and a joint decoder to fully capture both fine-grained and coarse-grained features during both encoding and decoding.

2.3 Self-supervised Learning

Self-supervised learning has attracted significant interest, pretext tasks [11,33, 61] and contrastive learning [5,14] based methods have been dominant recently with significant performance. For text recognition, SeqCLR [1] extends the Sim-CLR [5] with some modifications including the instances construction and data augmentations. [3] tries the MoCo [14] and RotNet [11] for text recognition. Inspired by the Masked Language Modeling in NLP, some works try masked image modeling for pre-training. MAE [13] and SimMIM [51] mask several patches in the image and then reconstruct the missing pixels. Such a simple task brings an impressive performance gain for vision tasks. In this paper, we first explore masked image modeling for text recognition and improve the MAE with a focusing mechanism for text recognition.

3 Method

3.1 Pipeline

As shown in Fig. 2, our proposed method adopts a fully Transformer based architecture with an encoder-decoder framework. The encoder aims to extract abundant visual and contextual features with a two-stages structure and the decoder is to predict the target texts by integrating the attention-based and CTC-based decoding. To improve the extraction of fine-grained visual features, we borrow the idea from MAE and extend it with a focusing mechanism. The details of the proposed techniques will be described as follows.

3.2 Two-Stage Encoder

The two-stage ViT encoder is proposed to capture both fine-grained visual features and coarse-grained contextual information, which are both essential for text recognition. As shown in Fig. 2, the first stage takes 4×4 fine-grained square patches as inputs after a patch embedding layer. The patch size of 4 is from the observation that CNNs adopted by most text recognition hold a $4\times$ downsampling rate. Owing to the fine-grained patches, more distinguishable visual features can be extracted based on several Transformer units. Denoted the inputs as x_i, where i indicates the outputs of i-th of N units, the forward within a Transformer unit are as follows:

$$x'_{i+1} = Dropout(MHSA(LN(x_i))) + x_i,$$
$$x_{i+1} = Dropout(FFN(LN(x'_{i+1}))) + x'_{i+1}. \tag{1}$$

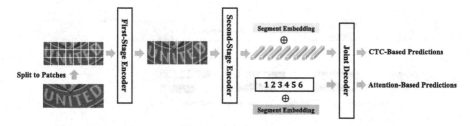

Fig. 2. The pipeline of our proposed method includes a two-stage ViT encoder and a joint decoder. The two-stage ViT encoder targets different grained features with different sizes of patches. The joint decoder involves both CTC-based and attention-based decoding with a deep interaction of corresponding grained features.

where MHSA is short for multi-head self-attention, LN represents the layer normalization, and the FFN denotes the feed-forward network. The operation of the MHSA and FFN is the same as the original Transformer.

The second stage of ViT focuses on the sequence contextual features based on the outputs x_N of the first stage. First, we recover the feature map with the same size as the input image from x_N. Next, we re-split the feature map into vertical patches with the size of $H \times 4$, where H is the height of the input image. With such a size of patches, the inputs of the second stage are converted to a 1D sequence, which focuses on sequence modeling. Same as the first stage, the patches will be input to another M Transformer units, and the outputs of the final units are regarded as the outputs of the encoder, denoted as x_{N+M}. Compared to the square patches, the strategy of vertical patches reduces computational complexity from $O((\lceil \frac{H}{P} \times \frac{W}{P} \rceil)^2 D)$ to $O((\lceil \frac{W}{P} \rceil)^2 D)$, where W is the width of the image, P is the patch size and D is the dimension of embedding.

3.3   Joint Decoder

The CTC-based decoding maps the feature of each patch to a corresponding character in a fine-grained manner, and the attention-based decoding predicts each character based on several correlated patches. To integrate the two-grained decoding features, we propose a joint decoder with two decoding techniques. Different from simple joint training with two separate branches, our joint decoder brings a deep interaction into the two decoding processes as shown in Fig. 2.

The inputs of the joint decoder contain two parts, the outputs of the encoder, and the embedding of the reading order [55,56], where the two inputs correspond to the features for two decoding processes. To help the model distinguish the two types of inputs, an additional learnable segment embedding is introduced. The inputs can be illustrated formally as follows:

$$y_0 = [O + S_o; X_{N+M} + S_x] + PE. \tag{2}$$

where y_0 represents the inputs of the decoder, O is denoted as the embedding of reading order. S_o and S_x are segment embedding for two inputs respectively, and PE is the position encoding. [;] indicates the operation of concatenation.

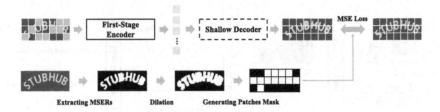

Fig. 3. The framework of the pre-training model with the process of generating patches mask. The encoder is the same as the first stage ViT encoder of text recognizer, and the decoder is introduced to reconstruct the images only existing during pre-training.

Next, the y_0 will be input to the L Transformer units to capture the characters-aware features for the final prediction, the process is similar to Eq. 1. Note that, the deep interaction conducted by each Transformer unit is bi-directional where the two types of inputs are both aware of each other. Finally, a linear layer is adopted to map the outputs y_L to the probability of each character. The probabilities are then split into two parts, which will be supervised by different objective functions. In this way, the first half of y_L contains fine-grained decoding features due to the CTC-based supervision, and the second half targets the coarse-grained features from attention-based decoding accordingly. When inference, the outputs of any decoder can be regarded as the final predictions.

3.4   MAE with Focusing Mechanism

MAE [13] shows that a masking and reconstruction pre-training with a high mask ratio can benefit the downstream visual tasks, here we explore the effectiveness of MAE in the field of text recognition. For our two-stage ViT encoder, we just involve the first stage in the pre-training, which is responsible for fine-grained visual feature extraction and may benefit from pixel construction learning. We follow the most setting of MAE, where the encoder is the same as our first-stage ViT, and the decoder is shallower with fewer embedding dimensions compared with the encoder, as shown in Fig. 3.

Text images can be divided into foreground and background, the former is the areas containing characters and the other areas are the background. There is no doubt that masking and reconstructing the pixels belonging to the areas of characters can better help the model focus on the details of characters and further improve downstream text recognition. To overcome this issue, we propose a focusing mechanism with an additional character-aware 0-1 mask for the reconstruction objective function. In this way, the model will focus on the reconstruction of the character-related pixels. However, we can not touch the specific localization of each character in a self-supervised manner. Inspired by traditional image processing, we adopt MSER to get the pixels inner a character.

As shown in the bottom branch of Fig. 3, we first extract the MSERs from the input image. To avoid too few or no MSERs, we constrained the ratio $r =$

$\frac{area_{msers}}{area_{image}} > 0.15$. If not, the mask will be set all to 1, in other words, the model will reconstruct all pixels. An additional dilation operation is supplemented to expand the area of MSERs, which will include pixels near the edges of the characters, which are essential to distinguish the foreground and background. Finally, we convert the dilated MSER maps to the patches mask under the standard that the patches intersecting with MSERs are regarded as foreground ones. The patches mask will be multiplied to the reconstruction loss, in this way, the model will more focus on predicting foreground pixels.

3.5   Objective Functions and Training Strategies

The objective functions adopted in this paper include pre-training and fine-tuning. For pre-training, a mean squared error (MSE) is adopted to supervise the reconstruction of the missing pixels. The patch masks proposed in Sect. 3.4 are multiplied to the loss to let the model focus on the reconstruction of the character-related pixels:

$$L_{pt} = Mask * MSE(p_{re}, p_{img})   \tag{3}$$

where, the p_{re} represents reconstructed pixels and p_{img} is the pixels in the target image. When fine-tuning, the objective function contains two parts focusing on the attention-based prediction and CTC-based prediction:

$$L_{ft} = CELoss(y_{attn}, label) + CTCLoss(y_{ctc}, label)   \tag{4}$$

the attention predictions y_{attn} are supervised by the standard cross-entropy loss and CTC predictions y_{ctc} adopt the corresponding CTC loss. Note that when fine-tuning, all parameters are under optimization in an end-to-end manner.

4   Experiments

4.1   Datasets

We adopt two synthetic datasets Synth90K (90K) [18] and SynthText (ST) [12] for training, and evaluates our method on six public benchmarks including IIIT5K-Words (IIIT5K) [31], ICDAR2013 (IC13) [20], ICDAR2015 (IC15) [19], Street View Text (SVT) [45], Street View Text Perspective (SVTP) [36] and CUTE80 (CUTE) [37]. Specifically, the IIIT5K, IC13, and SVT are used to evaluate the performance of the model on regular text, and IC15, SVTP, and CUTE focus on irregular text.

206   Z. Qiao et al.

Table 1. Comparison of effectiveness and efficiency with other works on public benchmarks, where previous works are divided into CNN-based (first part) and ViT-based (second part). The word-level accuracy is adopted as the evaluation metric, the **bold** represents the best performance, and underline represents the second best performance.

Methods	IIIT5K 3000	SVT 647	IC13 857	IC13 1015	IC15 1811	IC15 2077	SVTP 645	CUTE 288	Average	Params ×10⁶	Time ms/image
Wang et al. [46]	94.3	89.2	–	93.9	–	74.5	80.0	84.4	86.9	–	–
Qiao et al. [35]	93.8	89.6	–	92.8	80.0	–	81.2	83.6	88.4	–	80.1
Wan et al. [44]	93.9	90.1	–	92.9	79.4	–	84.3	83.3	88.6	57.0	56.8
Yue et al. [56]	95.3	88.1	–	94.8	–	77.1	79.5	90.3	88.2	–	–
Mou et al. [32]	94.4	92.3	–	95.0	–	82.2	84.3	85.0	89.8	–	61.3
Zhang et al. [58]	95.2	90.9	–	94.8	82.8	79.5	83.2	87.5	90.4	–	–
Yu et al. [55]	94.8	91.5	95.5	–	82.7	–	85.1	87.8	90.4	49.3	26.9
Qiao et al. [34]	95.2	91.2	95.2	93.4	83.5	81.0	84.3	84.4	90.5	–	28.4
Fang et al. [10]	95.4	93.2	**96.8**	–	84.0	–	87.0	88.9	91.5	32.8	19.5
Yan et al. [52]	95.6	94.0	96.4	–	83.0	–	87.6	91.7	91.5	–	67.4
Wang et al. [48]	95.8	91.7	95.7	–	83.7	–	86.0	88.5	91.2	–	–
Bhunia et al. [4]	95.2	92.2	–	**95.5**	84.0	–	85.7	89.7	91.2	–	–
He et al. [16]	95.8	**94.1**	**96.8**	–	84.6	–	87.9	**92.3**	92.1	42.1	18.8
Atienza et al. [2]	88.4	87.7	93.2	92.4	78.5	72.6	81.8	81.3	85.6	85.8	**9.8**
Tao et al. [41]	95.1	93.8	–	95.2	84.8	81.1	**88.1**	85.1	91.5	68.0	16.2
Ours	**96.1**	92.1	96.7	**95.5**	**86.3**	**83.3**	**88.1**	89.6	**92.4**	**13.2**	14.5

4.2 Implementation Details

Model Settings. The input images are resized to 32 × 128 without keeping ratio. The first stage of the encoder is with a patch size 4 × 4, and the second stage adopts the patch size of 32 × 4. The encoder contains 12 Transformer units, where 6 blocks for each stage and 4 units are used for the decoder. 38 symbols are covered for recognition, including digits, lower-case characters and the additional ⟨EOS⟩ token for attention-based decoding and ⟨Blank⟩ token for CTC-based decoding.

Pre-training. We sample 2 millions images from the two synthetic datasets 90K and ST for pre-training in total. The mask ratio is set to 75%, and the pre-training lasts 1600 epochs.

Fine-Tuning. To fairly compare with previous works, we train our model only on two synthetic datasets during fine-tuning. We adopt the Adam [21] as the optimizer with the initial learning rate of 10^{-4}. The model tends to converge in about 30 epochs, more epochs further improve the performance and we achieve our best model with epochs of 50.

4.3 Comparisons with State-of-the-Arts

We compare the performance of our method with other state-of-the-art methods on six benchmarks. As shown in Table 1, our method achieves the five best accuracies, and the best accuracy on average. The parameter amount of our method is only 13.2×10^6, which is the least compared with other methods. For inference speed, our method achieves a short inference time with only 14.5 ms, owing to the high parallelization of the Transformer.

Compared with CNN-based ABINet [10], our method gets 0.7% and 2.3% improvements on IIIT5K and IC15 respectively without any additional language-based modules. For the segmentation-based method S-GTR [16], our method works better on three benchmarks and comparable results on others. Both ABI-Net and S-GTR are non-autoregressive methods with a short inference time, our method further improves the inference speed of nearly 20% and the number of the parameters reduces by 60%. The advantages of our method in effectiveness and efficiency are verified.

For other ViT-based methods, our method achieves much better performance compared to ViTSTR [2] with fewer parameters. Compared with TRIG [41], our method works better on five benchmarks, especially 2.2% and 4.5% improvements on IC15 and CUTE. Without rectification module and GRU-based autoregressive decoder adopted by TRIG, our method also shows superiority on the number of parameters and inference speed.

4.4 Ablation Studies

Due to the limitation of the resources, the ablation studies are conducted with training epochs of 30.

The Effectiveness of Two-Stage ViT Encoder. To analyze the effectiveness of our two-stage ViT encoder, we conduct experiments to compare it with two single-grained encoders adopting all 4×4 patches and all 32×4 vertical patches respectively. As shown in Table 2, the ViT with the size of 32×4 patches works poorly on all six benchmarks, since the fine-grained visual features are ignored. Compared with single-stage ViT with all 4×4 patches, our proposed two-stage ViT works better on three benchmarks, and the superiority in FLOPs is also verified. However, the unsatisfactory performance of our proposed two-stage ViT encoder on CUTE indicates that there is still some room to improve the fusion between fine-grained and coarse-grained features, which will further help the model with curved texts. Taken overall, our two-stage encoder balances the effectiveness and efficiency.

Table 2. The comparison of accuracy with different patch sizes.

Patch Size	IIIT5K	SVT	IC13	IC15	SVTP	CUTE	FLOPs
4×4	95.9	91.9	**94.8**	**85.3**	86.0	**86.8**	2.84G
32×4	94.8	91.2	93.8	84.5	83.5	83.0	**0.45G**
$4 \times 4 + 32 \times 4$ (Ours)	**96.0**	**92.1**	94.4	**85.3**	86.5	85.1	1.58G

The Effectiveness of the Joint Decoder. As shown in Table 3, both CTC-based and attention-based predictions of the joint decoder are better in general compared with the respective single decoder. The performances between the two decoding techniques inner the joint decoder are similar, which illustrates the effectiveness of the deep interaction between different grained features.

Table 3. The comparison of accuracy with different decoders.

Methods	IIIT5K	SVT	IC13	IC15	SVTP	CUTE
CTC-based	95.9	91.3	93.4	84.7	86.2	85.1
Attention-based	95.8	91.2	94.9	85.7	86.9	87.2
Joint Decoder (CTC) (Ours)	**96.4**	**92.4**	94.8	85.9	86.7	**87.8**
Joint Decoder (Attention) (Ours)	96.1	92.3	**95.0**	**86.0**	**87.0**	86.8

The Effectiveness of the Self-supervised Pre-training. We analyze the effectiveness of the MAE-based pre-training and our proposed focusing mechanism for MAE. As shown in Table 4, the MAE-based pre-training improves the performances, especially 1.2% on IC13 and 0.6% on SVTP. Compared with the original MAE, our proposed focusing MAE works better on four benchmarks.

Table 4. The comparison of accuracy with different pre-training strategies.

Methods	IIIT5K	SVT	IC13	IC15	SVTP	CUTE
Scratch	96.0	92.1	94.4	85.3	86.5	85.1
MAE	95.8	91.9	**95.6**	85.4	**87.1**	86.1
Focusing MAE	**96.1**	**92.3**	95.0	**86.0**	87.0	**86.8**

We also visualize the reconstruction results in Fig. 4, the model is verified to reconstruct the original text images even with a large mask ratio. The reconstruction process will benefit the model with the awareness of the target characters and further improve the recognition results. The reconstruction results of our proposed focusing MAE seem better with more sharpened characters since it helps the model focus on the reconstruction of the pixels related to characters.

Fig. 4. The qualitative analysis of the results of reconstructions. For each quadruple, the masked image (left), MAE reconstruction (left middle), Focusing MAE reconstruction (right middle), and the original image (right) are shown. The red dashed rectangles help identify differences, and the reconstructed images with focusing MAE are more sharpened with more clear edges. (Color figure online)

In Fig. 5, we visualize some successful samples with pre-training. Here, the pre-training benefits the model with five cases: similar characters, blurred images, background interference, occluded characters, and curved text. We

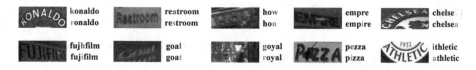

Fig. 5. Recognition results with (bottom) or without (top) pre-training.

explain that the pre-training of the pixels reconstruction can help the model focus on fine-grained differences of characters, distinguish between foreground and background, and model context dependencies.

4.5 Experiments on Occlusion Scene Text

VisionLAN [48] provides a new Occlusion Scene Text (OST) dataset to evaluate robust of the model on images with missing visual cues, the dataset is divided into two versions of weak and heavy according to the occluded degree. We also verify the effectiveness of our method on OST, as

Table 5. The comparison of accuracy on the occluded text images.

Methods	Average	Weak	Heavy
VisionLAN [48]	60.3	70.3	50.3
Ours	**74.3**	**81.3**	**67.4**

shown in Table 5, our method improves the accuracies significantly. Compared with VisionLAN, 14.0% improvement is achieved on average. We attribute the success to the powerful contextual dependencies modeling abilities of ViT, and the MAE pre-training also benefits the model with the occluded text images.

5 Conclusions

In this paper, we propose a novel ViT-based text recognition method with multi-grained encoding and decoding. For the encoder, we adopt a two-stage ViT structure with a different patching strategy, which extracts both fine-grained visual and coarse-grained sequence contextual features respectively, and improves the efficiency. Next, a joint decoder is proposed to integrate the CTC-based and attention-based decoding, and deep interaction between two-grained features corresponding to the two decoding processes is adopted, from which the two-grained features can benefit each other. Finally, the MAE-based pre-training is explored to improve the fine-grained feature extraction, and a focusing mechanism is proposed to let the model focus on reconstructing the characters-related pixels. Extensive experiments illustrate the effectiveness and efficiency of our method, four best results are achieved on six benchmarks with a faster inference speed and fewer parameters. In the future, we will further explore the techniques of unsupervised pre-training for text recognition and improve the structure of the ViT-based text recognizer.

Acknowledgments. This work was supported by National Key R&D Program of China, under Grant No. 2020AAA0104500.

References

1. Aberdam, A., et al.: Sequence-to-sequence contrastive learning for text recognition. In: CVPR, pp. 15302–15312 (2021)
2. Atienza, R.: Vision transformer for fast and efficient scene text recognition. In: Lladós, J., Lopresti, D., Uchida, S. (eds.) ICDAR 2021. LNCS, vol. 12821, pp. 319–334. Springer, Cham (2021). https://doi.org/10.1007/978-3-030-86549-8_21
3. Baek, J., Matsui, Y., Aizawa, K.: What if we only use real datasets for scene text recognition? toward scene text recognition with fewer labels. In: CVPR, pp. 3113–3122 (2021)
4. Bhunia, A.K., Sain, A., Kumar, A., Ghose, S., Chowdhury, P.N., Song, Y.Z.: Joint visual semantic reasoning: multi-stage decoder for text recognition. In: ICCV, pp. 14940–14949 (2021)
5. Chen, T., Kornblith, S., Norouzi, M., Hinton, G.E.: A simple framework for contrastive learning of visual representations. In: ICML, pp. 1597–1607. ACM (2020)
6. Cheng, Z., Bai, F., Xu, Y., Zheng, G., Pu, S., Zhou, S.: Focusing attention: towards accurate text recognition in natural images. In: ICCV, pp. 5076–5084. IEEE (2017)
7. Cheng, Z., Xu, Y., Bai, F., Niu, Y., Pu, S., Zhou, S.: AON: towards arbitrarily-oriented text recognition. In: CVPR, pp. 5571–5579. IEEE (2018)
8. d'Ascoli, S., Touvron, H., Leavitt, M.L., Morcos, A.S., Biroli, G., Sagun, L.: ConVit: improving vision transformers with soft convolutional inductive biases. In: ICML, pp. 2286–2296. ACM (2021)
9. Dosovitskiy, A., et al.: An image is worth 16x16 words: transformers for image recognition at scale. In: ICLR (2021)
10. Fang, S., Xie, H., Wang, Y., Mao, Z., Zhang, Y.: Read like humans: autonomous, bidirectional and iterative language modeling for scene text recognition. In: CVPR, pp. 7098–7107 (2021)
11. Gidaris, S., Singh, P., Komodakis, N.: Unsupervised representation learning by predicting image rotations. In: ICLR (2021)
12. Gupta, A., Vedaldi, A., Zisserman, A.: Synthetic data for text localisation in natural images. In: CVPR, pp. 2315–2324. IEEE (2016)
13. He, K., Chen, X., Xie, S., Li, Y., Dollár, P., Girshick, R.B.: Masked autoencoders are scalable vision learners. In: CVPR, pp. 16000–16009 (2022)
14. He, K., Fan, H., Wu, Y., Xie, S., Girshick, R.: Momentum contrast for unsupervised visual representation learning. In: CVPR, pp. 9726–9735 (2020)
15. He, P., Huang, W., Qiao, Y., Chen, C.L., Tang, X.: Reading scene text in deep convolutional sequences. In: AAAI, pp. 3501–3508. AAAI (2016)
16. He, Y., et al.: Visual semantics allow for textual reasoning better in scene text recognition. In: AAAI. AAAI (2021)
17. Hu, W., Cai, X., Hou, J., Yi, S., Lin, Z.: GTC: guided training of CTC towards efficient and accurate scene text recognition. In: AAAI, pp. 11005–11012. AAAI (2020)
18. Jaderberg, M., Simonyan, K., Vedaldi, A., Zisserman, A.: Reading text in the wild with convolutional neural networks. IJCV 116(1), 1–20 (2016)
19. Karatzas, D., et al.: ICDAR 2015 competition on robust reading. In: ICDAR, pp. 1156–1160. IEEE (2015)
20. Karatzas, D., et al.: ICDAR 2013 robust reading competition. In: ICDAR, pp. 1484–1493. IEEE (2013)
21. Kingma, D.P., Ba, J.: ADAM: a method for stochastic optimization. In: ICLR (2015)

22. Li, H., Wang, P., Shen, C., Zhang, G.: Show, attend and read: a simple and strong baseline for irregular text recognition. In: AAAI, pp. 8610–8617. AAAI (2019)
23. Li, M., et al.: TrOCR: transformer-based optical character recognition with pretrained models. arXiv preprint arXiv:2109.10282 (2021)
24. Liao, M., et al.: Scene text recognition from two-dimensional perspective. In: AAAI, pp. 8714–8721. AAAI (2019)
25. Litman, R., Anschel, O., Tsiper, S., Litman, R., Mazor, S., Manmatha, R.: SCATTER: selective context attentional scene text recognizer. In: CVPR, pp. 11962–11972. IEEE (2020)
26. Liu, Z., et al.: Swin Transformer: hierarchical vision transformer using shifted windows. In: ICCV, pp. 10012–10022 (2021)
27. Long, J., Shelhamer, E., Darrell, T.: Fully convolutional networks for semantic segmentation. In: CVPR, pp. 3431–3440 (2015)
28. Lu, N., et al.: Master: multi-aspect non-local network for scene text recognition. Pattern Recogn. **117**, 107980 (2021)
29. Luo, C., Jin, L., Sun, Z.: MORAN: a multi-object rectified attention network for scene text recognition. Pattern Recogn. **90**, 109–118 (2019)
30. Matas, J., Chum, O., Urban, M., Pajdla, T.: Robust wide baseline stereo from maximally stable extremal regions. In: BMVC, pp. 36.1-36.10. BMVA Press (2002)
31. Mishra, A., Alahari, K., Jawahar, C.: Scene text recognition using higher order language priors. In: BMVC. BMVA (2012)
32. Mou, Y., et al.: PlugNet: degradation aware scene text recognition supervised by a pluggable super-resolution unit. In: Vedaldi, A., Bischof, H., Brox, T., Frahm, J.-M. (eds.) ECCV 2020. LNCS, vol. 12360, pp. 158–174. Springer, Cham (2020). https://doi.org/10.1007/978-3-030-58555-6_10
33. Noroozi, M., Favaro, P.: Unsupervised learning of visual representations by solving jigsaw puzzles. In: Leibe, B., Matas, J., Sebe, N., Welling, M. (eds.) ECCV 2016. LNCS, vol. 9910, pp. 69–84. Springer, Cham (2016). https://doi.org/10.1007/978-3-319-46466-4_5
34. Qiao, Z., et al.: PIMNet: a parallel, iterative and mimicking network for scene text recognition. In: MM, pp. 2046–2055. ACM (2021)
35. Qiao, Z., Zhou, Y., Yang, D., Zhou, Y., Wang, W.: SEED: semantics enhanced encoder-decoder framework for scene text recognition. In: CVPR, pp. 13525–13534. IEEE (2020)
36. Quy Phan, T., Shivakumara, P., Tian, S., Lim Tan, C.: Recognizing text with perspective distortion in natural scenes. In: ICCV, pp. 569–576. IEEE (2013)
37. Risnumawan, A., Shivakumara, P., Chan, C.S., Tan, C.L.: A robust arbitrary text detection system for natural scene images. ESA **41**(18), 8027–8048 (2014)
38. Shi, B., Bai, X., Yao, C.: An end-to-end trainable neural network for image-based sequence recognition and its application to scene text recognition. TPAMI **39**(11), 2298–2304 (2016)
39. Shi, B., Yang, M., Wang, X., Lyu, P., Yao, C., Bai, X.: ASTER: an attentional scene text recognizer with flexible rectification. TPAMI **41**(9), 2035–2048 (2018)
40. Su, B., Lu, S.: Accurate recognition of words in scenes without character segmentation using recurrent neural network. Pattern Recogn. **63**, 397–405 (2017)
41. Tao, Y., Jia, Z., Ma, R., Xu, S.: TRIG: transformer-based text recognizer with initial embedding guidance. arXiv preprint arXiv:2111.08314 (2021)
42. Touvron, H., Cord, M., Douze, M., Massa, F., Sablayrolles, A., Jégou, H.: Training data-efficient image transformers & distillation through attention. arXiv preprint arXiv:1503.02531 (2021)

43. Vaswani, A., et al.: Attention is all you need. In: NeurIPS, pp. 5998–6008 (2017)
44. Wan, Z., He, M., Chen, H., Bai, X., Yao, C.: TextScanner: reading characters in order for robust scene text recognition. In: AAAI, pp. 12120–12127. AAAI (2020)
45. Wang, K., Babenko, B., Belongie, S.: End-to-end scene text recognition. In: ICCV, pp. 1457–1464. IEEE (2011)
46. Wang, T., et al.: Decoupled attention network for text recognition. In: AAAI, pp. 12216–12224. AAAI (2020)
47. Wang, W., et al.: Pyramid vision transformer: a versatile backbone for dense prediction without convolutions. In: ICCV, pp. 568–578 (2021)
48. Wang, Y., Xie, H., Fang, S., Wang, J., Zhu, S., Zhang, Y.: From two to one: a new scene text recognizer with visual language modeling network. In: ICCV, pp. 14194–14203 (2021)
49. Wu, H., et al.: CvT: Introducing convolutions to vision transformers. In: ICCV, pp. 22–31 (2021)
50. Xie, Z., Huang, Y., Zhu, Y., Jin, L., Liu, Y., Xie, L.: Aggregation cross-entropy for sequence recognition. In: CVPR, pp. 6538–6547. IEEE (2019)
51. Xie, Z., et al.: SimMIM: a simple framework for masked image modeling. In: CVPR, pp. 9653–9663 (2022)
52. Yan, R., Peng, L., Xiao, S., Yao, G.: Primitive representation learning for scene text recognition. In: CVPR, pp. 284–293 (2021)
53. Yang, M., et al.: Symmetry-constrained rectification network for scene text recognition. In: ICCV, pp. 9146–9155. IEEE (2020)
54. Yin, X.C., Yin, X., Huang, K., Hao, H.W.: Robust text detection in natural scene images. TPAMI 36(5), 970–983 (2014)
55. Yu, D., et al.: Towards accurate scene text recognition with semantic reasoning networks. In: CVPR, pp. 12110–12119. IEEE (2020)
56. Yue, X., Kuang, Z., Lin, C., Sun, H., Zhang, W.: RobustScanner: dynamically enhancing positional clues for robust text recognition. In: Vedaldi, A., Bischof, H., Brox, T., Frahm, J.-M. (eds.) ECCV 2020. LNCS, vol. 12364, pp. 135–151. Springer, Cham (2020). https://doi.org/10.1007/978-3-030-58529-7_9
57. Zhan, F., Lu, S.: ESIR: end-to-end scene text recognition via iterative image rectification. In: CVPR, pp. 2059–2068. IEEE (2019)
58. Zhang, C., et al.: SPIN: structure-preserving inner offset network for scene text recognition. In: AAAI, pp. 3305–3314. AAAI (2021)
59. Zhang, C., Gupta, A., Zisserman, A.: Adaptive text recognition through visual matching. In: Vedaldi, A., Bischof, H., Brox, T., Frahm, J.-M. (eds.) ECCV 2020. LNCS, vol. 12361, pp. 51–67. Springer, Cham (2020). https://doi.org/10.1007/978-3-030-58517-4_4
60. Zhang, H., Yao, Q., Yang, M., Xu, Y., Bai, X.: AutoSTR: efficient backbone search for scene text recognition. In: Vedaldi, A., Bischof, H., Brox, T., Frahm, J.-M. (eds.) ECCV 2020. LNCS, vol. 12369, pp. 751–767. Springer, Cham (2020). https://doi.org/10.1007/978-3-030-58586-0_44
61. Zhang, R., Isola, P., Efros, A.A.: Colorful image colorization. In: Leibe, B., Matas, J., Sebe, N., Welling, M. (eds.) ECCV 2016. LNCS, vol. 9907, pp. 649–666. Springer, Cham (2016). https://doi.org/10.1007/978-3-319-46487-9_40

Spatial Attention and Syntax Rule Enhanced Tree Decoder for Offline Handwritten Mathematical Expression Recognition

Zihao Lin[1], Jinrong Li[2], Fan Yang[1], Shuangping Huang[1,3]($\boxtimes$),
Xu Yang[4]($\boxtimes$), Jianmin Lin[2], and Ming Yang[2]

[1] South China University of Technology, Guangzhou, China
eehsp@scut.edu.cn
[2] CVTE Research, Guangzhou, China
{lijinrong,linjianmin,yangming}@cvte.com
[3] Pazhou Lab, Guangzhou 510330, China
[4] GRGBanking Equipment Co. Ltd., Guangzhou, China
yxu8@grgbanking.com

Abstract. Offline Handwritten Mathematical Expression Recognition (HMER) has been dramatically advanced recently by employing tree decoders as part of the encoder-decoder method. Despite the tree decoder-based methods regard the expressions as a LaTeX tree and parse 2D spatial structure to the tree nodes sequence, the performance of existing works is still poor due to the inevitable tree nodes prediction errors. Besides, they lack syntax rules to regulate the output of expressions. In this paper, we propose a novel model called **S**patial Attention and **S**yntax Rule Enhanced **T**ree **D**ecoder (SS-TD), which is equipped with spatial attention mechanism to alleviate the prediction error of tree structure and use syntax masks (obtained from the transformation of syntax rules) to constrain the occurrence of ungrammatical mathematical expression. In this way, our model can effectively describe tree structure and increase the accuracy of output expression. Experiments show that SS-TD achieves better recognition performance than prior models on CROHME 14/16/19 datasets, demonstrating the effectiveness of our model.

Keywords: Offline handwritten mathematical expression recognition · Tree decoder · Spatial attention · Syntax

1 Introduction

Offline handwritten mathematical expression recognition (HMER) has many applications like searching for questions in education, recording payment in finance, and many other fields. However, it is a challenge to recognize handwritten mathematical expression (HME) due to ambiguity of handwritten symbols and complexity of spatial structures in the expression. It means HMER

© The Author(s), under exclusive license to Springer Nature Switzerland AG 2022
U. Porwal et al. (Eds.): ICFHR 2022, LNCS 13639, pp. 213–227, 2022.
https://doi.org/10.1007/978-3-031-21648-0_15

should not only correctly recognize the symbols but also analyze the relationship between them, which put forward great difficulties in the recognition performance.

With the rapid advancement of deep learning-based methods for HMER, some studies have proposed string decoder [13,19,21,22,28] with handwritten mathematical expression images as input and LaTeX strings of expression as predicted targets. However, their works emphasize symbol recognition but seldom consider structure information, which leads to the unsatisfactory result in the recognition of expressions with complex structures like $\sqrt{2 + \sqrt{2}}$.

Moreover, in recent years, research based on tree decoder [24,26,27] which focuses on the structure was emerging in the field of HMER. The methods based on tree decoder inherently represent expression as tree structures which is more natural for HMER. Among them, Zhang et al. [27] proposed a tree decoder (DenseWAP-TD) to predict the parent-child relationship of trees during decoding procedure, which achieved the best recognition performance at that time. Although the tree decoder models the tree structure of HMEs explicitly, the structure prediction accuracy may decrease as they only use the node symbol information to predict the node structure. What's more, as shown in Fig. 1(c), the previous methods output an ungrammatical expression in which the symbol '2' can not have a 'Subscript' relation with the symbol '$\sum$'. It means the previous tree decoders may lack the syntactic rules to generate the grammatical expressions.

In order to solve the above problems, inspired by the DenseWAP-TD model [27] , we proposed a novel tree decoder (SS-TD) integrating spatial attention and syntax rules which is shown in Fig. 2. First, we introduce spatial attention mechanism to predict the parent node, which effectively improves the structure accuracy. What's more, for purpose of reducing the generation of ungrammatical mathematical expressions, we transform the grammar of the relationship between expression symbols into the syntax mask and introduce it into the process of relation prediction. The comparison results shown in Fig. 1 illustrate that our model can deal with the problems of the previous method. To further confirm the effectiveness of our model, we conduct the experiments on CROHME dataset. The results show that our model consistently achieves higher recognition rates over many other methods, demonstrating the effectiveness of our model for offline HMER.

The main contributions of this paper are as follows:

- We proposed a new tree decoder (SS-TD) which can effectively adapt encoders of other methods to achieve higher recognition performance.
- To improve the structure accuracy, we introduce spatial attention enhancement to predict node information. Besides, we introduce syntax masks to constrain the generation of ungrammatical mathematical expressions.
- We demonstrate the advantage of SS-TD on the CROHME14/16/19 dataset by experiments and achieve very competitive results.

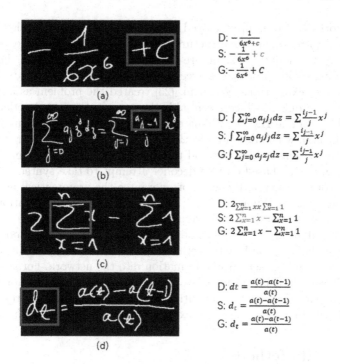

D: $-\frac{1}{6x^6+c}$

S: $-\frac{1}{6x^6}+c$

G: $-\frac{1}{6x^6}+C$

(a)

D: $\int \sum_{j=0}^{\infty} a_j j_j dz = \sum \frac{i_{j-1}}{j} x^j$

S: $\int \sum_{j=0}^{\infty} a_j j_j dz = \sum \frac{i_{j-1}}{j} x^j$

G: $\int \sum_{j=0}^{\infty} a_j z_j dz = \sum \frac{i_{j-1}}{j} x^j$

(b)

D: $2_{\sum_{x=1}^{n} xx \sum_{x=1}^{n} 1}$

S: $2\sum_{x=1}^{n} x - \sum_{x=1}^{n} 1$

G: $2\sum_{x=1}^{n} x - \sum_{x=1}^{n} 1$

(c)

D: $dt = \frac{a(t)-a(t-1)}{a(t)}$

S: $d_t = \frac{a(t)-a(t-1)}{a(t)}$

G: $d_t = \frac{a(t)-a(t-1)}{a(t)}$

(d)

Fig. 1. Comparison with the previous method DenseWAP-TD. 'D','S','G' stand for DenseWAP-TD, SS-TD and Ground-Truth. (a) and (b) denote parent node prediction error of previous method. (c) and (d) denote the relation prediction error of previous method. Symbols in red are the prediction errors. Symbols in green is the corresponding prediction (Color figure online)

2   Related Works

Early traditional methods [7,17,18] employed sequential approaches, which implemented symbol recognition and structural analysis separately. Recognition errors might be accumulated constantly in these methods. On the other hand, some studies [2,3,5] handled it as a global optimization to settle the problem of error accumulation, but made the process more inefficient.

Recently, deep learning-based encoder-decoder approaches were investigated by many studies on offline HMER. Some of them used string decoder to generate LaTeX strings. WAP [28] proposed by Zhang et al. employs a fully convolutional network encoder and an attention-equipped decoder, which achieved state-of-the-art at that time on CROHME 14/16 datasets. Then, Zhang et al. [25] proposed a DenseNet encoder to improve the WAP, which is also employed in our model. Wu et al. [22] proposed an adversarial learning strategy to overcome the difficulty about writing style variation. However, these studies lack structure relationship awareness, which leads to inevitable errors in some complex mathematical expression recognition.

Some works [1,4,6,9,10] employed tree decoder for various tasks, which is more natural to represent the structure. In this regard, some studies employed tree decoders to model the structure of mathematical expression. For online HMER, Zhang et al. proposed a tree decoder method SRD [26] to generate tree sequence of expressions. Wu et al. [23] treated the problem as a graph-to-graph learning problem, which essential is the same as the tree decoder. On the other hand, for offline HMER, Zhang et al. proposed a tree-structured decoder (DenseWAP-TD), which decomposes the structure tree into a sub-tree sequence to predict the parent-child nodes and the relationship between them. However, the existing methods based on tree decoder attempt to take syntactic information into account, which still could not guarantee the grammatical accuracy of the output expression. Yuan et al. [24] incorporated syntax information into an encoder-decoder network, which gave us a lot of inspiration.

As described above, although some tree decoders have been proposed, there is still a great challenge on how to further improve the accuracy in offline HMER and effectively combine semantic information into the network. For the purpose of solving these problems, we integrate spatial attention mechanism and syntax rules into the tree decoder to deal with complex mathematical expressions and adapt to the syntactic rationality of expressions as much as possible.

3   Proposed Method

Our model takes the input HME images and yields the symbol tree structure as shown in Fig. 2. We adopt an encoder to extract the feature vector sequence of the expression images. The decoder takes the feature sequence as input and then yields sequential triples step by step. Each triple which represents the sub-tree structure includes a child node, a parent node, and the relationship between parent-child nodes, represented as $(y_t^c, y_t^p, y_t^{rel})$. The child node y_t^c is composed of the predicted symbol and its tree sequence order. The predicted parent node y_t^p is one of the previous child nodes $y_{1,2...,t-1}^c$ which has best match with current child node y_t^c in the semantic and spatial information. The connection relationship y_t^{rel} between current parent and child nodes is represented as 6 forms: Above, Below, Sup, Sub, inside, and Right.

Just like the previous methods based on encoder-decoder [25] , we also employ DenseNet to encode the HME images $I \in R^{H \times W \times C}$ and extract the feature vector sequence $E = \{h_i\} \in R^{H' \times W' \times D}$, where $h_i \in R^D$. Then, the feature sequence is input into the child decoder and the parent decoder respectively to recognize the symbol step by step. For the purpose of modeling the expression structure, the decoder of SS-TD consists of three prediction modules: 1)the child node prediction module shown as ChildN.P in Fig. 2 and Fig. 3 to predict the current child node information y_t^c; 2)the parent node prediction module shown as ParentN.P in Fig. 2 and Fig. 3 to predict the parent node y_t^p from the previous child nodes which has best match with current child node; 3)the relationship prediction module shown as Relation.P in Fig. 2 and Fig. 3 to predict the relationships y_t^{rel} between parent and child nodes which are mentioned above. Consequently,

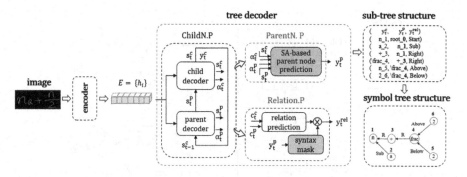

Fig. 2. Overview of SS-TD. The encoder is a DenseNet. 'SA-based parent node prediction' denotes the parent node prediction module equipped spatial attention (SA). 'Syntax mask' denotes transformed syntax rules introduced in our model. The details of the tree decoder are illustrated in Fig. 3

the prediction modules yield the triples which can be transformed to the symbol tree structure. At last, we traverse the tree structure to obtain the LaTeX strings.

3.1   Child Node Prediction Module

Like the previous tree decoder-based method DenseWAP-TD, we also adopt a parent decoder and a child decoder to recognize the child node symbol in sequence as shown in Fig. 3. These two decoders generate a tuple including the symbol and its recognition order, which is defined as child node information.

The parent decoder consists of two Gated Recurrent Units (GRU) layers and an attention block as shown in the left part of Fig. 3. It takes the previous child node y_{t-1}^c and its hidden state s_{t-1}^c as input, and outputs the parent context vector c_t^p and its hidden state s_t^p, which is also the input of the child decoder and the other two modules ParentN.P and Relation.P which shown in the right of Fig. 3:

$$c_t^p = \sum_{i=1}^{L} a_{ti}^p h_i \tag{1}$$

$$s_t^p = GRU_2^p(c_t^p, \hat{s}_t^p) \tag{2}$$

where h_i is the i-th element of feature map E. $L = H' \times W'$. s_t^p is the prediction of current parent hidden state. $a_t^p = \{a_{ti}^p\}$ is the current parent attention probabilities which is computed as follows:

$$\hat{s}_t^p = GRU_1^p(y_{t-1}^c, s_{t-1}^p) \tag{3}$$

$$a_t^p = f_{att}^p(\hat{s}_t^p, a_{1,2...,t-1}^p) \tag{4}$$

where f_{att}^p represents attention function followed the DenseWAP-TD [27] .

The structure of child decoder is almost the same as the parent decoder. The decoder inputs the previous parent node y_{t-1}^p and its hidden state s_t^p, and outputs the child context vector c_t^c and its hidden state s_t^c.

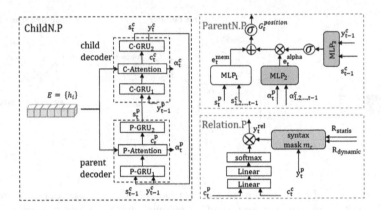

Fig. 3. The main prediction module in the decoder of SS-TD. 'ChildN.P' represents the child node prediction module. 'ParentN.P' represents the parent node prediction module. 'Relation.P' represents the relation prediction module.

We take the previous parent node y_{t-1}^p, the child node hidden state s_t^c and its context vector c_t^c as input to compute the probability of each output predicted child node $p(y_t^c)$:

$$p(y_t^c) = softmax(W_{out}^c(W_e y_{t-1}^p + W_h s_t^c + W_c c_t^c)) \qquad (5)$$

where $W_{out}^c \in R^S$ is a full connection layer parameter. S is the size of the recognition symbol set.

The classification loss of child node prediction module is:

$$L_c = -\sum_{t=1}^{T} log(p(y_t^c)) \qquad (6)$$

During the test process, we use the previous one-hot vector of the parent decoder and the child decoder in greedy algorithm to predict the child node:

$$\hat{y}_t^c = \arg\max_{y^c} p(y_t^c) \qquad (7)$$

where y^c represents all the symbols in the symbol set.

3.2   Spatial Attention-Based Parent Node Prediction Module

We treat predicting the parent node as finding previous child nodes which have the best match with current child node. We define the order of previous child nodes as the parent node position $\hat{y}^{pos}$, which is the predicted target in this module.

We first use the semantic information representing the hidden state of the nodes to predict the parent nodes. We employ Multilayer Perceptron (MLP) architecture to obtain the semantic energy factor e_{ti}^{mem} which is shown as the MLP$_1$ block in Fig. 3:

$$e_{ti}^{mem} = v_{mem}^T \tanh(W_{mem}s_t^p + U_{mem}s_{1,2...,t-1}^c) \tag{8}$$

where e_{ti}^{mem} is the semantic energy of $i - th$ element at decoding step t. s_t^p is the hidden state of parent decoder. $s_{1...t-1}^c$ are the hidden states of child nodes in history.

However, only using semantic information to predict the parent node may lead to inevitable error especially when facing the identical symbols of expression. Thus, we introduce spatial attention mechanism to alleviate the prediction error of parent node.

The spatial information of the parent node and previous child nodes is represented as the attention distribution a_t^p and $a_{1,2...,t-1}^c$ respectively. The spatial energy factor is computed as follows:

$$e_{ti}^{alpha} = v_{alpha}^T \tanh(W_{alpha}p(a_t^p) + U_{alpha}p(a_{1,2...,t-1}^c)) \tag{9}$$

where e_{ti}^{alpha} is the spatial energy of $i - th$ element at decoding step t. $p(\cdot)$ represents adaptive pooling. The size of its output vector is 4×32.

We further employ the gate mechanism to control the input of updated spatial position information as shown in the ParentN.P block of Fig. 3. The gate mechanism is computed as follows:

$$g_t = \sigma(W_{yg}y_{t-1}^c + U_{sg}s_{t-1}^c) \tag{10}$$

$$e_{ti}^{position} = e_{ti}^{mem} + g_t \odot e_{ti}^{alpha} \tag{11}$$

$$G_{ti}^{position} = \sigma e_{ti}^{position} \tag{12}$$

where $\odot$ means the element-wise product.

The loss of the parent node prediction module is defined as a cross entropy loss:

$$L_{pos} = -\sum_{t-1}^{T}\sum_{i=1}^{L}[\bar{G}_{ti}^{position}log(G_{ti}^{position}) + (1-\bar{G}_{ti}^{position})log(1-G_{ti}^{position})] \tag{13}$$

where $\bar{G}_{ti}^{position}$ represent the ground-truth of the parent node. If $i - th$ child node in history is the current parent node, $\bar{G}_{ti}^{position}$ is 1, otherwise 0.

In the testing stage, we compute the parent node position $\hat{y}^{pos}$:

$$\hat{y}^{pos} = \arg\max_g G^{position} \tag{14}$$

And then the prediction of parent node $\hat{y}^p$ is obtained by using the $\hat{y}^{pos}$ which also represents the child node order. After that, we put it into the child decoder to predict the child node.

3.3   Syntax Rule-Based Relation Prediction Module

In order to predict the grammatical expression, we introduce syntax rules into the relation prediction module. Two crucial syntax rules related to conjunction relations are summarized as follows:

- Different mathematical symbols have different restrictions of connection relation, like the '$\sum$' can not have the 'Inside' relation with other symbols.
- Each mathematical symbol cannot have repeated connection relations at the same time, like one symbol can not have 'Right' relation with two other symbols.

Table 1. Static syntax mask. The connection relationships of the symbol are transformed into the Static syntax mask, which is designed in the sequence of [Right, Sup, Sub, Above, Below, Inside]

	Right	Sup	Sub	Above	Below	Inside	Symbol
$\frac{\square}{\square}$	1	1	0	1	1	0	\frac
$\sqrt{\square}$	1	0	0	0	0	1	\sqrt
Σ, Π	1	1	1	1	1	0	[\sum, \prod]
$\lim_{\square}$	1	0	0	0	1	0	\lim
Letter	1	1	1	0	0	0	the Greek letter, English letter, the ending symbols...
Number	1	1	0	0	0	0	number, Trigonometric functions
Bin Symbol	1	0	0	0	0	0	the beginning symbols, operator...

Then, to integrate these rules into the model training process, we creatively put forward the syntax mask representing the mathematical connection relationship. According to the two syntax rules, we define the syntax mask of mathematical symbols. If the mathematical symbols can be connected to its related symbol in a relation of Right, Sup, Sub, Above, Below, Inside, the corresponding syntax mask value is 1, otherwise 0.

For the first syntax rule which is regarded as the prior knowledge, the mask of it could be obtained in advance as shown in Table.1. For example, the mask of the symbol '+' is '100000' since it only has one relation 'Right'. Besides, we group some symbols which have the same syntax mask, such as the Letter as shown in Table 1. In addition, we deal with the symbols which can be represented as different syntax masks by the logical 'OR' operator. For example, the mask of symbol 'e' is '111000' when it represents the lowercase letter, while the mask is '110000' when it represents the irrational constant. To sum up, the mask of 'e' is '111000‖110000=111000', where '‖' is the logical 'OR' operator.

We present the static syntax mask matrix as $R_{static} \in R^{C \times 6}$, and the static syntax mask of parent node as $m_t^s \in R^6$:

$$m_t^s = y_t^p R_{static} \tag{15}$$

where y_t^p is the one-hot vector of the symbol in the current parent node.

On the other hand, for the second syntax rule, we need to acquire information about the previous relations between parent-child nodes. To solve the problem, we propose a dynamic mask matrix $R_{dynamic}^t$ which is initialized into an all-zero matrix to store the historical relations in step t. For example, when the parent node and its relationship are predicted to be '\sqrt' and 'Inside', the value of the corresponding row of '\sqrt' and the corresponding column of 'Inside' in $R_{dynamic}^t$ should be updated as '1'. The dynamic syntax mask of parent node is represented as $m_t^d \in R^6$:

$$m_t^d = y_t^p R_{dynamic}^{t-1} \tag{16}$$

$$m_t = m_t^s \otimes m_t^d \tag{17}$$

where $\otimes$ is the logical exclusive or operator. The update of dynamic mask matrix $R_{dynamic}^t$ is computed as follows:

$$R_{dynamic}^t = R_{dynamic}^{t-1} + y_t^{pT} y_t^{rel} \tag{18}$$

where y_t^{pT} represents the transposition of the parent node y_t^p.

As shown in Fig. 3, We compute the probabilities of the relation prediction $p(y_t^{rel})$ as follows:

$$p(y_t^{rel}) = softmax(m_t \odot (W_{out}^{rel}(W_{cp}c_t^p + W_{cc}c_t^c))) \tag{19}$$

where $\odot$ represents multiplication element by element.

The classification loss of relation prediction module is computed as follows:

$$L_{rel} = -\sum_{t=1}^{T} log(p(y_t^{rel})) \tag{20}$$

In the testing stage, we take the relation of maximum probability as final prediction relation:

$$\hat{y}^{rel} = \arg\max_{y^{rel}} p(y_t^{rel}) \tag{21}$$

where y^{rel} represent the relations set.

3.4   Total Loss

We use an attention self-regularization loss to speed up network convergence. Specifically, a Kullback-Leibler divergence is employed to measure the difference in the distribution of attention generated by parent and child decoder:

$$L_{alpha} = \sum_{t=1}^{T} \hat{a}_t^p log(\frac{\hat{a}_t^p}{a_t^p}) \tag{22}$$

Thus, the training objective of the SS-TD is to minimize the loss as follows:

$$L = \lambda_1 L_c + \lambda_2 L_{pos} + \lambda_3 L_{rel} + \lambda_4 L_{alpha} \tag{23}$$

where, λ is the weight of each term.

4   Experiments

4.1   Dataset

We verified our proposed model on the CROHME dataset [14–16] which is the most widely used for HMER. The CROHME training set contains 8835 HMEs, 101 math symbol classes. There are 6 common spatial math relations (Above, Below, Right, Inside, Superscript and Subscript) in our implementation. We evaluate our model on CROHME 14/16/19 test set. Among them, the CROHME 2014 test set contains 986 HMEs, which is evaluated by most advanced model. And CROHME 2016 and 2019 test sets are collected and labeled for research after that, which contain 1147 and 1119 HMEs.

4.2   Implementation Details

For the training loss, we set $\lambda_1 = \lambda_2 = \lambda_3 = 1$ in our experiment indicating the same importance for the modules, and set $\lambda_4 = 0.1$.

Following the DenseNet-WAP [25] , We employ DenseNet as the encoder, which is composed of three DenseBlocks and two Transition layers. Each Dense-Block contains 48 convolution layers. The convolution kernel size is set to 11×11 for computing the coverage vector.

Our model follows the previous tree decoder-based method DenseWAP-TD using two decoders to recognize the symbols, which has redundant parameters. We first simplify the parameters to optimize the model, where the dimensions of child attention, parent attention and memory attention are set to 128 instead of 512 and the embedding dimensions of both child node and parent node are set to 128 instead of 256. As shown in the second row of Table 2, the parameters are reduced by near half and the inference speed increases by more than 50 ms without affecting the recognition performance. Regarding this, we take the same parameters into SS-TD to pursue better performance. Due to the introduced syntax masks, the complexity of our model is quite more than the simplified

Table 2. Performance on CROHME 2014 of our model versus DenseWAP-TD and Simplified DenseWAP on Expression Rate (ExpRate) (in %). Inference speed is represented as Speed. The number of parameters is shown in the last column.

Model	Exp.Rate$_{latex}$(%)	Speed	Params
DenseWAP-TD	49.10	144.9	7.93M
Simplified DenseWAP-TD	49.32	**92.56**	**4.68M**
SS-TD	**52.48**	94.48	4.77M

DenseWAP-TD, in which the gap is still smaller than that between DenseWAP-TD and simplified DenseWAP-TD.

We utilized the Adadelta algorithm for optimization and conduct on experiments. Besides, the framework was implemented in Pytorch. Experiments were conducted on 2 NVIDIA GeForce RTX 3090.

4.3   Ablation Experiment

In order to verify the performance improvement of SS-TD brought by the integration of spatial attention and syntax rules, we conduct ablation experiments on CROHME 14 datasets.

Table 3. Results of Ablation Experiment in % on CROHME 2014. 'Exp.Rate$_{tree}$' denotes the expression recognition rate of tree structure. 'Exp.Rate$_{latex}$' denotes the expression recognition rate of LaTeX string. 'WER$_{pos}$' denotes the word error rate of parent node prediction. 'WER$_{rel}$' denotes the word error rate of predicted relationship between parent-child nodes.

Spatial information	Static syntax mask	Dynamic syntax mask	CROHME 2014			
			Exp.Rate$_{tree}$ ↑	Exp.Rate$_{latex}$ ↑	WER$_{pos}$ ↓	WER$_{rel}$ ↓
			49.93	50.14	8.64	6.53
√			51.47	51.47	7.12	6.23
√	√		52.02	52.10	6.98	5.56
√	√	√	**52.48**	**52.48**	**6.72**	**5.12**

As shown in Table 3, after we introduce spatial attention mechanism, the Exp.Rate$_{tree}$ increases by 1.54% and the WER$_{pos}$ decreases by 1.52%. To further prove the improvement is due to the spatial attention mechanism itself instead of the attention parameters, we compare the precision(%)/Params(M) of attention employed and attention not employed, in which the value of attention employed is 10.79 and 0.12 higher than attention not employed. In addition, as shown in Fig. 4, the result predicted by only using semantic information is the

(a)Parent node Prediction: 'n'   (b)Parent node Prediction: '1'

Fig. 4. The visualization of spatial attention distribution. (a) denotes the correct parent node prediction by integrating spatial information. (b) denotes the error prediction by only using semantic information. Blocks in blue denote the attention distribution. (Color figure online)

symbol '1' which is incorrect obviously. And the attention distribution of child node symbols correctly indicate the right symbol 'n'. Hence the spatial attention enhancement achieved improvements and this demonstrates the importance of spatial information for the parent node position.

On the other hand, as illustrated in the third row and the forth row of Table 3, the performance has a great improvement on the whole after adding static and dynamic syntax masks. It is worth noting that the gap between $Exp.Rate_{tree}$ and $Exp.Rate_{latex}$ becomes smaller, which indicates the grammar accuracy of the output tree structures has great improvement. In general, after the enhancement we introduce into the model, $Exp.Rate_{tree}$ and $Exp.Rate_{latex}$ increase by 2.55% and 2.34%, WER_{pos} and WER_{pos} decrease by 1.92% and 1.41%, respectively, which validates the effectiveness of the spatial attention and syntax rules.

4.4 Performance Comparison

The comparison among our model and the early algorithms on CROHME 14/16/19 is listed in the Table 4. All the experiment results in Table 4 are selected from published papers. Our model is implemented without using data enhancement and beam search strategies. Obviously, the $Exp.Rate_{latex}$ of SS-TD is 52.48% on CROHME 2014, 55.29% on CROHME 2016, and 54.32% on CROHME 2019. Specifically, compared with the baseline model DenseWAP-TD, our model is 3.33% better on CROHME 2014 and 6.68% better on CROHME 2016 than DenseWAP-TD which demonstrates the effectiveness of our improvement.

However, the competitive performance of our method is still lower than some string decoder-based methods, such as [20] and [8] . The former uses an additional weakly supervised branch to achieve 53.35% on CROHME 14, the latter uses large scale input images from data augmentation and beam search decoding to achieve 57.72%/61.38% ExpRate on CROHME 16/19 which take longer to decode iteratively. Indeed, our method pursues the expression spatial structure resolution too much and loses part of the recognition performance of the symbols, which is the main problem that we want to break through in the future.

Table 4. Evaluation of offline HMER on CROHME 14/16/19 (%)

System	CROHME 14 Exp.Rate$_{latex}$	CROHME 16 Exp.Rate$_{latex}$	CROHME 19 Exp.Rate$_{latex}$
WAP [28]	40.4	37.1	37.1
DenseWAP [25]	43.0	40.1	41.7
PGS [12]	48.78	45.60	–
DenseWAP-TD [27]	49.1	48.5	51.4
PAL [22]	39.66	49.61	–
PAL-v2 [11]	48.88	51.53	–
WS-WAP [19]	**53.65**	**51.96**	–
SS-TD	52.48	51.29	**54.32**

5 Conclusion

In this paper, we proposed a new tree decoder model integrating spatial attention and syntax rules, which not only improve the structure prediction accuracy, but also decrease the apparent grammar error of mathematical expressions. We illustrate the great performance of SS-TD for offline HMER through the ablation study and comparisons with the advanced methods on CROHME datasets. In the future work, we will further improve the generalization and performance of SS-TD.

Acknowledgement. This work has been supported by the National Natural Science Foundation of China (No.62176093, 61673182), the Key Realm Research and Development Program of Guangzhou (No.202206030001), and the GuangDong Basic and Applied Basic Research Foundation (No.2021A1515012282).

References

1. Alvarez-Melis, D., Jaakkola, T.S.: Tree-structured decoding with doubly-recurrent neural networks. In: 5th International Conference on Learning Representations, ICLR 2017, Toulon, France, 24–26 April 2017, Conference Track Proceedings. OpenReview.net (2017). https://openreview.net/forum?id=HkYhZDqxg
2. Álvaro, F., Sánchez, J.A., Benedí, J.M.: An integrated grammar-based approach for mathematical expression recognition. Pattern Recogn. **51**, 135–147 (2016)
3. Awal, A., Mouchère, H., Viard-Gaudin, C.: A global learning approach for an online handwritten mathematical expression recognition system. Pattern Recogn. Lett. **35**, 68–77 (2014). https://doi.org/10.1016/j.patrec.2012.10.024
4. Chakraborty, S., Allamanis, M., Ray, B.: Tree2tree neural translation model for learning source code changes. CoRR abs/1810.00314 (2018). http://arxiv.org/abs/1810.00314

5. Chan, K., Yeung, D.: Elastic structural matching for online handwritten alphanumeric character recognition. In: Jain, A.K., Venkatesh, S., Lovell, B.C. (eds.) Fourteenth International Conference on Pattern Recognition, ICPR 1998, Brisbane, Australia, 16–20 August 1998, pp. 1508–1511. IEEE Computer Society (1998), https://doi.org/10.1109/ICPR.1998.711993

6. Chen, X., Liu, C., Song, D.: Tree-to-tree neural networks for program translation. In: Bengio, S., Wallach, H.M., Larochelle, H., Grauman, K., Cesa-Bianchi, N., Garnett, R. (eds.) Advances in Neural Information Processing Systems 31: Annual Conference on Neural Information Processing Systems (NeurIPS) 2018, pp. 2552–2562 (2018). https://proceedings.neurips.cc/paper/2018/hash/d759175de8ea5b1d9a2660894f-Abstract.html

7. Cortes, C., Vapnik, V.: Support-vector networks. Mach. Learn. **20**(3), 273–297 (1995)

8. Ding, H., Chen, K., Huo, Q.: An encoder-decoder approach to handwritten mathematical expression recognition with multi-head attention and stacked decoder. In: Lladós, J., Lopresti, D., Uchida, S. (eds.) ICDAR 2021. LNCS, vol. 12822, pp. 602–616. Springer, Cham (2021). https://doi.org/10.1007/978-3-030-86331-9_39

9. Dyer, C., Kuncoro, A., Ballesteros, M., Smith, N.A.: Recurrent neural network grammars. In: Knight, K., Nenkova, A., Rambow, O. (eds.) NAACL HLT 2016, The 2016 Conference of the North American Chapter of the Association for Computational Linguistics: Human Language Technologies, pp. 199–209. The Association for Computational Linguistics (2016). https://doi.org/10.18653/v1/n16-1024

10. Harer, J., Reale, C.P., Chin, P.: Tree-transformer: a transformer-based method for correction of tree-structured data. CoRR abs/1908.00449 (2019). http://arxiv.org/abs/1908.00449

11. Le, A.D.: Recognizing handwritten mathematical expressions via paired dual loss attention network and printed mathematical expressions. In: Proceedings of the IEEE/CVF Conference on Computer Vision and Pattern Recognition (CVPR) Workshops (2020)

12. Le, A.D., Indurkhya, B., Nakagawa, M.: Patten generation strategies for improving recognition of handwritten mathematical expression. Pattern Recogn. Lett. **128**, 255–262 (2019). https://doi.org/10.1016/j.patrec.2019.09.002

13. Li, Z., Jin, L., Lai, S., Zhu, Y.: Improving attention-based handwritten mathematical expression recognition with scale augmentation and drop attention. In: 17th International Conference on Frontiers in Handwriting Recognition, ICFHR, pp. 175–180. IEEE (2020). https://doi.org/10.1109/ICFHR2020.2020.00041

14. Mahdavi, M., Zanibbi, R., Mouchere, H., Viard-Gaudin, C., Garain, U.: ICDAR 2019 CROHME + TFD: competition on recognition of handwritten mathematical expressions and typeset formula detection. In: 2019 International Conference on Document Analysis and Recognition ICDAR, pp. 1533–1538. IEEE (2016). https://doi.org/10.1109/ICDAR.2019.00247

15. Mouchere, H., Viard-Gaudin, C., Zanibbi, R., Garain, U.: ICFHR 2016 CROHME: competition on recognition of online handwritten mathematical expressions. In: 2016 15th International Conference on Frontiers in Handwriting Recognition ICFHR, pp. 607–612. IEEE (2019). https://doi.org/10.1109/ICFHR.2016.0116

16. Mouchère, H., Zanibbi, R., Garain, U., Viard-Gaudin, C.: Advancing the state of the art for handwritten math recognition: the CROHME competitions, 2011–2014. Int. J. Docu. Anal. Recog. (IJDAR) **19**(2), 173–189 (2016). https://doi.org/10.1007/s10032-016-0263-5

17. Okamoto, M., Imai, H., Takagi, K.: Performance evaluation of a robust method for mathematical expression recognition. In: 6th International Conference on Document Analysis and Recognition ICDAR 2001), 10–13 September 2001, Seattle, WA, USA, pp. 121–128. IEEE Computer Society (2001). https://doi.org/10.1109/ICDAR.2001.953767

18. Qian, R.J., Huang, T.S.: Optimal edge detection in two-dimensional images. IEEE Trans. Image Process. **5**(7), 1215–1220 (1996). https://doi.org/10.1109/83.502412

19. Truong, T., Nguyen, C.T., Phan, K.M., Nakagawa, M.: Improvement of end-to-end offline handwritten mathematical expression recognition by weakly supervised learning. In: 17th International Conference on Frontiers in Handwriting Recognition, ICFHR 2020, Dortmund, Germany, September 8–10, 2020, pp. 181–186. IEEE (2020). https://doi.org/10.1109/ICFHR2020.2020.00042

20. Truong, T., Ung, H.Q., Nguyen, H.T., Nguyen, C.T., Nakagawa, M.: Relation-based representation for handwritten mathematical expression recognition. In: ICDAR 2021: 16th International Conference Document Analysis and Recognition, pp. 7–19 (2021). https://doi.org/10.1007/978-3-030-86198-8_1

21. Wang, J., Sun, Y., Wang, S.: Image to latex with densenet encoder and joint attention. In: Bie, R., Sun, Y., Yu, J. (eds.) 2018 International Conference on Identification, Information and Knowledge in the Internet of Things, IIKI 2018, Beijing, China, 19–21 October 2018. Procedia Computer Science, vol. 147, pp. 374–380. Elsevier (2018). https://doi.org/10.1016/j.procs.2019.01.246

22. Wu, J., Yin, F., Zhang, Y., Zhang, X., Liu, C.: Handwritten mathematical expression recognition via paired adversarial learning. Int. J. Comput. Vis. **128**(10), 2386–2401 (2020). https://doi.org/10.1007/s11263-020-01291-5

23. Wu, J., Yin, F., Zhang, Y., Zhang, X., Liu, C.: Graph-to-graph: towards accurate and interpretable online handwritten mathematical expression recognition. In: Proceedings of the AAAI Conference on Artificial Intelligence, pp. 2925–2933 (2021). https://doi.org/10.1609/aaai.v35i4.16399

24. Yuan, Y., et al.: Syntax-aware network for handwritten mathematical expression recognition. CoRR abs/2203.01601 (2022). https://doi.org/10.48550/arXiv.2203.01601

25. Zhang, J., Du, J., Dai, L.: Multi-scale attention with dense encoder for handwritten mathematical expression recognition. In: 24th International Conference on Pattern Recognition, ICPR 2018, Beijing, China, 20–24 August 2018, pp. 2245–2250. IEEE Computer Society (2018). https://doi.org/10.1109/ICPR.2018.8546031

26. Zhang, J., Du, J., Yang, Y., Song, Y., Dai, L.: SRD: a tree structure based decoder for online handwritten mathematical expression recognition. IEEE Trans. Multim. **23**, 2471–2480 (2021). https://doi.org/10.1109/TMM.2020.3011316

27. Zhang, J., Du, J., Yang, Y., Song, Y., Wei, S., Dai, L.: A tree-structured decoder for image-to-markup generation. In: Proceedings of the 37th International Conference on Machine Learning, ICML 2020, 13–18 July 2020, Virtual Event. Proceedings of Machine Learning Research, vol. 119, pp. 11076–11085. PMLR (2020). http://proceedings.mlr.press/v119/zhang20g.html

28. Zhang, J.: Watch, attend and parse: an end-to-end neural network based approach to handwritten mathematical expression recognition. Pattern Recogn. **71**, 196–206 (2017). https://doi.org/10.1016/j.patcog.2017.06.017

Handwriting Recognition
and Understanding

FPRNet: End-to-End Full-Page Recognition Model for Handwritten Chinese Essay

Tonghua Su[✉][ID], Hongming You, Shuchen Liu, and Zhongjie Wang[ID]

School of Software, Harbin Institute of Technology,
Harbin, People's Republic of China
{thsu,rainy}@hit.edu.cn

Abstract. Handwritten Chinese Essay Recognition (HCER) is a special branch of handwritten Chinese text recognition with great interest. In a naive way, it can be firstly segmented into text lines or even characters, followed by a text line or character recognition step. Instead, we propose an end-to-end recognition model named FPRNet which directly runs on full-page images in light of the segmentation-free strategy. Our well-designed model can extract text from a full-page image only supervised with text labels and adapt better to authentic noisy images. Besides, we propose an effective dimensionality reduction mechanism based on reshape operation to bridge features between 2D and 1D without information loss. Moreover, we propose an order-align strategy to mitigate the decoding confusion caused by skewness. Experiments are conducted on real-world essay images. Our model achieves a 5.83% character error rate (CER), which is comparable with the state-of-the-art approaches.

Keywords: Handwritten text recognition · Deep learning · Text and symbol recognition

1 Introduction

Handwritten Chinese Text Recognition (HCTR) has been a long-term research task in pattern recognition and document analysis [1,2]. In general, the goal of the HCTR task is to transcribe handwritten text images into a machine-encoded form so that downstream tasks can freely use the text. With the rise and popularization of deep learning technology, the HCTR system is widely used in office automation, forms automation, and AI-assisted education.

Typically, HCTR research focuses on line-level texts. Full-page text recognition is usually achieved in two steps [3,4]. It is segmented at the line level firstly. Then a recognition model will transcribe text from the segmented images. However, as a matter of fact, segmentation error will seriously affect the final recognition performance. Producing segmentation labels on massive data is also too costly [16].

U. Porwal et al. (Eds.): ICFHR 2022, LNCS 13639, pp. 231–244, 2022.
https://doi.org/10.1007/978-3-031-21648-0_16

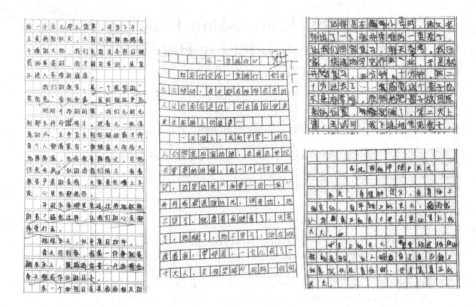

Fig. 1. Examples of authentic handwritten Chinese essay images. Left: samples with correction and noises. Middle: samples with scanning. Right: samples with modifications and different character scales.

In many real-world scenarios, handwritten Chinese essay recognition (HCER) system based on HCTR faces more complex challenges. In the case of education, there exists miswriting, writing separately, scribble marks, complex grid backgrounds, distorting scanning, and so on (Fig. 1). In addition, we need to consider various types of composition paper templates, different line spacing, and different character scales. Using language models like most state-of-the-art works [5] generally to enhance the model results is also not allowed because this might correct students' miswritten unintentionally.

This paper aims at providing a full-page recognition model (FPRNet) to meet the mentioned challenges. We construct a novel encoder-decoder architecture driven by CTC loss [6] lending recent advances in full-page text recognition works.

Our proposed model employs the encoder-decoder structure. As for the encoder, we design an architecture integrating gating mechanism and multi-branch structure. This architecture also refers to the MobileNetV2 [7] and reduces computational consumption by using pointwise convolution and depthwise convolution. To mitigate the mapping gap between 2D image and 1D text string, we leverage the reshape operation to align the output text and feature maps more precisely. Furthermore, we propose an order-align strategy to alleviate decoding confusion caused by skewness. The resultant method demonstrates a comparable performance with traditional methods.

This paper is organized as follows. Section 2 gives an overview of related work on the field. Section 3 describes our architecture design in detail. Section 4 describes the experiments and presents its result. Finally, concluding remarks are given in Sect. 5.

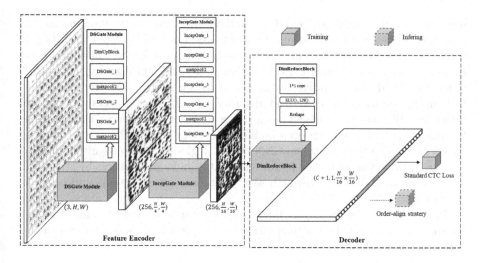

Fig. 2. Overview of the proposed FPRNet.

2   Related Works

We classify the previous works into two categories. Those requiring segmentation and ground-truth in line-level are called segmentation-based approaches. The other segmentation-free approach, which only requires ground-truth in full-page level, has been a recent hot research topic.

2.1   Segmentation-Based Approaches

The segmentation-based approach has been studied for a long time. The models proposed in [8,9] use a region proposal network and region of interest pooling to obtain bounding boxes for each word or line in the input image.

Subsequently, a single-line text recognition model transcribes the text from the boxes. A typical single-line text recognition model uses a CTC-based approach to convert text images into segmented sequences, and then decodes the sequences using a dynamic programming approach. Many kinds of architectures have been proposed for single-line text recognition. Multi-Dimensional Long-Short Term Memory(MDLSTM) [10] is used to recognize Chinese character lines. [11] proposes a separable MDLSTM-RNN model, which is able to extract contextual information in different directions. [12] proposes a general framework based on Gated CNN, which achieves higher computational efficiency.

2.2   Segmentation-Free Approaches

A segmentation-free approach is dedicated to achieving detection and recognition of multi-line images simultaneously or considering detection only as an implicit task without considering intermediate detection results. [13] proposes a segmentation-free approach based on MDLSTM, where a two-dimensional feature map is obtained by encoding. [14] illustrates a new SFR model consisting of three modules, in which the line head detection module locates the initial position of a text line by regression calculation of the horizontal and vertical offset of the line head point relative to the center of the lattice. [15] proposes a unified end-to-end model using hybrid attention to iteratively process a paragraph image line by line. [16] directly performs a segmentation-free and annotation-free one-stage end-to-end recognition model for multi-line handwritten text. The authors extended the traditional CNN+CTC single-line text recognition method by transforming the multi-line feature map into single-line features by stretching it in the vertical direction through several bilinear interpolations and convolutional layers.

In this work,we suggest combining the advantages of efficient single-line text recognition and multi-line segmentation-free approach to design an end-to-end full-page recognition method.

3 Architecture

Our proposed Full-Page Recognition Network (FPRNet) consists of two major procedures. One is an encoder based on depthwise separable convolution, Inception block and gate block. The other is a decoder based on dimensionality reduction block that enables the decoder to focus on the most relevant encoded features at each decoding time step. We also propose an order-align strategy to improve the orderliness of recognition result on skew images. Figure 2 illustrates the overview of the proposed model architecture.

3.1 Encoder

As shown in Fig. 2, the encoder extracts features f from input image X. It consists of a DSGate module and an IncepGate module, being mainly made up of a sequence of depthwise separable gate block (DSGate) and Inception gate block (IncepGate) respectively.

DSGate Block depicted in Fig. 3(a) is a residual block. Residual connections with element-wise sum operation enable strengthening the parameters update during backpropagation. It can be written as

$$\tilde{x} = x + F(x) \tag{1}$$

DSGate is motivated by the Inverted Residual Block in MobileNet V2 [7]. It can be factored into three parts: mapping $M(\cdot)$ from input space $\mathbf{R}^{H \times W \times C}$ to the lower dimensional depthwise convolution space $\mathbf{R}^{H \times W \times \frac{c}{2}}$, extracting $G(\cdot)$ image feature in the low dimensional space, and projecting $H(\cdot)$ back to the original dimensional input space. Then, Eq. 1 can be written as:

$$\tilde{x} = x + H(G(M(x))) \tag{2}$$

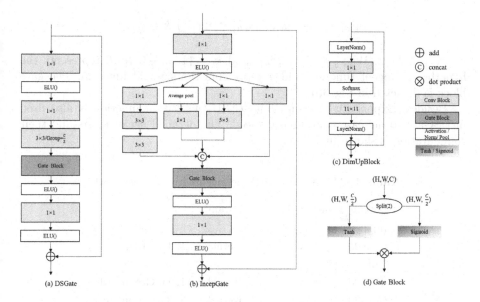

Fig. 3. The structure of primary building blocks in FPRNet. a) DSGate is an efficient block widely used in the encoder. b) IncepGate can learn spatial patterns at different scales. c) DimUpBlock is used in the front of FPRNet. d) the attention gates used to control the information flow and enhance model representation.

In DSGate block, $M(\cdot)$ and $H(\cdot)$ are both a 1×1 convolution. $G(\cdot)$ consists of a depthwise separable convolution and a gate block. Depthwise separable Convolutions are a key building block for many efficient neural network architectures [17,18]. Standard convolution takes an $h_i \times w_i \times d_i$ input tensor and produces an $h_i \times w_i \times d_j$ output tensor applying convolution kernel $K \in R^{k \times k \times d_i \times d_j}$. By splitting a full convolution into two separate layers, including a depthwise convolution and a 1×1 pointwise convolution, the computational cost drop from $h_i \times w_i \times d_i \times d_j \times k^2$ to $h_i \times w_i \times d_i \times (d_j + k^2)$. With the help of depthwise separable convolution, our architecture is trained at a faster convergency speed and lower computational cost without significantly reducing recognition accuracy.

The gates block, which controls the information flow has shown the strength in filtering-out unimportant background noise and sharpening signals [19]. Inspired by typical works as [12,20]. The attention gate structure with both tanh and sigmoid is also adopted by newer publications such as [21]. Let x of dimension $H \times W \times C$ be the input of the gate block. We propose to split the feature maps into two parts along with channel C. The first one refers to a traditional *tanh* activation, while the second one refers to the gating operation modeled using a *sigmoid* activation and acts as a self-attention mechanism over the first part. Let i, j and k be the index in the feature map, each pixel of output o can be then computed:

$$o[i, j, k] = tanh(x[i, j, k] \times \sigma(x[i, j, k + \frac{C}{2}])) \tag{3}$$

where $i \in \{1, ..., W\}$, $j \in \{1, ..., H\}$,and $k \in \{1, ..., \frac{C}{2}\}$. $tanh$ represents the hyperbolic tangent function and σ represents the sigmoid function. The gate block is depicted in Fig. 3(d).

Besides, Exponential Linear Units (ELU) activation function is also used in DSGate to increase convergence speed. The detailed list of layers and their configuration is shown in Table 1.

Table 1. ARCHITECTURE details used in DSGate block

Layers	Output shape	Param.
Input	(c, h, w)	
Conv2d	$(\frac{c}{2}, h, w)$	Kernel = (1, 1), Group = 1
ELU	$(\frac{c}{2}, h, w)$	
Conv2d	(c, h, w)	Kernel = (1, 1), Group = 1
Conv2d	(c, h, w)	Kernel = (3, 3), Group =c
Split	$(\frac{c}{2}, h, w)(\frac{c}{2}, h, w)$	dim = 1
Tanh	$(\frac{c}{2}, h, w)$	
Sigmoid	$(\frac{c}{2}, h, w)$	
Mul	$(\frac{c}{2}, h, w)$	
Conv2d	(c, h, w)	Kernel = (1, 1), Group = 1
ELU	$(\frac{c}{2}, h, w)$	

IncepGate Block is depicted in Fig. 3(b), which is similar to DSGate. The most significant difference is changing depthwise separate convolution to an Inception block which is proposed in GoogLeNet [22]. It allows for the utilization of varying convolutional filter sizes to learn spatial patterns at different scales. In the adopted Inception block, the input feature map is processed by four branches, and the results are jointly entered into the following blocks to get the final output. The mapping, projecting, activation, and residual connections are fully reserved.

Now, we describe more details in the encoder. It takes an image $X \in R^{H \times W \times 3}$ as input. Firstly, the input is fed to DimUpBlock, which is presented in Fig. 3(c). DimUpBlock starts with a layer normalization, which directly estimates the normalization statistics from the summed inputs within a single layer without introducing any new dependencies between training cases. The input features are then projected to a high dimension with a 1×1 convolutional layer. After normalizing by softmax, the feature will be fed to a depth-wise convolution with a 11×11 filter. Following a layer normalization layer and residual connection, the output of DimUpBlock is given to three continuous DSGate blocks and five IncepGate blocks. Noteworthy, the max pooling layer, whose stride is 2, is used after every two blocks to reduce feature map size. After this step, we have computed the feature $f \in R^{\frac{H}{16} \times \frac{W}{16} \times 256}$.

3.2 Decoder

As described above, we cannot use language models to correct typos in essay recognition tasks because students may write incorrectly, which is a factor in essay scoring. Thus, we train our model with the standard Connectionist Temporal Classification(CTC) loss function [6], which is termed as an alignment-free approach to sequence transduction. Before feeding the encoder-extracted feature f to the CTC function, we must devise methods to serialize the two-dimensional features to one-dimensional features.

OrigamiNet [16] proposes to use a series of up-scaling operations followed by convolution to unfold the 2D signal into single-line features. This approach achieves good results on multi-line English handwriting recognition, but the HCER task needs to face thousands of characters, multi-layer convolutions after up-scaling bring in a vast number of parameters. Through experiments, it was found that reducing the number of up-scaling layers and convolutions would seriously weaken the unfolding effect. In addition, the down-sampling operation performed on the feature map in the horizontal direction appears to be severely information-deficient, and the model cannot retain the implicit spatial information of the text effectively.

Owing to the above problems, we propose the dimensionality reduction block (DimReduceBlock) to unfold feature maps. Let N be the size of the charset. The encoder-extracted feature f is fed into a single convolution with 1×1 kernel and changes to $N + 1$ dimensions. Afterward, we reassemble 2D feature maps from top to bottom, left to right, using the reshape operations. Moreover, ELU activation and layer normalization are also used after convolution. The architecture of DimReduceBlock is shown in Fig. 2.

Similar to [25], the proposed dimensionality reduction mechanism based on reshape operation better preserves the encoder-extracted features and the independence between tokens. Each location in the serialized feature map focuses more on the visual information within the field of perception. In turn, a more accurate implicit alignment between the output text and the original image is achieved. The advantages of the decoder will be further demonstrated visually in the Sect. 5.

3.3 Order-Align Strategy

One of the critical internal problems of full-page text recognition is the unknown number of characters to be recognized. So we must guarantee that the length of the sequence output from the encoder is greater than the maximum value of the sequence length in all data. CTC can handle such unaligned sequences relatively well. However, many blank tokens lead to degradation of the decoding effect.

238　　T. Su et al.

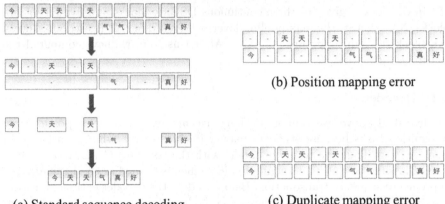

(b) Position mapping error

(a) Standard sequence decoding

(c) Duplicate mapping error

Fig. 4. Two common mapping problems during decoding. a) A standard decoding process with duplicate removal and blank character removal; b) text is incorrectly mapped to an adjacent token; c) text is repeatedly predicted by multiple tokens.

Especially when dealing with skewed text or images with large line spacing, problems such as those in Fig. 4 may occur. Words on the image may be mapped into adjacent tokens in the sequence or be duplicated. To solve these problems, we propose an order-align strategy to improve the orderliness of recognition results in the inference.

The execution flow of order-align strategy is described as follows. First, we iterate through each row of the 2D sequence and find the first non-zero element in each row, and once found use the current row and column as starting markers. Since there may be more than one empty line separating two rows, we traverse each element of the current row forward from the start marker, and if its upper and lower adjacent elements are not zero, we overwrite its value to the row where the start marker is located. Then, we mark the row where the non-zero element is located as the current row and continue traversing forward. Also, if we encounter the same character in the top and bottom adjacent elements, we keep only one character left.

The details are described in the *Algorithm* 1 below. In this way, we can move all elements on the upper and lower lines to the center line and then decode them to get an ordered string. This strategy can effectively improve the sequence orderliness in the inference stage. The effectiveness will be verified through experiments presented in Sect. 4.3.

Algorithm 1. Order-align Strategy

Input: *TextIndex*: the 1D-sequence of characters' index infered from model.
Output: *OrderedIndex*: ordered index sequence of characters
1: **function** ORDER-ALIGN(*TextIndex*)
2:　　reshape *TextIndex* to 2D with the shape of (h, w) as sq
3:　　**for** $i = 0 \to h - 1$ **do**
4:　　　　Iterate the row to find the first non-zero element,

5:    if found then record it's position $curRow(=i)$, $curCol$;
       otherwise continue.
6:    **for** $j = curCol \rightarrow w - 1$ **do**
7:        Check the upper and lower rows of $curRow$ at j
8:        If lower row at j is not zero, set $sq[i,j] = sq[curRow+1,j]$,
           set curRow and lower row at j to 0, set $curRow$ to $curRow + 1$.
9:        If upper row at j is not zero, set $sq[i,j] = sq[curRow-1,j]$,
           set curRow and upper row at j to 0,
           and set $curRow$ to $curRow - 1$.
10:       If character at $(curRow, j)$ is not zero, set $sq[i,j]$ to $sq[curRow,j]$
           and set $sq[curRow,j]$ to 0.
11:   **end for**
12:   **end for**
13:   reshape the 2D-sequence sq to 1D with the shape of $(w*h)$ as $OrderedIndex$
14:   **return** OrderedIndex
15: **end function**

---

4   Experiments and Results

4.1   Dataset

The FPRNet is proposed to solve the handwritten Chinese essay recognition. We specially collected thousands of essay images by school-aged students from the four provinces of Heilongjiang, Jilin, Hebei, and Jiangsu. The collected images include different grid backgrounds, criticism marks, and captured using different methods. Some details of the experimental data is shown in the Table 2.

Table 2. The distribution of experimental dataset

Region	Capture method	Type	Count
Jilin	Scan	Junior	133
Hebei	Scan	Senior	621
Jiangsu	Scan	Senior	317
Jilin	MobilePhone	Junior	101
Hebei	MobilePhone	Primary	66
Jilin	DocumentCamera	Junior	402
Hebei	DocumentCamera	Senior	211
Hebei	DocumentCamera	Senior	90
Heilongjiang	DocumentCamera	Junior	78
Total			**2019**

Since the images were collected in different ways and vary greatly in color, lighting, and size. We needed a uniform pre-processing of the data. The maximum value filter and minimum value filter were executed first, and all images were unified by padding to 1440 pixels height and 720 pixels width. We collected a total of 2019 authentic images, using 1718 of them as the training set and the remaining 301 for testing. Statistically, the dataset contains a total of 3464 different characters, which covers almost all common Chinese characters.

In addition to the complete dataset described above, the state-of-the-art approach [16] we compare requires square multi-line text as input. Therefore, we select 938 images from Hebei and Jiangsu, slice them by paragraph, and padding them to 800 pixels height and 800 pixels width. We obtained a total of 2336 paragraph images, 2024 in the training set and 312 in the testing set.

Most of our experiments are carried out on the full-page dataset, and the experiments for comparison with baseline will be carried out on the paragraph dataset. The specific experimental results will be shown in Sect. 4.3.

4.2 Experimental Setup

Our proposed FPRNet model works with the PyTorch framework and is trained on a single GeForce GTX TITAN X(12G) GPU. In all experiments, the networks are trained with the Adam [23] optimizer, with an initial learning rate of 10^{-2}, exponentially decay to 0.001 over 180k steps. The weight decay factor is the same for all the convolution layers in FPRNet. We use the same default data augmentation module as in Origaminet for a fair comparison.

Three different metrics have been used in the experimental evaluation of the proposed method. One is the Character Error Rate (CER), using the Levenshtein Edit Distance algorithm. It is the ratio between insertion, deletion, and substitution errors and the total number of characters. The formula is as shown in Eq. 4

$$CER = \frac{\sum_{i=1}^{k} d_{lev}(\hat{y}_i, y_i)}{\sum_{i=1}^{k} y_{len_i}} \tag{4}$$

where k is the number of images in the data set. We also use the CER normalized by label length (nCER) to measure the recognition accuracy for a single image. In addition, We introduce BLEU, an indicator reflecting sentence fluency in machine translation. The higher the BLEU value, the better the fluency of the sentence. The calculation formula is Eq. 5,

$$BLEU = BP \times \exp(\sum_{n=1}^{N} w_n \log p_n) \tag{5}$$

where p_n represents the n-gram accuracy after text block correction, which reflects the accuracy of translation, and BP represents the penalty factor.

4.3 Experimental Results

In this section, we present the comparison with other approaches on the full-page Handwritten Chinese essay datasets and paragraph sub-datasets, which are introduced in the Sect. 4.1. We compare our model with the OrigmaniNet [16] under different data inputs. The image in the paragraph datasets is an aspect ratio of 1, the same as the original paper. For encoders, we choose the recommended GTR-8 [24] to train. The other hyperparameters are the same as those set in the original paper. We also experimented with both baseline and FPR-Net on the full-page dataset with an image aspect ratio of 2, which is a more suitable input specification for the HCER task. We also compare the recognition results with those of the mainstream handwritten OCR API service provided by Tencent. The results were shown in Table 3.

From the results, we observe that our proposed FPRNet model achieves better performances with fewer parameters. Compared to the baseline model, our approach reaches a CER of 6.98% better than 10.25% on the paragraph datasets. On the full-page datasets, our approach also achieves a better performance with a CER of 7.58% comparing with the CER of 10.25% on baseline approach. As expected, the CER decreases from 7.58% to 5.83% along with the addition of order-align strategy, which directly demonstrates the effectiveness of the strategy.

In Table 3, we also make a comparison between our proposed method and an influential handwritten OCR provided by Tencent, an industry-leading cloud service vendor, which employs the classic two-stage approach and trains on multiple levels of Chinese handwritten data. Evaluation on the full-page essay dataset yields a CER of 7.42%. Achieving comparable results without line annotations demonstrates the advantages of our proposed segmentation-free approach.

Table 3. Comparison of the FPRNet with the baseline approaches and Cloud API on Handwritten Chinese Essay dataset

Model	CER (%)	nCER (%)	BLEU	Para#
Paragraph datasets				
[16]	10.25	10.17	80.17	4.1M
Ours	6.98	7.32	84.56	2.5M
Full-page datasets				
[16]	15.41	17.65	75.01	
Ours	7.58	8.09	83.78	
Ours + Order-align	**5.83**	**6.02**	**86.67**	
Tencent Cloud API	7.42			

We further visualize the advantage of the proposed architecture and discuss the reasons for the experimental result. Figure 5 shows the saliency map of the same characters under our proposed method and OrigamiNet.By comparision,it can be found that the region of interest for each character in the baseline method is concentered around this line, while FPRNet achieves a better alignment to correspond the output text and its corresponding position.

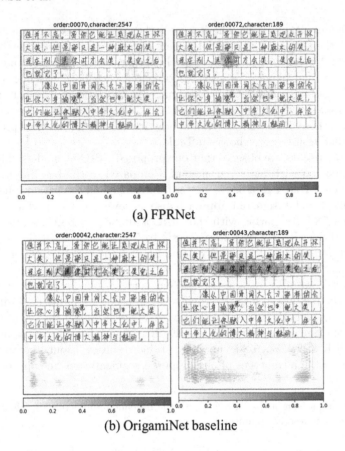

(a) FPRNet

(b) OrigamiNet baseline

Fig. 5. Saliency maps of FPRNet and OrigamiNet under the same sample.

5   Conclusion

In this paper, we propose an end-to-end full-page recognition network named FPRNet, performing well on the handwritten Chinese essay recognition (HCER) task. This method brings several advantages. First, the encoder incorporates new ideas from multi-resolution structure, which bring the ability to learn spatial pattern at different scales. Second, the reshape based dimensionality reduction mechanism applying to decoder achieves a better alignment to correspond the output text and its corresponding position. Finally, the proposed order-align strategy greatly guarantees the sequence orderliness of the inference stage.. On the HCER task, FPRNet achieves better results using fewer parameters than the baseline segmentation-free method. Comparable performance was also achieved compared to traditional two-stage methods.

Acknowledgment. Great thanks to Jifeng Wang. This work was supported by the National Key Research and Development Program of China (Grant No. 2020AAA0108003) and National Natural Science Foundation of China (Grant No. 62277011 and 61673140).

References

1. Su, T., Zhang, T., Guan, D., Huang, H.: Off-line recognition of realistic Chinese handwriting using segmentation-free strategy. Pattern Recogn. **42**(1), 167–182 (2009)
2. Wang, Q., Yin, F., Liu, C.: Handwritten Chinese text recognition by integrating multiple contexts. IEEE TPAMI **34**(8), 1469–1481 (2011)
3. Toledo, J., et al.: Handwriting recognition by attribute embedding and recurrent neural networks. In: ICDAR (2017)
4. Lei, K., et al.: Convolve, attend and spell: an attention-based sequence-to-sequence model for handwritten word recognition. In: German Conference on Pattern Recognition (2018)
5. Wu, Y., Yin, F., Liu, C.: Improving handwritten Chinese text recognition using neural network language models and convolutional neural network shape models. Pattern Recogn. **65**, 251–264 (2017)
6. Graves, A., Fernández, S., Gomez, F., et al.: Connectionist temporal classification: labelling unsegmented sequence data with recurrent neural networks. In: Proceedings of the 23rd International Conference on Machine Learning, pp. 369–376 (2006)
7. Sandler, M., Howard, A., Zhu, M., et al.: Mobilenetv 2: inverted residuals and linear bottlenecks. In: CVPR (2018)
8. Chung, J., Delteil, T.: A computationally efficient pipeline approach to full page offline handwritten text recognition. In: ICDAR Workshop (2019)
9. Carbonell, M., Fornés, A., Villegas, M., et al.: A neural model for text localization, transcription and named entity recognition in full pages. Pattern Recogn. Lett. **136**, 219–227 (2020)
10. Messina, R., Louradour, J.: Segmentation-free handwritten Chinese text recognition with LSTM-RNN. In: ICDAR (2015)
11. Wu, Y., Yin, F., et al.: Handwritten Chinese text recognition using separable multi-dimensional recurrent neural network. In: ICDAR (2017)
12. Yousef, M., Hussain, K.F., Mohammed, U.S.: Accurate, data-efficient, unconstrained text recognition with convolutional neural networks. Pattern Recogn. **108**, 107482 (2020)
13. Bluche, T.: Joint line segmentation and transcription for end-to-end handwritten paragraph recognition. In: NeuIPS (2016)
14. Wigington, C., Tensmeyer, C., Davis, B., et al.: Start, follow, read: end-to-end full-page handwriting recognition. In: ECCV (2018)
15. Coquenet, D., Chatelain, C., Paquet, T.: End-to-end handwritten paragraph text recognition using a vertical attention network. IEEE Trans. Pattern Anal. Mach. Intell. (2022). (Accepted)
16. Yousef, M., Bishop, T.: OrigamiNet: weakly-supervised, segmentation-free, one-step, full page text recognition by learning to unfold. In: CVPR (2020)
17. Howard, A., Zhu, M., Chen, B., et al.: MobileNets: efficient convolutional neural networks for mobile vision applications. ArXiv:1704.04861 (2017)
18. Chollet, F.: Xception: deep learning with depthwise separable convolutions. In: CVPR (2017)

19. Wang, F., Jiang, M., Qian, C., et al.: Residual attention network for image classification. In: CVPR (2017)
20. Srivastava, R.K., Greff, K., Schmidhuber, J.: Highway networks. ArXiv:1505.00387 (2015)
21. Coquenet, D., Chatelain, C., Paquet, T.: Recurrence-free unconstrained handwritten text recognition using gated fully convolutional network. In: International Conference on Frontiers in Handwriting Recognition, pp. 19–24 (2020)
22. Szegedy, C., Liu, W., Jia, Y., et al.: Going deeper with convolutions. In: CVPR (2015)
23. Kingma, D., Ba, J.: Adam: a method for stochastic optimization. ArXiv:1412.6980 (2014)
24. Yousef, M., Hussain, K., Mohammed, U.: Accurate, data-efficient, unconstrained text recognition with convolutional neural networks. Pattern Recogn. **108**, 107482 (2020)
25. Coquenet, D., Chatelain, C., Paquet, T.: SPAN: a simple predict and align network for handwritten paragraph recognition. In: International Conference on Document Analysis and Recognition (2021)

Active Transfer Learning for Handwriting Recognition

Eric Burdett[1], Stanley Fujimoto[1], Timothy Brown[2], Ammon Shurtz[2], Daniel Segrera[2], Lawry Sorenson[2], Mark Clement[2](✉) ⓘ, and Joseph Price[2] ⓘ

[1] Ancestry Corporation, Lehi, UT 84043, USA
{eburdett,sfujimoto}@ancestry.com
[2] Brigham Young University, Provo, UT 84602, USA
clement@cs.byu.edu

Abstract. With the advent of deep neural networks, handwriting recognition systems have recently achieved remarkable performance. Unfortunately, to achieve high-quality results, these models require large amounts of labeled training data, which is difficult to obtain. Various methods have been proposed to reduce the volume of training data required. We propose a framework for fitting new handwriting recognition models that joins both active and transfer learning into a unified framework. Empirical results show that our method performs better than traditional active learning, transfer learning, and standard supervised training methods.

Keywords: Handwriting recognition · Active learning · Transfer learning

1 Introduction

Handwriting recognition promises to reduce the human effort of manually transcribing handwritten text. To perform recognition on new sets of documents, however, human annotators must manually label a subset of the documents to train a handwriting recognition model which can then transcribe the content of remaining documents. These new sets of documents could be in another human language or in a new domain (death records vs birth records). Unfortunately, for high quality results, large amounts of hand labeled data is required and because this labeling process is expensive and time consuming, many digitized records have not been transcribed. Reducing the annotation effort will allow handwriting recognition to be applied to more records.

In attempting to reduce this manual annotation effort, researchers have proposed solutions utilizing transfer learning and domain adaptation techniques [12–14], synthetic data creation [2,4,6,7,14], self-supervision [1,14], and active learning [18,22]. These methods often sacrifice accuracy and there is rarely a substitute for well labeled training data. While each method has been extensively studied, little research has explored the advantages of using multiple methods in tandem.

© The Author(s), under exclusive license to Springer Nature Switzerland AG 2022
U. Porwal et al. (Eds.): ICFHR 2022, LNCS 13639, pp. 245–258, 2022.
https://doi.org/10.1007/978-3-031-21648-0_17

Transfer learning refers to training an existing handwriting recognition model, which was designed to transcribe documents from a previous data set, to recognize documents from a new set. This allows new models to utilize previously learned knowledge on new untranscribed data. Active learning involves selecting the documents that will provide the most knowledge to the model, labeling them, and then using them to train.

Combining active and transfer learning allows us to utilize previously learned knowledge from similar document sets while also providing a method for selecting the most important samples to label for further training. This approach helps to overcome the weaknesses of both transferred models and models obtained through active learning.

This approach both minimizes the annotation effort and provides adjustable accuracy, dependant on the number of labeled samples. Empirical results show the effectiveness of the active transfer framework compared to traditional active learning, transfer learning, and standard supervised training schemes.

2   Related Work

Significant research has been conducted to reduce the manual annotation effort for handwriting recognition. These methods usually fall under the realm of transfer learning, active learning, synthetic data creation, and self-supervision. We will briefly discuss each of these methods and consider their impact on reducing manual annotation.

2.1   Transfer Learning

Transfer learning is a broad term used to describe the concept of utilizing past knowledge learned on previous tasks for the current task. In the simplest case, a model trained on one dataset is used to make predictions on a different dataset. Fine-tuning a pre-trained model with additional labeled data from the target domain is also a fairly common computer vision practice [24].

Other transfer learning techniques involve training mechanisms that use labeled data from a source domain and unlabeled data from the target domain to condition the model to be more receptive to target data. For example, Domain Adversarial Networks [8] use a training scheme to encourage domain confusion using a domain classification network and a gradient-reversal layer.

When used in conjunction with standard supervised training in the source domain, the process encourages the feature-extraction layers to become domain-invariant. This method was applied to handwriting recognition where a synthetic dataset was used as the source domain and real handwritten text as the target domain [13]. The results were impressive with substantial reductions in error rates. However, error rates were still too high to be used in practice. Others applying domain adversarial networks to handwriting recognition include [14] and [12], with the latter seeking to disentangle content from style in handwritten images.

2.2   Active Learning

In contrast to selecting random instances for a training set, active learning is the process of selecting the most informative samples that will lead to superior model performance. The key element to active learning is the selection of an uncertainty function ϕ that will determine which samples will be annotated. Common examples of ϕ include using the magnitude of the most probable labeling, the margin between the first and second most probable labelings, and Shannon entropy [22].

Active learning has not been a well studied topic in the realm of handwriting recognition and it is limited to a small set of papers. Researchers have used n-best list entropy and the global entropy reduction for the uncertainty functions and were able to reduce the annotation effort by eighteen percent [22]. Most recently, [18] took an active learning approach by using derivational entropy to calculate ϕ. Their results indicated a small reduction of annotation effort, but their recognition error rates were still high due to their use of Hidden Markov Models rather than state-of-the-art deep neural networks. In both cases, the authors considered active learning in the context of a single dataset without utilizing previously trained models, labeled data from other datasets, or applying any transfer learning techniques.

2.3   Active Transfer Learning

To the best of our knowledge, the use of active learning and transfer learning in a single framework has not been studied for handwriting recognition; however, it has been studied in other areas of computer vision and natural language processing. To avoid the cold-start problem often seen in standard active learning, researchers [11] used a set of pre-trained model weights to initialize the active learner. Others have built upon this framework by using domain separation classifiers and clustering methods for inputting labels to provide slight increases in accuracy with the same amount of data [5,17,19].

In most cases, active learning and transfer learning are performed in two separate distinct steps. More recently, however, researchers [20] developed a loss function integrating both active and transfer learning techniques. This framework uses a domain adversarial network [8] with the addition of labeled data from the target domain being actively selected and trained in a supervised manner concurrently with domain adversarial training.

This research will apply Active Transfer Learning specifically for handwriting recognition.

3   Methodology

We propose a framework for fitting handwriting recognition models to new datasets by combining active and transfer learning. Rather than starting with a random weights initialization, we use a set of weights pre-trained on another dataset. Then, instead of randomly selecting images to label, we actively select

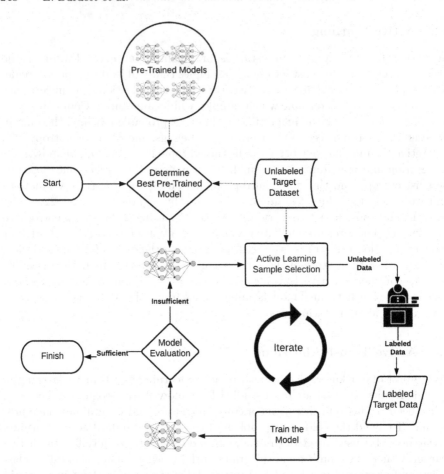

Fig. 1. System architecture for fitting a new dataset. We start by determining which of our pre-trained models will be best suited for the new dataset. The pre-trained model along with the unlabeled target data are given to the active learner for sample selection. The chosen samples are annotated and added to the pool of labeled target data. The model is trained and immediately evaluated. If it is determined after training that the model has achieved sufficient recognition accuracy, we are finished. Otherwise, we repeat the process until the desired accuracy is obtained.

images that will be the most helpful for training. The labeled images are added to the pool of labeled target data and the model is subsequently trained. This process of actively selecting labels and training the model with the entire pool of labeled data is repeated iteratively until the desired accuracy is achieved or until the model has fully converged and no improvement is observed.

The proposed framework is designed to be modular, where pre-trained weights and active learning algorithms can easily be inserted and replaced. The following provides a high-level overview of each step of the framework and the implementation we used to achieve our results. Figure 1 provides a visual representation of the ideas presented in this section.

3.1 Model Weights Initialization

The first step in our framework is determining which set of pre-trained model weights will provide the best initialization for the model that will be fine-tuned on data from the target domain. We explore using the weights from models that have been trained on other datasets that are somewhat similar to the target domain. The use of a self-supervision technique such as SeqCLR [1] for pre-training under this framework remains as an alternative method for future research.

Pre-trained Recognition Models. If multiple models exist that have been trained on data that are somewhat similar to the target domain, we can evaluate these models in two ways. First, we can evaluate the weights using a test set already created for the target domain. If a test set has not been created, we can use the average confidence score given by the recognition model when provided with a random sample of images from the target set.

For a confidence measure, we make use of the Connectionist Temporal Classification (CTC) loss function [9]. CTC provides a loss between the model's output, which is a two-dimensional matrix representing the probability of each character occurring for each timestep in the sequence, and the provided ground truth. For generating a confidence score, we can utilize this same loss function by using the model's best prediction as the ground truth, which is commonly the best-path (greedy) decoding of the model's output [9]. We can then transform the CTC loss value, given as the negative log likelihood, back to a probability by applying an exponential and removing the negative. Equation 1 provides the probability that the model's prediction is correct with L being the result of CTC loss when providing the best-path decoding of the model's prediction as the ground truth.

$$p = e^{-L} \tag{1}$$

Using this equation, we can provide a confidence score representing the general performance of the model. We do this by taking n random samples from the dataset and calculating the mean of all confidence scores. The model that provides the highest mean confidence over a random sample is used. The weights from this model will then be used as the starting initialization for the model in the active transfer framework.

3.2 Active Learning Sample Selection

The active learning step involves making predictions on samples from the unlabeled target set and scoring them with an uncertainty function, ϕ. Although various uncertainty functions can be inserted here, we choose to use standard entropy adapted for using with Connectionist Temporal Classification [9]. Because we can obtain a probability score for each prediction as given in Eq. 1, we can calculate the standard Shannon entropy to be used as the uncertainty function.

$$\phi = -p * \log(p) \tag{2}$$

Incorporating the CTC loss as given in Eq. 1 results in the following equation:

$$\phi = -e^{-L} * \log(e^{-L}) \tag{3}$$

The above equation simplifies to produce our final uncertainty function:

$$\phi = e^{-L} * L \tag{4}$$

Using this criterion, we then take the top k images with the highest uncertainty scores to give to the annotator for labeling. These images are added to the set of labeled data in the target domain, which is subsequently used for standard supervised training. For all of our experiments, we use a value of $k = 1000$ and rather than performing the actual annotation, we retrieve the provided transcription from the training set of that particular dataset.

3.3   Supervised Training

With the labeled data given by the annotator, we train the model using standard supervised training with CTC loss [9]. With each iteration of the active transfer learning process, we train with the entire pool of labeled target data. This pool is increased each iteration by the value of k as described in Sect. 3.2.

For each iteration, the model is trained for an unlimited number of epochs with early stopping used when there is no improvement on the validation set. The set of model weights that achieved the lowest loss on the validation set is carried on to the next step of the framework. Training ends after 10 epochs without improvement on the validation set. Note that performing early stopping requires a small subset of data from the target domain to be labeled beforehand. If this extra step of labeling is not possible, the model can be trained for a fixed number of epochs in each iteration. We resize all images to a height and width of (32, 128) and apply the random grid warp augmentation [23] during the training process.

Recognition Model. For all of our experiments, we use the recognition model described in [21]. The model consists of an encoder with several gated convolutional layers [3] and a decoder containing several layers of bidirectional GRUs. The goal of this work is not necessarily to achieve state-of-the-art performance on certain benchmarks, but rather to show the impact of active transfer learning for handwriting recognition compared to the traditional active, transfer, and standard supervised learning methods. Any recognition model can be easily inserted into the active transfer framework.

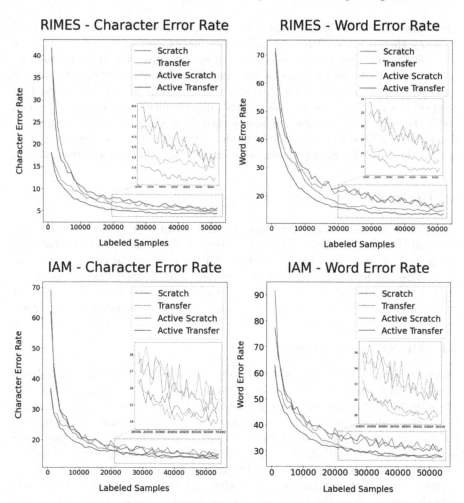

Fig. 2. Methods comparison on the RIMES dataset (top) and IAM dataset (bottom). The recognition models for the *Transfer* and *Active Transfer* methods were initialized with weights from a pre-trained model on the IAM and RIMES datasets respectively. For the RIMES dataset, the Active Transfer method achieved better error rates regardless of the number of labeled samples given for training. For the IAM dataset, the Active Transfer method achieved only comparable performance with the Active method likely because the pre-trained RIMES model provided little to no advantage when training on IAM.

3.4   Model Evaluation

Each iteration after the model has been trained, we evaluate its performance with a labeled test set in the target domain using the character error rate and word error rate. At times, it can be difficult to determine how much labeled data is needed to train a recognition model for a specific task. By evaluating the model each iteration, we can acquire enough labeled data to achieve the level of accuracy that is required for the task while not expending the effort and cost of labeling more data than is necessary. In essence, this step provides tunable accuracy, where the user can decide how much additional labeled data to provide depending on the current model performance. Once the model has achieved an acceptable level of accuracy or the model's improvement has plateaued, we can break the loop and consider the active transfer training process complete. If the model is still converging and the model performance has not yet crossed an acceptable threshold, we can continue to iterate until that point is reached.

4   Results

To validate our methodology, we test our framework on the IAM [16], RIMES [10], and CVL [15] datasets. They were selected because they are somewhat similar in format, but vary slightly in style. All are modern datasets and contain black/blue handwriting with white backgrounds. All trained models are evaluated at the word-level on their respective train-test splits. We evaluate each of the training setups using the standard character error rate and word error rate.

4.1   Methods Comparison

To evaluate the active transfer framework, we compare the results of training a handwriting recognition model using each of the following formats: Scratch, Transfer, Active Scratch, Active Transfer. Scratch is the typical supervised training method where the model is trained from scratch with a random weights initialization and a random selection of labeled training images. In Transfer the model weights are initialized from a pre-trained recognition model. Training images are randomly sampled from the target domain. In Active Scratch the model weights are trained from scratch with a random initialization, but data is actively selected from the target domain. In Active Transfer the model weights are initialized from a pre-trained recognition model. Training images are actively selected from the target domain.

When comparing these results, we observe the character error rate and word error rate when given the specified number of labeled samples. Figure 2 provides the results of the IAM and RIMES datasets across all four methods. The model for the RIMES dataset is initialized with a set of weights pre-trained on the IAM dataset. The model for the IAM dataset uses weights from RIMES.

For the RIMES dataset, the *Active Transfer* method was the clear winner and achieved better error rates compared to all other methods regardless of the number of labeled training samples. On the other hand, the *Active Transfer* method on the IAM dataset performed comparable or slightly worse when judged against the *Active* method. This is likely due to the fact that the pre-trained RIMES model weights were not as helpful to the IAM dataset, which is clearly seen when comparing the *Scratch* and *Transfer* methods. Thus, the weights initialization plays a large role in determining performance. Section 4.3 shows

Table 1. Best error rates and required samples to achieve fully-trained *Scratch* performance on the RIMES, IAM, and CVL datasets. This table is meant to show the reduction in labeled samples to achieve the same performance as a fully-trained *Scratch* model. For example, on the RIMES dataset, the *Active Transfer* method achieved an error rate of 5.20% with only 18,000 samples as opposed to 49,000 if the model was trained from *Scratch*. This results in a reduction of 31,000 samples. Note that the RIMES, IAM, and CVL datasets used pre-trained weights from the IAM, RIMES, and IAM datasets respectively.

RIMES Dataset

Method	CER	WER	Samples (CER)	Samples (WER)
Scratch	5.20%	15.87%	49,000	49,000
Transfer	4.82%	15.19%	48,000	48,000
Active Scratch	4.97%	14.40%	29,000	29,000
Active Transfer	**4.33%**	**13.21%**	**18,000**	**20,000**

IAM Dataset

Method	CER	WER	Samples (CER)	Samples (WER)
Scratch	14.63%	29.96%	48,000	53,000
Transfer	15.06%	30.26%	> 53,838	> 53,838
Active Scratch	**13.76%**	**27.34%**	**32,000**	27,000
Active Transfer	14.05%	27.81%	35,000	**23,000**

CVL Dataset

Method	CER	WER	Samples (CER)	Samples (WER)
Scratch	18.38%	29.86%	83,000	83,000
Transfer	13.73%	29.78%	**3,000**	87,000
Active Scratch	15.78%	25.62%	17,000	15,000
Active Transfer	**13.62%**	**25.27%**	**3,000**	**12,000**

that the proposed evaluation metrics can predict how a set of model weights will perform in this framework.

These results indicate a substantial reduction in the amount of labeled samples that are necessary to achieve performance comparable to that of a model trained using the standard *Scratch* method using all available data. For the RIMES dataset, the *Scratch* method achieved a character error rate of 5.20% using 49,000 labeled samples. To achieve the same character error rate using the *Active Transfer* method, only 18,000 labeled samples are required. If we estimate it takes 15 s to label each image, we can assume it takes roughly 204 h to annotate 49,000 samples. Reducing the required samples to 18,000 requires a much lower annotation effort of 75 h, which is approximately a 63% reduction.

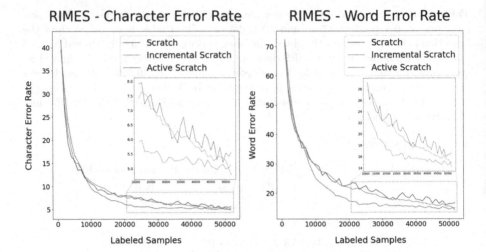

Fig. 3. Results of the incremental training method compared to active and scratch training on the RIMES dataset. Actively selecting training samples obviously provides a significant advantage when using fewer labeled samples. The *Incremental Scratch* method seems to be fairly comparable with the *Scratch* method throughout most of the training process. When all available data is used, the method achieves similar performance to *Active Scratch*. In this example, incremental training alone usually does not produce better results, but the combination of incremental training with active sampling can provide higher model performance.

We provide the full results of all the methods for the RIMES, IAM, and CVL datasets in Table 1. Although labeling reductions are significant, the reader should keep in mind that such reductions may not be possible when the source dataset for the pre-trained weights initialization is a substantially different style compared to the target dataset (i.e. handwriting style, parchment color, size of text, stroke thickness, etc.). An example of this is with the IAM dataset in Table 1, where little improvement is observed with or without the pre-trained weights initialization. With this in mind, the *Active Transfer* method seems to be an effective method for training handwriting recognition models, especially when the right pre-trained model is used to initialize weights.

It's also worthy to note the interesting results of the CVL dataset in Table 1. Using the pre-trained IAM model as the weights initialization was extremely beneficial as the *Transfer* method achieved similar performance to *Active Transfer* with the character error rate. However, the *Active Transfer* method was clearly superior to *Transfer* in reducing the word error rate. This is likely due to the fact that our uncertainty function for actively selecting samples provides an uncertainty for an entire word. Thus, words underrepresented in the dataset used for the pre-trained model weights will likely receive a greater uncertainty value and will be selected for labeling before other samples. Our uncertainty function inherently favors uncertain words, rather than uncertain characters. In the case

Table 2. Model weight evaluation metrics and final character error rates for the RIMES dataset using various pre-trained models for the weights initialization. *Confidence Eval* is calculated using Eq. 1. *CER Eval* provides the character error rate on the target domain's test set using the pre-trained model weights from the source domain. Both weight evaluation metrics are obtained before any fine-tuning is performed on the model. *Samples(CER)* is the minimum number of samples needed to reach the *Scratch* method's best character error rate, while *Final CER* is the character error rate on the test set after training a model using the *Active Transfer* method. The best character error rate achieved on the RIMES dataset using all available training data was 5.20%.

Model Weight Selection Metrics on the RIMES Dataset

Source Weights	Confidence Eval	CER Eval	Samples (CER)	Final CER
CVL	25.04%	74.65%	30,000	4.84%
IAM	**49.46%**	**36.30%**	**18,000**	4.33%
CVL + IAM	42.29%	38.96%	20,000	**4.24%**

of the CVL dataset in Table 1, the dramatic difference in word error rate between the *Transfer* and *Active Transfer* methods is likely due to the small amount of unique words available in the CVL dataset. Thus, if we can actively sample for labeling those unique or commonly used words, we should be able to reduce our word error rate sooner.

4.2 Incremental Iterative Training

It is interesting to note that for all the datasets in Fig. 2, the *Active* and *Active Transfer* methods achieved better error rates compared to the *Scratch* and *Transfer* methods even when all labeled samples were used. Our results empirically show the advantage of an incremental training approach where k samples are added to the training pool and the model is iteratively trained with the pool of labeled data whose size is increased by a value of k each iteration.

To ensure that the *Active* and *Active Transfer* methods were not obtaining better results solely due to an incremental training approach, we incrementally trained a model from scratch with k random samples (as opposed to k actively selected samples) being added to the pool of labeled data each iteration. For this experiment, we again used a value of $k = 1000$ and incrementally trained this model on the RIMES dataset. Figure 3 provides the results of this method compared to the *Active Scratch* and *Scratch* methods. Our results show the advantage of actively selecting labeled samples using an uncertainty function as opposed to an incrementally trained model using random samples, especially when using fewer labeled samples. For most of the training process, the *Scratch* and *Incremental Scratch* methods achieve fairly comparable performance, although an incremental training procedure may help stabilize training and allow for easier convergence.

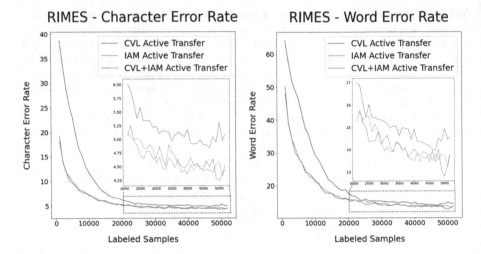

Fig. 4. Comparison of the *Active Transfer* method on the RIMES dataset using three different source weight initializations: CVL, IAM, CVL + IAM. The set of weights used for initialization seems to have a fairly large impact on the overall performance of the *Active Transfer* method.

4.3 Selection of Pre-trained Model Weights

Figure 4 provides the results of the *Active Transfer* method on the RIMES dataset while varying the source of the pre-trained weights. We observe the superiority of the IAM dataset compared to the CVL and combination of IAM and CVL datasets.

To ensure the two evaluation methods previously described for selecting weights for initialization will be good indications on how the model will perform overall, we provide these metrics in Table 2. Empirical results show that both evaluation metrics seem to be good indicators of superior pre-trained model weights when incorporated in the active transfer framework. The IAM dataset was superior to the CVL dataset and the model selection evaluation metrics agreed. The combined CVL + IAM dataset seemed to achieve comparable performance to IAM alone. The final error rate for CVL + IAM was slightly better, but IAM alone took slightly fewer samples to achieve comparable *Scratch* performance.

5 Conclusion

We have proposed an active transfer learning solution for fitting handwriting recognition models to new datasets. Empirical results showed a significant reduction in the manual annotation effort required to achieve comparable performance of a model trained from scratch and superior performance to traditional active or transfer learning techniques. This was especially evident when a strong pre-trained model is used. This method is especially useful for new collection of handwritten records with a similar style or format to those with an existing recognition model. The proposed implementation only considers the use of pre-trained model weights for transfer learning with an incremental active learning training process. Future work should consider additional, more sophisticated transfer learning methods described in Sect. 2 such as domain adversarial learning, synthetic data creation, and self-supervision. As the active transfer solution is modular, this work can be easily implemented. In addition, the experiments conducted only considered a few datasets with a similar style; the implications of transferring model weights from a substantially different source domain should be explored.

References

1. Aberdam, A., et al.: Sequence-to-sequence contrastive learning for text recognition. arXiv preprint arXiv:2012.10873 (2020)
2. Alonso, E., Moysset, B., Messina, R.: Adversarial generation of handwritten text images conditioned on sequences. In: 15th IAPR International Conference on Document Analysis and Recognition (ICDAR) (2019)
3. Bluche, T., Messina, R.: Gated convolutional recurrent neural networks for multilingual handwriting recognition. In: 2017 14th IAPR International Conference on Document Analysis and Recognition (ICDAR), pp. 646–651 (2017). https://doi.org/10.1109/ICDAR.2017.111
4. Chang, B., Zhang, Q., Pan, S., Meng, L.: Generating handwritten Chinese characters using cyclegan. In: 2018 IEEE Winter Conference on Applications of Computer Vision (WACV) (2018)
5. Chattopadhyay, R., Fan, W., Davidson, I., Panchanathan, S., Ye, J.: Joint transfer and batch-mode active learning. In: International Conference on Machine Learning (2013)
6. Davis, B., Tensmeyer, C., Price, B., Wigington, C., Morse, B., Jain, R.: Text and style conditioned Gan for generation of offline handwriting lines. In: The 31st British Machine Vision Conference (2020)
7. Fogel, S., Averbuch-Elor, H., Cohen, S., Mazor, S., Litman, R.: IEEE conference on computer vision and pattern recognition (CVPR). In: ScrabbleGAN: Semi-supervised Varying Length Handwritten Text Generation (2020)
8. Ganin, Y., et al.: Domain-adversarial training of neural networks. In: Csurka, G. (ed.) Domain Adaptation in Computer Vision Applications. ACVPR, pp. 189–209. Springer, Cham (2017). https://doi.org/10.1007/978-3-319-58347-1_10
9. Graves, A., Fernández, S., Gomez, F., Schmidhuber, J.: Connectionist temporal classification: labeling unsegmented sequence data with recurrent neural networks. In: International Conference on Machine Learning (2006)
10. Grosicki, E., Carré, M., Geoffrois, E., Prêteux, F.: Rimes evaluation campaign for handwritten mail processing. In: International Workshop on Frontiers in Handwriting Recognition (IWFHR 2006), pp. 231–235 (2006)
11. Kale, D., Liu, Y.: Accelerating active learning with transfer learning. In: 2013 IEEE 13th International Conference on Data Mining, pp. 1085–1090 (2013). https://doi.org/10.1109/ICDM.2013.160

12. Kang, L., Riba, P., Rusiñol, M., Fornés, A., Villegas, M.: Distilling content from style for handwritten word recognition. In: 2020 17th International Conference on Frontiers in Handwriting Recognition (ICFHR), pp. 139–144 (2020). https://doi.org/10.1109/ICFHR2020.2020.00035

13. Kang, L., Rusiñol, M., Fornés, A., Riba, P., Villegas, M.: Unsupervised adaptation for synthetic-to-real handwritten word recognition. In: 2020 IEEE Winter Conference on Applications of Computer Vision (WACV), pp. 3491–3500 (2020). https://doi.org/10.1109/WACV45572.2020.9093392

14. Keret, S., Wolf, L., Dershowitz, N., Werner, E., Almogi, O., Wangchuk, D.: Transductive learning for reading handwritten tibetan manuscripts. In: International Conference on Document Analysis and Recognition (ICDAR) (2019)

15. Kleber, F., Fiel, S., Diem, M., Sablatnig, R.: CVL-database: an off-line database for writer retrieval, writer identification and word spotting. In: 2013 12th International Conference on Document Analysis and Recognition, pp. 560–564 (2013). https://doi.org/10.1109/ICDAR.2013.117

16. Marti, U.V., Bunke, H.: The IAM-database: an English sentence database for offline handwriting recognition. Int. J. Doc. Anal. Recogn. 5, 39–46 (2002). https://doi.org/10.1007/s100320200071

17. Rai, P., Saha, A., Daumé, H., Venkatasubramanian, S.: Domain adaptation meets active learning. In: Proceedings of the NAACL HLT 2010 Workshop on Active Learning for Natural Language Processing, pp. 27–32. Association for Computational Linguistics, Los Angeles, California (2010). https://www.aclweb.org/anthology/W10-0104

18. Romero, V., Sánchez, J.A., Toselli, A.H.: Active learning in handwritten text recognition using the derivational entropy. In: 2018 16th International Conference on Frontiers in Handwriting Recognition (ICFHR), pp. 291–296 (2018). https://doi.org/10.1109/ICFHR-2018.2018.00058

19. Shi, X., Fan, W., Ren, J.: Actively transfer domain knowledge. In: Daelemans, W., Goethals, B., Morik, K. (eds.) ECML PKDD 2008. LNCS (LNAI), vol. 5212, pp. 342–357. Springer, Heidelberg (2008). https://doi.org/10.1007/978-3-540-87481-2_23

20. Singh, A., Chakraborty, S.: Deep active transfer learning for image recognition. In: 2020 International Joint Conference on Neural Networks (IJCNN), pp. 1–9 (2020). https://doi.org/10.1109/IJCNN48605.2020.9207391

21. de Sousa Neto, A.F., Bezerra, B.L.D., Toselli, A.H., Lima, E.B.: HTR-Flor++: a handwritten text recognition system based on a pipeline of optical and language models. In: Proceedings of the ACM Symposium on Document Engineering 2020. DocEng 2020, Association for Computing Machinery, New York, NY, USA (2020). https://doi.org/10.1145/3395027.3419603

22. Tsakalis, K.: Active learning for historic handwritten text recognition. Ph.D. thesis, University of Cambridge (2016)

23. Wigington, C., Stewart, S., Davis, B., Barrett, B., Price, B., Cohen, S.: Data augmentation for recognition of handwritten words and lines using a CNN-LSTM network. In: 2017 14th IAPR International Conference on Document Analysis and Recognition (ICDAR), vol. 01, pp. 639–645 (2017). https://doi.org/10.1109/ICDAR.2017.110

24. Zamir, A.R., Sax, A., Shen, W., Guibas, L., Malik, J., Savarese, S.: Taskonomy: disentangling task transfer learning. In: 2018 IEEE/CVF Conference on Computer Vision and Pattern Recognition, pp. 3712–3722 (2018). https://doi.org/10.1109/CVPR.2018.00391

Recognition-Free Question Answering on Handwritten Document Collections

Oliver Tüselmann[✉]⬤, Friedrich Müller⬤, Fabian Wolf⬤,
and Gernot A. Fink⬤

Department of Computer Science, TU Dortmund University, 44227 Dortmund,
Germany
{oliver.tueselmann,friedrich.mueller,fabian.wolf,
gernot.fink}@cs.tu-dortmund.de

Abstract. In recent years, considerable progress has been made in the research area of Question Answering (QA) on document images. Current QA approaches from the Document Image Analysis community are mainly focusing on machine-printed documents and perform rather limited on handwriting. This is mainly due to the reduced recognition performance on handwritten documents. To tackle this problem, we propose a recognition-free QA approach, especially designed for handwritten document image collections. We present a robust document retrieval method, as well as two QA models. Our approaches outperform the state-of-the-art recognition-free models on the challenging BenthamQA and HW-SQuAD datasets.

Keywords: Visual question answering · Information retrieval · Handwritten documents · Document understanding

1 Introduction

Question Answering (QA) is still an open and major research topic in a wide variety of disciplines [16,26,31]. Especially, the communities of Computer Vision (CV) and Natural Language Processing (NLP) focus on this task and made considerable progress [26,31]. Over the last few years, the Document Image Analysis (DA) community has shown an increasing interest in QA [15,16]. The majority of DA approaches tackle this task by adapting and using models from the NLP and CV communities [15,19,27]. Thereby, the text from a document image is transcribed and an answer is determined using a textual QA system [15,27]. This already leads to high performances for machine-printed document images with low recognition error rates [16]. However, the performances of these approaches decrease considerably on handwritten document images [15,16]. This is mainly due to the considerably reduced recognition accuracy, even though, substantial progress has been made in handwritten text recognition (HTR) over the last few years [8].

© The Author(s), under exclusive license to Springer Nature Switzerland AG 2022
U. Porwal et al. (Eds.): ICFHR 2022, LNCS 13639, pp. 259–273, 2022.
https://doi.org/10.1007/978-3-031-21648-0_18

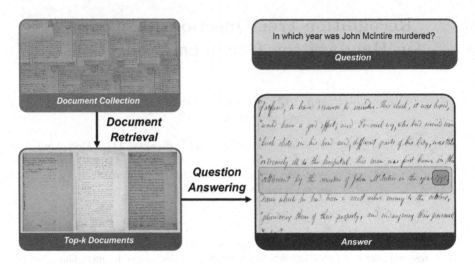

Fig. 1. An overview of the question answering pipeline on document image collections. Given a textual question and a document image collection, a document retriever identifies the k most relevant documents from the collection for answering the question. Finally, a word (blue) or line (green) region from one of these k document images is returned as the answer. (Color figure online)

Answering questions on handwritten document images requires models that are robust with respect to handwriting recognition errors or do not rely on textual input. This is particularly important for QA on unknown document collections, as training data is usually not available and therefore, high handwriting recognition error rates are expected. For developing and evaluating such approaches, Mathew et al. recently proposed the BenthamQA and HW-SQuAD datasets [15]. These datasets provide questions as strings in natural language and expect answers as image regions of a document image rather than a textual response. Finding an answer in a single document image is already a challenging task. However, it is even more complicated in real world scenarios as a large collection of document images is often given. Therefore, it is crucial to identify documents in the collection that are relevant to answer the question and afterwards extract the answer from these documents. Figure 1 provides an overview of this pipeline.

In this work, we propose a recognition-free approach for answering questions on handwritten document image collections. We present a robust document retriever as well as two QA approaches. The first QA model is based on the approach of Mathew et al. [15] and replaces their aggregation strategy with an attention based method. The second model is based on a QA architecture from the NLP domain and enables recognition-free QA on both word and line level. We compare our approach with recognition-free as well as recognition-based QA approaches on the challenging BenthamQA [15] and HW-SQuAD [15] datasets and are able to outperform state-of-the-art results by a large margin.

2   Related Work

QA on document collections usually requires a two-stage approach consisting of a document retriever and a QA model. We provide an overview on textual document retrieval (see Sect. 2.1) and QA in the visual, textual as well as document image domain (see Sect. 2.2).

2.1   Document Retrieval

Document retrieval is an information retrieval task that receives a textual request and returns a set of documents from a given document collection that best matches the query. Traditional approaches rely on counting statistics between query and document words [6,17]. Different weighting and normalization schemes over these counts lead to Term Frequency-Inverse Document Frequency (TF-IDF) models, which are still popular [6,17]. However, these models ignore the position of occurrences and the relationships with other terms in the document [6]. Therefore, models have been developed that can learn the relevance between questions and documents. Learning-To-Rank (LTR) [11] is a well-known document retrieval approach, which represents a query-document pair as a vector of hand-crafted features and trains a model to obtain similarity scores. Recently, deep neural ranking models outperform LTR models [18]. For a detailed overview on document retrieval, see [6,17].

2.2   Question Answering

QA is applied in various domains, leading to large variations among approaches. We present an overview on purely textual QA approaches as well as Visual Question Answering (VQA) models from the CV domain. Furthermore, we discuss recent progress for VQA on document images.

Textual Question Answering. The textual QA community is mainly focusing on the Machine Reading Comprehension (MRC) [30] and OpenQA [31] tasks. In MRC, only one document is given and the answer is a snippet of the document. There is also an extension for this task, whereby models have to decide whether a question is answerable based on the document [30]. Traditional MRC approaches are mainly implemented based on handcrafted rules or statistical methods [30]. Long Short-Term Memory (LSTM)-based models with attention [22] achieved further progress in this field. In recent years, Transformer models (e.g. BERT [4]) improved the results considerably [22]. These models benefit from largely pre-trained word embeddings, which encode useful semantic information between words. Currently, specialized transformer models (e.g. LUKE [28]) lead to state-of-the-art results. In contrast to MRC, OpenQA tries to answer a given question without any specified context. It usually requires the system to search for relevant documents in a large document collection and generate an answer based on the retrieved documents. OpenQA models are mainly a combination of document retrieval and MRC-based approaches [31]. For a detailed overview of textual QA, see [31].

Visual Question Answering. Given an image and a query in natural language, Visual Question Answering (VQA) tries to answer the question using visual elements of the image and textual information from the query [26]. Most approaches rely on an encoder-decoder architecture, which embed questions and images in a common feature space [5,12,21]. This allows learning interactions and performing inference over the question and the image contents. Practically, image representations are obtained with Convolutional Neural Networks (CNNs) pre-trained on object recognition [12,26]. Text representations are obtained with word embeddings pre-trained on large text corpora. RNNs are used to handle the variable size of questions. Further progress in this field has been made using attention [2]. The attention mechanism allows the model to assign importance to features from specific regions of the image. Recently, Transformer based architectures achieved state-of-the-art results on multiple VQA benchmarking datasets [9]. For a detailed overview of VQA, see [26].

Document Image Visual Question Answering. Mainly due to several new competitions [14,16] and datasets [14–16], there has been major progress in the area of answering questions on document images. These datasets provide MRC [14,16] as well as OpenQA tasks [15]. The approaches and datasets mainly focus on machine-printed documents, which contain visual and structural information (e.g. charts, diagrams) [15,19,27]. The layout is important for answering most of the questions [16,19,27]. The approaches are based on textual recognition results and adapt state-of-the-art QA systems from the NLP domain [19,27]. Recently, Mathew et al. [15] published a first dataset for QA on handwritten document collections. Furthermore, they proposed a recognition-free QA approach, which outperforms recognition-based QA models on handwritten datasets [15].

3 Method

In this section, we present our recognition-free approach for answering questions on document image collections. The approach consists of a document retriever (see Sect. 3.2) and a QA model (see Sect. 3.3). Both models are based on the robust Pyramidal Histogram of Characters (PHOC) attribute representation (see Sect. 3.1). Given a query and a collection of word and line-segmented document images, our document retrieval approach assigns a score to each document image, indicating its relevance to answer the query. For each of the K most relevant documents, our QA model determines answer snippets and an associated confidence score. Finally, the snippet with the highest score is returned as the answer.

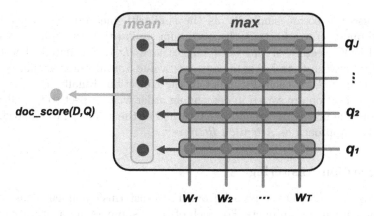

Fig. 2. Our attention based retrieval approach for calculating the similarity between a query $\mathbf{Q} = [\mathbf{q}_1, ..., \mathbf{q}_J]$ and a document or snippet $\mathbf{D} = [\mathbf{w}_1, ..., \mathbf{w}_T]$.

3.1 Query and Document Representation

To compute a similarity score between a document image and a query, as well as for question answering, the question words and the document images have to be transformed into a vector representation. Since we follow a recognition-free approach and the question is provided in a textual and the documents in a visual form, we use the Pyramidal Histogram of Characters (PHOC) representation that allows a robust mapping of words and images into the same space. A PHOC is a binary pyramidal representation of a character string and is used to represent visual attributes of a given word image. The embedding is successfully and widely used in the word spotting domain [1,10,23]. We use the TPP-PHOCNet [23] to realize a mapping from word images to a PHOC representation. The representations of the word images are finally stored in the order of their occurrences in the document image.

3.2 Retrieval

To determine the most relevant documents regarding a query in a given collection, we follow a similar approach as described in [15]. Hereby, they aggregate the question and the documents into vector representations of fixed size, by using the Fisher Vector framework [7]. Finally, the best matching documents are obtained by calculating the cosine similarity between the question and document embeddings. In contrast, our approach does not rely on such aggregation methods and instead uses the similarity between each word image from a document D and each question word from the pre-processed query Q as described in Eq. 1 and visualized in Fig. 2.

$$doc\_score(D, Q) = \frac{1}{|Q|} * \sum_{q \in Q} \max_{w \in D}[sim(w, q)] \qquad (1)$$

We use the cosine similarity as the similarity measure. For each PHOC encoded question word $q \in Q$, the maximum similarity between q and the predicted PHOC vectors of the word images $w \in D$ is calculated. The overall similarity between Q and D is the averaged value over all these similarity scores and is computed for each document in the collection. Finally, the documents from the collection are sorted in descending order with respect to the calculated scores and the first K documents are returned as the result. In the following, we denote this approach as *Attention-Retriever*.

3.3   Question Answering

The recognition-free QA approach from [15] transforms document images into a set of two-line image snippets. For each of these snippets, an aggregated vector representation is determined based on the corresponding word images and is used to compute a similarity score with respect to an aggregated query vector. The score represents the confidence of finding the answer in the corresponding document region and determines the final answer of the system. Even though this approach can correctly locate the answers for some questions, the intuition behind this method is fairly questionable. The approach does not learn any real relationship between context and question, but exploits the heuristic that question words often occur close to the answer. Therefore, the approach does not realize a classical QA system, but rather an adapted syntactic word spotting approach for snippets.

NLP models are mainly based on the successes in transfer learning, where contextualized word embedding models were pre-trained on very large text collections. Unfortunately, transfer learning on handwritten word images is currently difficult and a robust mapping of word images into a semantic space is challenging even for static semantic word embeddings [24]. Therefore, it is currently not straightforward to adapt state-of-the-art NLP approaches to this task. However, there are previous state-of-the-art QA models from the NLP domain that do not rely on contextualized word embeddings and still lead to high performances on most datasets.

We follow the approach of the textual Bidirectional Attention Flow for Machine Comprehension (BIDAF) [22] model from the NLP domain and adapt it to a recognition-free QA model working on line instead of word level (see Fig. 3). The architecture can be divided into the word embedding, phrase embedding, attention flow, modeling, line embedding and output parts.

In the word embedding layer, all word images from a given document as well as the textual question words are represented as PHOC vectors. Here, the word images from the documents are transferred into the PHOC representation using the TPP-PHOCNet [23]. We further encode the line correspondence of each word image in the document using the positional encoding strategy from [4] and append them to the corresponding PHOC representations. The phrase

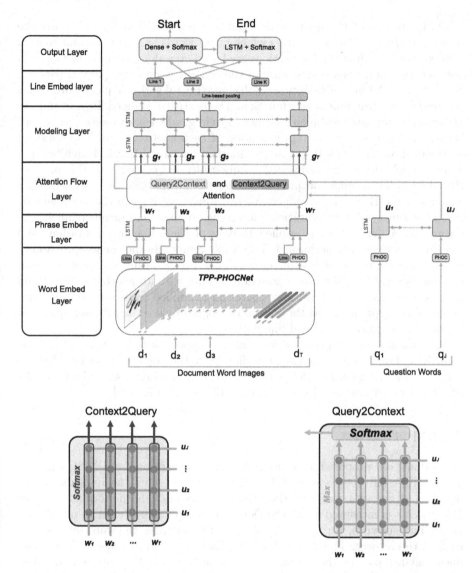

Fig. 3. The adapted BIDAF architecture for recognition-free question answering on line level.

embedding part uses a BLSTM to extract and model the temporal interactions between the word image representations from a given document as well as for the question word representations. Afterwards, the attention flow layer determines two types of attention scores between the obtained context $(\mathbf{w}_1, ..., \mathbf{w}_T)$ and question $(\mathbf{u}_1, ..., \mathbf{u}_J)$ vectors, namely Context2Query and Query2Context. Thereby, Context2Query signifies which context words have the closest similarity to one of the query words and are hence critical for answering the query and Query2Context signifies which query words are most relevant to each context word. Both of these attentions are based on a shared similarity matrix **S** between the contextual embeddings of the context **w** and the query **u**. Hereby, $S_{t,j}$ indicates the similarity between t-th context word and j-th query word and is computed by a trainable scalar function. The contextual embeddings **w** and the attention vectors are combined together to yield **g**, where each vector can be considered as the query-aware representation of each context word. The obtained representations serve as the input to another two-stage BLSTM architecture, which models the relationship between questions and contexts. In this process, the BLSTM has as many outputs as the number of words in the document. The outputs of the BLSTM are reduced to the number of lines in the document by summing the word representations according to their line membership in the document. A dense layer is applied to each of these line representations and a softmax operation is performed. The result represents the pseudo-probability distribution for the start line of the answer. For calculating a similar distribution for the end line, the line representations are fed into another BLSTM and a dense layer as well as a softmax operation is applied to its output. The confidence for the prediction is the sum of the values before the softmax operation for the predicted start and end line indices.

The architecture can be used for word-level predictions by removing the line embedding layer. In the following, we will refer to the line-level model as *BIDAF-Line* and the word-level model as *BIDAF-Word*. In addition, we will refer to the adapted recognition-free QA approach of [15] as *Attention-QA*.

4 Experiments

We evaluate our proposed recognition-free QA approach on the HW-SQuAD and BenthamQA datasets (see Sect. 4.1). Section 4.2 presents the implementation details and Sect. 4.3 discusses the evaluation results. The performance of the QA systems is measured using the Double Inclusion Score (DIS) as introduced in [15] and shown in Eq. 2. The Small Box (SB) includes the word images that contain the answer. The Large Box (LB) includes all word images from those lines that are part of the SB as well as those from the lines above and below it. The Answer Box (AB) includes the word images from the lines predicted by the QA system. A visual example of these box definitions is given in Fig. 4. The prediction is considered a correct answer if the score is above 0.8.

$$DIS = \frac{AB \cap SB}{|SB|} \times \frac{AB \cap LB}{|AB|} \tag{2}$$

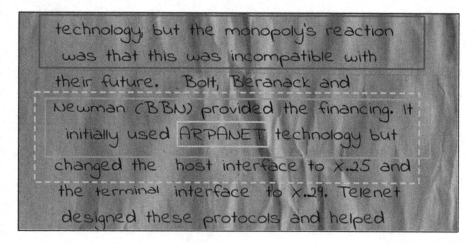

Fig. 4. An example of a correct (green) and an incorrect (red) predicted Answer Box for the question "Telnet used what interface technology?". Here, the rectangle around the word "ARPANET" represents the Small Box, whereas the dashed rectangle illustrates the Large Box. (Color figure online)

4.1 Dataset

We train and evaluate our models on the recently proposed HW-SQuAD and BenthamQA datasets [15] (see Fig. 5), which contain question-answer pairs on handwritten document image collections in the English language. The datasets vary considerably in their size and characteristics and include synthetically generated as well as real handwritten documents.

BenthamQA [15] is a small historical handwritten QA dataset where questions and answers were created using crowdsourcing. The historic dataset contains 338 documents written by the English philosopher Jeremy Bentham and shows some considerable variations in writing styles. The dataset provides only a test set consisting of 200 question-answer pairs on 94 document images. The remaining 244 documents from the collection are used as distractors.

HW-SQuAD [15] is a QA dataset on syntactically generated handwritten document images from the textual SQuAD1.0 [20] dataset. The textual dataset is actually defined for an MRC task and was adapted by Mathew et al. [15] to an OpenQA task. The synthetic dataset consists of 20963 document pages containing a total of 84942 questions. The official partitioning splits the dataset into 17007 documents for training, 1889 for validation and 2067 for testing. Thereby, the training, validation and test sets contain 67887, 7578 and 9477 questions respectively.

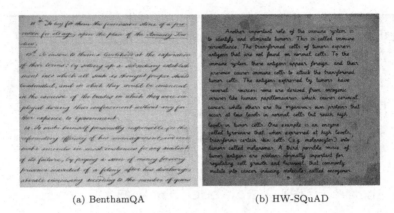

(a) BenthamQA (b) HW-SQuAD

Fig. 5. Example documents for the BenthamQA and HW-SQuAD datasets.

4.2 Implementation Details

Our proposed document retriever relies on pre-segmented word images and our QA approaches also need line annotations. For our experiments, we use the gold-standard word and line bounding boxes available with the datasets. Questions are split into words and stopwords are removed using NLTK [3]. For training the BIDAF architecture, we use the HW-SQuAD dataset. We do not change the proposed parameters from [22] and use a hidden layer size of 100 as well as dropout with probability 0.2 for the BLSTMs. For optimization, we use ADADELTA [29] with the Cross Entropy loss and a learning rate of 0.5. The positional line encoding produces a 30-dimensional vector using sine and cosine functions.

For word representation, we use a 504-dimensional PHOC vector consisting of lowercase letters (a–z), numbers (0–9) and the levels 2, 3, 4 and 5. We pre-train the TPP-PHOCNet on the HW-SQuAD [15] as well as IIIT-HWS [10] datasets and fine-tune the model on the IAM database [13]. We use a batch size of 40 and a momentum of 0.9. The parameters of the network are updated using the Stochastic Gradient Decent optimizer and the Cosine loss. The learning rate is set to 0.01 during pre-training and 0.001 while fine-tuning. It is divided by two if the loss has not decreased in the last three epochs. We binarize the word images to remove the background from text. This is especially important as the background of document images from BenthamQA largely differ from those in IIIT-HWS and IAM.

4.3 Results

In this section, we show the evaluation performances of our recognition-free QA approach on handwritten document image collections. We evaluate and compare the document retrieval approach in Sect. 4.3 and the three QA approaches in Sect. 4.3. Finally, we present and discuss the results on the combination of the retrieval and QA approaches in Sect. 4.3. For all subsequent experiments, we use the Attention-Retriever approach presented in Sect. 3.2.

Table 1. Top-5 accuracy (%) for document retrieval approaches.

	Approach	HW-SQuAD	BenthamQA
Pred.	Mathew et al. (rec-free) [15]	46.5	55.5
	Mathew et al. (rec-based) [15]	86.1	32.0
	Attention-Retriever	**86.2**	**92.5**
GT	Mathew et al. (rec-based) [15]	90.2	98.5
	Attention-Retriever	87.2	98.0

Retrieval. For answering a given question in a document collection, it is common to determine the five most relevant documents with respect to the query. To evaluate those retrieval models, we use the Top-5 accuracy as described in [15]. This score represents the percentage of questions from a given test set that have their associated answer document in the top five predicted retrieval results. In Table 1 we present the Top-5 accuracy scores for our document retrieval approach and compare it to the literature. We also show the results of NLP models working on ground truth transcriptions in this table.

The results show that we are able to improve the state-of-the-art Top-5 accuracy scores on both datasets. We clearly outperform the recognition-free approach from the literature. On the HW-SQuAD dataset, we perform marginally better compared to the recognition-based approach proposed in [15]. The recognition based model performs on nearly perfect recognition results (97.9% word accuracy). When the recognition performance becomes worse as in BenthamQA (23.2% word accuracy), the vulnerability of the approach to recognition errors is revealed and only a low performance can be achieved. The results from our model with ground truth and predicted PHOCs are almost identical, demonstrating the robustness of our approach. The differences between HW-SQuAD and BenthamQA can be explained by the lower prediction performance of the PHOC vectors for BenthamQA. In this case, the TPP-PHOCNet can achieve a query-by-string score of 98.5 on HW-SQuAD and 77.3 on BenthamQA. Interestingly, the NLP method working on ground truth transcriptions can only achieve marginally higher scores compared to our attention approach, demonstrating the capabilities of our model.

Question Answering. In order to evaluate the performance of our three proposed QA approaches without the influence of the retrieval model, we evaluate their performances on a MRC rather than the OpenQA task. Thus, the QA systems only work on the document that contains the answer to the question. Table 2 shows the results of our QA approaches as well as upper bounds using a state-of-the-art QA approach (BERT [4]) working on ground truth transcriptions.

Table 2. Machine reading comprehension. Performance measured in DIS.

	QA-Approach	HW-SQuAD	BenthamQA
Pred.	Attention-QA	47.5	38.5
	BIDAF-Word	57.2	28.0
	BIDAF-Line	**68.1**	**50.5**
GT	Attention-QA	47.7	39.0
	BIDAF-Word	57.7	47.0
	BIDAF-Line	68.7	62.0
	BERT [4]	94.4	88.0

The results show that our line-based BIDAF model can achieve higher scores compared to the word-based model and the attention based approach. A comparison with the literature is not possible, as Mathew et al. do not evaluate their approaches on this task. As already shown for the document retrieval, a similar relationship emerges between the performances of the models based on ground truth and predicted PHOCs. However, compared to the attention approach, the PHOC prediction errors have a stronger impact on the performances of the BIDAF models. In comparison to the line-based approach, the word-based BIDAF model seems to be quite sensitive to erroneous PHOC predictions. Presumably, the line-based model is less sensitive to the recognition errors due to the aggregation on line level. It should be noted, that the performances of our approaches show potential for improvement compared to NLP models working on textual annotations. This gap is likely due to the successful application of transfer learning in the textual domain.

End-to-End Question Answering. In the previous subsections, we have evaluated the individual components of our system. For answering questions in document collections, a combination of those is required. Table 3 shows the results for the combination of our document retriever and the line-based BIDAF model as well as approaches from the literature. For this evaluation, the document retrieval approaches extract the top five documents from the collections.

Our approach can clearly outperform the recognition-free method proposed by Mathew et al. [15]. The recognition-based system from [15], however, outperforms our approach on the HW-SQuAD dataset, but clearly fails on BenthamQA. This supports the common research outcomes, whereas the performances of textual NLP models are quite high on datasets with low recognition errors, but decrease considerably when the amount of recognition errors rise [25]. The results show that the PHOC prediction errors affect the performance of our model, however, it shows a considerably improved robustness compared to recognition-based models.

Table 3. End-to-end answer line snippet extraction. Performance measured in DIS.

	QA-Approach	HW-SQuAD	BenthamQA
Pred.	Mathew et al. (rec-free) [15]	15.9	17.5
	Mathew et al. (rec-based) [15]	**59.3**	2.5
	BIDAF-Line	45.0	**37.5**
GT	Mathew et al. (rec-based) [15]	74.8	74.0
	BIDAF-Line	45.3	55.0

5 Conclusions

In this work, we present a recognition-free question answering system for handwritten document image collections. The system consists of an attention based document retriever as well as a question answering approach. Our document retrieval model achieves new state-of-the-art scores for the retrieval task on all considered datasets. For question answering, textual approaches benefit from transfer learning methods and outperform recognition-free approaches on the HW-SQuAD dataset with low word error rates. Considering the desired application to historical datasets with presumably no annotated training material, error rates are usually significantly higher. As seen on BenthamQA, this leads to a considerable decrease of the QA performance for recognition-based models. Our experiments show the robustness of our proposed combination of recognition-free retrieval and QA system and that it is able to outperform recognition-free as well as recognition-based methods from the literature.

References

1. Almazán, J., Gordo, A., Fornés, A., Valveny, E.: Word spotting and recognition with embedded attributes. IEEE Transactions on Pattern Analysis and Machine Intelligence **36**(12), 2552–2566 (2014)
2. Anderson, P., He, X., Buehler, C., Teney, D., Johnson, M., Gould, S., Zhang, L.: Bottom-up and top-down attention for image captioning and visual question answering. In: Int. Conf. on Computer Vision and Pattern Recognition. pp. 6077–6086. Salt Lake City, UT, USA (2018)
3. Bird, S.: NLTK: The natural language toolkit. In: Annual Meeting of the Association for Computational Linguistics. pp. 69–72. Sydney, Australia (2006)
4. Devlin, J., Chang, M., Lee, K., Toutanova, K.: BERT: Pre-training of deep bidirectional transformers for language understanding. In: Annual Conf. of the North American Chapter of the Association for Computational Linguistics - Human Language Technologies. pp. 4171–4186. Minneapolis, MN, USA (2019)
5. Fukui, A., Park, D.H., Yang, D., Rohrbach, A., Darrell, T., Rohrbach, M.: Multimodal compact bilinear pooling for visual question answering and visual grounding. In: Proc. Conf. on Empirical Methods in Natural Language Processing. pp. 457–468 (2016)

6. Guo, J., Fan, Y., Pang, L., Yang, L., Ai, Q., Zamani, H., Wu, C., Croft, W.B., Cheng, X.: A deep look into neural ranking models for information retrieval. Information Processing and Management **57**(6), 102067 (2020)
7. Jaakkola, T.S., Haussler, D.: Exploiting generative models in discriminative classifiers. In: Int. Conf. on Neural Information Processing Systems. pp. 487–493. Denver, CO, USA (1998)
8. Kang, L., Toledo, J.I., Riba, P., Villegas, M., Fornés, A., Rusiñol, M.: Convolve, attend and spell: An attention-based sequence-to-sequence model for handwritten word recognition. In: Proc. German Conf. on Pattern Recognition. pp. 459–472. Stuttgart, Germany (2018)
9. Khan, A.U., Mazaheri, A., da Vitoria Lobo, N., Shah, M.: MMFT-BERT: Multimodal fusion transformer with BERT encodings for visual question answering. In: Proc. Conf. on Empirical Methods in Natural Language Processing. pp. 4648–4660. Online (2020)
10. Krishnan, P., Jawahar, C.V.: HWNet v2: An efficient word image representation for handwritten documents. Int. J. Doc. Anal. Recognit. **22**, 387–405 (2019)
11. Liu, T.: Learning to rank for information retrieval. Found. Trends Inf. Retr. **3**(3), 225–331 (2009)
12. Malinowski, M., Rohrbach, M., Fritz, M.: Ask your neurons: A neural-based approach to answering questions about images. In: Int. Conf. on Computer Vision. pp. 1–9. Santiago, Chile (2015)
13. Marti, U., Bunke, H.: The IAM-database: An English sentence database for offline handwriting recognition. Int. J. Doc. Anal. Recognit. **5**(1), 39–46 (2002)
14. Mathew, M., Bagal, V., Tito, R.P., Karatzas, D., Valveny, E., Jawahar, C.V.: Infographicvqa. CoRR abs/2104.12756 (2021)
15. Mathew, M., Gómez, L., Karatzas, D., Jawahar, C.V.: Asking questions on handwritten document collections. Int. J. Doc. Anal. Recognit. **24**, 235–249 (2021)
16. Mathew, M., Karatzas, D., Jawahar, C.V.: DocVQA: A dataset for VQA on document images. In: IEEE Workshop on Applications of Computer Vision. pp. 2199–2208. Waikoloa, HI, USA (2021)
17. Mitra, B., Craswell, N.: An introduction to neural information retrieval. Foundations and Trends in Information Retrieval **13**(1), 1–126 (2018)
18. Pang, L., Lan, Y., Guo, J., Xu, J., Xu, J., Cheng, X.: DeepRank: A new deep architecture for relevance ranking in information retrieval. In: Proc. ACM Int. Conf. on Information and Knowledge Management. pp. 257–266. Singapore (2017)
19. Powalski, R., Borchmann, Ł., Jurkiewicz, D., Dwojak, T., Pietruszka, M., Pałka, G.: Going Full-TILT boogie on document understanding with text-image-layout transformer. In: Proc. Int. Conf. on Document Analysis and Recognition. pp. 732–747. Lausanne, Switzerland (2021)
20. Rajpurkar, P., Zhang, J., Lopyrev, K., Liang, P.: SQuAD: 100, 000+ questions for machine comprehension of text. In: Proc. Conf. on Empirical Methods in Natural Language Processing. pp. 2383–2392. Austin, TX, USA (2016)
21. Saito, K., Shin, A., Ushiku, Y., Harada, T.: DualNet: Domain-invariant network for visual question answering. In: Int. Conf. on Multimedia and Expo. pp. 829–834. Ypsilanti, MI, USA (2017)
22. Seo, M., Kembhavi, A., Farhadi, A., Hajishirzi, H.: Bidirectional attention flow for machine comprehension. In: Int. Conf. on Learning Representations. Toulon, France (2017)
23. Sudholt, S., Fink, G.A.: PHOCNet: A deep convolutional neural network for word spotting in handwritten documents. In: Proc. Int. Conf. on Frontiers in Handwriting Recognition. pp. 277–282 (2016)

24. Tüselmann, O., Wolf, F., Fink, G.A.: Identifying and tackling key challenges in semantic word spotting. In: Proc. Int. Conf. on Frontiers in Handwriting Recognition. pp. 55–60. Dortmund, Germany (2020)
25. Tüselmann, O., Wolf, F., Fink, G.A.: Are end-to-end systems really necessary for NER on handwritten document images? In: Lladós, J., Lopresti, D., Uchida, S. (eds.) ICDAR 2021. LNCS, vol. 12822, pp. 808–822. Springer, Cham (2021). https://doi.org/10.1007/978-3-030-86331-9_52
26. Wu, Q., Teney, D., Wang, P., Shen, C., Dick, A.R., van den Hengel, A.: Visual question answering: A survey of methods and datasets. Computer Vision and Image Understanding **163**, 21–40 (2017)
27. Xu, Y., Xu, Y., Lv, T., Cui, L., Wei, F., Wang, G., Lu, Y., Florêncio, D.A.F., Zhang, C., Che, W., Zhang, M., Zhou, L.: LayoutLMv2: Multi-modal pre-training for visually-rich document understanding. In: Annual Meeting of the Association for Computational Linguistics and Int. Joint Conf. on Natural Language Processing. pp. 2579–2591. Bangkok, Thailand (2021)
28. Yamada, I., Asai, A., Shindo, H., Takeda, H., Matsumoto, Y.: LUKE: Deep contextualized entity representations with entity-aware self-attention. In: Proc. Conf. on Empirical Methods in Natural Language Processing. pp. 6442–6454. Online (2020)
29. Zeiler, M.D.: ADADELTA: An adaptive learning rate method. CoRR abs/1212.5701 (2012)
30. Zeng, C., Li, S., Li, Q., Hu, J., Hu, J.: A survey on machine reading comprehension: Tasks, evaluation metrics, and benchmark datasets. CoRR abs/2006.11880 (2020)
31. Zhu, F., Lei, W., Wang, C., Zheng, J., Poria, S., Chua, T.: Retrieving and reading: A comprehensive survey on open-domain question answering. CoRR abs/2101.00774 (2021)

Handwriting Recognition and Automatic Scoring for Descriptive Answers in Japanese Language Tests

Hung Tuan Nguyen[1]([✉]) [ID], Cuong Tuan Nguyen[1] [ID], Haruki Oka[2],
Tsunenori Ishioka[3] [ID], and Masaki Nakagawa[1] [ID]

[1] Wacom-TUAT Joint Research Laboratory, Tokyo University of Agriculture and Technology,
Tokyo, Japan
{fx7297,fx4102}@go.tuat.ac.jp, nakagawa@cc.tuat.ac.jp
[2] Recruit Co. Ltd., Tokyo, Japan
haruki_oka@r.recruit.co.jp
[3] The National Center for University Entrance Examinations, Tokyo, Japan
tunenori@rd.dnc.ac.jp

Abstract. This paper presents an experiment of automatically scoring handwritten descriptive answers in the trial tests for the new Japanese university entrance examination, which were made for about 120,000 examinees in 2017 and 2018. There are about 400,000 answers with more than 20 million characters. Although all answers have been scored by human examiners, handwritten characters are not labeled. We present our attempt to adapt deep neural network-based handwriting recognizers trained on a labeled handwriting dataset into this unlabeled answer set. Our proposed method combines different training strategies, ensembles multiple recognizers, and uses a language model built from a large general corpus to avoid overfitting into specific data. In our experiment, the proposed method records character accuracy of over 97% using about 2,000 verified labeled answers that account for less than 0.5% of the dataset. Then, the recognized answers are fed into a pre-trained automatic scoring system based on the BERT model without correcting misrecognized characters and providing rubric annotations. The automatic scoring system achieves from 0.84 to 0.98 of Quadratic Weighted Kappa (QWK). As QWK is over 0.8, it represents an acceptable similarity of scoring between the automatic scoring system and the human examiners. These results are promising for further research on end-to-end automatic scoring of descriptive answers.

Keywords: Handwritten Japanese answers · Handwriting recognition · Automatic scoring · Ensemble recognition · Deep neural networks

1 Introduction

Descriptive answers are better for evaluating learners' understanding and problem-solving ability. They encourage learners to think rather than select. However, scoring

H. Oka—Work done while at The University of Tokyo.

U. Porwal et al. (Eds.): ICFHR 2022, LNCS 13639, pp. 274–284, 2022.
https://doi.org/10.1007/978-3-031-21648-0_19

them requires large work and time. In recent years, it was proposed to add descriptive questions in the new university entrance common examinations in Japan as well as the current multiple-choice questions [1], but given up due to the short period of scoring handwritten answers and the anxiety about reliable scoring.

One approach is to score handwritten descriptive answers automatically and feedback scores to examinees and examiners to correct scoring errors. Another approach is to apply automatic scoring or to cluster them for human examiners to score them efficiently and reliably [2, 3]. For both of them, handwritten answers need to be recognized and scored.

A few datasets storing handwritten answers have been published and used in research on handwriting recognition, such as SCUT-EPT (Chinese handwritten answers) [4] and Dset-Mix, which is artificially prepared by synthesized handwritten math answers [5]. Note that these datasets are all fully labeled and ideally suited to train handwriting recognizers based on deep neural networks.

The National Center for University Entrance Examinations (NCUEE) conducted trial tests for the new university entrance common exams with 64,518 and 67,745 examinees in 2017 and 2018, respectively. Three descriptive questions were included in the Japanese language test for each trial test. All handwritten answers were scored by human examiners. The scanned images and scores by human are used in this research.

However, the offline images are only raw images and have not been segmented or labeled. It is infeasible to prepare labels for a whole set of scanned handwritten answers. On the other hand, automatic pattern recognition methods, especially well-known deep neural networks, require large-scale labeled data for training. Hence, we present normalization, segmentation, and handwriting recognition from this handwritten answer set and their automatic scoring. In particular, we focus mainly on training the handwriting recognizer to adapt to the actual data from the examinee's answers. In addition, we also incorporated the language model to re-rank the predicted candidates so that the ambiguous patterns are corrected by linguistic context. In summary, the main contributions are as follows:

- We present an ensemble deep neural network-based recognizer for offline Japanese handwritten answer recognition.
- We propose a training procedure with multiple steps (pre-training, fine-tuning and ensemble learning) to adapt the ensembled handwriting recognizer to real patterns.
- We evaluate the handwriting recognizer in combination with the latest automatic scoring system.

Note that the combined architecture does not try to correct of misrecognized characters or require rubric annotations since we expect that some small misrecognitions would not affect to automatic scoring model. In practice, the human examiners do not need to recognize all the characters in answers as they score answers based on the meaning of sentences or detection of keywords.

2 Related Works

Handwritten text recognition has been studied for many decades [6]. Offline recognition (Optical recognition from scanner or camera recently) followed by online recognition from tablet have been developed for reading postal addresses, bank checks, business forms and documents. Recently, more challenging problem of historical document recognition is being studied extensively [7, 8].

However, there is only a few studies on handwritten text recognition for handwritten answers. Handwritten answers are usually different from each other in terms of content on top of writing style so that the cost for completely labeling them is almost impractical. Recently, one of the first large-scale datasets of examination answers, the SCUT-EPT dataset, was introduced [4]. It contains 50,000 handwriting Chinese text line images provided by 2,986 volunteers.

The proposed SCUT-EPT dataset shows challenges, including character erasure, text line supplement, character/phrase switching, noised background, non-uniform word size, and unbalanced text length. The current advanced text recognition methods, such as convolutional recurrent neural network (CRNN), exhibits poor performance on the proposed SCUT-EPT dataset. According to the visualizations and error analyses, the authors reported that the performance of examiners is much better than the CRNN method. Furthermore, they compared three sequential labelling methods, including connectionist temporal classification (CTC), attention mechanism, and cascaded attention-CTC. Even though the attention mechanism has been demonstrated to be effective in English scene text recognition, its performance was far inferior to the CTC method in the case of the SCUT-EPT dataset with a large-scale character set.

This paper presents the first attempt to recognize and score large-scale unconstrained Japanese handwritten answers from high school students for the trial university entrance examinations.

3 Handwritten Japanese Answer Dataset

In 2017 and 2018, the National Center for University Entrance Examinations of Japan has conducted trial tests using descriptive questions for the university entrance examinations for the Japanese language and Math. At that time, nearly 38% of high schools in Japan attended these trial tests, but the students were not asked to participate in all subjects. For the Japanese language tests, 64,518 and 67,745 answer sheets were collected in 2017 and 2018, respectively. We call this dataset NCUEE-Handwritten Japanese Answers (NCUEE-HJA) with two sub-datasets NCUEE-HJA-2017 and NCUEE-HJA-2018.

Note that all answer images are not labeled except their scores, which is a major barrier to adopting the latest deep neural network-based handwriting recognition methods. Thus, we need to prepare a small but effective number of samples for training handwriting recognizers. In order to prepare these labeled samples, we first employed multiple segmentation stages to obtain the single characters from the answer images, as shown in Fig. 1. Then, the text-line label is manually prepared.

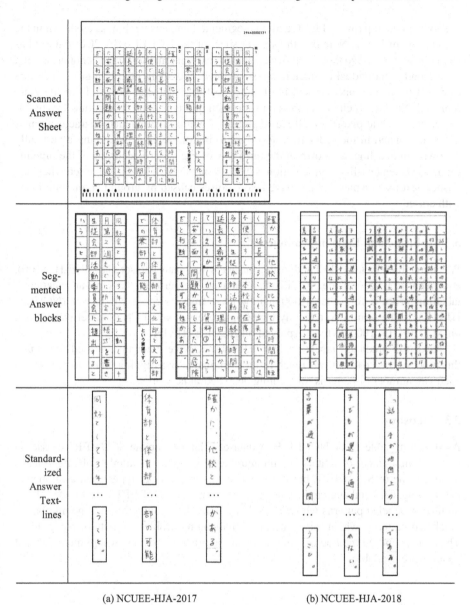

Scanned Answer Sheet		
Seg-mented Answer blocks		
Standard-ized Answer Text-lines		

(a) NCUEE-HJA-2017   (b) NCUEE-HJA-2018

Fig. 1. Samples in the NCUEE-HJA dataset and their processed results.

3.1 Handwritten Text-Line Segmentation

We used a pipeline of two steps to segment handwritten text-lines from answer sheet images. First, the answer block segmentation step is completed by horizontal and vertical histogram-based projections to detect answer blocks. For the NCUEE-HJA-2017 dataset, an answer image of an examinee consists of three answers for three descriptive questions, as shown in the top row of Fig. 1, and it is segmented into answer blocks as shown in the second row of Fig. 1. Note that the answer sheet of the NCUEE-HJA-2018 dataset has been pre-segmented so that the above answer block segmentation stage is not necessary.

Secondly, individual characters are segmented. We apply morphological operators to detect horizontal and vertical borderlines to form character boxes. Then, they are removed. Next, each character image inside each box is padded to a unique size of 64-by-64 using white pixels. Finally, a sequence of white-pixel padded character images are concatenated in the order from top right to bottom left for generating a vertically handwritten text-line, as shown in the third row of Fig. 1. The accuracy of the pipeline for handwritten text-line segmentation is 99.42% on the NCUEE-HJA dataset, where the unsuccessful segmented images have heavy noises caused by ground color stain made with eraser.

3.2 Splitting and Labeling Samples

We manually labeled 100 handwritten answers for each dataset (NCUEE-HJA100), equivalent to 0.05% of the NCUEE-HJA dataset. We performed a semi-automatically labelling process for other 1,000 handwritten answers for each dataset (NCUEE-HJA1K). In the semi-automatically labelling process, a small CNN model was trained on NCUEE-HJA100 and then deployed on NCUEE-HJA1K.

Next, the predicted labels on NCUEE-HJA1K were manually verified. Although this process was costly, it was much faster than manually assigning labels to prepare an adequate number of patterns for training deep neural networks.

3.3 Statistics

As shown in Table 1, the NCUEE-HJA dataset is larger than the SCUT-EPT dataset in terms of the number of writers and the number of samples. Its number of categories is less than one of SCUT-EPT, but it consists of many character types such as alphabet, kanji, hiragana, katakana, and numeric. Moreover, the SCUT-EPT dataset has a large number of labeled patterns while the NCUEE-HJA dataset only has a large number of unlabeled patterns, which raises the most challenging problem in pattern recognition. Thus, the proposed method is different from the previous studies, where many labeled patterns are available.

Table 1. Comparison between SCUT-EPT and NCUEE-HJA.

Database Characteristics	SCUT-EPT	NCUEE-HJA	
Language	Chinese	Japanese	
Collected Year	2018	2017	2018
Type	Offline Handwritten Text (labeled)	Offline Handwritten Text (unlabeled)	
No. of Categories	4,250	3,125	
No. of Writers	2,986 (2,986)	64,518 (1,000)	67,745 (1,000)
No. of Text-lines	50,000 (40,000)	193,554 (1,000)	>11 million (61,883)
No. of Characters	1,267,167 (1,018,432)	203,235 (1,000)	>10 million (53,370)

( ) *presents the value of the corresponding training set.*

4 Handwritten Answer Recognition and Automatic Scoring

Deep Neural Networks (DNNs) have achieved state-of-the-art results on pattern recognition challenges such as object detection, image classification, or even image transcription during the last decade. In this paper, we propose employing the different DNN models for two tasks: handwriting recognition and text classification, to perform automatic handwritten answer scoring as shown in Fig. 2.

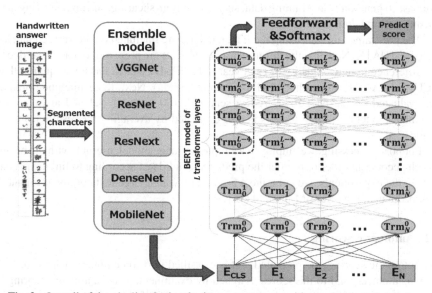

Fig. 2. Overall of the pipeline for handwritten answer recognition and automatic scoring.

4.1   Handwritten Answer Recognition

Generally, deep neural networks have become a state-of-the-art approach for image classification, especially Convolutional Neural Networks (CNNs). According to the results from [7], we employ a set of five well-known CNN models (VGG16, MobileNet24, ResNet34, ResNeXt50, and DenseNet121). These models are modified to adapt to the small input images as a single character image of 64-by-64 pixels. Their first convolutional layers are modified to use the kernel size of 3-by-3. Moreover, the final classification layers are modified to match the number of categories in the NCUEE-HJA dataset. The details of CNN models are presented as follows.

The first network is derived from VGGNet, which consists of 16 convolutional layers divided into four blocks, as presented in [9]. The second network, MobileNet24, with 24 convolutional layers [10], is proposed to work efficiently on smart devices with limited computational resources. The third and fourth networks based on the residual connections (ResNet34 and ResNeXt50) consist of 34 and 50 layers. These layers are grouped into multiple blocks with 3 convolutional layers and a residual connection in each block [11, 12]. The fifth network is DenseNet121, in which the extracted feature maps in lower-level layers are concatenated to obtain dense features for higher-level layers [13]. We reused the pre-trained parameters of the five models on ImageNet as provided in the PyTorch vision library[1] to have a generalized initialization.

For solving the challenge of a few labeled samples, good initialization of recognition models is important. It requires a large-scale dataset in a similar domain to the NCUEE-HJA dataset. Thus, we employed all handwritten character datasets from the ETL database [14]. We applied various transformations to training patterns from the ETL database to generalize the training data, such as rotation, shearing, shifting, blurring, and noising with random parameters.

The pre-trained models using the ETL database are finetuned on the verified labeled NCUEE-HJA1K. Note that the different trained models are prepared for each year due to the difference in image quality of each year (NCUEE-HJA-2018 has lower quality than NCUEE-HJA-2017 using the JPEG lossy compression). Next, these finetuned models are evaluated on the NCUEE-HJA test sets with 100 manually labeled answers. An ensemble model of all five models is also evaluated by averaging the results from the single models.

For post-processing, we employed a simple language model, n-gram, with the beam-search decoder in order to reorder the predicted candidates according to linguistic context. In the experiments, we compute a 5-g model from a large text corpus of Japanese Wikipedia using the KenLM method [15].

4.2   Automatic Scoring

For automatic scoring, the predicted text is classified into some ranks, such as [0, 1, 2, 3], which correspond to the scores assigned by examiners. Thus, automatic scoring is considered as text classification in this paper. We use BERT [16], which is pre-trained on Japanese Wikipedia. By adding the information of the handwritten answers to it, we can

---

[1] https://pytorch.org/vision/stable/models.html.

make reasonable estimates using the knowledge obtained from the large-scale model. The procedure is as follows:

1. Divide the recognized answers into morphemes. Morphemes are called tokens, and each is given an ID. A special token called CLS is given at the beginning of the sentence.
2. Convert the ID assigned to each token into a 768-dimensional vector from the result of pre-learning in Japanese Wikipedia.
3. Obtain the vector of CLS tokens in the final 4 layers of the 12 hidden layers in BERT.
4. Combine the extracted 768-dimensional vectors in the column direction. This creates a 768 × 4 matrix.
5. Dimension transformation of this matrix with a linear layer is then applied to the softmax function to output the category with the highest probability.

For the estimation by BERT, all the data were divided into 3:1:1, that is, 60%, 20%, 20%, used as training data, validation data, and test data, respectively. Adam is used for model optimization [17], and training was performed with a batch size of 16 and an epoch number of 5. Finally, the second-order weighted κ coefficient (QWK) [18] is used to measure the degree of agreement between scores by the automatic scoring system and those by human examiners. Generally, QWK values of 0.61–0.8 indicate substantial agreement, and > 0.8 suggest almost perfect agreement [19].

5 Experiment Results

The following sections present the performance of the recognition model and the automatic scoring model.

5.1 Performance of Recognition Model

The performance of the shallowest and deepest models as well as the ensemble of all five models, are shown in Table 2. Although the image quality of NCUEE-HJA in 2018 is lower than in 2017, the character recognition rates are improved due to fine-tuning and the n-gram context procession. Thus, the CNN models were able to extract discriminative features regardless of the image quality after the fine-tuning process. Moreover, even the shallowest model performed similarly to the deepest one. However, a single model might not perform well on a specific dataset due to the distribution of features and the network structure. In order to avoid this problem, we employed the ensemble model by averaging the predictions from single models. Note that the ensemble model does not outperform the VGG16 and DenseNet121 on the test sets of NCUEE-HJA 2017 and 2018, respectively. However, it is expected to perform well in a large number of writing styles.

5.2 Performance of Automatic Scoring Model

Table 3 shows the statistics for scoring a total of 6 questions, 3 for each of 2017 and 2018. The number of answers, the score range, the average, the standard deviation, the

Table 2. Character Recognition Accuracy (%) on the NCUEE-HJA-2017 and NCUEE-HJA-2018 test sets of 100 answers.

	Fine-tuned	Applied n-gram	NCUEE-HJA-2017	NCUEE-HJA-2018
VGG16	✗	✗	67.36	39.50
	✗	✓	74.55	49.06
	✓	✗	98.11	98.61
	✓	✓	**98.52**	98.27
Dense Net121	✗	✗	67.23	49.21
	✗	✓	73.87	53.71
	✓	✗	94.00	98.02
	✓	✓	96.08	**98.44**
Ensemble of five models	✗	✗	74.64	54.97
	✗	✓	84.15	65.92
	✓	✗	96.99	98.18
	✓	✓	97.75	97.97

number of allowed characters, and the QWK scores of different models are shown in order. We calculated the mean and standard deviation of the scores from a 4-point scale {a, b, c, d} by setting a → 3, b → 2, c → 1, d → 0.

From the score range and mean columns in Table 3, 2017#1 seems most straightforward, as implied by its highest mean score, while 2017#3 seems most difficult due to the lowest mean score. According to QWK scores of these questions, the worst case of 2017#3 is 0.82; while the worst cases of other questions are higher than 0.9. Thus, QWK scores seem to be correlated with the difficulty of the question. As mentioned in the recognition performance analysis, the ensemble model is expected to perform well on a large number of writing styles. In Table 3, the ensemble model outperforms the single models, such as VGG16 and DenseNet121, on all questions.

In order to investigate the robustness of our method with misrecognized text, we compare the ensemble model that combines fine-tuned models and uses n-gram with two fine-tuned single recognizers with n-gram. Moreover, the fine-tuned ensemble models with and without n-gram are compared. Table 3 also shows the degree of deterioration. Although the recognition results are not perfect as 100%, the predicted text seems usable for automatic scoring. For the single models, the QWK scores decreased by nearly 3 percentage points (pcps). For the ensemble model without a language model, the QWK scores are almost similar to the ensemble model with an n-gram language model except for the question 2018#3. Even when considering the language model, the performance did not improve. We found that the more difficult the question, the larger gap in the QWK scores between the worst and best cases.

Table 3. Statistics on scoring for each question.

Questions	No. of answers	Score range	Mean (Std. Div.)	# of allowed char	Quadratic Weighted Kappa (QWK)			
					VGG16	DenseNet 121	Ensemble model (no n-gram)	Ensemble model
2017#1	62,222	0–6	4.46 (1.67)	~50	0.977	0.974	0.975	**0.980**
2017#2	61,777	0–2	1.51 (0.87)	~25	0.957	0.952	0.957	**0.959**
2017#3	59,791	0–5	0.43 (1.10)	80–120	0.844	0.820	**0.847**	0.830
2018#1	67,332	0–3	2.51 (0.88)	~30	0.972	0.970	**0.973**	0.970
2018#2	66,246	0–3	1.87 (1.14)	~40	0.952	**0.953**	0.950	**0.953**
2018#3	58,159	0–3	0.76 (1.07)	80–120	0.933	0.935	0.937	**0.941**

6 Conclusions

This paper presented the first attempt to score handwriting descriptive answers for Japanese language tests automatically. The proposed pipeline consists of multiple state-of-the-art deep neural networks to recognize Japanese handwriting characters and automatically score the answers. The recognizers achieved a high character accuracy of over 97%, with only 0.1% labeled patterns from the dataset. Moreover, the automatic scoring model performed almost the same as examiners with QWK scores from 0.84 to 0.98, depending on the difficulty of the questions. These results suggest that an end-to-end automatic scoring system on descriptive answers, where the recognizer and automatic scoring models are combined as a single model, would be promising for further research.

Acknowledgement. This research is being partially supported by JSPS KAKENHI: Grant Number JP20H04300 and A-STEP: JST Grant Number JPMJTM20ML.

References

1. Takao, S.: Reform in articulation between high school and university as an urgent task of Japanese public policy. Ann. Public Policy Stud. **10**, 87–108 (2016)
2. Nakagawa, M., Hirai, Y.: "Peta-gogy" for future: toward computer - assisted and automated marking of descriptive answers. Inf. Process. Soc. Japan. **57**, 920–924 (2016)
3. Khuong, V.T.M., Phan, K.M., Ung, H.Q., Nguyen, C.T., Nakagawa, M.: Clustering of hand-written mathematical expressions for computer-assisted marking. IEICE Trans. Inf. Syst. **E104.D**, 275–284 (2021). https://doi.org/10.1587/TRANSINF.2020EDP7087

4. Zhu, Y., Xie, Z., Jin, L., Chen, X., Huang, Y., Zhang, M.: SCUT-EPT: new dataset and benchmark for offline Chinese text recognition in examination paper. IEEE Access. 7, 370–382 (2019). https://doi.org/10.1109/ACCESS.2018.2885398

5. Phan, K.M., Khuong, V.T.M., Ung, H.Q., Nakagawa, M.: Generating synthetic handwritten mathematical expressions from a LaTeX sequence or a MathML script. In: Proceedings of the 15th International Conference on Document Analysis and Recognition, pp. 922–927 (2019). https://doi.org/10.1109/ICDAR.2019.00152

6. Plamondon, R., Srihari, S.N.: Online and off-line handwriting recognition: a comprehensive survey. IEEE Trans. Pattern Anal. Mach. Intell. 22, 63–84 (2000). https://doi.org/10.1109/34.824821

7. Nguyen, H.T., Ly, N.T., Nguyen, K.C., Nguyen, C.T., Nakagawa, M.: Attempts to recognize anomalously deformed Kana in Japanese historical documents. In: Proceedings of the 4th International Workshop on Historical Document Imaging and Processing, pp. 31–36 (2017). https://doi.org/10.1145/3151509.3151514

8. Xu, Y., Yin, F., Wang, D.H., Zhang, X.Y., Zhang, Z., Liu, C.L.: CASIA-AHCDB: a large-scale Chinese ancient handwritten characters database. In: Proceedings of the 15th International Conference on Document Analysis and Recognition. pp. 793–798 (2019). https://doi.org/10.1109/ICDAR.2019.00132

9. Simonyan, K., Zisserman, A.: Very deep convolutional networks for large-scale image recognition. In: Proceedings of the 3rd International Conference on Learning Representations (2015)

10. Howard, A.G., et al.: MobileNets: efficient convolutional neural networks for mobile vision applications (2017)

11. He, K., Zhang, X., Ren, S., Sun, J.: Deep residual learning for image recognition. In: Proceedings of the 29th IEEE Conference on Computer Vision and Pattern Recognition, pp. 770–778 (2016). https://doi.org/10.1109/CVPR.2016.90

12. Xie, S., Girshick, R., Dollár, P., Tu, Z., He, K.: Aggregated residual transformations for deep neural networks. In: Proceedings of the 30th IEEE Conference on Computer Vision and Pattern Recognition, pp. 1492–1500 (2017). https://doi.org/10.1109/CVPR.2017.634

13. Huang, G., Liu, Z., Van Der Maaten, L., Weinberger, K.Q.: Densely connected convolutional networks. In: Proceedings of the 30th IEEE Conference on Computer Vision and Pattern Recognition, pp. 2261–2269 (2017). https://doi.org/10.1109/CVPR.2017.243

14. Saito, T., Yamada, H., Yamamoto, K.: On the database ETL 9 of handprinted characters in HIS Chinese characters and its analysis. Trans. IECE Jpn. J68-D(4), 757–764 (1986)

15. Heafield, K., Pouzyrevsky, I., Clark, J.H., Koehn, P.: Scalable modified Kneser-Ney language model estimation. In: Proceedings of the 51st Annual Meeting of the Association for Computational Linguistics, pp. 690–696 (2013)

16. Devlin, J., Chang, M.W., Lee, K., Toutanova, K.: BERT: pre-training of deep bidirectional transformers for language understanding. In: Proceedings of the North American Chapter of the Association for Computational Linguistics: Human Language Technologies, NAACL HTL2019, pp. 4171–4186 (2019)

17. Kingma, D.P., Ba, J.L.: Adam: a method for stochastic optimization. In: Proceedings of the 3rd International Conference on Learning Representations, San Diego (2015)

18. Cohen, J.: Weighted kappa: nominal scale agreement provision for scaled disagreement or partial credit. Psychol. Bull. 70, 213–220 (1968). https://doi.org/10.1037/H0026256

19. Landis, J.R., Koch, G.G.: The measurement of observer agreement for categorical data. Biometrics 33, 159–174 (1977). https://doi.org/10.2307/2529310

A Weighted Combination of Semantic and Syntactic Word Image Representations

Oliver Tüselmann[(⊠)] [iD], Kai Brandenbusch[iD], Miao Chen[iD],
and Gernot A. Fink[iD]

Department of Computer Science, TU Dortmund University,
44227 Dortmund, Germany
{oliver.tueselmann,kai.brandenbusch,miao.chen,
gernot.fink}@cs.tu-dortmund.de

Abstract. In contrast to traditional keyword spotting, semantic word spotting allows users to search not only for word images with the same transcription as the keyword, but also for concepts which are latent or hidden inside a query. However, it has been shown that mapping word images to semantic representations proves to be a difficult task. As semantic embeddings do not consider syntactic similarity, it is common to find search results with highly ranked semantically similar word images, while words with the same transcription as the search query appear in lower ranks. To counteract this problem, a combination of semantic and syntactic representations usually provides a good trade-off w.r.t. semantic and syntactic metrics. In this work, we present methods for realizing a weighted combination of semantic and syntactic information. This allows users to focus more on semantic or syntactic aspects and thus provides new insights to their document collections. Thereby, our proposed methods are not limited to the use of word spotting, but also aim to address the optimization of recognition-free NLP downstream tasks.

1 Introduction

In recent years, great progress has been made in the field of handwriting recognition [18]. However, the transformation of handwritten document images into digital text remains a difficult task, especially for small as well as historical datasets [17]. For this reason, there is a sustained interest in and use of recognition-free word spotting approaches [7]. The goal of word spotting is to retrieve word images from a collection of document images that are similar w.r.t. a given query. Traditional approaches define similarity based on visual properties and avoid an explicit recognition step. This leads to search results in which word images with syntactically close transcriptions to the query are highly ranked. Thereby, two words are syntactically similar if their string edit distance is small. Over the last few years, deep learning based approaches achieved remarkable results on most benchmark datasets in this domain [13,21,24].

© The Author(s), under exclusive license to Springer Nature Switzerland AG 2022
U. Porwal et al. (Eds.): ICFHR 2022, LNCS 13639, pp. 285–299, 2022.
https://doi.org/10.1007/978-3-031-21648-0_20

An extension of this task is semantic word spotting [14,22,24], where semantic information is taken into account during retrieval. This allows users to not only search for syntactically similar words, but also for concepts which are latent or hidden inside a query. A concept could be a similar meaning (e.g. great and huge) or a categorical relationship (e.g. animal and cat). Searching for semantically similar occurrences of words in a given document collection offers users a new way to efficiently explore their collections. However, from a technical perspective, such a search poses major challenges [14,22]. This is mainly due to the high variability of handwriting, which often leads to an incorrect prediction of semantic information [14,22,24]. Another limiting factor is the challenging prediction of semantic information for words that have not been seen during training [14,22]. To overcome these limitations, Krishnan et al. have recently shown that a stacking of semantic and syntactic representations can yield promising results in terms of both semantic and syntactic evaluation measures [14].

However, a simple stacking of these embeddings often leads to a semantic or syntactic bias caused by the different characteristics of the representations (e.g. dimensions and dynamic ranges). In this work, we show how syntactic and semantic representations can be combined with equal importance. Thereby, we present methods for combining both semantic and syntactic information into a single representation by using a weighted combination. The adjustment of the weighting parameter is task-dependent and enables users to prioritize semantic or syntactic aspects and thus obtain new insights into their document collections.

The remainder of this paper is organized as follows. Section 2 introduces the basics and related work in the field of word spotting and word embeddings. In Sect. 3, we present the approaches used for predicting semantic and syntactic embeddings from word images and how to combine them appropriately. We then evaluate the weighted combinations quantitatively as well as qualitatively on three handwriting datasets in Sect. 4. Finally, we summarize the results in Sect. 5.

2   Related Work

In this section, we provide an overview on traditional as well as semantic word spotting. We further introduce an outline of syntactic and semantic word representations used in the context of word spotting.

2.1   Traditional Word Spotting

Word spotting is a retrieval-based method for determining regions in a document image that are similar w.r.t. a given query. The approaches avoid an explicit text recognition and use visual features to determine the similarity between a word image and a query. The similarities are finally used to determine a retrieval list of the most similar word images w.r.t. a given query.

Word spotting approaches can be divided into multiple categories. There exists a variety of different query types with Query-by-Example (QbE) and

Query-by-String (QbS) being the most prominent ones. In QbE applications, the query is a word image whereas in QbS it is a textual string representation. Furthermore, word spotting can be divided into segmentation-free (i.e. entire document images are used without any segmentation) and segmentation-based (i.e. a word level segmentation is required) approaches. There is a wide range of methods in this area covering Bag-of-Feature representations, sequence models, Support Vector Machines and Neural Networks [7]. Recently, approaches based on Convolutional Neural Networks (CNNs) have achieved remarkable results for most benchmark datasets in this area [13,21]. For a more detailed overview of word spotting, see [7].

2.2 Semantic Word Spotting

Semantic word spotting realizes a semantic word image retrieval and can be seen as an extension of the traditional word spotting approach. The aim of this task is to retrieve all word images with the same transcription as the query, followed by semantically similar ones. Even if there are multiple ways to achieve a semantic word image retrieval, we only consider approaches that predict semantic information directly from a word image without transcribing it.

The first approaches of semantic word spotting rely on ontology-based knowledge [8,11] and are thus limited to a small set of human labeled semantic relationships. To overcome this limitation, Wilkinson et al. [24] use a two-stage CNN-based approach to map word images into a textually pre-trained semantic space. Tueselmann et al. [22] show that the semantic space used in [24] encodes only few semantic relations and is close to a syntactic embedding in many parts. They suggest to use a pre-trained FastText [3] embedding and present an optimized two-stage architecture for mapping word images into a semantic space. Recently, Krishnan et al. [14] proposed an end-to-end approach for mapping word images into a semantic space and explored word normalization methods from the Natural Language Processing (NLP) domain, such as Lemmatization and Stemming. Furthermore, they showed that a stacking of semantic and syntactic representations is able to tackle major problems of just using semantic word image embeddings.

2.3 Word Embeddings

Word embeddings convert strings into vector representations. They are often used in word spotting systems to enable a comparison between word images and strings [2,21,24]. The Pyramidal Histogram of Characters (PHOC) embedding is the most commonly used embedding in the field of word spotting [2,13,21]. A PHOC is a binary pyramidal representation of a character string and is used to represent visual attributes of a given word image. Besides PHOC, the Discrete Cosine Transform of Words (DCToW) [24] and the Deep Embedding approach by Krishnan et al. [13] provide state-of-the-art results for many word spotting benchmark datasets.

For realizing a semantic word spotting approach, it is necessary to encode semantic relationships between word images. There are several approaches from the NLP domain that provide this kind of information for textual inputs [3,5,19,20]. The approaches can be divided into context-based and static word embeddings. Context-based methods [5,20] calculate a word representation for a given word based on the context in which it appears. Whereas, static methods [3,19] always output the same embedding for a word, regardless of the context in which it occurs. Since words in different contexts often have different meanings, context-based models are preferable in most applications [6]. However, if there is no context information as in segmentation-based word spotting, static embeddings are preferable.

The combination of embeddings has already been successfully applied in the NLP domain [1,4,23]. The motivation for this combination is that different embeddings encode different semantic aspects and thereby, their combination leads to performance gains in many tasks [1,23]. In contrast to our work, the concatenated embeddings just serve as input to neural architectures and the equal importance of both embeddings is ignored.

3 Method

In this section, we present methods for realizing a weighted combination of semantic and syntactic word image representations. In Sect. 3.1, we introduce the representations used and show how they are predicted from word images using CNNs. Then, we present two approaches for a weighted combination in Sect. 3.2. Finally, we motivate in Sect. 3.3 that a normalization of semantic and syntactic embeddings is necessary before stacking them, due to their different properties in dimensionality and ranges.

3.1 Word Image Representation

For realizing a weighted combination of semantic and syntactic information, suitable word image representations are needed. Similar to previous works in the semantic word spotting domain [14,22], we use FastText as the semantic representation. Due to the state-of-the-art performance in traditional word spotting and the ability to adjust the dimensionality of the embedding size, we use the HWNetv2 [13] for obtaining our syntactic representation.

We use the proposed networks from Krishnan et al. [14] for mapping word images to word representations. Thereby, the HWNetv2 [13] is used for predicting syntactic and a modified ResNet (Attribute-ResNet) for semantic features. The Attribute-ResNet uses a ResNet34 architecture [10] for feature extraction, whereby the global average pooling layer at the end of the network is replaced with a Temporal Pyramid Pooling (TPP) layer. The output of the TPP layer is transferred into a 3-layer Fully-Connected Network (FCN). This FCN has as many neurons in the last layer as there are dimensions in the word representation to be predicted (e.g. FastText = 300). Except for the final layer, the ReLU activation function is applied to the output of all layers in the network.

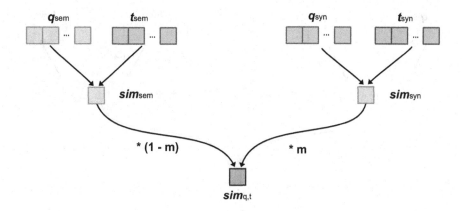

Fig. 1. An overview of the indirect weighting approach.

3.2 Weighted Combination Approaches

We propose two approaches for a weighted combination of semantic and syntactic embeddings. Both approaches receive a query, a word image from the test set, and a weighting factor $m \in [0, 1]$ as input and output a similarity score. In the following, the proposed methods are described in more detail.

We obtain the semantic (q_{sem}) and syntactic (q_{syn}) representations for a given query q as already shown in the previous section. The query can be either a word image or a text string. Similarly, for an element of the test set t, the representations are denoted by t_{syn} and t_{sem}. As a metric we employ the cosine similarity which is commonly used in retrieval tasks.

Indirect Approach. The indirect approach follows an intuitive realization of a weighted combination. After computing the semantic and syntactic similarities separately, the weighted similarity score is obtained by using a linear interpolation (see Fig. 1). Technically, the cosine similarity sim_{syn} of q_{syn} and t_{syn}, as well as sim_{sem} of q_{sem} and t_{sem} are computed respectively. Finally, sim_{syn} and sim_{sem} are weighted by a parameter m and $(1 - m)$ respectively, resulting in its new similarity score $sim_{q,t}$ (see Eq. 1).

$$sim_{syn} = similarity(q_{syn}, t_{syn})$$
$$sim_{sem} = similarity(q_{sem}, t_{sem})$$
$$sim_{q,t} = m \cdot sim_{syn} + (1 - m) \cdot sim_{sem} \quad (0 <= m <= 1)$$

(1)

Direct Approach. In contrast to the indirect approach, the semantic and syntactic vectors for both the query and the test image are first weighted element-wise. Afterwards, these weighted representations are concatenated and finally, the similarity between the concatenated representations of the query and test

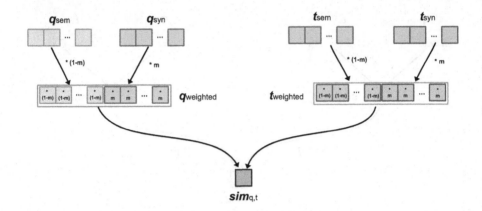

Fig. 2. An overview of the direct weighting approach.

element is calculated (see Fig. 2). A fundamental benefit of this approach is that the concatenated representations can be used not only for obtaining similarity scores, but also for recognition-free downstream tasks in the NLP domain. Technically, each element of the syntactic and semantic embeddings is multiplied with m and $(1 - m)$ respectively. The weighted syntactic and semantic embeddings of q and t are then stacked, resulting in new representations $q_{weighted}$ and $t_{weighted}$. Finally, the cosine similarity of these two embeddings provides their new similarity score $sim_{q,t}$ (see Eq. 2).

$$q_{weighted} = concat((1 - m) \cdot q_{sem}, m \cdot q_{syn})$$
$$t_{weighted} = concat((1 - m) \cdot t_{sem}, m \cdot t_{syn}) \quad (0 <= m <= 1) \qquad (2)$$
$$sim_{q,t} = similarity(q_{weighted}, t_{weighted})$$

3.3   Normalization

For the indirect combination of semantic and syntactic embeddings we can compute the distances in the original embedding spaces. For our approach of stacking the embeddings, however, we have to take the differences of the embedding spaces into consideration. First, the dimensionality of FastText (300-D) differs from the dimensionality of the HWNet embedding (2048-D). Second, the dynamic range of the features in the embeddings differ as shown in Fig. 3a.

In order to mitigate these problems, we apply different normalization methods to the embeddings before the concatenation. The effects of the different normalization approaches to the statistics of the concatenated embeddings can be observed in Fig. 3. Following the approach of Krishnan et al. [14], we normalize the embeddings to have zero mean and unit variance (standardization, see Figs. 3c and 3d). In contrast to their work, we standardize both embeddings instead of only FastText. Another common approach is normalizing the embeddings to have an L2-norm of 1 (see Figs. 3e and 3f). While this normalization

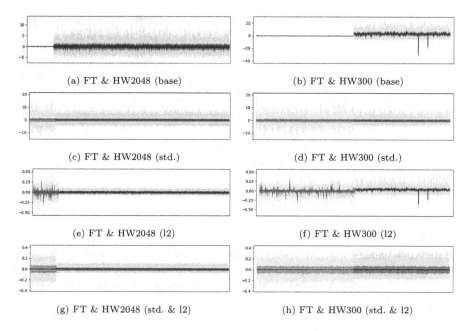

(a) FT & HW2048 (base)   (b) FT & HW300 (base)

(c) FT & HW2048 (std.)   (d) FT & HW300 (std.)

(e) FT & HW2048 (l2)   (f) FT & HW300 (l2)

(g) FT & HW2048 (std. & l2)   (h) FT & HW300 (std. & l2)

Fig. 3. Mean, standard deviation and dynamic range of the different embedding components for combinations of FastText (FT, blue) and HWNet (orange) with 2048 (HW2048) and 300 (HW300) dimensions. We applied standardization (std.), L2-normalization (l2) and combinations of the both to the embeddings before concatenation. The embeddings were computed on the IAM test set. (Color figure online)

has no effect on the cosine similarity of the individual embeddings, changing the length of the vectors before stacking changes the orientation and thus the cosine similarity of the combined embeddings. Additionally, we explore the combination of standardization and length normalization (see Figs. 3g and 3h). In order to examine the influence of the difference in dimensionality of the embedding spaces, we trained the HWNet embeddings to have the same number of dimensions as FastText (300 instead of 2048). The statistics of the combined embeddings using less dimensions are shown in the right column of Fig. 3.

4   Experiments

We evaluate our proposed approaches quantitatively as well as qualitatively on three handwriting datasets (see Sec. 4.1). We supply implementation details in Sect. 4.2 and present in Sect. 4.3 the metrics used. In Sect. 4.4, the impact of our proposed normalization steps before concatenating the semantic and syntactic embeddings is evaluated. Finally, we show and discuss the quantitative as well as qualitative results in Sect. 4.5.

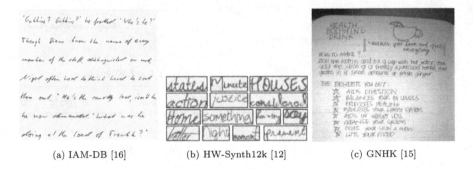

(a) IAM-DB [16]   (b) HW-Synth12k [12]   (c) GNHK [15]

Fig. 4. Example images for the datasets used.

4.1 Datasets

In our experiments, we train and evaluate our approaches on three English hand-writing datasets (see Fig. 4). The datasets vary considerably in their size and characteristics and include synthetically generated as well as real handwritten documents.

IAM-DB. The IAM Database [16] is a popular benchmark for handwriting recognition and word spotting. The documents contain modern English sentences and were written by a total of 657 different people. The database includes 1539 text pages containing a total of 13353 text lines and 115320 words. The official writer independent partitioning splits the database in 6161 lines for training, 1840 for validation and 1861 for testing. The pages contain text from a diverse set of categories (e.g. press, religion, fiction).

HW-Synth12k. The HW-Synth12k (HW) dataset is an in-house version of the HW-Synth dataset proposed in [12]. The dataset consists of synthetically rendered word images from handwritten fonts. The word images contain the 12000 most common words from the English language. For each word, a writer independent split of 50 training and 4 test images is generated. The font is randomly sampled from over 300 publicly available True Type fonts that resemble handwriting.

GNHK. The GoodNotes Handwriting Kollection (GNHK) dataset [15] includes unconstrained camera-captured images of English handwritten text. It consists of 687 documents containing a total of 9363 text lines and 39026 words. The official partitioning divides the data into training and test sets with a ratio of 75% and 25%, respectively.

4.2 Implementation Details

In all of our semantic experiments, we use the pre-trained FastText model proposed by [9]. Similar to Krishnan et al. [14], we normalize the FastText representations to have zero mean and unit variance. For the HWNetv2 model, the PHOC representation consists of layers $2, 3, 4, 5$ and an alphabet with characters $a - z$ and $0 - 9$. The syntactic network follows the proposed optimization and hyper-parameter settings as described in [14].

The semantic network is optimized using the Mean Squared Error Loss and the Stochastic Gradient Descent (SGD) algorithm. The SGD algorithm uses a momentum of 0.9 and a batch size of 64. The networks are first pre-trained on the HW-Synth12k dataset and then fine-tuned on the IAM or GNHK dataset respectively. A learning rate of 0.01 is used during pre-training and 0.001 while fine-tuning. Furthermore, a warm-up strategy is used to train the networks. The images are scaled and padded to a fixed size of 128×384, while keeping the aspect ratio.

4.3   Evaluation Protocol

We use mean Average Precision (mAP) for evaluating the syntactic quality of our approach. MAP is a metric for evaluating retrieval tasks and is de-facto the standard quality measure in the word spotting domain [2,13,21]. In our experiments, we perform evaluation under both segmentation-based QbE and QbS setting. Thereby, we follow the protocol proposed in [2] and discard stop words as queries for the IAM-DB.

Similar to Krishnan et al. [14], we use Word Analogy (WA) for evaluating the semantic quality of our approach. In the WA task, three words a, b and c are given and the goal is to infer the fourth word d that satisfies the following condition: a is to b as c is to d. For example, *Berlin* is to *Germany* as *Paris* is to *France*. In our evaluation, we use the collection of human-defined WA examples proposed in [19]. Note, that questions which contain words that are not part of the test corpus of a dataset are excluded from the evaluation. The accuracy of correctly predicted analogies is used as the final semantic evaluation score.

4.4   Normalization

Figure 5 shows the influence of the weight parameter for the direct combination on the evaluation metrics when applying different normalization methods as explained in Sect. 3.3. The goal of our approach is to allow a user to interpolate between a semantic and syntactic representation as intuitive as possible by adjusting the weight parameter m. Therefore, the course of the metrics should be smooth and robust to small changes in the weighting. When combining the embeddings without any normalization (Fig. 5a), we can observe that FastText is only taken into account for very small values of m. For larger m, the metrics are solely determined by the HWNet embedding. Applying standardization to zero mean and unit variance (Fig. 5b) helps slightly smoothing the metrics as a function of the weight parameter. A further improvement can be achieved by applying L2-normalization after the standardization (Fig. 5c). Reducing the number of dimensions of the HWNet embedding to match the 300 dimensions of FastText allows for a smoother weighting than just combining the original

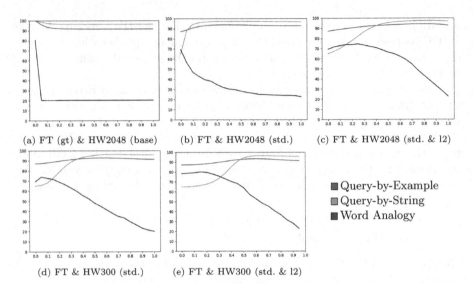

(a) FT (gt) & HW2048 (base)   (b) FT & HW2048 (std.)   (c) FT & HW2048 (std. & l2)

(d) FT & HW300 (std.)   (e) FT & HW300 (std. & l2)

■ Query-by-Example
▢ Query-by-String
■ Word Analogy

Fig. 5. Impact of the weighted combination for different normalization steps. Results are given in accuracy for WA and mAP for QbE and QbS on the IAM dataset. The values on the x-axis represent the weighting factor m from purely semantic (0) to purely syntactic (1). (Color figure online)

HWNet with 2048 dimensions and FastText (Fig. 5d). However, this combination is still dominated by the HWNet embedding for most choices of m. Therefore, we apply standardization and normalization on these embeddings as well (Fig. 5e), achieving a well balanced influence of both embeddings.

A similar effect of the normalization strategies can be observed in Fig. 6 where similarities between example words are depicted depending on the weight parameter. For this example, we chose three syntactically similar (edit distance of 1 or 2) but semantically unrelated words and three semantically similar words which do not share a single character. As a baseline, Fig. 6f shows the indirect approach which is a linear interpolation of the semantic and the syntactic similarity. Figure 6a shows the aforementioned problem of highly unbalanced contributions of the different embeddings towards the similarity between the stacked embeddings when no normalization is applied. Applying standardization (Figs. 6b & 6d) and the combination of standardization followed by L2-normalization (Figs. 6c & 6e) leads to an S-shaped course of the similarity score approximating the desired behaviour of the linear interpolation in the indirect combination. In contrast to the indirect embedding, however, the combined embeddings can be used for other downstream tasks.

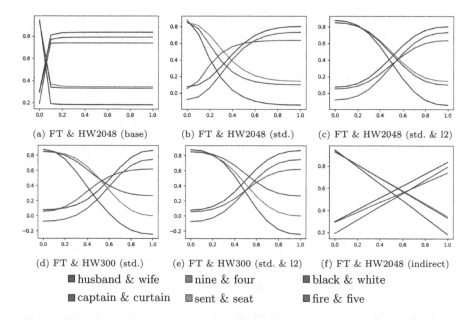

(a) FT & HW2048 (base)   (b) FT & HW2048 (std.)   (c) FT & HW2048 (std. & l2)

(d) FT & HW300 (std.)   (e) FT & HW300 (std. & l2)   (f) FT & HW2048 (indirect)

■ husband & wife   ■ nine & four   ■ black & white

■ captain & curtain   ■ sent & seat   ■ fire & five

Fig. 6. Combined similarities for either syntactically or semantically similar word pairs. The values on the x-axis represent the weighting factor m from purely semantic (0) to purely syntactic (1). (Color figure online)

4.5 Results

We evaluate our method quantitatively based on the semantic and syntactic metrics described in Sect. 4.3. Since the benefit for a human adaptation of the weighting cannot be easily expressed in a score, we further provide a qualitative evaluation where we show representative retrieval lists for different weightings.

Quantitative Evaluation. We provide the performances of the direct and indirect weighting approaches in Fig. 7. The plots show the course of the syntactic (QbE and QbS) and semantic (WA) metrics at different weights from purely semantic (0) to purely syntactic (1). Interestingly, the results are very similar for both approaches on all datasets. However, the S-shaped course for the direct method provides a qualitatively improved and more natural ranking. This is also supported by the plots in Fig. 6. In general, the QbS score is strictly monotonically increasing and rises most strongly for weights in the range between 0 and 0.5. After that, only small improvements can be achieved. Furthermore, the QbE score remains fairly constant. The WA score shows a surprising course on the IAM and GNHK datasets, as it rises in the range from 0 to 0.2, drops slightly from 0.2 to 0.5 and finally drops sharply. The increase at the beginning can be explained by the fact that there are a few syntactic analogies in the WA task and they may benefit from the syntactic information.

Table 1 compares our proposed approaches with the recently published stacking method by Krishnan et al. [14]. The main focus of this evaluation is on the

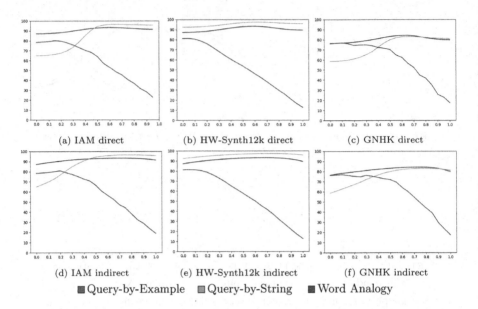

(a) IAM direct   (b) HW-Synth12k direct   (c) GNHK direct

(d) IAM indirect   (e) HW-Synth12k indirect   (f) GNHK indirect

■ Query-by-Example   □ Query-by-String   ■ Word Analogy

Fig. 7. Impact of the weighted combination for the proposed direct and indirect combination approaches. Results are given in accuracy for WA and mAP for QbE and QbS. The values on the x-axis represent the weighting factor m from purely semantic (0) to purely syntactic (1). (Color figure online)

Table 1. Performances on the three evaluated datasets using accuracy for the WA task (semantic) and mAP for QbE and QbS word spotting (syntactic).

Approach	HW-Synth12k			IAM-DB			GNHK		
	WA	QbE	QbS	WA	QbE	QbS	WA	QbE	QbS
FastText GT [3]	82.6	—	—	87.4	—	—	91.3	—	—
Syntactic (HW300)	16.4	89.4	95.8	18.9	91.3	**95.7**	17.4	79.9	**81.0**
Semantic (FastText)	**81.2**	87.3	92.3	**78.4**	87.1	65.0	**76.0**	76.3	58.7
Krishnan et al. (Stacked) [14]	—	—	—	61.5	90.6	94.3	—	—	—
Ours (Indirect)	57.7	**92.8**	**96.8**	66.6	**92.8**	94.3	72.7	82.6	78.5
Ours (Direct)	57.1	92.2	96.3	64.1	**92.8**	94.3	70.3	**83.1**	79.8

comparison of syntactically and semantically combined methods. However, for a better interpretation of the results, we also provide the scores of the purely syntactic and semantic models. Unfortunately, a comparison with [14] is not straightforward, as it is trained and evaluated on an unpublished version of the HW-Synth dataset and no scores have been published for the GNHK dataset. Therefore, we can only compare our optimized concatenation approaches on the IAM dataset. The results show that both of our models outperform the approach of Krishnan et al. [14] w.r.t. the WA and QbE metrics and achieve identical scores for QbS. Our methods achieve a proper trade-off between the

Fig. 8. The top-10 of the retrieval lists for the query *jacky* and different weightings of combination on the IAM-DB test set.

syntactic and semantic metrics and can enhance the QbE scores remarkably on all datasets. Compared to the textual FastText model, working on the ground truth (GT) annotations of the datasets, the pure semantic model is able to achieve fairly high scores on the IAM and GNHK datasets. However, these are still several points behind the textual approach. For the HW-Synth12k dataset the WA scores are close to the annotation-based system, which is presumably due to the lower variability of the synthesized data and that all words of the test set have been seen during training.

Even though achieving high scores on these metrics is beneficial, the interpretability of the quantitative results is limited. Since the focus of this work is on the realization of an appropriate weighting, the insights from the trajectories given in Fig. 5 and the following qualitative evaluation are increasingly important.

Qualitative Evaluation. Even though the quantitative results for the combination of syntactic and semantic embeddings could show their ability to perform well on both semantic and syntactic scores, the benefit for the user in exploring document collections cannot be captured that easily. Hence, Fig. 8 and 9 show two examples for the query words *jacky* and *hotel* on the IAM-DB test set with the weightings of (0, 0.25, 0.5, 0.75, 1). The figures illustrate the problem of a purely semantic embedding (0), whereby semantically similar words w.r.t. the query appear in the top 10 of the result lists, but word images with the same transcription as the query are often missing. When syntactic information is taken into account (0.25, 0.5), these words are highly ranked and the list also contains many semantically relevant word images. When the focus shifts more to the syntactic representation (0.75, 1), the words with the same transcription as the query are highly ranked, however, the semantic information is lost. The examples show that a weighted exploration of datasets has many advantages, for example a user searching for the word *hotel* is likely more interested in occurrences of the words *hostel*, *stay* or *room* then to words like *motor* or *heel*.

298   O. Tüselmann et al.

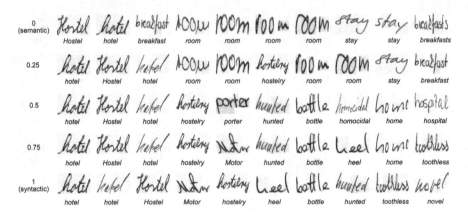

Fig. 9. The top-10 of the retrieval lists for the query *hotel* and different weightings of combination on the IAM-DB test set.

5 Conclusions

In this work, we present and evaluate two approaches for realizing an user-adaptable weighted combination of semantic and syntactic word image representations. Our experiments show both qualitatively and quantitatively that a weighted combination of such representations offers many advantages and yields interesting results especially in the area of semantic word spotting. Although the primary objective of this approach is on the manual adjustment of the weighting parameter by the user, the approach could be extended by an automatic recommendation of this parameter. In future work, we plan to investigate the effect of this combination in the context of recognition-free NLP tasks.

References

1. Akbik, A., Bergmann, T., Blythe, D., Rasul, K., Schweter, S., Vollgraf, R.: FLAIR: an easy-to-use framework for state-of-the-art NLP. In: NAACL-HLT, Minneapolis, MN, USA, pp. 54–59 (2019)
2. Almazán, J., Gordo, A., Fornés, A., Valveny, E.: Word spotting and recognition with embedded attributes. IEEE Trans. Pattern Anal. Mach. Intell. **36**(12), 2552–2566 (2014)
3. Bojanowski, P., Grave, E., Joulin, A., Mikolov, T.: Enriching word vectors with subword information. Trans. ACL **5**, 135–146 (2017)
4. Dang, H., Le-Hong, P.: A combined syntactic-semantic embedding model based on lexicalized tree-adjoining grammar. Comput. Speech Lang. **68**, 101202 (2021)
5. Devlin, J., Chang, M., Lee, K., Toutanova, K.: BERT: pre-training of deep bidirectional transformers for language understanding. In: ACL, Minneapolis, MN, USA, pp. 4171–4186 (2019)
6. Ethayarajh, K.: How contextual are contextualized word representations? comparing the geometry of BERT, ELMo, and GPT-2 embeddings. In: Proceedings of Conference on Empirical Methods in Natural Language Processing, Hong Kong, pp. 55–65 (2019)

7. Giotis, A.P., Sfikas, G., Gatos, B., Nikou, C.: A survey of document image word spotting techniques. Pattern Recogn. **68**, 310–332 (2017)

8. Gordo, A., Almazán, J., Murray, N.: LEWIS: latent embeddings for word images and their semantics. In: ICCV, Santiago, Chile (2015)

9. Grave, E., Bojanowski, P., Gupta, P., Joulin, A., Mikolov, T.: Learning word vectors for 157 languages. In: Proceedings of International Conference on Language Resources and Evaluation, Miyazaki, Japan (2018)

10. He, K., Zhang, X., Ren, S., Sun, J.: Deep residual learning for image recognition. In: CVPR, Las Vegas, NV, USA, pp. 770–778 (2016)

11. Krishnan, P., Jawahar, C.V.: Bringing semantics in word image retrieval. In: ICDAR, Washington, DC, USA, pp. 733–737 (2013)

12. Krishnan, P., Jawahar, C.V.: Generating synthetic data for text recognition. CoRR abs/1608.04224 (2016)

13. Krishnan, P., Jawahar, C.V.: HWNet v2: an efficient word image representation for handwritten documents. Int. J. Doc. Anal. Recogn. (IJDAR) **22**, 387–405 (2019)

14. Krishnan, P., Jawahar, C.: Bringing semantics into word image representation. Pattern Recogn. **108**, 107542 (2020)

15. Lee, A.W.C., Chung, J., Lee, M.: GNHK: a dataset for English handwriting in the wild. In: ICDAR, Lausanne, Switzerland, pp. 399–412 (2021)

16. Marti, U., Bunke, H.: The IAM-database: an English sentence database for offline handwriting recognition. Int. J. Doc. Anal. Recogn. (IJDAR) **5**(1), 39–46 (2002)

17. Mathew, M., Gomez, L., Karatzas, D., Jawahar, C.V.: Asking questions on handwritten document collections. Int. J. Doc. Anal. Recogn. (IJDAR) **24**(3), 235–249 (2021). https://doi.org/10.1007/s10032-021-00383-3

18. Memon, J., Sami, M., Khan, R.A., Uddin, M.: Handwritten optical character recognition (OCR): a comprehensive systematic literature review. IEEE Access **8**, 142642–142668 (2020)

19. Mikolov, T., Chen, K., Corrado, G., Dean, J.: Efficient estimation of word representations in vector space. In: ICLR, Scottsdale, AZ, USA (2013)

20. Peters, M.E., et al.: Deep contextualized word representations. In: NAACL, New Orleans, LA, USA, pp. 2227–2237 (2018)

21. Sudholt, S., Fink, G.A.: PHOCNet: a deep convolutional neural network for word spotting in handwritten documents. In: ICFHR, Shenzhen, China, pp. 277–282 (2016)

22. Tüselmann, O., Wolf, F., Fink, G.A.: Identifying and tackling key challenges in semantic word spotting. In: ICFHR, Dortmund, Germany, pp. 55–60 (2020)

23. Wang, X., et al.: Automated concatenation of embeddings for structured prediction. In: ACL/IJCNLP, Bangkok, Thailand, pp. 2643–2660 (2021)

24. Wilkinson, T., Brun, A.: Semantic and verbatim word spotting using deep neural networks. In: ICFHR, Shenzhen, China, pp. 307–312 (2016)

Combining Self-training and Minimal Annotations for Handwritten Word Recognition

Fabian Wolf$^{(\boxtimes)}$ (D) and Gernot A. Fink (D)

Department of Computer Science, TU Dortmund University,
44227 Dortmund, Germany
{fabian.wolf,gernot.fink}@cs.tu-dortmund.de

Abstract. Handwritten Text Recognition (HTR) relies on deep learning to achieve high performances. Its success is substantially driven by large annotated training datasets resulting in powerful recognition models. Performances suffer considerably when applied to document collections with a distinctive style that is not well represented by training data. Applying a recognition model to a new data collection poses a tremendous annotation effort, which is often out of scope, for example considering historic collections. To overcome this limitation, we propose a training scheme that combines multiple data sources. Synthetically generated samples are used to train an initial model. Self-training offers the possibility to exploit unlabeled samples. We further investigate the question of how a small number of manually annotated samples can be integrated to achieve maximal performance with limited annotation effort. Therefore, we add labeled samples at different stages of self-training and propose two criteria, namely confidence and diversity, for the selection of samples to annotate. In our experiments, we show that the proposed training scheme is able to considerably close the gap to fully-supervised training on the designated training set with less than ten percent of the labeling demand.

Keywords: Handwritten text recognition · Self-training · Semi-supervised learning

1 Introduction

Deep learning has transformed all areas of computer vision and machine learning. Across many domains, we have observed the diminishing use of designed structural model components. While at the beginning of deep learning, networks usually relied on convolutional layers and recurrent components [6,17,28], recent developments such as transformer networks do not rely on these forms of structural regularization [11,19]. In order to counter the lack of structural design, novel architectures have a tremendous number of parameters and need to learn these from data. Given enough annotated training data, this approach is extremely successful for all kinds of applications. Handwritten text recognition

(HTR) has seen a similar development. Early models built upon Hidden Markov Models account for the sequential nature of text and allow for the definition of character and word models [22]. HMMs have been mostly replaced by convolutional networks combined with LSTMs or similar recurrent components [14,28]. Removing structural model design and relying on highly overparameterized neural architectures resulted in performance gains across benchmarks.

The removal of structural biases seems to be beneficial from a pure performance perspective, but introduces a severe drawback. Optimizing a highly overparameterized deep learning model requires a tremendous amount of manually annotated data. Today, a major source of handwritten text to be analyzed by HTR technology comes from historic documents. Writing styles are usually quite specific and models trained on modern handwriting often yield poor performances. In order to train a state-of-the-art model for such a collection, annotations need to be created by scholarly experts. This makes the application of a deep learning HTR model extremely costly and often out of scope for the application to a new document collection.

In this work, we investigate the question of how to train an HTR model for maximum performance with limited annotation effort. Therefore, we rely on different types of data and levels of supervision. Synthetic data allows to pretrain the recognizer to derive an initial model. To further adapt the model to a target data collection, we apply a self-training approach to exploit unlabeled data. While most works either train on a representative training dataset or do not use any labeled data, we aim at maximizing the performance gains from small numbers of labeled samples. Therefore, we experiment on how to most efficiently integrate a small number of labeled samples into a self-training process. As self-training relies on a large dataset of pseudo-labeled samples, it is an open question if a model may benefit from the addition of comparably small numbers of manually annotated samples. This also implies the question of how to choose samples for labeling in order to maximize the value of manual annotations. We propose two selection schemes based on a confidence measure and introduce a method to ensure diversity among selected samples. In our experiments, we show that the proposed training scheme is highly efficient and we are able to significantly close the performance gap to fully-supervised models with only 1%, 5% or 10% of the labeling demand.

The remainder of this paper is organized as follows. Section 2 discusses related works on text recognition, semi-supervised learning and self-training in document analysis. The proposed method relies on a sequence-to-sequence based HTR model that is trained via self-training with additional manually annotated samples. The underlying architecture, the proposed training scheme and selection criteria are presented in Sect. 3. Finally, we present experiments on three considerably different benchmark datasets in Sect. 4.

2   Related Work

Handwritten Text Recognition (HTR) is a central problem in document analysis and has a long standing history of being tackled with machine learning techniques [22]. Early models are based on hand designed features in combination with Hidden Markov Models [22]. As the LeNet essentially solves an HTR problem, the research area was at the brink of the uprise of convolutional neural networks and deep learning [17]. Today, HTR vastly exploits the capabilities of deep neural networks [5,11,14,19,28]. The de facto state of the art relies on a deep convolutional architecture that is combined with recurrent components such as LSTMs or GRUs. In order to optimize the tremendous amount of parameters, the model is trained using Connectionist Temporal Classification (CTC) [6,14,28]. The CTC-Loss reduces the framewise character predictions to a recognition sequence and serves as the optimization criterion. Lately, different recognition approaches received increasingly more attention due to their success in other areas of deep learning. Sequence-to-sequence approaches exploit attention mechanisms to learn the mapping between sequences of different length [11,26]. With the recent popularity of transformer architectures in many application areas of deep learning, they were investigated for the task of HTR and showed competitive performances [9,19].

Text recognition has observed a steady increase of performances driven by increasingly complex models. The application of deep models that are highly overparameterized requires a large amount of annotated training data. For many application scenarios such as the analysis of historic documents, annotations from the domain are essential for high performances. The high and costly demand for in-domain annotations encourages the investigation of methods that do not require full supervision. Transfer learning describes a technique to train a model on related data followed by a phase of finetuning on a limited training set. Several works show that this reduces the demand of labeled data from the target domain considerably and increases performances [8,21]. As handwritten text and documents in general are highly regularized by script and language constraints, it is straightforward to design synthesis pipelines that allow for the automatic generation of samples. Several works use synthetic data to pretrain models that are then further finetuned with labeled samples [7,15]. Although models trained purely on synthetic data usually perform poorly, labeling demand can be reduced considerably. Motivated by the success of self-supervised training for visual representation learning, [1] showed that it is beneficial to pretrain a recognition model with a self-supervised contrastive learning strategy. Other works investigate adaptation strategies either from related or synthetic data that do not rely on any manually annotated samples [10,30,31]. In [10,31], unsupervised domain adaptation strategies are proposed that exploit an adversarial loss to align extracted features from synthetic and real samples.

Another popular method common in semi-supervised learning is the concept of self-training [2,18,24]. In order to exploit unlabeled samples, an initial model, which is either trained on a synthetic or on a small labeled training set, makes predictions for unlabeled data. These predictions are considered to be correct and used as labels for further training. Self-training is usually combined with a confidence based selection of samples and consistency regularization techniques [2,24]. This approach has also been heavily investigated for document analysis. In

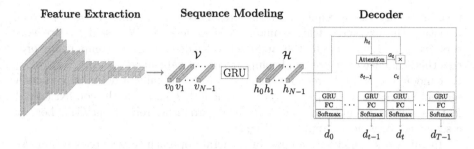

Fig. 1. Sequence-to-Sequence handwritten text recognition architecture with an attention based decoder.

[23,25], it was shown that self-training allows for increased performances when used in a transductive learning scenario. In this case, pseudo-labeled samples from a test set are used during training. In [27], a self-training approach allows to adapt an HTR model trained on a dataset in one language to another one using a language model as additional supervision. Self-training has been shown to be an efficient way to adapt OCR models with already high initial performances [4,12]. Despite the rather poor performances of the initial models, [29,30] showed that it is feasible to train models solely by means of self-training starting from synthetic data. This allows to train models with high performances without any manually labeled samples from the target domain.

3   Method

In this work, we investigate the combination of synthetic and unlabeled data with a small number of manually annotated samples to efficiently train a handwritten word recognizer. Our method is built upon a state-of-the-art handwritten word recognizer based on a sequence-to-sequence architecture (Sect. 3.1). Section 3.2 presents the proposed training scheme that allows to combine pretraining on synthetic data and self-training on pseudo-labels with supervision from manually annotated samples. Integrating labeled samples into a semi-supervised training method poses the question of which samples to select and annotate. While a random selection is common in the literature, Sect. 3.3 discusses two criteria namely confidence and diversity to be taken into account.

3.1   Text Recognition

Over the past years, a plethora of handwriting recognition models has been proposed. As in most areas of document analysis and computer vision, models almost exclusively rely on neural architectures. Text recognition approaches commonly share a general architecture that has been proven to achieve state-of-the-art performances. First, a convolutional neural network extracts two dimensional feature maps $\mathcal{V}$. For languages that rely on an alphabetic writing system, text inherently has a sequential structure. This is taken into account by representing the extracted feature maps as a sequence of N visual feature vectors $(v_0, v_1, \ldots, v_{N-1})$. On the architectural side, recurrent components process this sequence of vectors and compute contextual features $(h_0, h_1, \ldots, h_{N-1})$ that model a certain degree of sequential context. Finally, a decoder predicts a sequence of characters $(y_0, y_1, \ldots, y_{T-1})$ of length T based on the contextualized feature vectors. Common decoders in this regard either rely on the CTC Loss or a sequence-to-sequence approach (Fig. 1).

In this work, we closely follow this general approach to text recognition. As the focus of our work is not the design of a novel text recognition architecture but the development of an efficient training scheme, we chose model components closely comparable to [10, 30]. In this case, a VGG-19-BN architecture serves as the feature extractor of the model. Two layers of Bidirectional Gated Recurrent Units model the sequence of contextualized feature vectors. A local attention mechanism allows to combine several feature vectors into a context vector. Based on the current decoder state and the context, fully connected layers with softmax compute the final prediction. At each decoding step, the decoder predicts the pseudo-probabilities d_i over the set of characters. During training, the model is optimized using a Cross-Entropy loss between the character annotation and the predicted pseudo-probability.

3.2   Training Scheme

In order to optimize the model, we aim at combining several training techniques. Training on synthetic data allows to derive an initial model that is then further adapted to the target data. Self-training allows to exploit unlabeled data by using the current weights of the model to predict pseudo-labels which are then considered as training samples. In this work, we investigate the question if adding a small portion of manually labeled samples increases performances. Furthermore, it is not obvious at which training stage manually labeled samples are most efficiently added to the training set.

Let $S = \{(s_n, y_n); n \in (1, \ldots, N_s)\}$ be a batch of synthetic samples of size N_s with labels y. $U = \{(u_n, \hat{p}_{n,i}); n \in (1, \ldots, N_u)\}$ denotes the unlabeled dataset with predicted pseudo-labels $\hat{p}$ at self-training cycle i. Annotating a limited number of samples K with their actual label results in a labeled set $L = \{(x_n, p_n); n \in (1, \ldots, K)\}$. For training, we combine pseudo-labeled and labeled data resulting in a combined training set $T = U \setminus \{(u_n, \hat{p}_{n,i}); u_n \in L\} \cup L$. In this work, we investigate multiple training schemes that integrate labeled data at different points of training. First, we train an initial model on the synthetic dataset. A simple approach, which is common in the literature on semi-supervised learning, is to further finetune the model exclusively on a small labeled dataset.

As discussed in [30], significant performance gains may be achieved by training on a pseudo-labeled dataset. We follow the same strategy and generate a set of pseudo-labels for the unlabeled dataset. We exploit the confidence measure

---

Algorithm 1: Semi-supervised training procedure

Input : Synthetic Data $S = \{(s_n, y_n); n \in (1, \ldots, N_s)\}$, Unlabeled Data
$U = \{(u_n, \hat{p}_{n,i}); n \in (1, \ldots, N_u)\}$, Labeled Data $L = \{\}$, number of
self-training cycles M, number of manually labeled samples K,
confidence measure $c(\cdot)$, threshold τ

1   Train initial model on S;
2   if *Add Before* then
3   $\quad \lfloor$ $L = LABEL(U, K)$

4   if *Self-Training* then
5   $\quad \mid$ for $i \leftarrow 0$ to M do
6   $\quad \quad \mid$ Estimate Pseudo Labels for U;
7   $\quad \quad \mid$ Keep confident samples U_{conf} with $c(u_b) > \tau$ for each $u_b \in U$;
8   $\quad \quad \mid$ if *Add During* then
9   $\quad \quad \quad \lfloor$ $L = L \cup LABEL(U, K/M)$;
10 $\quad \quad \lfloor$ $T = U_{conf} \cup L$, Train on T;

11 if *Add After* then
12 $\quad \lfloor$ $L = LABEL(U, K)$, Train on L;

---

introduced in Sect. 3.3 to remove unconfident samples from the set of pseudo-labeled samples with a confidence threshold of τ. After training an epoch on the training set, pseudo-labels are reestimated. Combining self-training with annotated samples gives several options on when to include labeled samples. After training the initial model, a set of labeled samples may be added to the combined training set. Training is then performed on the combination of labeled and pseudo-labeled samples, while pseudo-labels are iteratively reestimated. This allows the model to exploit labeled data for pseudo-labeling from the beginning of the training procedure. Alternatively, the model may be first trained solely by means of self-training. After adaptation further finetuning is performed by training exclusively on labeled data. Instead of integrating an entire labeled dataset at once, an active learning strategy can be pursued. At each self-training cycle, we add a limited number of labeled samples to the current training set.

Algorithm 1 summarizes the entire training scheme. Labels are either added before, after or during self-training. Despite the difference at which point to add labeled samples, a selection strategy is required. As presented in Sect. 3.3, the *LABEL* function can either select samples randomly, maximally unconfident or unconfident with additionally ensured diversity.

3.3   Confidence and Diversity

The performance of a model is strongly determined by the extent to which the training data is representative for a test sample. As the training set can naturally only represent an approximation of the test distribution, individual samples are differently well represented. Depending on the training distribution

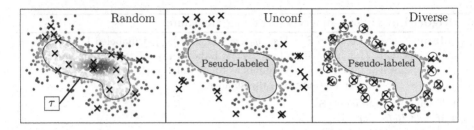

Fig. 2. Toy example for different sampling strategies. Red indicates a low confidence value. Pseudo-labeled samples with $c(u_i) < \tau$ are used during self-training. × represents samples selected for annotation, ○ correspond to cluster centroids. (Color figure online)

and the generalization capability of the model, predictions underly a degree of uncertainty. The degree of uncertainty is often measured by finding a numeric estimation on how confident a network is in its prediction. Several works exploit such a confidence measure to select samples in semi-supervised training schemes. Despite the observation that neural networks are often overconfident, the use of the activation of output neurons is capable to identify erroneous predictions.

To derive a confidence estimate for the recognition model, we follow the same approach as [30]. For each word image, the recognizer predicts a sequence of pseudo-probabilities d_i. The confidence of the character prediction is determined by its corresponding activation. Character confidences are then aggregated to a resulting confidence estimate c for the entire sequence. We then select samples for annotation with minimal confidence:

$$\text{LABEL}(\mathbf{U}, K) = \underset{\mathbf{U}}{\arg\min} \sum_{i=0}^{K-1} c(u_i), \qquad c(u_i) = \frac{1}{T} \sum_{t=0}^{T-1} \max\left(d_t\right) \qquad (1)$$

A potential drawback of selecting unconfident samples to annotate and include into semi-supervised training is their lack of diversity. Visually similar samples that are not well represented by the training distribution share low confidence values. Additionally, datasets often include degraded samples that are not well suited as additional training material. To counter the effect of a pure confidence based selection, we propose to ensure diversity among unconfident samples. First, all unlabeled samples are thresholded based on confidence. For each sample, we derive a feature representation by aggregating the contextual feature vectors $(h_0, \ldots, h_{N-1})$. K-means clustering is performed over the resulting embedding. We select samples with minimal cosine similarity d_{cos} to the cluster centroids c_i to ensure a diverse selection:

$$\{c_0, \ldots, c_{K-1}\} = \text{kmeans}(f(\mathbf{U} \setminus \mathbf{U}_{\text{conf}}), K), \qquad f(u_i) = mean(\{h_{0,i}, \ldots, h_{N-1,i}\}) \qquad (2)$$

$$\text{LABEL}(\mathbf{U}, K) = \underset{\mathbf{U}}{\arg\min} \sum_{i=0}^{K-1} d_{cos}(f(u_i), c_i) \qquad (3)$$

Table 1. Datasets

Dataset	Train	Validation	Test
George Washington [16]	2404	1291	1164
IAM Database [20]	47981	7554	20305
CVL [13]	12146	–	87756

In this case, the number of clusters K corresponds to the number of samples that are annotated and included into training. Figure 2 visualizes a toy example for the different selection approaches.

4   Experiments

In our evaluations, we conduct experiments on three different datasets that vary considerably in size, characteristics and the ratio of labeled to unlabeled data, see Sect. 4.1. We report confidence intervals to account for the random nature of the empirical evaluations, see Sect. 4.2. In Sect. 4.3, we present the results of our experiments which we compare to other results from the literature in Sect. 4.4.

Most training parameters are adopted from [30]. Initial models are pretrained for one epoch on the synthetic datasets proposed in [30] that were generated with true type fonts using the Gutenberg Project as a textual database. All models are trained using ADAM optimization and a batch size of 32. In all experiments, we train for 50 epochs on the respective training set, which can either exclusively contain labeled samples or a combination of pseudo-labeled samples with actual labeled ones. Pseudo-labels are generated for all samples independent from their corresponding partition and thresholded at a confidence value of $\tau = 0.55$. Note that a threshold of $\tau = 0.55$ is surpassed only by quite confident samples, as the training target is numerically defined as 0.6 due to label smoothing. To allow for a fair comparison, labeled samples are only selected from the designated training partitions of the datasets. We evaluate our model in terms of character (CER) and word error rates (WER).

4.1   Datasets

We train and evaluate the model on three different datasets. See Table 1 for an overview of the number of samples and respective partitions. The George Washington dataset [16] is a historic collection written by a single writer with rather homogeneous appearance. Overall, training and test data is highly limited. Note, that training with 1%, 5% or 10% of the training data corresponds to adding only 24, 120 or 240 labeled samples. The IAM database [20] is the de facto standard benchmark for evaluating handwriting recognition models. We use the RWTH partitioning for word recognition. The IAM database contains handwritten samples from 657 different writers and is considerably bigger in size compared to GW. Additionally, we conduct experiments on the CVL dataset

[13]. In contrast to the other two benchmarks, the multiwriter CVL database is not exclusively written in the English language but also contains German samples. In our experiments, we do not make any modifications in this regard meaning that unigrams which do not occur in English will not be predicted by the model. Furthermore, the test set is significantly larger than the training partition. Adding 1%, 5% or 10% of the training set results to an extremely low number of labeled samples (120, 600, 1200) compared to the total number of unlabeled data of 99902.

4.2 Confidence Intervals

In order to account for the random nature of our experiments, we report confidence intervals for the presented results. Whenever an experiment relies on a random label selection, we conducted the experiment five times and report mean and standard deviation. In all other cases, the reported intervals rely on the assumption that the recognition process can be considered a Bernoulli process [3]. With k being the number of correctly recognized and n the total number of patterns, the recognition rate can be expressed as $p = k/n$. A confidence interval is calculated by solving the following quadratic equation:

$$(n + c^2)p^2 - (2k + c^2)p + \frac{k^2}{n} = 0 \qquad (4)$$

For $c = 1.96$, the actual error rate falls in the confidence interval with a probability of 95%.

4.3 Results

Finetuning. In a first series of experiments, we investigate the performance of the model when trained without self-training. Therefore, we first pretrain the model for one epoch on the synthetic dataset and then perform 50 epochs of finetuning on a small labeled dataset. Labeled data is either selected randomly, based on confidence estimations or with ensured diversity. Figure 3 presents examples that are selected under the different sampling strategies. Qualititively, we observe that a unconfident selection results in short sequences of one or two characters for IAM, where recognition is almost impossible due to the lack of context. For GW and CVL, these artifacts do not exist in the databases and we see a high focus on single words such as the abbreviation "G.W." or the word "effect". The proposed diversity criterion avoids this characteristic and leads to a visually more diverse set of samples.

Table 2 presents the results for finetuning with 1%, 5% and 10% of the original training set. Our model is able to adapt to the target data already with an extremely limited number of samples. Note, that taking 1% of the George Washington training set results in only 24 samples. Considering the different selection strategies, we observe that taking the most unconfident samples is harmful compared to a random selection for the smallest training portions. This indicates

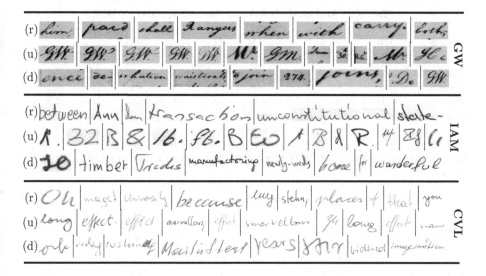

Fig. 3. Examples for random (r), unconfident (u) and diverse (d) selection.

Table 2. Experiments on finetuning from synethtic data.

Method		GW						IAM						CVL					
		1%	±	5%	±	10%	±	1%	±	5%	±	10%	±	1%	±	5%	±	10%	±
Random	CER	16.6	0.4	10.2	1.1	8.3	0.5	13.3	0.2	11.1	0.1	10.0	0.1	21.7	0.3	16.9	0.2	14.2	0.3
Unconf		21.5	–	9.0	–	8.3	–	14.8	–	11.7	–	10.1	–	26.3	–	17.1	–	13.6	
Diverse		14.1	–	9.1	–	7.3	–	12.7	–	10.4	–	10.1	–	21.3	–	16.1	–	13.7	
Random	WER	41.4	1.2	28.3	1.7	24.3	1.3	31.6	0.4	27.1	0.4	25.5	0.4	48.0	0.4	40.2	0.3	34.5	0.6
Unconf		52.0	2.9	26.1	2.5	24.2	2.5	33.7	0.7	28.1	0.6	25.0	0.6	52.9	0.3	41.4	0.3	34.3	0.3
Diverse		38.0	2.8	27.0	2.5	22.1	2.4	31.1	0.6	26.0	0.6	25.7	0.6	47.2	0.3	38.5	0.3	33.7	0.3

Initial Models (0%) in CER (WER): GW: 25.4 (62.1) IAM: 21.8 (48.1) CVL: 27.5 (56.07)

that in these cases training focuses too much on unrepresentative samples. For higher portions of training data unconfident selections achieve slightly better results especially under the diversity criterion.

Add Labels Before Self-training. An intuitive approach to add labeled samples to training is to include a small number of samples at the beginning. This allows the model to exploit knowledge learned from labeled data while predicting pseudo-labels. We, therefore, combine small portions of labeled samples from the designated training sets with unlabeled data. Then, we conduct 50 cycles of self-training on iteratively estimated pseudo-labels. At each cycle, we train for one epoch over the combined set of labeled and pseudo-labeled samples. Independent from the selection strategy, the integration of labeled samples increases performances for IAM and GW, see Table 3. In case of CVL performance gains can only be observed for 5% and 10% of labeled training data in combination with a

Table 3. Add labels before the first self-training cycle.

Method		GW						IAM						CVL					
		1%	±	5%	±	10%	±	1%	±	5%	±	10%	±	1%	±	5%	±	10%	±
Random	CER	10.3	0.6	8.9	0.9	8.2	0.4	9.7	0.2	8.6	0.9	7.3	0.3	8.3	0.7	8.1	0.6	7.7	0.7
Unconf		9.9	–	7.9	–	7.2	–	11.6	–	8.2	–	7.3	–	8.1	–	7.3	–	7.1	
Diverse		10.6	–	8.3	–	7.5	–	9.9	–	9.3	–	7.6	–	9.6	–	7.6	–	7.2	
Random	WER	27.2	1.3	23.8	2.2	22.4	1.6	26.9	0.5	21.7	0.6	19.5	1.0	23.5	1.6	23.2	1.1	23.2	1.9
Unconf		25.6	2.5	22.0	2.4	20.7	2.3	31.8	0.6	22.9	0.6	21.0	0.6	24.3	0.3	23.5	0.3	23.2	0.3
Diverse		28.0	2.6	22.9	2.4	20.7	2.3	27.1	0.6	26.3	0.6	21.4	0.6	26.6	0.3	23.5	0.3	22.8	0.3

No added labeles (0%) in CER (WER): GW: 11.6 (29.5) IAM: 10.7 (28.5) CVL: 8.1 (24.8)

Table 4. Add labels after self-training.

Method		GW						IAM						CVL					
		1%	±	5%	±	10%	±	1%	±	5%	±	10%	±	1%	±	5%	±	10%	±
Random	CER	10.1	0.5	8.3	0.1	7.4	0.4	9.0	0.3	8.2	0.1	7.8	0.1	6.5	0.4	4.7	0.4	3.6	0.3
Unconf		9.7	–	7.4	–	7.4	–	10.0	–	9.8	–	8.6	–	8.2	–	5.3	–	4.5	
Diverse		10.0	–	7.5	–	5.9	–	10.2	–	9.2	–	8.1	–	6.2	–	5.2	–	4.9	
Random	WER	25.9	1.5	21.3	0.3	19.5	0.8	22.9	0.7	21.1	0.4	20.1	0.2	18.0	1.4	11.4	0.4	8.6	0.4
Unconf		26.2	2.5	21.3	2.4	21.9	2.4	27.6	0.6	27.2	0.6	22.6	0.6	24.9	0.3	13.7	0.3	10.7	0.3
Diverse		27.2	2.6	20.5	2.3	17.7	2.2	28.1	0.6	24.9	0.6	21.6	0.6	17.2	0.2	13.2	0.3	11.9	0.3

No added labeles (0%) in CER (WER): GW: 11.6 (29.5) IAM: 10.7 (28.5) CVL: 8.1 (24.8)

confidence based selection strategies. In this case, the ratio of labeled samples to pseudo-labels is especially unfavorable, due to the exceptionally large unlabeled dataset. The observed difference when using confidence based selections indicates that for such a dataset it is increasingly important to select samples that lie outside of the distribution that is well modeled by pseudo-labeled samples.

Add Labels After Self-Training. Instead of adding labeled samples at the beginning of training process, adaptation may be first performed solely via self-training. After the model has been optimized by training on pseudo-labels, sample selection and annotation is performed. Summarizing, we first train the pre-trained model for 50 self-training cycles and then perform finetuning exclusively on labeled samples for another 50 epochs. In general, we observe that our model is able to benefit from increasing numbers of added labeled samples, see Table 4. For GW, choosing diverse unconfident samples improves performances especially for the largest portion of labeled samples of 10%. For the other datasets, performances appear to be generally independent from the selection as the model already benefit from a random sampling strategy.

Active Sampling. Adding labels before or after the self-training cycles, does not allow the model to iteratively adapt sample selection and pseudo-label prediction based on knowledge learned from small portions of labeled samples during

Table 5. Active label selection.

Method		GW				IAM				CVL			
		5%	±	10%	±	5%	±	10%	±	5%	±	10%	±
Random	CER	9.7	0.7	9.7	0.5	9.7	0.6	8.6	0.1	9.2	0.7	8.1	0.6
Unconf		8.6	–	7.6	–	8.5	–	7.9	–	8.2	–	6.8	
Diverse		8.5	–	7.6	–	8.8	–	8.6	–	7.4	–	6.8	
Random	WER	26.3	1.8	26.5	1.6	26.5	1.9	22.5	0.5	25.4	1.6	23.1	2.3
Unconf		25.5	2.5	22.6	2.4	23.9	0.6	22.3	0.6	25.9	0.3	22.5	0.3
Diverse		23.4	2.4	22.0	2.4	25.3	0.6	24.1	0.6	24.4	0.3	21.2	0.3

No added labeles (0%) in CER (WER): GW: 11.6 (29.5) IAM: 10.7 (28.5) CVL: 8.1 (24.8)

Table 6. Comparison of the proposed training approaches.

Method		GW			IAM			CVL		
		1%	5%	10%	1%	5%	10%	1%	5%	10%
Finetune	CER	14.1	9.1	7.3	12.7	10.4	10.1	21.3	16.1	13.7
Add First		10.6	8.3	7.5	**9.9**	9.3	**7.6**	9.6	7.6	7.2
Add Last		**10.0**	**7.5**	**5.9**	10.2	9.2	8.1	**6.2**	**5.2**	**4.9**
Active		–	8.5	7.6	–	**8.8**	8.6	–	7.4	6.8
Finetune	WER	38.0	27.0	22.1	31.1	26.0	25.7	47.2	38.5	33.7
Add First		28.0	22.9	20.7	**27.1**	26.3	**21.4**	26.6	23.5	22.8
Add Last		**27.2**	**20.5**	**17.7**	28.1	**24.9**	21.6	**17.2**	**13.2**	**11.9**
Active		–	23.4	22.0	–	25.3	24.1	–	24.4	21.2

self-training. Therefore, we propose an active learning strategy in which labels are added during self-training. Every second self-training cycle a small number of n samples is added to the labeled dataset. Pseudo-labels and the current set of labeled samples then constitute the training set for which the model is trained for one epoch. n is chosen such that the total number of labeled samples corresponds to either 5% or 10% of the training dataset. Despite the addition of labeled samples improves performance upon exclusive self-training, our experiments indicate that the model does not scale well with additional training data when samples are selected randomly, see Table 5. In case of an active learning approach, selecting unconfident samples seems increasingly important. This might be due to the extremely small number of added samples at each step which only provide too limited novelty as they are already represented well by pseudo-labels.

4.4 Comparison

In this section, we compare the different training approaches and their performances to other works from the literature that train HTR models with no or

Table 7. Comparison to the state of the art.

Method		GW				IAM				CVL			
		0%	5%	10%	100%	0%	5%	10%	100%	0%	5%	10%	100%
Ours	CER	11.6	**7.5**	**5.9**	**4.0**	10.7	**9.2**	**8.1**	**6.3**	**8.1**	**5.2**	**4.9**	**2.6**
Ours*		25.4	–	–	4.2	21.8	–	–	7.2	27.5	–	–	5.1
Kang [10]		16.3	–	–	4.6	14.1	–	–	6.9	19.2	–	–	3.6
Wolf [30]		**10.5**	–	–	–	**8.8**	–	–	–	8.2	–	–	–
Ours	WER	31.1	**20.5**	**17.7**	**12.8**	28.5	**24.9**	**21.6**	**16.5**	**24.8**	**13.2**	**11.9**	**5.4**
Ours*		62.1	–	–	13.7	48.1	–	–	18.7	56.1	–	–	12.2
Aberdam [1]		–	–	–	–	–	59.7	47.7	20.1	–	26.9	25.2	22.2
Kang [10]		39.9	–	–	13.5	34.9	–	–	17.5	44.3	–	–	7.8
Wolf [30]		**30.7**	–	–	–	**25.3**	–	–	–	25.7	–	–	–

(∗) no self-training

limited training data. Table 6 summarizes the results of the different training schedules. We consider the experiments with a confidence based selection with ensured diversity, as this yields robust results across all datasets and training portions. Exploiting self-training and pseudo-labels improves performances in almost all cases. In our experiments, adding labeled samples after the completion of the self-training process resulted in the best performances on all datasets except the IAM database. This indicates that it is most effective to first learn which data is well represented by synthetic and pseudo-labeled samples before the addition of manually labeled data.

Table 7 compares our experiments to results from the literature. Kang et al. [10] and Wolf et al. [30] report results for recognition with the exact same architecture and hyperparameters as ours. Both works report performances for a scenario where no training data is available. In [10], an adversarial learning strategy adapts the model from synthetic to real data, while Wolf et al. [30] use a similar self-training approach to ours that is additionally combined with a consistency regularization method. [1] reports results for different amounts of available training data. Their work does not rely on synthetic samples but exploits a self-supervised contrastive learning strategy to pretrain a model that is later finetuned with labeled data. In general, we observe that a self-training approach is able to benefit from small numbers of manually labeled data, and scales fairly well with the availability of annotated samples. The performance gap to a model trained only on the designated training data can be closed considerably with only ten percent of the labeling demand. The self-supervised strategy proposed in [1] is clearly outperformed for all portions of data. This shows the effectiveness of using synthetic data in combination with self-training that also allows to efficiently select samples used for finetuning.

Even though the focus of our work is the application in limited data scenarios, combining self-training with labeled data also benefits performances compared to exclusively training on the presumably representative training set.

5   Conclusions

In this work, we investigated the question of how to train an HTR model with limited manual annotation effort. We showed that the strategy of adapting a model from synthetic data to the application domain with self-training benefits from the integration of small numbers of annotated samples. Despite the tremendously higher number of pseudo-labeled samples, manual annotation still benefits performances. Our experiments indicate that adding labeled samples late in the process specifically after a self-training phase is most efficient. We further propose two sampling strategies for selecting samples to annotate. Generally, it seems beneficial to take confidence estimates into account when selecting labels. Ensuring diversity by clustering increases robustness as the selected samples are less prone to dataset artifacts or to a potentially bad random selection.

References

1. Aberdam, A., et al.: Sequence-to-sequence contrastive learning for text recognition. In: Proceedings of IEEE Conference on Computer Vision and Pattern Recognition, Nashville, TN, USA, pp. 15302–15312 (2021)
2. Berthelot, D., Carlini, N., Goodfellow, I.J., Papernot, N., Oliver, A., Raffel, C.: MixMatch: a holistic approach to semi-supervised learning. In: Proceedings of International Conference on Neural Information Processing Systems, Vancouver, BC, Canada, pp. 5050–5060 (2019)
3. Brown, L.D., Cai, T.T., DasGupta, A.: Interval estimation for a binomial proportion. Stat. Sci. **16**(2), 101–133 (2001)
4. Das, D., Jawahar, C.V.: Adapting OCR with limited supervision. In: Proceedings of International Workshop on Document Analysis Systems, Wuhan, China, pp. 30–44 (2020)
5. Diaz, D.H., Qin, S., Ingle, R.R., Fujii, Y., Bissacco, A.: Rethinking text line recognition models. CoRR abs/2104.07787 (2021). https://arxiv.org/abs/2104.07787
6. Graves, A., Fernández, S., Gomez, F.J., Schmidhuber, J.: Connectionist temporal classification: Labelling unsegmented sequence data with recurrent neural networks. In: Proceedings of International Conference on Machine Learning, Pittsburgh, PA, USA, vol. 148, pp. 369–376 (2006)
7. Gurjar, N., Sudholt, S., Fink, G.A.: Learning deep representations for word spotting under weak supervision. In: Proceedings of International Workshop on Document Analysis Systems, Vienna, Austria, pp. 7–12 (2018)
8. Jaramillo, J.C.A., Murillo-Fuentes, J.J., Olmos, P.M.: Boosting handwriting text recognition in small databases with transfer learning. In: Proceedings of International Conference on Frontiers in Handwriting Recognition, Niagara Falls, NY, USA, pp. 429–434 (2018)
9. Kang, L., Riba, P., Rusiñol, M., Fornés, A., Villegas, M.: Pay attention to what you read: non-recurrent handwritten text-line recognition. Pattern Recogn. **129**, 108766 (2022)
10. Kang, L., Rusinol, M., Fornés, A., Riba, P., Villegas, M.: Unsupervised writer adaptation for synthetic-to-real handwritten word recognition. In: Winter Conference on Applications of Computer Vision, Snowmass Village, Co, USA, pp. 3502–3511 (2020)

11. Kang, L., Toledo, J.I., Riba, P., Villegas, M., Fornés, A., Rusiñol, M.: Convolve, attend and spell: an attention-based sequence-to-sequence model for handwritten word recognition. In: German Conference on Pattern Recognition, Stuttgart, Germany, vol. 11269, pp. 459–472 (2018)

12. Kiss, M., Benes, K., Hradis, M.: AT-ST: self-training adaptation strategy for OCR in domains with limited transcriptions. In: Proceedings of International Conference on Document Analysis and Recognition, Lausanne, Switzerland, vol. 12824, pp. 463–477 (2021)

13. Kleber, F., Fiel, S., Diem, M., Sablatnig, R.: CVL-database: an off-line database for writer retrieval, writer identification and word spotting. In: Proceedings International Conference on Document Analysis and Recognition, Washington, DC, USA, pp. 560–564 (2013)

14. Krishnan, P., Dutta, K., Jawahar, C.V.: Word spotting and recognition using deep embedding. In: Proceedings of International Workshop on Document Analysis Systems, Vienna, Austria, pp. 1–6 (2018)

15. Krishnan, P., Jawahar, C.V.: HWNet v2: an efficient word image representation for handwritten documents. Int. J. Doc. Anal. Recogn. **22**(4), 387–405 (2019)

16. Lavrenko, V., Rath, T.M., Manmatha, R.: Holistic word recognition for handwritten historical documents. In: International Workshop on Document Image Analysis for Libraries, Palo Alto, CA, USA, pp. 278–287 (2004)

17. LeCun, Y., Bottou, L., Bengio, Y., Haffner, P.: Gradient-based learning applied to document recognition. Proc. IEEE **86**(11), 2278–2324 (1998)

18. Lee, D.: Pseudo-label: the simple and efficient semi-supervised learning method for deep neural networks. In: ICML Workshop on Challenges in Representation Learning, Atlanta, GA, USA (2013)

19. Li, M., et al.: Trocr: transformer-based optical character recognition with pre-trained models. CoRR abs/2109.10282 (2021). https://arxiv.org/abs/2109.10282

20. Marti, U., Bunke, H.: The IAM-database: an English sentence database for offline handwriting recognition. Int. J. Doc. Anal. Recogn. **5**(1), 39–46 (2002)

21. Nair, R., Sankaran, N., Kota, B., Tulyakov, S., Setlur, S., Govindaraju, V.: Knowledge transfer using neural network based approach for handwritten text recognition. In: Proceedings of International Workshop on Document Analysis Systems, Vienna, Austria, pp. 441–446 (2018)

22. Plötz, T., Fink, G.A.: Markov models for offline handwriting recognition: a survey. Int. J. Doc. Anal. Recogn. **12**(4), 269–298 (2009)

23. Retsinas, G., Sfikas, G., Nikou, C.: Iterative weighted transductive learning for handwriting recognition. In: Proceedings of International Conference on Document Analysis and Recognition, Lausanne, Switzerland, vol. 12824, pp. 587–601 (2021)

24. Sohn, K., et al.: FixMatch: simplifying semi-supervised learning with consistency and confidence, vol. 33, pp. 596–608 (2020)

25. Stuner, B., Chatelain, C., Paquet, T.: Self-training of BLSTM with lexicon verification for handwriting recognition. In: Proceedings of International Conference on Document Analysis and Recognition, Kyoto, Japan, pp. 633–638 (2017)

26. Sueiras, J., Ruíz, V., Sánchez, Á., Vélez, J.F.: Offline continuous handwriting recognition using sequence to sequence neural networks. Neurocomputing **289**, 119–128 (2018)

27. Tensmeyer, C., Wigington, C., Davis, B.L., Stewart, S., Martinez, T.R., Barrett, W.: Language model supervision for handwriting recognition model adaptation. In: Proceedings of International Conference on Frontiers in Handwriting Recognition, Niagara Falls, NY, USA, pp. 133–138 (2018)

28. Wigington, C., Stewart, S., Davis, B.L., Barrett, B., Price, B.L., Cohen, S.: Data augmentation for recognition of handwritten words and lines using a CNN-LSTM network. In: Proceedings of International Conference on Document Analysis and Recognition, Kyoto, Japan, pp. 639–645 (2017)
29. Wolf, F., Fink, G.A.: Annotation-free learning of deep representations for word spotting using synthetic data and self labeling. In: Proceedings of International Workshop on Document Analysis Systems, Wuhan, China, pp. 293–308 (2020)
30. Wolf, F., Fink, G.A.: Self-training of handwritten word recognition for synthetic-to-real adaptation. CoRR abs/2206.03149 (2022). https://arxiv.org/abs/2206.03149
31. Zhang, Y., Nie, S., Liu, W., Xu, X., Zhang, D., Shen, H.T.: Sequence-to-sequence domain adaptation network for robust text image recognition. In: Proceedings of IEEE Conference on Computer Vision and Pattern Recognition, Long Beach, CA, USA, pp. 2740–2749 (2019)

Script-Level Word Sample Augmentation for Few-Shot Handwritten Text Recognition

Wei Chen, Xiangdong Su$^{(\boxtimes)}$, and Haoran Zhang

College of Computer Science, Inner Mongolia University, China National & Local Joint Engineering Research Center of Intelligent Information Processing Technology for Mongolian, Hohhot, China
cssxd@imu.edu.cn

Abstract. The variety of handwriting styles and the scarcity of training data often result in poor performance of character recognizer. Rather than tedious data collection and annotation, researchers prefer to use low-cost data augmentation to improve the robustness of the recognizer. However, most existing data augmentation methods treat handwritten text as ordinary images and generate new samples through holistic transformation, which greatly limits the diversity of generated samples. To solve the problem, this paper proposes a script-level handwritten text augmentation method, where each component is treated as a Bézier curve. Specifically, we first segment the character into components based on skeleton detection. Then, we move the control points of each component according to the prior knowledge of languages. Finally, we transform the component by Bézier curves and assemble them into new samples. Our method is simple, controllable, and friendly to few-shot handwritten text. Experiments on four datasets in different languages show that the proposed script-level augmentation method performs better than the holistic augmentation methods. Apart from it, we also modify Affine transformation, a commonly used augmentation method, from a holistic to script-level way. Experimental results demonstrate that script-level affine can achieve better performance than holistic affine in the character recognition task. Our code is available at https://github.com/ IMU-MachineLearningSXD/script-level_aug_ICFHR2022.

Keywords: Data augmentation · Script-level transformation · Multilingual handwritten text · Few-shot OCR

1 Introduction

Text recognition models [1–4] based on deep learning have gained great attention in recent years. The high recognition rate and strong robustness of these models depend on large-scale training data. In the handwritten text recognition task, the handwriting styles of different writers vary greatly. Hence, the recognition model needs diverse handwritten samples to improve its fitting ability. Unfortunately,

© The Author(s), under exclusive license to Springer Nature Switzerland AG 2022
U. Porwal et al. (Eds.): ICFHR 2022, LNCS 13639, pp. 316–330, 2022.
https://doi.org/10.1007/978-3-031-21648-0_22

collecting enough training data is difficult for languages with few users. As a result, the model will not perform satisfactorily. Instead of tedious data collection and annotation, researchers prefer to use low-cost data augmentation methods to solve the problem. However, most data augmentation methods are general at the image level, and they are not designed for handwritten text.

Previous data augmentation methods are mainly divided into two categories: rule-based and deep learning-based. Rule-based approaches are general at the image level, but are not designed specifically for handwritten text, such as Affine [6], Add noise [8], Hide and Seek [7], etc. These methods treat handwritten text as a normal image for holistic processing. Figure 1 illustrates their augmentation effects. Although partially effective, they do not fully exploit the characteristics of handwritten text at the script-level. Additionally, methods based on deep learning have become widely accepted in recent years, such as GAN-based methods [11–14]. By using sufficient training data or pre-training, these methods can generate good samples. On the other hand, their huge data requirement is also their greatest weakness. Many minority languages, such as Mongolian, suffer from the problem of a lack of data, which makes deep learning methods ineffective on few-shot datasets.

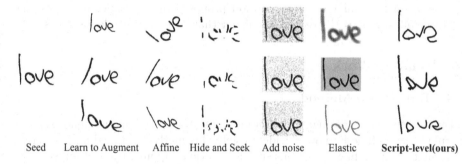

Seed   Learn to Augment   Affine   Hide and Seek   Add noise   Elastic   **Script-level(ours)**

Fig. 1. Comparison of augmented samples of previous methods and script-level method.

To deal with the weakness of holistic augmentation, InAugment [9] proposes an augmentation method based on local deformation. It deforms the local area of the image separately, and then pastes the deformed area back into the original image to create an augmented sample. In light of this, we propose a script-level handwritten word augmentation method that can generate large-scale training samples by making local variations to script in word images, improving the performance of handwritten text recognition. In our approach, each component is regarded as a quadratic Bézier curve [10]. At first, we extract the skeleton of each handwritten word and segment the whole word into several components with the skeleton and eight-neighborhood information. At second, we move the three control points of each component according a empirical formula, since each component is controlled by Bézier control points and making script variation is based on the control point movement. Finally, we transform new script from the control points after movement by quadratic Bézier curves [10] and assemble them together to form a new word sample.

To demonstrate the effectiveness and generality of our method in multiple languages, we conduct extensive experiments on four handwritten datasets about English, Arabic, Russian and Mongolian. Experimental results show that our method achieves better performance than the baselines. Moreover, we prove that the script-level word augmentation paradigm is better than the holistic word augmentation paradigm with the most commonly used augmentation method Affine [6]. That is, we do word augmentation on the holistic word and script-level individually.

The main contributions of our work can be summarized as follow:

- We propose a script-level handwritten text augmentation method for few-shot text recognition, in which each component is treated as a quadratic Bezier curve.
- We use our script-level innovations to upgrade a frequently used augmentation method, Affine [6]. We demonstrate that Affine (script-level) outperforms Affine (holistic).
- The experimental results on datasets of four different languages prove that the script-level augmentation method proposed in this paper improves the recognition rate of the recognizer higher than that of multiple baselines.

2   Related Work

2.1   Rule-Based Method

There are two types of augmentation methods applied to handwritten text, one is rule-based. Affine [6] now is widely used. This method is based on geometric deformation, including operations such as rotation, flipping, translation, scaling, and stretching of pictures. It is a holistic operation on handwritten text, and it has strong versatility. However, Affine [6] does not take into account the script-level deformation of handwritten text, which brings obvious macro changes to the picture but lacks micro changes. Krishna et al. [7] proposed Hide-and-Seek, which is a method of hiding image patches. Hide-and-Seek is similar to Random Erasing [5], forcing the network to improve the generalization ability without knowing the full picture of the data. It works in weak supervision tasks and can simulate data corruption in real scenarios. In addition, Jonghoon et al. [8] added different kinds of noise to the input image. They made the model robust in adversarial noise training and not sensitive to noise scenarios. Finally, Color transformation [15] is often used in image classification and scene text recognition tasks. Its operations include changing the hue, saturation, contrast, etc. of the image. This method is effective in data with rich color combinations. But handwritten texts often have a single color tone, which is not applicable. The above methods are all processed at the image level, with strong regularity and generality, and can augment high-quality samples with very few samples. However, they are not all applicable to handwritten text. The reason is that they focus on the whole picture, while the main information of handwritten text is concentrated on the author's writing script. On the contrary, the script-level augmentation method proposed in this paper is specially designed for handwritten text. Similarly, it is also rule-based and can operate on few-shot data. Our generality is reflected in multilingualism.

2.2   Deep Learning-Based Method

Another category is based on deep learning. Luo et al. [14] proposed a method combining geometric deformation and joint adversarial training. The generator in the network can provide good deformation parameters to generate adversarial samples, which makes the recognizer gradually improve the robustness during adversarial training. Unfortunately, this method cannot generate samples directly but parameters. Recently, Gan-based methods [16–19] have been vigorously discussed. Fogel et al. [11] proposed ScrabbleGAN, which introduced semi-supervision to alleviate the problem of GAN requiring a large amount of training data and generated variable-length handwritten text in the form of scrabble games. Ji et al. [12] proposed HIGAN, which can generate handwritten text or words according to a given variable-length text label. Its new samples are realistic and handwriting styles are also diverse. Apart from it, to alleviate two key issues in handwritten text, style representation, and content embedding, Luo et al. [13] proposed SLOGAN, it used a style bank to parameterize specific handwriting styles and fed it into the generator as prior knowledge. The samples generated by this method can have new styles not included in the training data. The above-mentioned deep learning-based methods all have such a fatal problem, that is, the network must be trained or pre-trained with a sufficient amount of training data, otherwise it will not be effective.

Both the rule-based method and the deep learning-based method have their own advantages and disadvantages. The rule-based method is very traceable and controllable, and does not require network training, but the generated data cannot have a new style outside the rules. On the contrary, deep learning-based methods just make up for this shortcoming. Such methods can be applied to datasets with sufficient data, but the performance on few-shot datasets is unsatisfactory.

3   Methodology

This section describes the steps of the script-level augmentation method, including script segmentation, script control point movement and script transformation and assembling. Figure 2 illustrates the proposed method as well as a sample.

3.1   Script Segmentation

Since our method deforms each component individually, it is necessary to segment each word into components. First, we use the image thinning algorithm

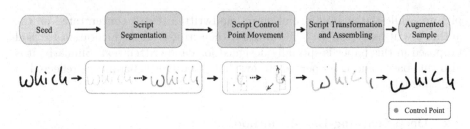

Fig. 2. Illustration of the workflow of the proposed method and an augmented word sample.

in Zhang et al. [24] to extract single-pixel skeletons of word images. This algorithm removes redundant pixels in the image through two sub-iterations. Next, we detect the corners of the skeleton. The corner points will be used to split the script and act as control points for the ends of the components. We compute the sum of the eight neighbor pixel counts for each pixel of the single-pixel skeleton. The points that conform to Eq. 1 are the end points of the skeleton and the multi-branch points, which are classified as corner points.

$$\sum_{i=1}^{n=8} P_i = 1 \ or \ \sum_{i=1}^{n=8} P_i \geq 3 \tag{1}$$

Finally, to distinguish different components, we hide the corners in the skeleton and use the classic Two-Pass algorithm in [25] to detect connected domains. The algorithm can classify the pixels in the image during the two scans and assign labels to them. We adopt the eight-neighborhood mode in the algorithm, which can effectively segment the skeleton into multiple components. Each component will then be reconstructed as a Bézier curve [10].

In fact, the above method can not fully meet the segmentation of special script, such as shapes similar to S and O. We also have additional corner completion steps, the specific description can be seen in the code we provide.

3.2 Script Control Point Movement

Our method transforms each segmented component as a quadratic Bézier curve [10]. The shape of each curve is determined by three control points, among which two control points $P_a(x_a, y_a)$ and $P_b(x_b, y_b)$ are the ends of components obtained in Subsect. 3.1. According to the definition of [10], we determine the third control point $P_c(x_c, y_c)$ as follows. First, we use the distance formula to find the farthest pixel point $P_t(x^*, y^*)$ of each script from the line connecting the control points at both ends, as shown in Eq. 2.

$$P_t(x^*, y^*) = \arg\max_{x,y} \left(\frac{(x_b - x_a) \times y + (x_b - y_a) \times x + x_b y_a - x_a y_b}{(x_b - x_a)^2 + (y_b - y_a)^2} \right) \tag{2}$$

where x, y represent the coordinates of the remaining points on the component. Then, we use $P_t(x^*, y^*)$ to calculate the position of the third control point $P_c(x_c, y_c)$, as shown in Eq. 3.

$$P_c(x_c, y_c) = 2P_t(x^*, y^*) - \frac{1}{2}(P_a(x_a, y_a) + P_b(x_b, y_b)) \tag{3}$$

At this point, we get three control points $P_a(x_a, y_a)$, $P_b(x_b, y_b)$, $P_c(x_c, y_c)$ of each component. Next, we define the movement area for the control points. To avoid introducing erroneous word samples in script transformation, we make the position of the control points constrained by one another. For the movement area of the component endpoints, we use the closest point as the limit. Suppose the endpoint is $P_m(x_m, y_m)$ and the closest point is $P_n(x_n, y_n)$, we compute the horizontal movement range W_{range} and vertical movement range H_{range} of the endpoints as

$$W_{range} = 2k \times |x_m - x_n|, \ k \in [0, 1] \tag{4}$$

$$H_{range} = 2k \times |y_m - y_n|, \ k \in [0, 1] \tag{5}$$

where k is a coefficient used to control the size of the moving area. That is, the endpoints $P_a(x_a, y_a)$ and $P_b(x_b, y_b)$ are moved in the rectangle defined by W_{range} and H_{range}.

For the third control point $P_c(x_c, y_c)$, its movement area S_{area} is a square area which is centered on it. We make an constraint of S_{area} with $P_a(x_a, y_a)$, $P_b(x_b, y_b)$ and k. The equation of S_{area} is shown in 6.

$$S_{area} = 2k \times \left(\frac{(x_b - x_a) \times y_c + (y_b - y_a) \times x_c + x_b y_a - x_a y_b}{(x_b - x_a)^2 + (y_b - y_a)^2} \right) \tag{6}$$

Similarly, the third control point $P_c(x_c, y_c)$ are moved in the square defined by S_{area}. By randomly moving the control points P_a, P_b, P_c, we get three new control points P'_a, P'_b, P'_c, respectively. These three new control points will be used to transform the script.

3.3   Script Transformation and Assembling

In Subsect. 3.2, we get three new control points P'_a, P'_b, P'_c for each component. According to the definition of quadratic Bézier curve [10], we generate a new component with these three points. The calculation of new component is shown as Eq. 7. P_t represents the point on the new component.

$$P_t = (1 - t)^2 P'_a + 2t(1 - t)P'_c + t^2 P'_b \qquad t \in [0, 1] \tag{7}$$

Finally, we assemble all the newly generated components and place them on an adaptive sized background. New handwriting samples are generated as a result. Our method not only allows script-level variation, but also customizes script thickness, which is not possible with other baseline methods.

4 Experiments

4.1 Datasets

To verify the generalization of our method, we conduct experiments on the following four datasets. The language of each dataset is different. Meanwhile, to fit few-shot handwriting recognition, we adjust the settings of the dataset. Table 1 describes the language of the dataset, the number of unique words, and the size of the training and test sets in the experiments. The last column also reports the recognition accuracy of the CRNN model [1] on different datasets without any data augmentation. By the way, the augmented samples we use in all experiments are randomly generated by the method itself and not selected, which ensures fairness.

- **Mongolian-Database.** The offline handwriting dataset is provided by Inner Mongolia University. We pick 10,000 unique words from this dataset, with only two samples per word. The selected data are equally divided into training and test sets.
- **CVL-Database** [20]. It is a public database for writer retrieval, writer identification, and word spotting. The database consists of 7 different handwritten texts (1 German and 6 English Texts). A total of 311 authors contributed to the writing. We change the settings of the original dataset, swapping the training and test sets.
- **IFN/ENIT-Database** [21]. The IFN/ENIT-database contains material for training and testing Arabic handwriting recognition software. It has binary images of over 2200 handwritten sample tables from 411 authors. The dataset is divided into 4 disjoint subsets. We only extracted two subsets as a training set and test set.
- **HKR** [22]. HKR is used for handwriting recognition on Russian and Kazakh databases (with about 95% of Russian and 5% of Kazakh words/sentences respectively). The database consists of more than 1400 filled forms. There are approximately 63000 sentences, and more than 715699 symbols produced by approximately 200 different writers. We split the training and test sets of the dataset in a 1:10 ratio.

Table 1. Information about the datasets in the experiments. The last column indicates the word recognition accuracy without data augmentation.

Dataset	Language	Label	Size (train/test)	Accuracy (no aug)
Mongolian-Database	Mongolian	10000	10k/10k	0.0639
CVL-Database	English & German	395	12k/88k	0.4041
IFN/ENIT-Database	Arabic	2259	7k/7k	0.3799
HKR	Russian & Kazakh	2808	6k/59k	0.1509

4.2   Baseline

To verify the effectiveness of the proposed method, we select the following five widely used data augmentation methods as baselines. Some baselines fuse multiple operations under the same class. In addition, our experimental scenario is few-shot handwriting recognition, which render deep learning-based methods ineffective. Therefore, we do not compare with deep learning based augmentation methods.

- **Affine** [6] is a commonly used augmentation method not only in handwriting recognition but also in computer vision. Its operations include rotation, scaling, translation and stretching of the entire image. Not only do we use it as a baseline, but we also apply this holistic operation to the script-level. This will help us to verify that the script-level augmentation method is better than the holistic processing method under the same method.
- **Color and Blur** [15]. This type of method mainly changes the hue, saturation, contrast, etc. of the picture, as well as the operation of blurring the picture and changing the grayscale. Such methods are currently mainly used in image classification and scene text recognition tasks.
- **Advanced** [5,8]. The method consists of random masking of the image and the operation of adding adversarial noise to the image. Noise includes Gaussian noise and salt and pepper noise. This augmented sample can simulate real-world data corruption, forcing the network to be robust without knowing the full picture of the data.
- **Hide and Seek** [7]. It first divides the image into patches of the same size, then masks these patches according to the probability value. It is similar to Random Erasing [5], forcing the recognizer to use limited information to improve the recognition rate.
- **Learn to Augment** [14]. This method first divides the picture into N blocks, and the four corners on each block are controlled by the base point. The base point will move in random directions and distances, driving the image to deform. The authors combine it with joint adversarial training. The generative network will predict the optimal distance R to move these base points and the value of N to split into blocks. At the same time, the method can also be used off the network. We set the value of N to 3 and R to 10 as suggested by the paper.

4.3   Evaluation Method

In order to compare the pros and cons of our proposed script-level augmentation method and the baseline, we use word recognition accuracy as the evaluation metric. The recognizer is a combination of CRNN [1] and CTC [23], an end-to-end image model. The overall process is not complicated. The CNN layer first extracts the image convolution features of text pictures. Then the LSTM further extracts the sequence features. Finally, the recognizer introduces CTC loss to solve the problem that characters cannot be aligned during training.

5   Results and Discussion

5.1   Comparison with Baselines

We compare our method with baselines in this subsection. The selection of k in our method refers to the ablation experiments. Meanwhile, to demonstrate the multilingual generality of the proposed method, we conduct experiments on multiple datasets in different languages. We set the number of augmentations to [20, 50] with a step size of 10, which is to study the difference in the improvement effect of different augmentation methods on the recognizer under different augmentation times. In addition, we manually tune the parameters in the baseline method as better or optimal. The experimental results are shown in Table 2. The following is our analysis and conclusions.

Table 2. Comparison of recognition accuracy between our method and baselines.

Dataset	Method	Augmentation times			
		20	30	40	50
Mongolian-Database	Affine	0.2679	0.2633	0.2711	0.2686
	Color and Blur	0.0955	0.0958	0.1029	0.0994
	Advanced	0.0842	0.0723	0.0751	0.0735
	Hide and Seek	0.0692	0.0655	0.0575	0.0550
	Learn to Augment	0.2981	0.3234	0.3337	0.3493
	Ours	**0.3442**	**0.3638**	**0.3734**	**0.3747**
CVL-Database	Affine	0.7848	0.7922	0.7909	0.7902
	Color and Blur	0.6351	0.6422	0.6430	0.6464
	Advanced	0.6586	0.6647	0.6559	0.6629
	Hide and Seek	0.6172	0.6211	0.6245	0.6112
	Learn to Augment	0.8155	0.8261	0.8303	0.8291
	Ours	**0.8265**	**0.8330**	**0.8379**	**0.8411**
IFN/ENIT-Database	Affine	0.6967	0.7034	0.6997	0.7045
	Color and Blur	0.5809	0.5945	0.6025	0.6015
	Advanced	0.5979	0.6098	0.6077	0.6191
	Hide and Seek	0.5866	0.5951	0.5933	0.5927
	Learn to Augment	0.7320	0.7411	0.7368	0.7379
	Ours	**0.7498**	**0.7469**	**0.7534**	**0.7469**
HKR	Affine	0.6773	0.6910	0.6961	0.6986
	Color and Blur	0.3857	0.4308	0.4571	0.4713
	Advanced	0.4823	0.5135	0.5349	0.5441
	Hide and Seek	0.5053	0.5365	0.5474	0.5554
	Learn to Augment	0.7375	0.7707	**0.7780**	0.7736
	Ours	**0.7623**	**0.7726**	0.7779	**0.7807**

First, in the process of increasing the number of augmentations from 20 to 40, almost every method has a positive improvement in recognition accuracy.

This shows that in few-shot handwriting recognition, data augmentation can effectively alleviate the starvation problem of the model under the appropriate augmentation times. However, as the number of augmentations increases from 40 to 50, the recognition accuracy declines or increases slowly. This indicates that there is a problem of insufficient diversity of augmented samples and misleading models with erroneous samples. Therefore, the use of data augmentation should not be indiscriminate. It is necessary to choose an appropriate number of augmentations according to the data scarcity of the dataset. Conversely, excess data will cause the recognizer to overfit and wrong samples will mislead the recognizer.

Secondly, from the perspective of the improvement effect brought by the baselines, well-performed are Affine [6] and Learn to Augment [14], the latter of which is better. This shows that the method that can deform the handwritten text can improve the recognition accuracy more directly. The poor performers are Color and Blur [15], Advanced [5,8] and Hide and Seek [7]. This proves that simply changing the color of the image or adding noise and randomly masking the image is not completely suitable for the augmentation of handwritten text. In a word, the diversity of handwriting styles is the main contradiction that augmentation methods should focus on.

Finally, experimental data show that our method outperforms baseline methods in almost all experiments. Compared to the best performing baseline - Learn to Augment [14], our method improves the most on the Mongolian-Database. Our improvements in different augmentation times on the Mongolian-Database are 15.6%, 12.5%, 11.9%, 7.3%, with an average of 11.8%. Meanwhile, the improvement on the four datasets averages 3.9%. This proves that our method has multilingual generality. Apart from it, our method is only slightly 0.01% inferior to Learn to Augment [14] on HKR [22] with 40 augmentation times. The data demonstrate that our proposed script-level augmentation method can bring a more positive effect to the recognizer than the baseline, generate better sample diversity, and introduce fewer erroneous samples.

5.2   Ablation Experiment

In this subsection, we discuss the k values mentioned in Subsect. 3.2. k is used to define the maximum range of control point movement. A too small range of motion may limit the diversity of samples augmented by our method. If the range is too large, it may cause the control points to be shifted to inappropriate positions, resulting in erroneous samples. Apart from it, Data sets in different languages may be adapted to different k values. Therefore, an ablation experiment on the k value is necessary.

We limit k to $[0.3, 0.8]$ in our experiments and set the step size to 0.1. At the same time, in order to explore the adaptation of k value in different languages, we conduct experiments on four datasets. Finally, the selection of the k value will be based on the recognition accuracy of the word. Again, the number of augmentations is set to 20.

The results are shown in Table 3. When the k value is set to 0.6 on the three datasets of Mongolian-Database, CVL-Database [20] and HKR [22], the word recognition accuracy reaches the best. The optimal k value on the IFN/ENIT-Database [21] is 0.7. As the value of k increases, the recognition accuracy first increases and then decreases. The results confirm our previous guess and suggest that an inappropriate value of k will limit the performance of augmented samples. Therefore, we recommend setting the value of k at 0.6 when using our method.

Table 3. The effect of k value on the recognition accuracy. The augmentation times is 20.

k	0.3	0.4	0.5	0.6	0.7	0.8
Mongolian-Database	0.2980	0.3243	0.3246	**0.3442**	0.3344	0.3152
CVL-Database	0.7990	0.8146	0.8238	**0.8265**	0.8261	0.8210
IFN/ENIT-Database	0.7314	0.7431	0.7455	0.7477	**0.7498**	0.7407
HKR	0.7135	0.7401	0.7513	**0.7623**	0.7582	0.7530

5.3 Improvement for Affine

To demonstrate that in handwritten text augmentation, script-level processing is better than holistic, we retrofit Affine [6], a data augmentation method that has been widely used. The processing level of Affine [6] is holistic, which mainly includes operations such as rotation, scaling, and translation of images. We adapt this approach to script-level processing. The detailed operation is similar to the process shown in Fig. 2. We only convert the original deformation method using Bézier curves [10] to the deformation operation using Affine [6] on the Script Control Point Movement module. Finally each deformed component is assembled to form new augmented samples.

In order to compare the difference between Affine [6] before and after improvement, we conduct experiments on four datasets. We use these two methods to augment the dataset. Every sample in the training set is augmented 20 times. The original samples and the augmented samples are used as a new training set to train the recognizer. We report the experimental results in Table 4.

Compared with Affine (holistic) [6], Affine (script-level) obtains the most obvious improvement on the Mongolian-Database, with an increase of 13.7%. The average improvement on the four datasets is 7.4%. It proves that Affine (script-level) is significantly better than Affine (holistic) [6] when only changing the processing level. We analyze that script-level processing makes different changes to each component. Conversely, holistic processing is equivalent to setting the same parameters for all script variations. This will greatly limit the diversity of augmented samples.

Table 4. The effect of Affine (holistic) and Affine (script-level) on word recognition accuracy when the augment times is 20.

Method	Affine (holistic)	Affine (script-level)
Mongolian-Database	0.2679	**0.3045**
CVL-Database	0.7848	**0.8147**
IFN/ENIT-Database	0.6967	**0.7297**
HKR	0.6773	**0.7264**

Language	Seed	Augmented Samples

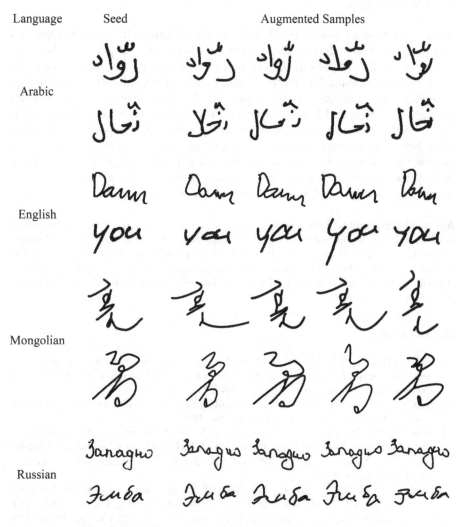

Arabic

English

Mongolian

Russian

Fig. 3. Samples generated by script-level augmentation.

5.4 Diversity of Augmented Samples

Figure 3 shows examples generated by our proposed script-level augmentation method. The first two columns are the original images waiting to be augmented and the languages they represent. The remaining columns are new samples. Different from the holistic augmentation method, our method does not perform holistic operations such as rotation and scaling of handwritten text. In contrast, the handwriting style of our augmented samples is reflected in script-level changes. Moreover, the deformation of each component is independent. Script-level deformation can fit more different handwriting styles of authors. It can greatly improve the recognition accuracy of the recognizer.

6 Conclusion

In this paper, we propose a rule-based script-level handwritten word augmentation method. It first divides the handwritten word into components, then uses Bézier curves [10] to deform the component respectively, and finally assembles them into new samples. Our method works on few-shot handwritten datasets and augments tons of effective new samples. At the same time, it can be used for multiple languages and has the advantage of bringing more variation at script-level to handwritten words. Experiments show that the proposed method can bring better improvement to the recognizer than the baselines. In addition, we use the idea of script-level to improve traditional Affine [6]. Experiments show that Affine (script-level) outperforms Affine (holistic), which again proves the advantages of script-level augmentation proposed in this paper.

Acknowledgements. This work was funded by National Natural Science Foundation of China (Grant No. 61762069), Key Technology Research Program of Inner Mongolia Autonomous Region (Grant No. 2021GG0165), Key R&D and Achievement Transformation Program of Inner Mongolia Autonomous Region (Grant No. 2022YFHH0077), Big Data Lab of Inner Mongolia Discipline Inspection and Supervision Committee (Grant No. 21500-5206043).

References

1. Shi, B., Bai, X., Yao, C.: An end-to-end trainable neural network for image-based sequence recognition and its application to scene text recognition. IEEE Trans. Pattern Anal. Mach. Intell. **39**(11), 2298–2304 (2017)
2. Luo, C., Jin, L., Sun, Z.: MORAN: a multi-object rectified attention network for scene text recognition. Pattern Recogn. **90**, 109–118 (2019)
3. Wan, Z., He, M., Chen, H., Bai, X., Yao, C.: TextScanner: reading characters in order for robust scene text recognition. In: AAAI Conference on Artificial Intelligence, New York, vol. 34, pp. 12120–12127. AAAI Press (2020). https://doi.org/10.1609/aaai.v34i07.68
4. Baek, J., Matsui, Y., Aizawa, K.: What if we only use real datasets for scene text recognition? Toward scene text recognition with fewer labels. In: IEEE/CVF Conference on Computer Vision and Pattern Recognition (CVPR), pp. 3113–3122. IEEE, Virtual (2021)

5. Zhong, Z., Zheng, L., Kang, G., Li, S., Yang, Y.: Random erasing data augmentation. In: AAAI Conference on Artificial Intelligence, New York, vol. 34, pp. 13001–13008. AAAI Press (2020). https://doi.org/10.1609/aaai.v34i07.7000
6. Jaderberg, M., Simonyan, K., Zisserman, A.: Spatial transformer networks. In: Neural Information Processing Systems (NeurIPS), Montréal, Canada, pp. 2017–2025. NeurIPS (2015)
7. Singh, K.-K., Yu, H., Sarmasi, A., Pradeep, G., Lee, Y.-J.: Hide-and-seek: a data augmentation technique for weakly-supervised localization and beyond. arXiv preprint arXiv:1811.02545 (2018)
8. Jin, J., Dundar, A., Culurciello, E.: Robust convolutional neural networks under adversarial noise. arXiv preprint arXiv:1511.06306 (2015)
9. Arar, M., Shamir, A., Bermano, A.: InAugment: improving classifiers via internal augmentation. In: IEEE/CVF International Conference on Computer Vision (ICCV) Workshops 2021, pp. 1698–1707. IEEE, Virtual (2021)
10. Forrest, A.-R.: Interactive interpolation and approximation by Bézier polynomials. Comput. J. **15**(1), 71–79 (1972)
11. Fogel, S., Averbuch-Elor, H., Cohen, S., Mazor, S., Litman, R.: ScrabbleGAN: semi-supervised varying length handwritten text generation. In: IEEE Computer Society Conference on Computer Vision and Pattern Recognition (CVPR), USA, pp. 4323–4332. IEEE (2020)
12. Gan, J., Wang, W.: HiGAN: handwriting imitation conditioned on arbitrary-length texts and disentangled styles. In: AAAI Conference on Artificial Intelligence, pp. 7484–7492. IEEE, Virtual (2021)
13. Luo, C., Zhu, Y., Jin, L., Li, Z., Peng, D.: SLOGAN: handwriting style synthesis for arbitrary-length and out-of-vocabulary text. IEEE Trans. Neural Netw. Learn. Syst. (TNNLS) (2022)
14. Luo, C., Zhu, Y., Jin, L., Wang, Y.: Learn to augment: joint data augmentation and network optimization for text recognition. In: IEEE Computer Society Conference on Computer Vision and Pattern Recognition (CVPR), USA, pp. 13743–13752. IEEE (2020)
15. Atienza, R.: Data augmentation for scene text recognition. In: IEEE/CVF International Conference on Computer Vision (ICCV) Workshops, pp. 1561–1570. IEEE, Virtual (2021)
16. Zhao, C., Yen, G.G., Sun, Q., Zhang, C., Tang, Y.: Masked GAN for unsupervised depth and pose prediction with scale consistency. IEEE Trans. Neural Netw. Learn. Syst. (TNNLS) **32**(12), 5392–5403 (2020)
17. You, H., Cheng, Y., Cheng, T., Li, C., Zhou, P.: Bayesian cycle-consistent generative adversarial networks via marginalizing latent sampling. IEEE Trans. Neural Netw. Learn. Syst. (TNNLS) **32**(10), 4389–4403 (2020)
18. Yeo, Y.J., Shin, Y.G., Park, S., Ko, S.J.: Simple yet effective way for improving the performance of GAN. IEEE Trans. Neural Netw. Learn. Syst. (TNNLS) **33**(4), 1811–1818 (2021)
19. Peng, X., Tang, Z., Yang, F., Feris, R.S., Metaxas, D.: Jointly optimize data augmentation and network training: adversarial data augmentation in human pose estimation. In: IEEE Conference on Computer Vision and Pattern Recognition (CVPR), Salt Lake City, pp. 2226–2234. IEEE (2018)
20. Kleber, F., Fiel, S., Diem, M., Sablatnig, R.: CVL-database: an off-line database for writer retrieval, writer identification and word spotting. In: 2013 12th International Conference on Document Analysis and Recognition (ICDAR), Washington, pp. 560–564. IEEE (2013). https://doi.org/10.1109/ICDAR.2013.117

21. Pechwitz, M., Maddouri, S.S., Märgner, V., Ellouze, N., Amiri, H.: IFN/ENIT-database of handwritten Arabic words. In: Proceedings of CIFED, vol. 2, pp. 127–136. Citeseer (2002)
22. Nurseitov, D., Bostanbekov, K., Kurmankhojayev, D., Alimova, A., Abdallah, A., Tolegenov, R.: Handwritten Kazakh and Russian (HKR) database for text recognition. Multimed. Tools Appl. **80**(21), 33075–33097 (2021)
23. Graves, A., Fernandez, S., Gomez, F., Schmidhuber, J.: Connectionist temporal classification: labelling unsegmented sequence data with recurrent neural networks. In: 23rd International Conference on Machine Learning (ICML), Pittsburgh, PA, USA, pp. 369–376 (2006). https://doi.org/10.1145/1143844.1143891
24. Zhang, T.-Y., Suen, C.-Y.: A fast parallel algorithm for thinning digital patterns. Commun. ACM **27**(3), 236–239 (1984)
25. Rosenfeld, A., Pfaltz, J.L.: Sequential operations in digital picture processing. J. ACM (JACM) **13**(4), 471–494 (1966)

Towards Understanding and Improving Handwriting with AI

Suman Bhoi[1]([⊠])(iD) and Suman Sourav[2](iD)

[1] National University of Singapore, Singapore, Singapore
sumanbhoi@u.nus.edu
[2] Singapore University of Technology and Design, Singapore, Singapore
suman_sourav@sutd.edu.sg

Abstract. What makes a handwriting good? If the aesthetic judgment of handwriting follows implicit rules, can those rules be recovered by observing good and bad examples? To answer these questions, we apply explainability techniques to the classification of good and bad handwriting. We show that it is indeed possible to recover these inherent rules. We develop an AI system that uses a modified version of LIME Image Explainer and generates images containing suggestions for improvement. We use single-character and word-level datasets labelled with binary labels generated via accepted rules for handwriting classification. We discuss the possible improvements to the current system as well as where this research could be applied, such as user-specific auto-suggestions.

Keywords: Handwriting analysis · Explainability · Feature attribution

1 Introduction

Analysis of handwriting is a well studied and classic problem in computer vision. Early image recognition tools were used to classify handwritten numbers [11], whereas more recent studies have even shown that autism and medical issues can be detected through an individual's handwriting [8].

While two sets of handwritten material may convey the same objective meaning, some handwriting will be visually pleasing and satisfying to look at, while others might appear unsatisfactory to some, or be close to illegible. In this study, we work towards quantifying this subjective human perception of "good" and "bad" in handwriting, using machine learning techniques. Essentially, *"What makes some handwriting visually pleasing, while others not so much?"*.

While previous studies have scored the legibility and conformity of handwriting, we create a system to interpret the features that lead to the handwriting being classified as good or bad. Learning *why* handwriting is classified as bad, can enable us to understand the necessary aspects to change for making it better. The results of this research can aid schools when teaching handwriting, and also

© The Author(s), under exclusive license to Springer Nature Switzerland AG 2022
U. Porwal et al. (Eds.): ICFHR 2022, LNCS 13639, pp. 331–344, 2022.
https://doi.org/10.1007/978-3-031-21648-0_23

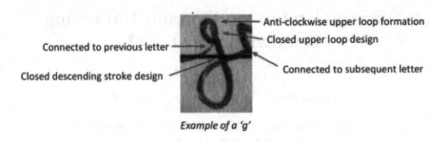

Example of a 'g'

Fig. 1. Construction features for a character [2]

aid in typographic design. Additionally, individuals can have increased awareness of how their handwriting could be deficient, and assist in self-improvement.
The research questions that we try to answer in this paper are the following:

RQ1: Can an explainer be created to infer how a user should change their
handwriting to make their handwriting better, including what aspects to
retain and what aspects that require change?

RQ2: What are the key differences between a "good" and a "bad" handwriting?

To answer **RQ1**, an image classification algorithm is trained on datasets of
"bad" and "good" handwriting according the rules in Sect. 3.1. Once text is
presented and classified, we use LIME [14], an explainer for machine learning
classifiers, to extract the segments in order to quantify why some text is classified
as "bad" while others are classified as "good". Qualitative analysis is then used
to infer the differences between these two classes.

To answer **RQ2**, we generate a data set which covers the extreme examples
of "bad" and "good" handwriting (detailed in Sect. 3.1). This is to test if the
classifier can capture the features that humans tend to evaluate.

2   Related Work

There has been some existing analysis work on aesthetic features of handwriting.
Agius et al. use logistic regression, classification and regression tree (CRT) [3]
to model the relationship between the nationality of the writer and handwriting
features of 7 characters (e.g. the letter "g" in Fig. 1). However, the framework
presented in the paper is based on numerical data, which is manually coded.
Computer vision is not taken into account to automatically extract the construction features for the testing data. It also does not consider explicit features,
such as slant and relative character sizes, which could contribute to the classification results. Handwriting beautification of digital ink has been done on stroke
level leveraging on recurrent neural network (RNN) cells [4]. The stroke samples
rely on a large database of writings from a user [22].

Additionally there have been a number of studies to score the handwriting
of individuals. Falk et al. [7] conducts a study where students use a digital pen

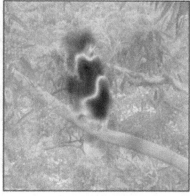

Fig. 2. Image of a cockatoo (left) and evidence for classification (right). Areas highlighted in red show evidence for the classification, and blue show evidence against the classification [21] (Color figure online)

to write a series of words. The centre of mass of each letter is then used to compute the size, space and alignment between words, and students are scored according to the Minnesota Handwriting Assessment [13]. Bouillon et al. use a neural network to compare teacher's ideal letters to those of the students and rate similarity percentage [6]. While these studies feedback to the user a score for their handwriting, they do not say *why* it was scored as good or bad.

Shrikumar et al. propose DeepLIFT (Deep Learning Important FeaTures) [18], which explains a neural network's output classification for a specific input, by "back-propagating the contributions of all neurons in the network to every feature of the input". Zintgraf et al. develop the prediction difference analysis method for deep neural networks [21], which provides visual explanations to how classification is reached for an input image. As seen in Fig. 2, areas highlighted in red show positive evidence for the classification, while areas in blue show evidence against the classification. A similar visual explainability technique was developed by Sundararajan et al. [20]. Such techniques could be used to explain why handwritten text is classified the way it is.

3   Technical Approach

3.1   Dataset

To build and verify our classification model, we use an existing well studied dataset within computer vision, as well as word-based datasets which closely match our ideal use case. Both datasets follow the "chalkboard" convention of white writing on black background. We label these datasets using a set criteria, explained further in each case below.

MNIST. We use the MNIST handwritten digits dataset as it is a classic data set within the image recognition community, and consists of thousands of samples of handwritten numbers. Due to its high adoption, there are pre-existing examples of frequently incorrectly classified MNIST samples within literature for reference [5,19].

Due to the unavailability of an extensively labelled data-set of "good" and "bad" numbers, we label the sixes within MNIST using the criteria for "bad" samples as shown in Fig. 3.

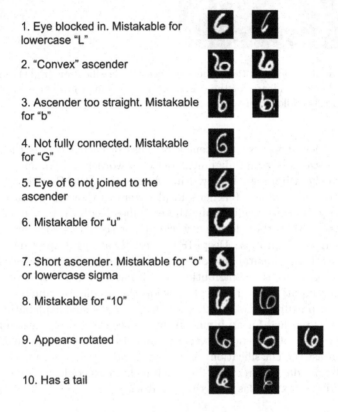

1. Eye blocked in. Mistakable for lowercase "L"

2. "Convex" ascender

3. Ascender too straight. Mistakable for "b"

4. Not fully connected. Mistakable for "G"

5. Eye of 6 not joined to the ascender

6. Mistakable for "u"

7. Short ascender. Mistakable for "o" or lowercase sigma

8. Mistakable for "10"

9. Appears rotated

10. Has a tail

Fig. 3. Rules and samples of MNIST sixes labelled "bad"

We label 4,100 of the 6,875 samples of the number 6 within the MNIST data set. For completeness, examples of sixes labelled "good" are shown in Fig. 4.

Handwritten Word Dataset. We train a model using multiple samples of a single handwritten word. To label words as "good" or "bad", we adhere to the Minnesota Handwriting Assessment (MHA) [13], as it is used widely within the education sector [10,14,16]. MHA guidelines[1] when assessing handwriting are as follows:

---

[1] We exclude "rate" as we do not have this information.

Fig. 4. Examples of MNIST samples labelled "good"

Legibility: Is the word readable?
Form: Do the letters look as they are supposed to? Are lines connected and curved/straight in the correct way?
Alignment: Are the characters aligned level?
Size: Are the characters all the same size?
Spacing: Are the characters spaced evenly?

We generate a dataset of the word "studio" according to the following criteria in Fig. 5. This handwritten dataset constitutes of 981 samples containing 435 good samples and 483 bad samples. Samples are consequently denoised by Gibbs sampling, as well as adjusted to black and white images.

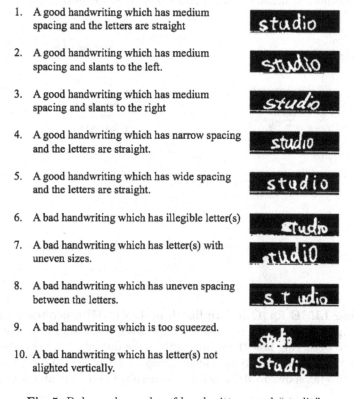

1. A good handwriting which has medium spacing and the letters are straight
2. A good handwriting which has medium spacing and slants to the left.
3. A good handwriting which has medium spacing and slants to the right
4. A good handwriting which has narrow spacing and the letters are straight.
5. A good handwriting which has wide spacing and the letters are straight.
6. A bad handwriting which has illegible letter(s)
7. A bad handwriting which has letter(s) with uneven sizes.
8. A bad handwriting which has uneven spacing between the letters.
9. A bad handwriting which is too squeezed.
10. A bad handwriting which has letter(s) not alighted vertically.

Fig. 5. Rules and samples of handwritten word "studio"

3.2   Training Our Model

For each data set, we train a generic convolutional neural network due to their accuracy in classification, well-supported libraries and use for explainability within computer vision. The overview of our system is shown in Fig. 6. In each case, the training and validation accuracies of the classifiers are all within 89–91%.

3.3   Explainer: LIME

Local Interpretable Model-Agnostic Explanations (LIME) [15] is used to explain our trained models. In each case, we aim to capture the top regions, or superpixels, that contribute positively or negatively to the two classes via the explainer. LIME fits a local linear decision boundary by perturbing the interpretable components around the sample's neighborhood. For image explanation, the interpretable components are the segmented superpixels, which are turned on and off during perturbation. There are 4 choices of segmentation algorithms: Felzenszwalb's method, Quickshift, SLIC, and Compact watershed [17].

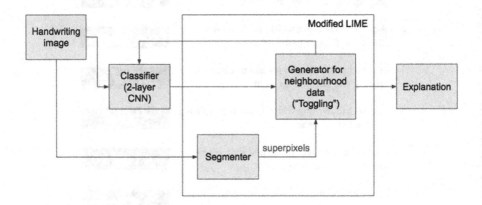

Fig. 6. High level structure of our handwriting explainer

Modifying LIME for Our Application. LIME [15] generates explanations by toggling superpixels in the input image with some alternative (e.g. black pixels), and observing how the change affects the output of the classifier. LIME does this by generating a replacement image, for looking up alternative pixel values. Conventionally, LIME takes in a superpixel and generates a mean in all three channels (RGB), then writes this mean to the corresponding pixels in the replacement image.

The images in our application are binary i.e., our segments are not in color and do not contain discriminative details within their boundaries. Thus the conventional mean replacement image generated by LIME contain identical segments and hence do not generate a good explanation.

Therefore, to improve on the existing options offered by LIME, we make a small change in the LIME source code to handle two other options, "reverse" and "reverse_nearest". In the "reverse" option, we take the image to be tested and create a negative of that image. Thus, our replacement image is now black ink and white background, opposite to the original image. This ensures that all the pixels will be toggled to their opposite, to see the effect on the classifier. This allows us to treat the handwriting problem as a binary problem for each segment - "Should this part be written to, or not?" The explainer can then test for changes in the classification output when the black segments are turned white, and when the white segments are turned black.

The "reverse" option offers more information, as it tests whether to keep or write to background segments, as well as whether to keep or erase written segments. However, this option also tests if background pixels far from the written sections should be written to or not, even if such edits are not of interest for improving handwriting.

Therefore, we create a "reverse_nearest" option that uses a Gaussian filter to blur the inverted image, and sets all pixels not affected by the blur to the background colour, black. This new replacement image does not store the opposites of pixels that are too far from the actual ink. All the pixels that are far from the writing are black, as in the original image, and thus toggling them causes no effect. In a way, this is imposing domain knowledge to guide the classifier to generate explanations only in the vicinity of the ink. Thus, we focus on the pixels that contain the letters and the pixels around it as this is the area we expect to contain explanations about the presence and absence of good and bad features. We use a σ of 0.8 for the blur, which is empirically found to yield the best results. This value can vary with stroke thickness.

Segmentation. The segmenter is responsible for generating the superpixels for the toggling and vital for generating interpretable explanations. This can be thought of as a way of "expert guidance" for the system to look for the nature and size of the superpixels and roughly the number of superpixels in the image.

SLIC [1] is found to be ideal for segmentation of our MNIST dataset. SLIC uses a k-means style algorithm for generating the superpixels. The algorithm initializes K cluster centers, where K is a user input, and creates roughly same size superpixels around these cluster centers. The proximity measure controls the compactness of the superpixels. Higher values make the superpixels squarer as the proximity is weighted heavily.

Figure 7 shows the segmentation of different segmenters on a sample from MNIST. We find that the best number of superpixels used and their compactness are very dependent on the size of the image, as well as the thickness of the strokes. For the low resolution MNIST image (28×28), the segmentation works well with high compactness of 50 (square superpixels are good as the image does not have many high dimensional features) and a moderate number of segments.

For the much higher resolution (40×73), we choose the Felzenszwalb segmenter instead (min_size = 2, scale = 0.01, $\sigma = 0.8$), for the following reasons:

- SLIC creates segments of nearly equal sizes. There exists edges and especially curves that SLIC may not segment correctly at larger segment sizes. Furthermore, with smaller segments, too many small squarish segments are created in the background. Felzenszwalb is able to create small segments near the high-detail areas and large segments in the background.
- Felzenszwalb is much faster, as the clustering process of the SLIC algorithm takes longer for a larger image.
- Felzenszwalb produces thin curved segments that wrap the letter strokes, and toggling these can correspond to shifting the strokes slightly for better spacing or slant.

Figure 8 shows the segmentation of different segmenters on a handwritten word sample.

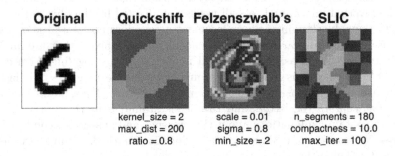

Original	Quickshift	Felzenszwalb's	SLIC
	kernel_size = 2 max_dist = 200 ratio = 0.8	scale = 0.01 sigma = 0.8 min_size = 2	n_segments = 180 compactness = 10.0 max_iter = 100

Fig. 7. Comparison of segmenters for the MNIST sample

Generating Explanations. For any sample image, we can generate an explanation by deploying LIME with the desired segmenter and replacement image settings, as well as the number of perturbations to try. In general, using 10,000 sample perturbations returns interpretable results for both digit and word images, within 5–30 s on a i5 1.5 GHz CPU with 8 GB RAM. The generated segments are sorted by their absolute contribution to the label, i.e. their influence on that outcome.

How the explanation result is displayed has a significant effect on its interpretability. We choose to visualise three aspects of the explanation, 1) Top contributing segments to the predicted label, 2) Most influential segments for both

Original   Quickshift   Felzenszwalb's   SLIC

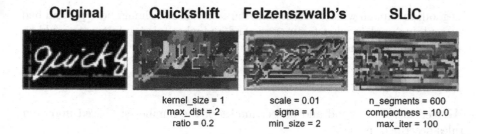

kernel_size = 1	scale = 0.01	n_segments = 600
max_dist = 2	sigma = 1	compactness = 10.0
ratio = 0.2	min_size = 2	max_iter = 100

Fig. 8. Comparison of segmenters for the handwritten word "quickly" sample

labels, 3) Result of suggested changes and segments to keep, based on most influential segments.

In particular, the most influential segments are colour-coded as follows (see Fig. 9):

- **Red (Good to change):** segments that contribute positively to the "bad" label, and segments that contribute negatively to the "good" label
- **Green (Good to keep):** segments that contribute positively to the "good" label, and segments that contribute negatively to the "bad" label

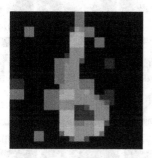

Fig. 9. A six from MNIST which looks like the letter "b" and is classified "bad". Red areas are to change, and green areas are to keep. (Color figure online)

To visualise the top contributing segments, we can either opt to show the top number of segments by absolute contribution value (this can be filtered to show positive contributions only), or set a threshold that selects all segments with contributions above a minimum weight.

4   Results

We aim to offer explanation about *why* writing is classified as good or bad, and how to improve bad writing. Therefore we give our classifier examples classified as good, to explain what features contribute to this and should be kept, and bad examples to explain what features are bad and what missing features should be added to improve the writing.

4.1 MNIST

We feed in several good and bad examples of the number six based upon our rules described in Fig. 3.

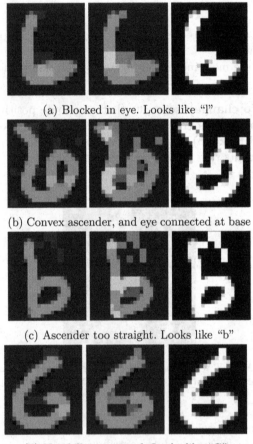

(a) Blocked in eye. Looks like "1"

(b) Convex ascender, and eye connected at base

(c) Ascender too straight. Looks like "b"

(d) Not fully connected. Looks like "G"

Fig. 10. MNIST Sixes classified as "bad". The first column presents the evidence for the classification, the second column highlights the favourable aspects (green) and the unfavourable aspects (red), and the third column gives the suggested change. (Color figure online)

How Should I Change My Bad Six? Figure 10 shows different examples of "bad" sixes according to our rules laid out in Fig. 3. All were correctly classified as "bad". Figure 10a and 10b suggest to add an eye while extending the ascender and connecting the eye leftward while adjusting the ascender respectively. Similarly, Fig. 10c and 10d suggest to add curvature to the ascender and connect, close the eye respectively. From these examples we observe that the explainer corrected the mistakes identified in our labelling stage.

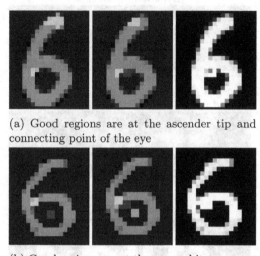

(a) Good regions are at the ascender tip and connecting point of the eye

(b) Good regions are at the eye and its connecting point

Fig. 11. MNIST Sixes classified as "good". The first column presents the evidence for the classification, the second column highlights the favourable aspects (green) and the unfavourable aspects (red), and the third column gives the suggested change. (Color figure online)

Why is My Six So Great? Additionally, it is interesting to know why your handwriting is classified as good, to assist in maintaining good habits.

Figure 11 shows an examples of sixes classified as "good", where the explainer seems to infer that it is good to keep the curved tip of the ascender, keep the eye of the six empty, and connect the eye properly.

4.2 Handwritten Word Dataset

The handwritten word dataset has greater levels of complexity than MNIST, but has far smaller number of samples.

Figure 12 shows an explanation for a good sample, where the explainer suggests to straighten the letter "d" and add the dot for letter "i". There's also a slight indication on closing the loop of letter "o". These suggestions are relevant to the form of the letters. There are also suggestions to add lines below the floating "u" and "o", which are relevant to alignment. The above results match the form and alignment requirements of the Minnesota Handwriting Assessment. The applied changes cause the handwriting score for the "good" class to increase from 74.6325% to 99.7178%.

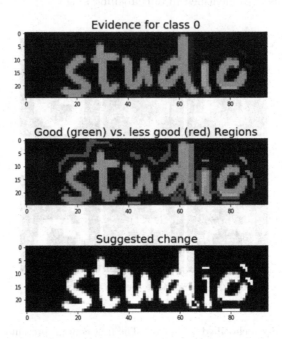

Fig. 12. Explanation of a "good" sample for "studio"

5   Limitations and Future Work

The dataset is a major limitation to our system since it is quite small and thus cannot represent all possible cases. Due to the shortcomings of the dataset both in quantity and quality, the classifier could not be tuned accurately, which directly affected the output of the explainer. Additionally, existing datasets have a lot of artifacts and noise. To overcome these limitations, a considerably large and curated dataset can be created. For labelling, we can use input from several people (or use Mechanical Turk) and follow majority consensus to reduce personal bias.

Currently we are using raw pixel values as features for the classifier. It would be an interesting future direction to exploit feature engineering techniques to train the classifier. Aspects such as slant, spacing, stroke curvature, and relative size of characters, could help the classifier to better learn the rules mentioned in Sect. 3.1.

The current segmenter choices do not create segments that follow the stroke contours and extend strokes smoothly. Interlacing the current LIME structure with a segmenter which better follows the contours and curvature of the characters in a text, such as SCALP [9] could be a probable future work. Furthermore, exploring an alternative explainer to LIME such as DeepLift [18], SHAP [12] can be helpful to improve the system performance.

6 Conclusion

We create a system that aims to explain the rules learnt by the machine learning model which distinguishes between good and bad handwriting. One application area of this research could be the education sector where this system can help children to improve their handwriting, enabling educators to focus on other academic aspects of the individuals. Explaining the working of the classifier while assessing handwriting will help users or administrators to establish faith in the system. Given enough varied data, it is possible for such a system to generate improvements for personalized styles of handwriting. Thus, an explainable and transparent AI that can give user-specific suggestions for improvements can be created, enabling seamless and easier learning transitions.

References

1. Achanta, R., Shaji, A., Smith, K., Lucchi, A., Fua, P., Süsstrunk, S.: SLIC superpixels compared to state-of-the-art superpixel methods. IEEE Trans. Pattern Anal. Mach. Intell. **34**(11), 2274–2282 (2012)
2. Agius, A., et al.: Dataset of coded handwriting features for use in statistical modelling. Data Brief **16**, 1010–1024 (2018)
3. Agius, A., et al.: Using handwriting to infer a writer's country of origin for forensic intelligence purposes. Forensic Sci. Int. **282**, 144–156 (2018)
4. Aksan, E., Pece, F., Hilliges, O.: DeepWriting: making digital ink editable via deep generative modeling. In: SIGCHI Conference on Human Factors in Computing Systems, CHI 2018. ACM, New York (2018)
5. Belongie, S., Malik, J., Puzicha, J.: Shape matching and object recognition using shape contexts. IEEE Trans. Pattern Anal. Mach. Intell. **24**(4), 509–522 (2002)
6. Bouillon, M., Anquetil, E.: Handwriting analysis with online fuzzy models. In: 17th Biennal Conference of the International Graphonomics Society (IGS) (2015)
7. Falk, T.H., Tam, C., Schellnus, H., Chau, T.: On the development of a computer-based handwriting assessment tool to objectively quantify handwriting proficiency in children. Comput. Methods Programs Biomed. **104**(3), e102–e111 (2011)
8. Fuentes, C.T., Mostofsky, S.H., Bastian, A.J.: Children with autism show specific handwriting impairments. Neurology **73**(19), 1532–1537 (2009)
9. Giraud, R., Ta, V.T., Papadakis, N.: SCALP: superpixels with contour adherence using linear path. In: International Conference on Pattern Recognition (ICPR), pp. 2374–2379. IEEE (2016)

10. Kaiser, M.L., Albaret, J.M., Doudin, P.A.: Relationship between visual-motor integration, eye-hand coordination, and quality of handwriting. J. Occup. Ther. Sch. Early Interv. **2**(2), 87–95 (2009)
11. LeCun, Y., Bottou, L., Bengio, Y., Haffner, P.: Gradient-based learning applied to document recognition. Proc. IEEE **86**(11), 2278–2324 (1998)
12. Lundberg, S.M., Lee, S.I.: A unified approach to interpreting model predictions. In: Advances in Neural Information Processing Systems, pp. 4768–4777 (2017)
13. Reisman, J.: Minnesota Handwriting Assessment. Psychological Corporation, San Antonio (1999)
14. Reisman, J.E.: Development and reliability of the research version of the Minnesota Handwriting Test. Phys. Occup. Ther. Pediatr. **13**(2), 41–55 (1993)
15. Ribeiro, M.T., Singh, S., Guestrin, C.: Why should i trust you?: Explaining the predictions of any classifier. In: 22nd ACM SIGKDD International Conference on Knowledge Discovery and Data Mining, pp. 1135–1144. ACM (2016)
16. Roston, K.L., Hinojosa, J., Kaplan, H.: Using the Minnesota Handwriting Assessment and Handwriting Checklist in screening first and second graders' handwriting legibility. J. Occup. Ther. Sch. Early Interv. **1**(2), 100–115 (2008)
17. Scikit-image: comparison of segmentation and superpixel algorithms (2017). http://scikit-image.org/docs/dev/auto_examples/segmentation/plot_segmentations.html
18. Shrikumar, A., Greenside, P., Kundaje, A.: Learning important features through propagating activation differences. arXiv preprint arXiv:1704.02685 (2017)
19. Stuhlsatz, A., Lippel, J., Zielke, T.: Feature extraction with deep neural networks by a generalized discriminant analysis. IEEE Trans. Neural Netw. Learn. Syst. **23**(4), 596–608 (2012)
20. Sundararajan, M., Taly, A., Yan, Q.: Axiomatic attribution for deep networks. arXiv preprint arXiv:1703.01365 (2017)
21. Zintgraf, L.M., Cohen, T.S., Adel, T., Welling, M.: Visualizing deep neural network decisions: prediction difference analysis. arXiv preprint arXiv:1702.04595 (2017)
22. Zitnick, C.L.: Handwriting beautification using token means. ACM Trans. Graph. **32**(4), 53:1–53:8 (2013)

Chaco: Character Contrastive Learning for Handwritten Text Recognition

Xiaoyi Zhang[1], Tianwei Wang[1], Jiapeng Wang[1], Lianwen Jin[1,2](✉),
Canjie Luo[1], and Yang Xue[1]

[1] South China University of Technology, Guangzhou, China
lianwen.jin@gmail.com, yxue@scut.edu.cn
[2] SCUT-Zhuhai Institute of Modern Industrial Innovation, Zhuhai, China

Abstract. Current mainstream text recognition models rely heavily on large-scale data, requiring expensive annotations to achieve high performance. Contrast-based self-supervised learning methods aimed at minimizing distances between positive pairs provide a nice way to alleviate this problem. Previous studies are implemented from the perspective of words, taking the entire word image as model input. But characters are actually the basic elements of words, so in this paper, we implement contrastive learning from another perspective, i.e., the perspective of characters. Specifically, a simple yet effective method, termed ChaCo, is proposed, which takes the characters and strokes (called a character unit) cropped from the word image as model input. However, in the commonly used random cropping approach, the positive pairs may contain completely different characters, in which case it is unreasonable to minimize the distance between positive pairs. To address this issue, we introduce a Character Unit Cropping Module (CUCM) to ensure the positive pairs contain the same characters by constraining the selection region of the positive sample. Experiments show that our proposed method can achieve much better representation quality than previous methods while requiring fewer computation resources. Under the semi-supervised setting, ChaCo can achieve promising performance with an accuracy improvement of 13.1 points on the IAM dataset.

Keywords: Self-supervised learning · Text recognition · Contrastive learning

1 Introduction

Text recognition is a vital task of computer vision. Most existing text recognition methods [16,19,20,23] are based on a fully supervised manner, which means that there is a corresponding manual annotation for each input text image. However, manual annotation requires a considerable amount of time and cost. Moreover, because cognitive comprehension differs among individuals, certain labeling problems may exist. For example, the label can be incorrect, interfering

U. Porwal et al. (Eds.): ICFHR 2022, LNCS 13639, pp. 345–359, 2022.
https://doi.org/10.1007/978-3-031-21648-0_24

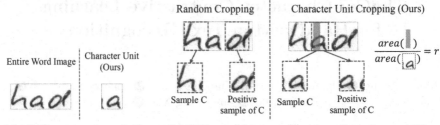

(a) Difference in Self-Supervised Model Input.

(b) Difference between Random Cropping and Character Unit Cropping.

Fig. 1. Differences between our method and other methods in model input and cropping approach.

with the model training process. Therefore, reducing manual annotation in text recognition has emerged as an important research topic [1,12,14].

Self-supervised learning [1,4–6,8,10] provides a method to learn from unlabeled data. In recent years, generative and contrastive self-supervised methods have been emerging [13], among which contrast-based methods have demonstrated significant potential. Contrast-based methods achieve self-supervised feature learning by minimizing the distance between positive pairs while maximizing the distance from negative ones [4,5,8,10]. The gap between self-supervised learning and fully supervised learning is gradually narrowing, with contrastive learning demonstrating promising generalization ability [18] in several fields, including image classification and object detection [5,8,21].

In the field of text recognition, only a few studies have been reported on contrast-based self-supervised learning. SeqCLR [1] considered the sequence relationship between the characters of a word in the self-supervised learning stage and proposed an instance mapping function for sequence modeling. PerSec [12] learned low- and high-level features of entire word images and proposed a dual-context perceiver to distinguish the sequential features within one image. All of these methods are implemented from the perspective of words, taking the entire word image as model input during the self-supervised stage. However, in the task of word recognition, the model generally first recognizes the characters in the image individually and then determines the words formed by these characters. Therefore, in word recognition, a more essential task is to recognize the features of characters. Thus, we adopt another perspective in self-supervised learning, i.e., the perspective of characters.

In this paper, we propose a new character contrastive learning approach named **ChaCo** that takes the characters and strokes cropped from the word image as model input. Figure 1(a) shows the difference in the self-supervised model input. Previous contrastive methods used the entire image as input, while our method takes the character unit as input. The character unit is an image containing some characters and strokes, cropped from the word image.

For the cropping approach, in previous contrastive learning methods [4,10], the position of the positive pairs obtained by the commonly used random cropping is irrelevant in the image. As shown in Fig. 1(b), the random cropping may result in sample C containing completely different characters from its positive sample, in which case it is unreasonable to minimize the distance between them in contrastive learning. Therefore, we introduce a Character Unit Cropping Module (CUCM) to obtain the positive pairs. The CUCM constrains the selection region of the positive samples so that the positive pairs contain a certain proportion of overlap. As shown in Fig. 1(b), the orange region represents the centrosymmetric region of sample C, and its area ratio to C is r. Both sample C and its positive sample contain the orange region. This approach allows the model to focus on mining the features of the same part of the positive sample and C, thereby distinguishing the characters of sample C from its neighboring characters and learning the features of characters in the self-supervised stage. In addition, the character unit is cropped from the entire word image, such that its size is much smaller than that of the entire word image, which can greatly reduce memory requirements in the self-supervised learning stage.

In the downstream text recognition stage, we follow the same experimental setup as existing methods [1,12] to validate the effectiveness of the proposed method. We take the entire word image as input and then perform the recognition in the "encoder-decoder" paradigm, where the encoder loads the encoder parameters obtained from the self-supervised stage. In representation-quality experiments, our method outperforms other methods by a large margin, surpassing the IAM dataset by 30.3 points. In semi-supervised experiments, our method achieves promising results with an accuracy improvement of 13.1 points on the IAM dataset.

In summary, the contributions of this paper are three-fold:

1) A character contrastive learning method called ChaCo is proposed. This simple yet effective method takes the character unit cropped from the word image as model input, which greatly reduces the demand for memory.
2) A Character Unit Cropping Module is proposed to obtain the positive pairs. The selection region of positive samples is constrained by the position of the original sample, so the positive pairs contain a certain proportion of overlap.
3) Experiments on three widely used handwritten datasets indicate that our method achieves better representation quality and promising results.

2 Related Work

2.1 Contrast-Based Self-supervised Framework

Self-supervised learning takes advantage of unlabeled data, significantly reducing the requirement for manual labeling. Contrastive learning has recently become a common method for self-supervised learning. Many studies have contributed to different aspects of the contrastive learning framework, such as the design of positive and negative sample pairs [4,5,10], asymmetric network structures [6,8], and the construction of loss functions [24].

He *et al.* [10] proposed a self-supervised architecture named MoCo, which utilizes the momentum mechanism to update the parameters of the momentum encoder to avoid forgetting previous negative samples. Chen *et al.* [4] proposed a self-supervised framework called SimCLR that used a large batch size to retain negative samples and used the projection head to achieve better performance. Inspired by SimCLR, Chen *et al.* [5] proposed MoCo v2, which added the projection head and more data augmentation based on MoCo to achieve better performance. Grill *et al.* [8] proposed an asymmetric network structure termed BYOL to eliminate the use of negative samples. Zbontar *et al.* [24] proposed a method named Barlow Twins, which avoided collapse by measuring the cross-correlation matrix between the outputs of two identical networks fed with distorted versions of a sample to render it as close to the identity matrix as possible. Chen *et al.* [6] proposed a simple siamese network called SimSiam, which was free of negative samples, large batches, or momentum encoders. They used a stop-gradient operation, which played an essential role in preventing collapse.

These contrastive learning methods are applied to the single-object image classification task. However, words consist of multiple characters, and the above methods are not suitable for multi-object images. Thus, we decompose word images into single-character images for contrastive learning.

2.2   Self-supervised Text Recognition

Recently, with the development of generative and contrastive self-supervised learning, self-supervised text-learning methods have emerged. In the generative self-supervised method, SimAN [14], designed for scene text, first divided a text image into two adjacent patches. Then, data augmentation was performed on one patch, and the other patch was used to guide the recovery to learn the text features.

In the contrastive self-supervised method, SeqCLR [1] used sequence-to-sequence contrastive learning and modeled the sequence relationship of text images at the feature map level using an instance mapping function. PerSec [12] learned the features of text images from low- and high-level respectively and proposed a dual-context perceiver to distinguish sequential features within one image. These contrastive methods took the entire word image as model input during the self-supervised stage. However, in word recognition, the model generally first recognizes the characters in the image individually and then determines the words formed by these characters. Therefore, a more essential task of word recognition is to recognize the features of characters. In this paper, we adopt the perspective of characters, taking the characters and strokes in the word image as model input.

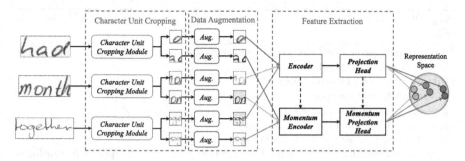

Fig. 2. The framework of ChaCo consists of three major parts: character unit cropping, data augmentation denoted by 'Aug.', and feature extraction.

3 Method

The proposed ChaCo framework, illustrated in Fig. 2, consists of three major parts: character unit cropping, data augmentation, and feature extraction. To verify the effectiveness of ChaCo, the downstream text recognition task is then performed in the text recognizer.

3.1 Character Unit Cropping

Among previous contrastive self-supervised text learning methods, PerSec [12] and SeqCLR [1] maintained the integrity of the entire word during the input process of the self-supervised stage. However, in the word recognition task, the model first recognizes the characters in the image individually and then determines the words formed by these characters. Thus, a more essential task in word recognition is character recognition. Therefore, in the self-supervised learning stage, the proposed method takes characters and strokes in the word image as model input to learn character features.

The idea that word image integrity is no longer maintained in model input has emerged in the generative self-supervised method SimAN [14]. It learned the features of text by splitting an image of a word into two non-overlapping patches and then recovering the augmented patch through another patch. Inspired by SimAN, we explore the idea of no longer maintaining word image integrity from a contrast-based method.

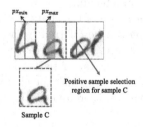

Fig. 3. Positive sample selection.

The character unit is defined as an image containing some characters and strokes. To get the character unit, we need to crop the word image. Random cropping has been a commonly used data augmentation method in contrast-based self-supervised methods [4,5,10]. However, the randomly cropped positive sample is not related to the position of the original

---

Algorithm 1. Pseudo-code of Character Unit Cropping Module in a Python-like style.

Input: Grayscale image Img, Width scale (w_{min}, w_{max}), Height scale (h_{min}, h_{max}), Center ratio r, Output size (cw, ch)
Output: Image C, Positive image PC
1: width,height=Img.shape
2: $w_{ratio} = $ random. uniform(w_{min}, w_{max})
3: $h_{ratio} = $ random. uniform(h_{min}, h_{max})
4: $w, h = w_{ratio} \times width, h_{ration} \times height$
5: $x, y = $ randint$(0, width - w),$ randint$(0, height - h)$ ▷ Upper left coordinate
6: $px_{min} = \max(x - \lfloor \frac{w}{2} \times (1 - r)\rfloor, 0)$
7: $px_{max} = \min(x + \lfloor \frac{w}{2} \times (1 - r)\rfloor, width - w)$
8: $px = $ randint(px_{min}, px_{max}) ▷ Upper left coordinate
9: $py = $ randint$(0, height - h)$
10: $C = crop(x, y, h, w)$ from Img
11: $PC = crop(px, py, h, w)$ from Img
12: $C, PC = $ resize$(C, (cw, ch)),$ resize$(PC, (cw, ch))$

---

sample in the image, which may result in the positive pairs having completely different characters. In this case, it is unreasonable to minimize the distance between them in contrastive learning. Therefore, building on the random cropping approach, a Character Unit Cropping Module (CUCM) is proposed to generate character units and their positive samples such that the position of the positive samples in the image is constrained by that of the original character unit.

The algorithm for the CUCM is presented in Algorithm 1. In the handwritten text, characters are generally combined horizontally into words. To obtain a character unit, a large-scale crop needs to be performed along the width, and a small-scale crop should be performed along the height. Therefore, the selection of (w_{min}, w_{max}) is relatively small, such as (0.1, 0.4), and the selection of (h_{min}, h_{max}) is relatively large, such as (0.8, 1.0). Thus, the initial height of the character unit is close to that of the original image $(h \approx height)$. Although the ordinate of the character unit and its positive sample may not be the same, $y \approx py$ can be inferred from Eq. 1. Eventually, the character unit differs significantly from the positive sample only along the abscissa; while still containing the same character parts. This positive sample selection method can drive the model to distinguish the characters of a character unit from its adjacent character strokes, thereby learning the features of the character.

$$\begin{cases} y = \text{randint}(0, height - h) \\ py = \text{randint}(0, height - h) \Rightarrow y \approx py \\ h \approx height \end{cases} \qquad (1)$$

where randint(a, b) represents an integer randomly selected from $[a, b)$.

As shown in Fig. 3, the orange region is the centrosymmetric region of sample C, and its area ratio to C is r. The positive sample of C contains the orange region, and the width and height are the same as those of C. Eventually, the selection region of the positive sample is shown as the red box. It is worth noting that when the center ratio r is larger, the selection region of the positive sample is smaller.

3.2 Data Augmentation

Data augmentation is used to increase the diversity of training samples. Many existing studies have shown that data augmentation is essential for improving the feature representation ability of a model [4,5].

The data augmentation method used in this paper includes distortion, brightness adjustment, blurring, and noise. These data augmentation types are similar to those used in the self-supervised process of SeqCLR [1]. Albumentation [3] is used to perform augmentation, which is a CPU-efficient toolkit.[1] The pseudo-code is shown in Appendix A. Figure 4 shows examples of data augmentation used for character units and their positive samples. It is worth noting that the data augmentation presented here is used during the self-supervised training stage. For a fair comparison, in the downstream text recognition stage, the data augmentation settings of SeqCLR [1] are adopted.

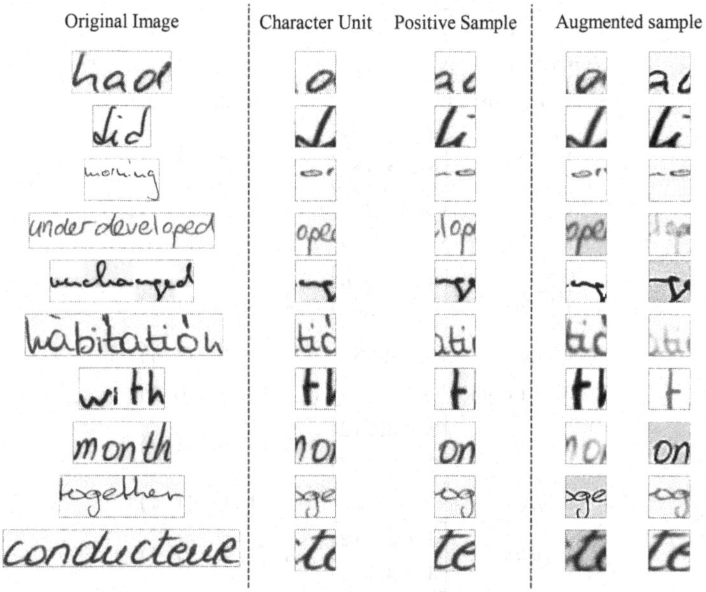

Fig. 4. Examples of data augmentation for character units and their positive samples.

---

[1] https://github.com/albumentations-team/albumentations.

3.3   Feature Extraction

For feature extraction, we follow MoCo v2 [5]. The feature extractor, illustrated in Fig. 2, consists of an encoder and a projection head, and the parameters are updated using backpropagation. The structure of the encoder is ResNet29, which is the same as that of SeqCLR [1]. The specific structure of the encoder is listed in Table 1. The projection head consists of two fully connected layers. The momentum feature extractor, consisting of a momentum encoder and a momentum projection head, has the same structure as the feature extractor. However, its parameters are updated with momentum according to the parameters of the feature extractor. Formally, denoting the parameters of the feature extractor as θ_q and those of the momentum feature extractor as θ_v, the parameter θ_v is updated by

$$m\theta_v + (1 - m)\theta_q \rightarrow \theta_v \tag{2}$$

where $m \in [0, 1)$ is a momentum coefficient.

Table 1. Architecture of ResNet29. *kernel*, *c*, *s*, and *p* represent the size of the convolution kernel, the dimension of the convolution kernel, stride, and padding, respectively.

Layer	Configuration		
	kernel, c	*s*	*p*
Conv1	$3 \times 3, 32$	$(1, 1)$	$(1, 1)$
Conv2	$3 \times 3, 64$	$(1, 1)$	$(1, 1)$
Maxpool1	2×2	$(2, 2)$	$(0, 0)$
Block1	$\begin{bmatrix} 3 \times 3, 128 \\ 3 \times 3, 128 \end{bmatrix} \times 1$	$(1, 1)$	$(1, 1)$
Conv3	$3 \times 3, 128$	$(1, 1)$	$(1, 1)$
Maxpool2	2×2	$(2, 2)$	$(0, 0)$
Block2	$\begin{bmatrix} 3 \times 3, 256 \\ 3 \times 3, 256 \end{bmatrix} \times 2$	$(1, 1)$	$(1, 1)$
Conv4	$3 \times 3, 256$	$(1, 1)$	$(1, 1)$
Maxpool3	2×2	$(2, 1)$	$(0, 1)$
Block3	$\begin{bmatrix} 3 \times 3, 512 \\ 3 \times 3, 512 \end{bmatrix} \times 5$	$(1, 1)$	$(1, 1)$
Conv5	$3 \times 3, 512$	$(1, 1)$	$(1, 1)$
Block4	$\begin{bmatrix} 3 \times 3, 512 \\ 3 \times 3, 512 \end{bmatrix} \times 3$	$(1, 1)$	$(1, 1)$
Conv6	$2 \times 2, 512$	$(2, 1)$	$(0, 1)$
Conv7	$2 \times 2, 512$	$(1, 1)$	$(0, 0)$

The feature extractor and momentum feature extractor perform feature extraction to generate the features of the positive pairs. Negative samples are obtained from a predefined queue. The features of the queue are initialized randomly. Subsequently, the features extracted by the momentum feature extractor are enqueued each time, and the earliest mini-batch features are dequeued.

The contrastive loss function InfoNCE [17] is formulated as

$$\mathcal{L} = -\log \frac{\exp\left(q \cdot k_+/\tau\right)}{\sum_{i=0}^{N} \exp\left(q \cdot k_i/\tau\right)} \tag{3}$$

where N is the queue length, τ is the temperature hyper-parameter [22], $k_i(i = 1 \cdots N)$ represents the feature vector in the queue, q and k_+ represent the feature vector extracted by the feature extractor and momentum feature extractor respectively from the same word image.

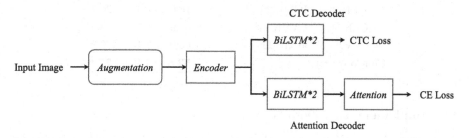

Fig. 5. Downstream text recognizer, consisting of an encoder and a decoder. There are two types of decoders: a CTC-based decoder and an attention-based decoder.

3.4 Downstream Text Recognizers

After self-supervised training, the effectiveness of the proposed method, ChaCo, is validated on the downstream text recognition task. As shown in Fig. 5, the text recognizer consists of an encoder and a decoder. The structure of the encoder is ResNet29. There are two types of text recognizers: CTC-based [7] and attention-based [2] recognizers. Note that this paper uses the same text recognizer as SeqCLR [1] for a fair comparison.

4   Experiments

4.1   Datasets and Metrics

The proposed method was evaluated on several widely used public benchmarks in handwritten text recognition studies. These datasets are IAM [15], RIMES [9], and CVL [11]. During the self-supervised and text recognition training stages, we only use the training set. For example, for the model on the IAM dataset, only the training samples of IAM are used for training, which is the same as the setting of SeqCLR [1].[2] To evaluate the performance of the text recognizer, we adopt the metrics of word-level recognition accuracy for all experiments in Sect. 4.

---

[2] We contacted the authors of SeqCLR to get the training details.

Table 2. Representation quality. The evaluation metric is word-level recognition accuracy. The encoder freezes the parameters obtained by self-supervised learning, and only the decoder is trained. **Bold** represents the best result. "MoCo v2* [5]" means using the same data augmentation, cropping range, and input size as ChaCo.

Method	Venue	Decoder	IAM	RIMES	CVL
Baseline	–	CTC	29.9	22.7	24.3
SimCLR [4]	ICML'20		4.0	10.0	1.8
SeqCLR [1]	CVPR'21		39.7	63.8	66.7
MoCo v2* [5]	arXiv'20		60.9	76.3	75.0
ChaCo (ours)	–		**70.0**	**79.4**	**75.6**
Baseline	–	Attention	33.9	28.7	35.0
SimCLR [4]	ICML'20		16.0	22.0	26.7
SeqCLR [1]	CVPR'21		51.9	79.5	74.5
MoCo v2* [5]	arXiv'20		65.0	82.4	76.8
ChaCo (ours)	–		**72.9**	**84.9**	**77.6**

4.2   Implementation Details

Self-supervised Stage. The scale range of the width is $(0.1, 0.4)$, and for the height, it is $(0.8, 1.0)$. The center ratio $r = 0.2$, and the size of the character unit is 32×32. We use a batch size of 512. The experiment required only 6000 Mb of memory on one A40 GPU, which shows the advantage of requiring fewer computational resources. Note that SeqCLR [1] used 4 V100 GPUs, and PerSec [12] used 32 A100 GPUs.

We train for 75K iterations for each dataset, which takes approximately 15 h. Specifically, the IAM and RIMES datasets are trained for 700 epochs and the CVL dataset is trained for 500 epochs. For the optimization, we use SGD as the optimizer with an initial learning rate of 0.03. Data augmentation adopts the augmentation in Sect. 3.2. The output dimension of the projection head is 128. Other self-supervised settings are the same as those in MoCo v2 [5].

It should be noted that the difference between MoCo v2 [5] and ChaCo is that MoCo v2 [5] uses random cropping, while ChaCo uses character unit cropping. So to verify the effectiveness of CUCM, MoCo v2 [5] uses the same data augmentation, cropping range, and input size as ChaCo. Specifically, "MoCo v2* [5]" means using data augmentation in Sect. 3.2, a cropping range of $(0.1, 0.4)$, and an input size of 32×32.

Downstream Text Recognition Task. For a fair comparison, we use the same experimental setup as that of SeqCLR [1]. The text recognition model is an "encoder-decoder" paradigm, using the entire word image as input, which needed to be scaled to 32×100. We use a batch size of 256 and the same data augmentation as SeqCLR [1], including light cropping, linear contrast, and Gaussian blur.

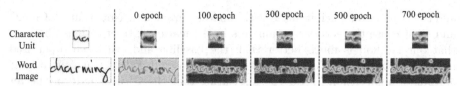

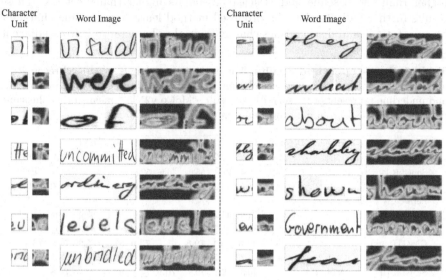

(a) Examples of feature extraction for 0th, 100th, 300th, 500th, and 700th epochs.

(b) Examples of feature extraction for the 700th epochs.

Fig. 6. Visualization of feature extraction of the encoder in the self-supervised training stage.

4.3   Representation Quality

To validate the effectiveness of the representations, we follow the evaluation criteria described in [1,12]. The encoder parameters are frozen, and only the decoder is trained.

Figure 6 shows the visualization of the feature extraction of the encoder in the self-supervised training stage. Note that the visualizations are performed on the IAM dataset. As shown in Fig. 6(a), as the number of training epochs increased, the encoder gradually learned the features of the characters in the character unit. For the entire word image, the encoder can still identify the features of each character in the word. Figure 6(b) provides more examples of feature extraction for the 700th epochs, which confirms that ChaCo effectively learns good feature representations in self-supervised training.

Table 2 shows the comparison of our method with the baseline, SimCLR [4], SeqCLR [1], and MoCo v2* [5]. For the baseline, the parameters of the encoder and decoder are initialized randomly. While for other methods, the encoder is initialized with the encoder parameters obtained from self-supervised training, and then the decoder is initialized with random parameters. Note that MoCo v2* and ChaCo use the encoder parameters in the feature extractor. It is observed that our method is much better than the baseline and other self-supervised methods on both the CTC-based and attention-based decoders. ChaCo is at least 0.6 points higher than MoCo v2*, indicating that the CUCM performs better than random cropping. In addition, the proposed method is at least 39 points higher than the baseline and at least 3.1 points higher than SeqCLR. These results further indicate that the proposed method learns outstanding character representations and distinguishes characters in the character unit from adjacent characters.

Table 3. Semi-supervised results. The evaluation metric is word-level recognition accuracy. The encoder and decoder are jointly trained. **Bold** represents the best result, and <u>underline</u> represents the second-best result. "MoCo v2* [5]" means using the same data augmentation, cropping range, and input size as ChaCo.

Method	Venue	Decoder	IAM			RIMES			CVL		
			5%	10%	100%	5%	10%	100%	5%	10%	100%
SimCLR [4]	ICML'20	CTC	15.4	21.8	65.0	36.5	52.9	84.5	52.1	62.0	74.1
SeqCLR [1]	CVPR'21		31.2	44.9	76.7	**61.8**	71.9	**90.1**	**66.0**	**71.0**	77.0
PerSec-CNN [12]	AAAI'22		–	–	77.9	–	–	–	–	–	<u>78.1</u>
Baseline	–		35.8	44.2	78.6	56.0	66.6	87.4	63.0	70.6	77.7
MoCo v2* [5]	arXiv'20		<u>43.0</u>	<u>51.7</u>	<u>78.8</u>	<u>61.4</u>	<u>72.1</u>	89.1	64.0	<u>70.6</u>	77.9
ChaCo (ours)	–		**44.3**	**53.1**	**79.8**	61.3	**73.0**	<u>89.2</u>	<u>65.6</u>	70.3	**78.1**
SimCLR [4]	ICML'20	Attention	22.7	32.2	70.7	49.9	60.9	87.8	59.0	65.6	75.7
SeqCLR [1]	CVPR'21		40.3	52.3	79.9	**70.9**	<u>77.0</u>	**92.5**	**73.1**	**74.8**	77.8
PerSec-CNN [12]	AAAI'22		–	–	<u>80.8</u>	–	–	–	–	–	**80.2**
Baseline	–		40.8	49.8	79.5	62.1	72.7	90.0	68.4	73.8	77.3
MoCo v2* [5]	arXiv'20		<u>43.9</u>	<u>53.4</u>	80.2	67.6	76.1	90.1	67.3	73.5	77.9
ChaCo (ours)	–		**46.7**	**55.5**	**81.4**	<u>68.4</u>	**77.0**	<u>90.6</u>	<u>68.9</u>	<u>74.0</u>	<u>78.2</u>

4.4   Semi-supervised Evaluation

In this setting, the encoder and decoder are jointly trained. The text recognizer initialization for each method is the same as in Sect. 4.3.

Table 3 shows the comparison of our method with the baseline, SimCLR [4], MoCo v2* [5], SeqCLR [1], and PerSec [12]. "5%", "10%", and "100%" in Table 3 represent training using 5%, 10%, and 100% of the training data, respectively. It can be observed that our method almost achieves the best or second-best results on all three datasets. Compared with the baseline, the pre-trained ChaCo model can achieve higher accuracy, reflecting the effectiveness of self-supervised contrastive learning. Our method outperforms MoCo v2* on most datasets, indicating the effectiveness of CUCM. Compared with the current self-supervised text methods (SeqCLR [1] and PerSec [12]), our method performs better on the IAM dataset, where the accuracy of the CTC-based decoder is 13.1 points higher than that of SeqCLR on 5% of the IAM data. ChaCo also achieves competitive results on the RIMES and CVL datasets. Based on these experiments, it can be seen that contrastive learning from the character unit perspective is effective.

Table 4. Ablation studies on the center ratio r.

r	0	0.2	0.4	0.6	0.8	1.0
Accuracy	81.3	**81.4**	81.3	80.4	80.9	80.5

Table 5. Ablation studies on the width scale.

(w_{min}, w_{max})	(0.1, 0.4)	(0.3, 0.6)	(0.5, 0.8)
Accuracy	**81.4**	81.3	81.1

5 Ablation Study

Because the center ratio r and the width scale (w_{min}, w_{max}) are critical parameters of the Character Unit Cropping Module, so we perform ablation studies on them. We conduct experiments on the IAM dataset in the semi-supervised setting. We use the attention-based decoder.

Table 4 shows the ablation studies for center ratio r. When the center ratio r is larger, the selection region of the positive sample in the word image is smaller. Therefore, when r is larger, the characters and strokes contained in the positive pairs are more similar, making it hard to view the adjacent characters of the character unit in the word image. Therefore, the model performs better when $r < 0.5$, and the best result is achieved when $r = 0.2$.

Table 5 shows the ablation studies for the width scale (w_{min}, w_{max}). The width scale (w_{min}, w_{max}) reflects the proportion of characters contained in character units to the word. Under the character unit size setting of 32×32, the proportion of characters below 50% is better, and the proportion of characters between 10% and 40% is the best.

6 Conclusions

In this paper, we propose a simple yet effective method named **ChaCo**. Different from previous contrast-based studies [1,12] which took the entire word image as model input, ChaCo learns from the character unit of the word image. The character units and their corresponding positive samples are obtained by the Character Unit Cropping Module (CUCM), which contain at least a certain proportion of the same parts. Consequently, the model can distinguish the character unit from its adjacent characters in the self-supervised stage to learn effective character features.

Extensive experiments show that the performance of the proposed method on handwritten text recognition is superior to or competitive with previous methods. In the future, the design of effective self-supervised learning tasks to learn more powerful discriminant features for text recognition is worthy of further investigation.

Acknowledgement. This research is supported in part by NSFC (Grant No. 61936003), GD-NSF (no. 2017A030312006, No. 2021A1515011870), Zhuhai Industry Core and Key Technology Research Project (no. ZH22044702200058PJL), and the Science and Technology Foundation of Guangzhou Huangpu Development District (Grant 2020GH17).

A Appendix

In this appendix, the pseudo-code of data augmentation in Sect. 3.2 is shown below for reference.

```
1  import albumentations as A
2  augmentation=A.Sequential([
3      # Distortion
4      A.OneOf([
5          A.ElasticTransform (alpha=30,sigma=5,alpha_affine=0,p=1),
6          A.Affine (scale=(0.8,1),shear=(-15,15),cval=255,p=1),
7          A.Perspective(p=1),
8          A.PiecewiseAffine(scale=(0.03, 0.1), nb_rows=4, nb_cols
           =4,cval=255,p=1),
9          A.GridDistortion(num_steps=3,p=1),]),
10     # Brightness Adjustment
11     A.OneOf([
12         A.ColorJitter(0.5, 0.5, 0.5, 0.5,p=1),
13         A.RandomBrightnessContrast(p=1),
14         A.Sharpen(p=1),]),
15     # Blurring
16     A.OneOf([
17         A.Blur(blur_limit=(3,5),p=1),
18         A.GaussianBlur(blur_limit=(3,5),p=1),
19         A.MedianBlur(blur_limit=(3,5),p=1),
20         A.MotionBlur(blur_limit=(3,5),p=1),]),
21     # Noise
22     A.OneOf([
23         A.GaussNoise(p=1),
24         A.MultiplicativeNoise(p=1),]),])
```

References

1. Aberdam, A., et al.: Sequence-to-sequence contrastive learning for text recognition. In: CVPR, pp. 15302–15312 (2021)
2. Bahdanau, D., Cho, K., Bengio, Y.: Neural machine translation by jointly learning to align and translate. In: ICLR (2015)
3. Buslaev, A., Iglovikov, V.I., Khvedchenya, E., Parinov, A., Druzhinin, M., Kalinin, A.A.: Albumentations: fast and flexible image augmentations. Information **11**(2), 125 (2020)
4. Chen, T., Kornblith, S., Norouzi, M., Hinton, G.: A simple framework for contrastive learning of visual representations. In: ICML, pp. 1597–1607 (2020)
5. Chen, X., Fan, H., Girshick, R., He, K.: Improved baselines with momentum contrastive learning (2020). arXiv preprint arXiv:2003.04297

6. Chen, X., He, K.: Exploring simple siamese representation learning. In: CVPR, pp. 15750–15758 (2021)
7. Graves, A., Fernández, S., Gomez, F., Schmidhuber, J.: Connectionist temporal classification: labelling unsegmented sequence data with recurrent neural networks. In: ICML, pp. 369–376 (2006)
8. Grill, J.B., et al.: Bootstrap your own latent-a new approach to self-supervised learning. In: NeurIPS, vol. 33, pp. 21271–21284 (2020)
9. Grosicki, E., Abed, H.E.: ICDAR 2009 handwriting recognition competition. In: ICDAR, pp. 1398–1402. IEEE Computer Society (2009)
10. He, K., Fan, H., Wu, Y., Xie, S., Girshick, R.: Momentum contrast for unsupervised visual representation learning. In: CVPR, pp. 9726–9735 (2020)
11. Kleber, F., Fiel, S., Diem, M., Sablatnig, R.: CVL-DataBase: an off-line database for writer retrieval, writer identification and word spotting. In: ICDAR, pp. 560–564 (2013)
12. Liu, H., et al.: Perceiving stroke-semantic context: hierarchical contrastive learning for robust scene text recognition. In: AAAI (2022)
13. Liu, X., et al.: Self-supervised learning: generative or contrastive. IEEE Trans. Knowl. Data Eng. 1 (2021)
14. Luo, C., Jin, L., Chen, J.: SimAN: exploring self-supervised representation learning of scene text via similarity-aware normalization. In: CVPR (2022)
15. Marti, U.V., Bunke, H.: The IAM-database: an English sentence database for offline handwriting recognition. Int. J. Doc. Anal. Recogn. 5(1), 39–46 (2002). https://doi.org/10.1007/s100320200071
16. Nguyen, N., et al.: Dictionary-guided scene text recognition. In: CVPR, pp. 7383–7392 (2021)
17. van den Oord, A., Li, Y., Vinyals, O.: Representation learning with contrastive predictive coding (2018). arXiv preprint arXiv:1807.03748
18. Tendle, A., Hasan, M.R.: A study of the generalizability of self-supervised representations. Mach. Learn. Appl. 6, 100124 (2021)
19. Wang, T., et al.: Decoupled attention network for text recognition. In: AAAI, vol. 34, pp. 12216–12224 (2020)
20. Wang, T., et al.: Implicit feature alignment: learn to convert text recognizer to text spotter. In: CVPR, pp. 5973–5982 (2021)
21. Wang, X., Zhang, R., Shen, C., Kong, T., Li, L.: Dense contrastive learning for self-supervised visual pre-training. In: CVPR, pp. 3024–3033 (2021)
22. Wu, Z., Xiong, Y., Yu, S.X., Lin, D.: Unsupervised feature learning via non-parametric instance discrimination. In: CVPR, pp. 3733–3742 (2018)
23. Yan, R., Peng, L., Xiao, S., Yao, G.: Primitive representation learning for scene text recognition. In: CVPR, pp. 284–293 (2021)
24. Zbontar, J., Jing, L., Misra, I., LeCun, Y., Deny, S.: Barlow twins: self-supervised learning via redundancy reduction. In: ICML, pp. 12310–12320 (2021)

Enhancing Indic Handwritten Text Recognition Using Global Semantic Information

Ajoy Mondal$^{(\boxtimes)}$ and C. V. Jawahar

International Institute of Information Technology, Hyderabad, India
{ajoy.mondal,jawahar}@iiit.ac.in

Abstract. Handwritten Text Recognition (HTR) is more interesting and challenging than printed text due to uneven variations in the handwriting style of the writers, content, and time. HTR becomes more challenging for the Indic languages because of (i) multiple characters combined to form conjuncts which increase the number of characters of respective languages, and (ii) near to 100 unique basic Unicode characters in each Indic script. Recently, many recognition methods based on the encoder-decoder framework have been proposed to handle such problems. They still face many challenges, such as image blur and incomplete characters due to varying writing styles and ink density. We argue that most encoder-decoder methods are based on local visual features without explicit global semantic information.

In this work, we enhance the performance of Indic handwritten text recognizers using global semantic information. We use a semantic module in an encoder-decoder framework for extracting global semantic information to recognize the Indic handwritten texts. The semantic information is used in both the encoder for supervision and the decoder for initialization. The semantic information is predicted from the word embedding of a pre-trained language model. Extensive experiments demonstrate that the proposed framework achieves state-of-the-art results on handwritten texts of ten Indic languages.

Keywords: Indic handwritten text · Encoder-decoder · Global semantic information · Word embedding · Language model · Indic language

1 Introduction

Optical Character Recognition (OCR) is the electronic or mechanical conversion of printed or handwritten document images into a machine-readable form. OCR is an essential component in the workflow of document image analytics. Typically, an OCR system includes two main modules (i) a text detection module and (ii) a text recognition module. A text detection module aims to localize all text blocks within the image, either at the word or line levels. The text recognition module

© The Author(s), under exclusive license to Springer Nature Switzerland AG 2022
U. Porwal et al. (Eds.): ICFHR 2022, LNCS 13639, pp. 360–374, 2022.
https://doi.org/10.1007/978-3-031-21648-0_25

aims to understand the text image content and transcribe the visual signals into natural language tokens. The problem of handwritten text recognition is more interesting and challenging than printed text due to the presence of uneven variations in handwriting style to the writers, content, and time. The handwriting of a person is always unique, and this uniqueness property creates motivation and interest among the researchers to work in this exigent and challenging field.

Among various languages around the world, many of them have disappeared as their usage is limited due to their presence in rural or geographically inaccessible parts of the globe. At this point, it is highly recommended to use technologies like OCR and natural language processing to stop the extermination of languages in the world. There are almost 7000 languages in the world[1]. However, handwritten OCR systems/tools are available only for a few. OCR systems are mostly available for languages that are of huge importance and strong economic value, like English [16,23,29], Chinese [28,37,38], Arabic [18,25], and Japanese [24,26]. Most of the languages derived from Indic script appear to be at the risk of vanishing due to the absence of efforts. So, there is an immense need for character recognition-related research for Indic scripts/languages.

Officially, there are 22 languages in India, many of which are used only for communication purposes. Among these languages, Hindi, Bengali, and Telugu are the top three languages in terms of the percentage of native speakers [21]. In many Indic scripts, two or more characters are often combined to form conjuncts which considerably increase the number of characters/symbols to be tackled by OCR systems [1]. These inherent features of Indic scripts make Handwritten Recognizer (HWR) more challenging as compared to Latin scripts. Compared to the 52 unique (upper case and lower case) characters in English, most Indic scripts have over 100 unique basic Unicode characters [27].

The problem of Indic handwritten text recognition, more generally text recognition, is formulated as a seq-2-seq prediction task where both the input and output are treated as a sequence of vectors. It aims to maximize the probability of predicting the output label sequence given the input feature sequence [2,10,11]. The encoder-decoder (e.g., CNN-RNN) framework is very popular to handwritten text recognition tasks [10,11,14]. The encoder extracts rich features and generates a context vector containing global information of the input text image. While the decoder converts the context vector to the target text.

Despite great effectiveness, the encoder-decoder framework has limited ability to generate context information that represents the whole input image [19]. Inspired by human visual attention, the researchers introduce the attention mechanism into the encoder-decoder framework, referred to as the attention-based encoder-decoder framework. Several works [19,34,35] exist in this direction. The attention mechanism helps the decoder to select the appropriate context at each decoding step resulting in accurate text recognition. It can resolve long-range dependency problems and the alignment between the encoder and decoder module. Kass and Vats [19] show that the attention mechanism improves accuracy more than a simple encode-decode framework for recognizing handwrit-

---

[1] https://www.ethnologue.com/guides/how-many-languages.

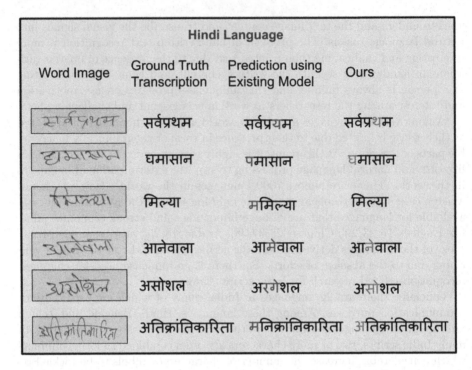

Fig. 1. Shows a comparison of our method with the existing encoder-decoder framework [14]. The first column shows examples of some challenging Hindi word images. The second column is ground truth transcription. The third column is the results of method [14]. The fourth column gives the predictions of our approach. The Red colored characters are wrongly recognized by method [14]. The Green colored characters are correctly recognized by the proposed method. (Color figure online)

ten English text. This type of framework works accurately for most of the scenarios except for low-quality images. In the case of handwritten text, due to varying handwriting styles and the density of the ink, text may be distorted, blurred, and have incomplete characters. Global semantic information is an alternative feature to handle these problems.

In this work, we enhance the performance of Indic handwritten text recognizers using global semantic information. Inspired by the work [30], we propose an Indic handwritten text recognizer based on an attention-based encoder-decoder framework with an additional semantic information module to predict global information. The semantic information is used to initialize the decoder and it has two main advantages (i) it can be supervised by a word embedding in the natural language processing field and (ii) it can reduce the gap between the encoder focusing on the visual feature and the decoder focusing on the language information. The training goal is to reduce the difference between the predicted semantic information and word embedding from a pre-trained language model. This way, the semantic module predicts richer semantic information, guides the decoder during the decoding process, and improves decoder performance. Some examples are shown in Fig. 1. However, our framework can correct it with the global semantic information. In other words, semantic information works as an "intuition", like a glimpse before people read a word carefully.

The main contributions of this work are as follows

- We propose an attention-based encoder-decoder framework with a semantic module to recognize Indic handwritten text. The semantic module predicts global semantic information, which guides the decoder to recognize text accurately. FastText, the pre-trained language model, supervises the predicted semantic information.
- Extensive experiments on public benchmark Indic handwritten datasets demonstrate that the proposed framework obtains state-of-the-art performance.

2 Related Work

There are three popular ways of building handwritten word recognizers for Indic scripts in the literature. The first one is to use segmentation-free, but lexicon-dependent methods train on representing the whole word [20,32,33]. In [33], Shaw et al. represent word images using a histogram of chain-code directions in the image strips scanning from left to right by a sliding window as the feature vector. A continuous density Hidden Markov Model (HMM) is proposed to recognize handwritten Devanagari words. Shaw et al. [32] discuss a novel combination of two different feature vectors for holistic recognition of offline handwritten word images in the same direction.

Another approach is based on segmentation of the characters within the word image and recognition of isolated characters using an isolated symbol classifier such as Support Vector Machine (SVM) [4], Artificial Neural Network (ANN) [3, 22]. In [31], Roy et al. segment Bengali and Devanagari word images into the upper, middle, and lower zones, using morphology and shape matching. The symbols present in the upper and lower zone are recognized using an SVM, while a Hidden Markov Model (HMM) was used to recognize the characters in the middle zone. Finally, the results from all three zones are combined. This category of approaches suffers from the drawback that we have to use a script-dependent character segmentation algorithm.

The third category of approaches treats word recognition as a seq-2-seq prediction problem where both the input and output are treated as a sequence of vectors. The aim is to maximize the probability of predicting the output label sequence given the input feature sequence [2,10,11]. Garain et al. [13] proposed a recognizer using Bidirectional Long short-term memory (BLSTM) with Connectionist Temporal Classification (CTC) layer to recognize unconstrained Bengali offline handwriting words. Adak et al. [2] used Convolutional Neural Network (CNN) integrated with an LSTM along with a CTC layer to recognize Bengali handwritten words. In the same direction, Dutta et al. proposed CNN-RNN hybrid end-to-end model to recognize Devanagari, Bengali [11], and Telugu [10] handwritten words. In this work [14], the authors use Spatial Transformer Network along with hybrid CNN-RNN with CTC layer to recognize word images in eight different Indic scripts such as Bengali, Gurumukhi, Gujarati, Odia, Kannada, Malayalam, Tamil, and Urdu. The authors use various data augmentation functions to improve recognition accuracy. This category of methods does not require character-level segmentation and is not bounded for recognizing a limited set of words.

3   Proposed Method

The proposed method for Indic handwritten text recognition tasks is discussed in detail. Figure 2 shows the proposed Indic handwritten text recognizer, which consists of four components: (i) the rectification module to rectify the irregular word image, (ii) the encoder to extract rich visual features, (iii) the semantic module to predict semantic information from the visual feature, and (iv) the decoder transcribes the final recognized text.

Most of the text recognizers are built on the encoder-decoder architecture with attention. The decoder focuses on specific regions of visual features extracted by the encoder and recognizes the corresponding characters step by step. This type of framework works accurately for most of the scenarios except for low-quality images. In the case of handwritten text, due to varying handwriting styles and the density of the ink, text may be distorted and blurred. Global semantic information is an alternative feature to handle these problems. As shown in Fig. 2, the semantic module extracts semantic information, which helps the decoder to predict accurate characters. The use of word embedding from a pre-trained language model acts as a supervisor to extract semantic information and help the decoder to improve its performance. In the following subsections, we discuss each module in detail.

3.1   Rectification Module

Generally, the diverse handwriting style present in handwritten data makes handwritten text recognizers more challenging. The rectification module learns to apply input-specific geometric transformations to rectify the image. By rectifying the input image, this module simplifies the recognition tasks. Normally Spatial Transformer Network (STN) [17] and its variants are used to rectify the input text image. In this work, the rectification module is based on STN. The STN models spatial transformation as the learnable parameters for the given input text image. The network first predicts a set of control points via its localization network. Then Thin-plate Spline (TPS) [7] transformation is calculated from the control points and passed to the grid generator and the sampler to generate the rectified image. Since the control points are predicted from the input text image, the rectification network takes no extra inputs other than the input image.

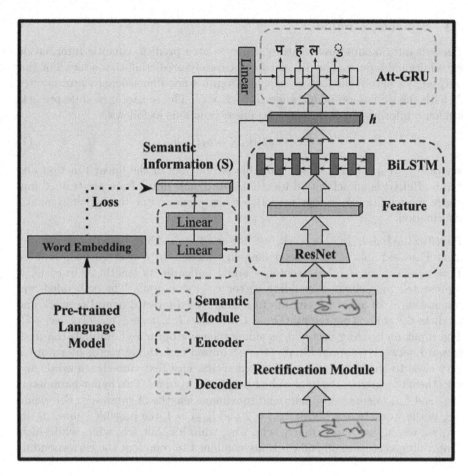

Fig. 2. Presents detail of the proposed Indic handwritten text recognizer. It consists of four main components: rectification module, encoder, semantic module, and decoder.

3.2 Encoder Module

The rectified image is forwarded to the encoder module consisting of a 45-layer ResNet based CNN similar to [34] followed by a 2-layer Bidirectional LSTM (BiLSTM) [15]. The output of the encoder module is a feature sequence $h = (h_1, ..., h_L)$ with the shape of $L \times C$, where L is the width of the last feature map in CNN and C is the depth.

3.3 Semantic Module

The semantic module uses feature sequence h to predict semantic information for additional input to the decoder for accurately predicting characters. For this purpose, we flatten the feature sequence h into a one-dimensional feature vector X with the dimension of K, where $K = L \times C$. The semantic module predicts semantic information S through two linear functions as follows

$$S = W_2\sigma(W_1 X + b_1) + b_2. \tag{1}$$

where W_1, W_2, b_1, and b_2 are trainable parameters of the linear function and σ (i.e., ReLU) is an activation function. Word embedding from pre-trained language model (e.g., FastText model) acts as a supervisor to predict semantic information.

FastText Model: In this work, we use word embedding based on skip-gram from FastText [6], a pre-trained language model. In a text corpus, suppose $T = w_i - 1, ..., w_i + 1$ be a sentence and l indicates its length. A word w_i is represented by a single embedding vector v_i in skip-gram. The embedded representation v_i of a word w_i inputs to a simple feed-forward neural network and predicts the context representation as $C_i = \{w_i - l, ..., w_i - 1, w_i + 1, ..., w_i + l\}$. The input embedding vector v_i is simultaneously optimized through the feed-forward network training. Finally, the optimized embedding vector of a word is very close to the words with similar semantics. FastText embeds subwords and uses them to generate the final embedding of the word w_i. Two hyper-parameters l_{min} and l_{max} denote minimum and maximum lengths of subwords. For example, in the word "which" with $l_{min} = 2$ and $l_{max} = 4$, the possible subwords are wh, wi, wc, hi, hc, hh, ic, ih, ch, whi, whc, whh, hic, hih, ich, whic, whih, hich. Embedding vectors of all subwords are combined to represent the corresponding word.

3.4 Decoder Module

The decoder consists of a single layer attentional GRU [8] with 512 hidden units and 512 attention units. The decoder adopts the Bahdanau-Attention mechanism [5]. The decoder is single-directional. The semantic information S is used to initialize the state of GRU. The decoder uses both local visual information h and global semantic information S to generate more accurate results.

3.5 Loss Functions and Training Strategy

The proposed model is trained in an end-to-end manner. We add supervision in both the semantic module and the decoder module. The loss function is defined as

$$L = L_r + \lambda L_e, \tag{2}$$

where L_r is the standard cross-entropy loss of the predicted probabilities with respective ground truths and L_e is the cosine embedding loss of the predicted semantic information S to the word embedding of the transcription label from the pre-trained FastText model. λ is a balancing parameter, and in this work, we set it as 1. Cosine embedding loss L_e is defined as $L_e = 1 - cos(S, E)$, where S is predicted semantic information through a semantic module, and E is the word embedding from the pre-trained FastText model.

3.6 Implementation Details

The proposed Indic handwritten text recognizer is implemented in PyTorch (base code is available in[2]). We use pre-trained FastText model[3] trained on common Crawl[4] and Wikipedia[5]. We use 45-layer ResNet architecture similar to [34] and 2-layer Bidirectional LSTM with 256 hidden units. We resize the input text image into 64×256 without keeping the ratio. We train the model for 50 epochs and use Adadelta [15] to minimize loss functions. We set the batch size to 64 and the learning rate to 1.0.

For inference, we resize the input images to the same size, similar to the training process. We use beam search for GRU decoding and set beam width to 5 for all experiments.

Data Augmentation: The works [9,10,14,36] highlight that the data augmentation strategies improve the handwritten text recognizer performance on Latin and Indic scripts. It enables the network to learn invariant features for a given task and prevents over-fitting. Similar to the work [14], we apply affine and elastic transformations on input text images to imitate the natural distortions and variations presented in handwriting text. We also use brightness and contrast augmentation on input text images to learn invariant features for text and background.

4 Experiments

4.1 Datasets

We used publicly available benchmark Indic handwriting dataset: IIIT-INDIC-HW-WORDS [10,12,14] for this experiment. Table 1 shows details of the used datasets containing word images of ten different languages.

---

[2] https://github.com/Pay20Y/SEED.
[3] https://fasttext.cc/docs/en/crawl-vectors.html.
[4] https://commoncrawl.org/.
[5] https://www.wikipedia.org.

Table 1. Shows the statistics of used datasets for this experiments. #: indicates number.

Script	#Writer	#Word instance	#Lexicon
Bengali	24	113K	11,295
Gujarati	17	116K	10,963
Gurumukhi	22	112K	11,093
Devanagari	12	95K	11,030
Kannada	11	103K	11,766
Odia	10	101K	13,314
Malayalam	27	116K	13,401
Tamil	16	103K	13,292
Telugu	11	120K	12,945
Urdu	8	100K	11,936

4.2 Evaluation Metric

Two popular evaluation metrics such as Character Recognition Rate (CRR) (alternatively Character Error Rate, CER) and Word Recognition Rate (WRR) (alternatively Word Error Rate, WER) are used to evaluate the performance of recognizers. Error Rate (ER) is defined as

$$ER = \frac{S + D + I}{N}, \tag{3}$$

where S indicates number of substitutions, D indicates the number of deletions, I indicates the number of insertions and N the number of instances in reference text. In case of CER, Eq. (3) operates on character levels, and in case of WER, Eq. (3) operates on word levels. Recognition Rate (RR) is defined as

$$RR = 1 - ER. \tag{4}$$

In the case of CRR, Eq. (4) operates on character levels and in the case of WRR, Eq. (4) operates on word levels.

4.3   Results

Ablation Study. We perform three different experiments for our ablation study such as (i) only encoder-coder framework, (ii) encoder-decoder with attention module, and finally (iii) encoder-decoder with attention and semantic modules. There are two steps to the semantic module in the proposed method such as (i) word embedding supervision and (ii) initialization of decoder with predicted semantic information. We evaluate these two steps separately by using IIIT-INDIC-HW-WORDS [14] as a training dataset. Table 2 shows the results of the handwritten word dataset of the Bengali script.

Table 2. Show the performance comparison between different strategies. Attn. represents attention module in the decoder. WES indicates word embedding supervision. INIT represents initializing the state of the GRU in the decoder. ↓ indicates the lower value corresponds to better performance.

Script	Method	Att.	WES	INIT	WER↓
Bengali	CNN-RNN	–	–	–	15.23
	CNN-RNN	✓			14.02
	CNN-RNN	✓	✓		13.48
	CNN-RNN	✓		✓	13.01
	CNN-RNN	✓	✓	✓	12.34

When attention is used in the decoder, the model reduces the WER by 1.21%. Only the model supervised with word embedding may not reduce much in WER (0.54%). Using predicted holistic features from the encoder to initialize the decoder reduces the WER by almost (1.01%). A combination of word embedding supervision and initialization of the decoder with predicted semantic information reduces most WER.

Comparison with State-of-the-Art. We compare our method with state-of-the-art methods on benchmark dataset—IIIT-INDIC-HW-WORDS [10,12,14]. The results are shown in Table 3. Our method reduces overall more than 2% average (average over ten languages) Word Error Rate (WER) as compared to the state-of-the-art. Among all languages, due to the complexity of Urdu script, the proposed method has the highest error rate (WER and CER) for Urdu languages. The proposed method achieves the minimum error rate in the Kannada language.

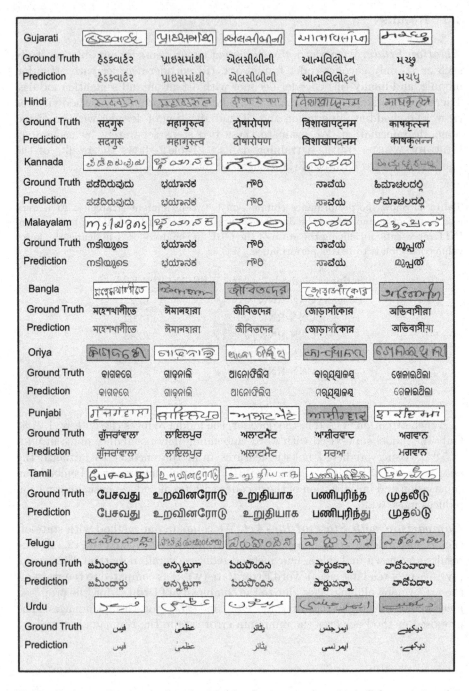

Fig. 3. Shows qualitative results obtained by the proposed method. Column 1, 2 and 3 present correctly recognized word images. Column 4 and 5 shows incorrect recognized word images. (Color figure online)

Table 3. Quantitative comparison with State-of-the-arts. ↓ indicates the lower value corresponds to better performance.

Script	Performance score			
	Method [14]		Ours	
	WER↓	CER↓	WER↓	CER↓
Bengali	14.77	4.85	12.34	2.35
Gujarati	11.39	2.39	09.21	1.19
Hindi	11.23	3.17	09.16	1.98
Kannada	10.93	1.79	08.57	1.01
Malayalam	11.34	1.92	09.37	1.12
Odia	14.97	3.00	12.32	1.32
Punjabi	12.78	3.42	10.77	2.10
Tamil	11.36	1.92	09.18	1.25
Telugu	13.98	3.18	12.11	2.15
Urdu	20.35	5.07	18.76	3.89

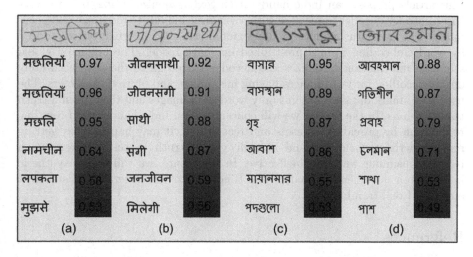

Fig. 4. Shows the visualization of cosine similarity between the predicted semantic information from the image and the word embedding of the words from lexicons. Larger value indicates more similar semantics.

Qualitative Results and Visualization. The visual results shown in Fig. 3 highlight that the proposed method obtains correct prediction for word images with incomplete characters and blurring characters due to varying ink density. We explain that semantic information will provide an effective global feature to the decoder, robust to the interference in the images. The first three columns of Fig. 3 show the correctly recognized text of ten languages. While last two columns of Fig. 3 shows wrongly recognized words by the proposed method. One or more wrongly recognized characters of the words are highlighted by red color. We perform experiments on the IIIT-INDIC-HW-WORDS dataset to visualize the validation of the predicted semantic information to the decoder for recognizing text. As presented in Fig. 4, we compute the cosine similarity between the predicted semantic information from the word image and the word embedding of each word from lexicons (50 words for each image). In Fig. 4, the predicted semantic information is related to word images which have similar semantics. Words of similar meanings have a large cosine similarity value, while words of different meanings have less cosine similarity value. For example, in the case of Fig. 4(a), three words from the top have a similar meaning, resulting in a higher cosine similarity score, While the remaining words are different meanings indicating a lower cosine similarity score. A similar observation is found for other word images shown in Fig. 4. With the help of a semantic module, the model tries to distinguish words very easily.

5   Conclusions and Future Works

This article proposes an Indic handwritten text recognizer using an attention-based encoder-decoder framework with a semantic module to recognize text accurately. The semantic module predicts global semantic information supervised by word embedding from a pre-trained language model. The predicted global semantic information initializes the decoder to recognize the text accurately during decoding for word images having incomplete and blurred characters. The used benchmark dataset contains only word-level images and their ground truth transcriptions. In the future, we will concentrate on Indic handwritten text line recognition by providing datasets and recognizers. It may happen that getting real handwritten documents and manually ground truth transcription generation is time-consuming and cost-ineffective. In the future, we will create synthetic handwritten documents to reduce the time and cost of generating a sufficient amount of real training data.

References

1. Script Grammar. For Indian languages. http://language.worldofcomputing.net/grammar/script-grammar.html. Accessed 26 Mar 2020
2. Adak, C., Chaudhuri, B.B., Blumenstein, M.: Offline cursive Bengali word recognition using CNNs with a recurrent model. In: ICFHR, pp. 429–434 (2016)
3. Alonso-Weber, J.M., Sesmero, M., Sanchis, A.: Combining additive input noise annealing and pattern transformations for improved handwritten character recognition. Expert Syst. Appl. **41**(18), 8180–8188 (2014)
4. Arora, S., Bhattacharjee, D., Nasipuri, M., Malik, L., Kundu, M., Basu, D.K.: Performance comparison of SVM and ANN for handwritten Devnagari character recognition. arXiv (2010)
5. Bahdanau, D., Cho, K., Bengio, Y.: Neural machine translation by jointly learning to align and translate. arXiv (2014)

6. Bojanowski, P., Grave, E., Joulin, A., Mikolov, T.: Enriching word vectors with subword information. Trans. Assoc. Comput. Linguist. **5**, 135–146 (2017)
7. Bookstein, F.L.: Principal warps: thin-plate splines and the decomposition of deformations. IEEE Trans. Pattern Anal. Mach. Intell. **11**(6), 567–585 (1989)
8. Cho, K., et al.: Learning phrase representations using RNN encoder-decoder for statistical machine translation. arXiv (2014)
9. Dutta, K., Krishnan, P., Mathew, M., Jawahar, C.V.: Improving CNN-RNN hybrid networks for handwriting recognition. In: ICFHR, pp. 80–85 (2018)
10. Dutta, K., Krishnan, P., Mathew, M., Jawahar, C.V.: Towards spotting and recognition of handwritten words in Indic scripts. In: ICFHR, pp. 32–37 (2018)
11. Dutta, K., Krishnan, P., Mathew, M., Jawahar, C.V.: Towards accurate handwritten word recognition for Hindi and Bangla. In: Rameshan, R., Arora, C., Dutta Roy, S. (eds.) NCVPRIPG 2017. CCIS, vol. 841, pp. 470–480. Springer, Singapore (2018). https://doi.org/10.1007/978-981-13-0020-2_41
12. Dutta, K., Krishnan, P., Mathew, M., Jawahar, C.: Offline handwriting recognition on Devanagari using a new benchmark dataset. In: DAS, pp. 25–30 (2018)
13. Garain, U., Mioulet, L., Chaudhuri, B.B., Chatelain, C., Paquet, T.: Unconstrained Bengali handwriting recognition with recurrent models. In: ICDAR (2015)
14. Gongidi, S., Jawahar, C.V.: IIIT-INDIC-HW-WORDS: a dataset for Indic handwritten text recognition. In: Lladós, J., Lopresti, D., Uchida, S. (eds.) ICDAR 2021. LNCS, vol. 12824, pp. 444–459. Springer, Cham (2021). https://doi.org/10.1007/978-3-030-86337-1_30
15. Graves, A., Liwicki, M., Fernández, S., Bertolami, R., Bunke, H., Schmidhuber, J.: A novel connectionist system for unconstrained handwriting recognition. IEEE Trans. Pattern Anal. Mach. Intell. **31**(5), 855–868 (2008)
16. Graves, A., Schmidhuber, J.: Offline handwriting recognition with multidimensional recurrent neural networks. In: NIPS (2008)
17. Jaderberg, M., Simonyan, K., Zisserman, A., et al.: Spatial transformer networks. In: Advances in Neural Information Processing Systems, vol. 28 (2015)
18. Jemni, S.K., Ammar, S., Kessentini, Y.: Domain and writer adaptation of offline Arabic handwriting recognition using deep neural networks. Neural Comput. Appl. **34**, 2055–2071 (2022). https://doi.org/10.1007/s00521-021-06520-7
19. Kass, D., Vats, E.: AttentionHTR: handwritten text recognition based on attention encoder-decoder networks. In: Uchida, S., Barney, E., Eglin, V. (eds.) DAS 2022. LNCS, vol. 13237, pp. 507–522. Springer, Cham (2022). https://doi.org/10.1007/978-3-031-06555-2_34
20. Kaur, H., Kumar, M.: On the recognition of offline handwritten word using holistic approach and AdaBoost methodology. Multimed. Tools Appl. **80**, 11155–11175 (2021). https://doi.org/10.1007/s11042-020-10297-7
21. Krishnan, P., Jawahar, C.V.: HWNet v2: an efficient word image representation for handwritten documents. IJDAR **22**, 387–405 (2019). https://doi.org/10.1007/s10032-019-00336-x
22. Labani, M., Moradi, P., Ahmadizar, F., Jalili, M.: A novel multivariate filter method for feature selection in text classification problems. Eng. Appl. Artif. Intell. **70**, 25–37 (2018)
23. Li, M., et al.: TrOCR: transformer-based optical character recognition with pretrained models. arXiv (2021)
24. Ly, N.T., Nguyen, C.T., Nakagawa, M.: Training an end-to-end model for offline handwritten Japanese text recognition by generated synthetic patterns. In: ICFHR (2018)

25. Maalej, R., Kherallah, M.: Improving the DBLSTM for on-line Arabic handwriting recognition. Multimed. Tools Appl. **79**, 17969–17990 (2020). https://doi.org/10.1007/s11042-020-08740-w

26. Nguyen, K.C., Nguyen, C.T., Nakagawa, M.: A semantic segmentation-based method for handwritten Japanese text recognition. In: ICFHR (2020)

27. Pal, U., Chaudhuri, B.: Indian script character recognition: a survey. Pattern Recogn. **37**(9), 1887–1899 (2004)

28. Peng, D., et al.: Recognition of handwritten Chinese text by segmentation: a segment-annotation-free approach. arXiv, 1–14 (2022)

29. Pham, V., Bluche, T., Kermorvant, C., Louradour, J.: Dropout improves recurrent neural networks for handwriting recognition. In: ICFHR (2014)

30. Qiao, Z., Zhou, Y., Yang, D., Zhou, Y., Wang, W.: SEED: semantics enhanced encoder-decoder framework for scene text recognition. In: CVPR, pp. 13528–13537 (2020)

31. Roy, P.P., Bhunia, A.K., Das, A., Dey, P., Pal, U.: HMM-based Indic handwritten word recognition using zone segmentation. Pattern Recogn. **60**, 1057–1075 (2016)

32. Shaw, B., Bhattacharya, U., Parui, S.K.: Combination of features for efficient recognition of offline handwritten Devanagari words. In: ICFHR, pp. 240–245 (2014)

33. Shaw, B., Parui, S.K., Shridhar, M.: Offline handwritten Devanagari word recognition: a holistic approach based on directional chain code feature and HMM. In: ICIT, pp. 203–208 (2008)

34. Shi, B., Yang, M., Wang, X., Lyu, P., Yao, C., Bai, X.: ASTER: An attentional scene text recognizer with flexible rectification. IEEE Trans. Pattern Anal. Mach. Intell. **41**(9), 2035–2048 (2018)

35. Wang, T., et al.: Decoupled attention network for text recognition. In: Proceedings of the AAAI Conference on Artificial Intelligence, pp. 12216–12224 (2020)

36. Wigington, C., Stewart, S., Davis, B., Barrett, B., Price, B., Cohen, S.: Data augmentation for recognition of handwritten words and lines using a CNN-LSTM network. In: ICDAR, pp. 639–645 (2017)

37. Wu, Y.C., Yin, F., Chen, Z., Liu, C.L.: Handwritten Chinese text recognition using separable multi-dimensional recurrent neural network. In: ICDAR (2017)

38. Xie, Z., Sun, Z., Jin, L., Feng, Z., Zhang, S.: Fully convolutional recurrent network for handwritten Chinese text recognition. In: ICPR (2016)

Yi Characters Online Handwriting Recognition Models Based on Recurrent Neural Network: RnnNet-Yi and ParallelRnnNet-Yi

Zhixin Yin[1,2], Shanxiong Chen[2(✉)], Dingwang Wang[2], Xihua Peng[2], and Jun Zhou[2]

[1] College of Business, Southwest University, Chongqing 402460, China
[2] College of Computer and Information Science, Southwest University, Chongqing 400715, China
csxpml@163.com

Abstract. As the sixth largest minority language in China, the Yi language is used by 8 million people and records the development of human civilization. Deep learning has been widely used and effective in mainstream characters' recognition, but there are few achievements in recognition of Yi characters, particularly online handwriting recognition. Most of the Yi strokes are curved, and the writing is irregular. Consequently, there are some problems, such as arbitrarily change for the writing order, multiple strokes concatenated or abbreviated, stroke position offset, and so on. Due to different sample collection equipment, there is also dimensional diversity and sampling frequency diversity between sample collection devices, which will bring more significant interference to the identification. In this paper, we construct a Yi online handwriting recognition database and propose two Yi online handwriting recognition models based on different usage scenarios: RnnNet-Yi (for high accuracy requirements) and ParallelRnnNet-Yi (for resource-constrained lightweight requirements), merging deep learning and feature extraction methods. The experimental results verify the effectiveness of the models proposed in this paper in upgrading the accuracy and training speed of Yi online handwriting recognition, which fills the gap in Yi online handwriting recognition research.

Keywords: Online handwriting recognition · Gated recurrent unit · Yi character · Recurrent neural network

1 Introduction

Online handwriting recognition has made breathtaking progress in widely used scripts such as Chinese, Latin and Japanese, and this technology is also applied in practice, such as touch screen handwriting input method, verification of the electronic signature, etc. In recent years, some researchers have also carried out research work for online handwriting recognition in other language scripts, such as Mongolian [7], Tibetan [25], Indian [2], Thai [23], Arabic [1,18], etc. Generally, these scripts serve as the official scripts of the country, and there are many precise requirements for text recognition scenarios, so the recognition of online handwriting has received more attention. As the script used by the Yi people, which are the sixth largest minority nationality in China, the Yi script has well-established writing norms and standards. At present, some domestic and foreign research institutes and universities are conducting research work on the recognition of offline handwriting in Yi script during the process of Yi language documentation [29,32,33], while there are fewer studies related to the recognition of online handwriting in Yi script [4]. In southwestern China, where the Yi ethnic group is concentrated, Yi script information systems and the recording of Yi script materials require online Yi script handwriting recognition technology to process handwritten data and enhance Yi language informatization.

Since Yi is a pictographic script, it is closer to Chinese than other languages such as Tibetan or Mongolian. Meanwhile, the overall structure of the standardized Yi character is more concise than that of the Chinese character, so the online handwriting recognition technology for Chinese characters can provide a reference for recognition in Yi. However, there are still many problems and differences in practical application. For example, there are no radicals and indexing components in Yi, so the first stroke is usually used as the main stroke, and other strokes are added to different parts of the primary stroke, or the font structure is shifted to make a new Yi character. Since most strokes are arcs and the writing is arbitrary, there may be subjective factors of writers such as exchanging the order of the strokes, continuous or abbreviated writing of multiple strokes, shifting of the writing position, etc. The difference in the scale and sampling frequency between sample collection devices also causes many problems. These uncertainties bring enormous interference to recognition, so finding a suitable sample processing algorithm is necessary to reduce the problem caused by delays and the differences between samples.

Based on the above problems, this paper constructs a Yi Online Handwriting Database (YI-OLHWDB) that can be used for online handwriting recognition. We present a method for analyzing and processing the structure of Yi strokes based on the uniqueness of the structural features of the Yi character based on online handwriting recognition techniques for Chinese characters. We also propose two Yi online handwriting recognition models, RnnNet-Yi and ParallelRnnNet-Yi, for the server side with sufficient computational power and the mobile side with relatively limited resources.

2   Related Works

Recently, handwriting recognition based on deep learning (CNN and RNN, etc.) has achieved practical results in various language scripts. Online handwriting recognition in mainstream scripts has been implemented commercially [16,21,30] and integrated into many application scenarios [8,24], etc. Research on online handwriting recognition of other scripts based on deep learning has also been promoted. Suhan Chowdhury et al. [5] proposed a weighted finite-state transducer (WFST) based model for the problem of character sticking and character missing in Indic Script handwriting recognition, which improves recognition accuracy by introducing language information. Further, Bharath et al. [2] used HMM-Based Lexicon-Driven for recognition to achieve better results to solve the problem of online recognition of Devanagari and Tamil in Indic Scripts. S. Bhattacharya et al. [3] proposed a hybrid architecture consisting of CNN, RNN and CTC for the problems of unconstrained Devanagari and Bangla writing styles, variable character sizes and the presence of shape-similar characters. In online Arabic handwriting recognition, Maalej.R et al. [18] proposed a new system based on DBLSTM that performs well under the conditions of variety of writers, the large vocabulary and the diversity of style. It was reported to lower the error rate by 10.99% compared to the previous best-performing Arabic research. Fakhraddin.Alwajih F et al. [1] proposed an end-to-end AOHR based on bidirectional long short-term memory and the connectionist temporal classification (BLSTM-CTC) for Arabic online handwriting recognition called DeepOnKHATT, which is capable of performing recognition at the sentence level in real-time. The article [9] groups Farsi characters according to their delayed strokes and main bodies, finally improving the recognition accuracy of multi-styled Farsi by considering the handwriting styles and extracting samples with wrongly spelled or incorrect structures for correction. Nguyen. HT et al. [19] used a BLSTM network to recognize Vietnamese online handwritten characters and achieved 90% accuracy, despite many long sequences with several delayed strokes. Kim et al. [14] presented a stacked gated recurrent unit (GRU) based model with 86.2% accuracy for recognizing 2350 Hangul characters. It was shown that the potential for RNN models to learn high-level structural information from sequential data.

China has a rich cultural history and diverse minority ethnic groups and languages. In addition to a series of achievements in Chinese character recognition, fantastic progress has been made in the study of character recognition for ethnic minorities. Mayire Ibrayim et al. [12] first applied RNN and CTC to build end-to-end online Uyghur handwriting recognition. It is remarkable that Simayi. W et al. [20] presented an Uyghur online handwritten words database which was filtered 125,020 samples of 2030 words written by 393 people. Meanwhile, the researchers proposed 1D convolutional model for sequence feature extraction and used for recognition, with the highest classification accuracy of 94.95%. Wang.W et al. [25] established an online Tibetan Handwriting Database (NMU-OLTHWDB) and used feature extraction with pen-trajectory direction and spatial combination features, then dimensionality reduction of the feature with LDA and quadratic discriminant function (QMQDF) for classification, resulting in recognition accuracy of 75.2%. Wu Wei et al. [26] implemented online Mongolian handwriting recognition based on HMM and developed a Mongolian handwriting recognition system. The system can achieve a recognition rate of 92% under the condition of constraint on the stroke order.

Compared with other script handwriting recognition research, Yi started late and only some universities for nationalities have made some achievements in offline handwriting recognition. In recent years, the introduction of deep learning into Yi handwriting recognition has improved accuracy. Jia et al. [13] designed a recognition model based on CNN, which gets an accuracy of 99.65% on the dataset containing 100 classes. Chen.SX et al. [4] also used CNN to construct five Ancient Yi character recognition models and used Alpha-Beta divergence as a penalty term to implement the coding for output neurons, improving recognition performance.

However, research on online handwriting recognition is lagging. Southwest Minzu University and the Central Nationalities Language Translation Center have developed online Yi handwriting recognition, and Sino Voice Technology Company has also launched online Yi handwriting recognition. Still, the number of recognized characters is limited and recognition accuracy is not up to the requirements of organizing and recording Yi documents.

3   Yi Online Handwriting Recognition Models

We propose two recurrent neural network recognition models, RnnNet-Yi and ParallelRnnNet-Yi, by drawing on the network framework of the paper [31] and considering the online handwritten Yi characteristics. RnnNet-Yi is mainly used for the automatic recognition of handwritten characters with high accuracy, and ParallelRnnNet-Yi is a lightweight model specifically for handwritten input scenarios with limited computational resources such as embedded devices. Since GRU has a more straightforward structure and fewer training parameters relative to LSTM, it can effectively reduce the training time [6], and the performance of GRU and LSTM is close to each other in most cases. Therefore, this paper's recurrent neural networks are constructed based on GRU.

3.1   RnnNet-Yi

To enhance the recognition effect, we construct a six-dimensional input sequence according to the coordinates of the sample in the plane. Based on the original two-dimensional coordinates, four attributes of x-axis displacement, y-axis displacement, stroke start point, and stroke end point are added, where $\Delta x_i = x_{i+1} - x_i$ and $\Delta y_i = y_{i+1} - y_i$, when the stroke start point or end point is 1, representing it as the corresponding flag. Thus, each item of the sequence input to the recurrent neural network is a six-dimensional vector:

$$[x_i, y_i, \Delta x_i, \Delta y_i, S_i, E_i] \tag{1}$$

The network structure of RnnNet-Yi is shown in Fig. 1. The model inputs the pre-processed stroke data into the network for feature extraction, in the first GRU layer, the input at each time step is the information of the handwritten sample and the features in the hidden state of the previous time step; the input of each time step of the other GRU layers contains two parts: the hidden state of the same time step of the previous GRU layer and the features in the hidden state of the previous time step of the current layer. Then the feature matrix output from the final layer GRU is expanded into a one-dimensional vector by the Flatten Layer, and then a Fully Connected Layer is used for decision output to obtain a probability vector of 1165-dimensions, and the character with the highest probability value after SoftMax classification is the final network recognition result.

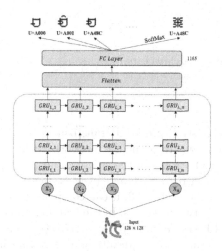

Fig. 1. RnnNet-Yi architecture.

3.2 ParallelRnnNet-Yi

Because the model of RnnNet-Yi is computationally intensive, it has limitations on the hardware resources of the device in practical use. Moreover, compared with other scripts, the number of strokes of a single character in Yi is smaller and more stable. Considering input scenarios with limited computational resources, we propose a more miniature model based on RnnNet-Yi: ParallelRnnNet-Yi.

ParallelRnnNet-Yi combines the ideas of recurrent neural networks and traditional stroke-based feature extraction, which centers on a series of small-scale recurrent neural networks to recognize each sequence of strokes in a sample. In this network, all strokes are separated and then recognized individually using parallel sub-networks. Since the sequence recognized by each sub-network must be a complete stroke, there is no need to use additional information to mark the beginning and end of the stroke, so the sequence input to the network can be streamlined to a four-dimensional vector:

$$[x_i, y_i, \Delta x_i, \Delta y_i] \tag{2}$$

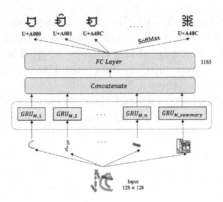

Fig. 2. ParallelRnnNet-Yi architecture. GRU_{M_s} uses the structural information of strokes as data augmentation, which is applied to assist in model decision making. (Color figure online)

Figure 2 shows the structure of ParallelRnnNet-Yi, $GRU_{M_1} \sim GRU_{M_n}$ denotes n independent recurrent neural networks, where n is determined by the maximum number of strokes in the sample, and $GRU_{M_{summary}}$ is the recurrent neural network used to receive and identify the information of the stroke structure extracted by the Sect. 4.2.2, it can compensate for some of the spatial-temporal correlations between strokes. In the training process, each stroke in the character sample will be input to $GRU_{M_1} \sim GRU_{M_n}$ separately for recognition and output feature vector, and the results of all strokes will be output through the Concatenate and Fully Connected Layers.

4   Database: YI-OLHWDB

To address the problem of no publicly available Yi online handwriting dataset, we construct a standard Yi online handwriting database: YI-OLHWDB, and proposed a pre-processing method applicable to Yi character samples to improve the accuracy of subsequent model training.

4.1   Database Collection

We collected data from mobile devices such as phone and iPad, and rotated and panned the samples to achieve sample expansion [11]. The final sample database was constructed with 442,768 samples in 1165 categories, and the coordinates of the samples were normalized to the standard interval of [0, 128].

Based on the feature of arc in Yi character, we remove the redundant points according to the paper [31], keep only the critical points of the beginning, turning point and end of the strokes, thus preserving the overall skeleton of the strokes while reducing the spatial and temporal overhead. In addition, all strokes in the sample are changed in order by the number of points after redundancy removal to achieve a normalized preprocessing of the sample.

4.2 Data Augmentation

4.2.1 DropStroke and MergeStroke Stroke omission is expected in the writing process. Inspired by the online handwritten Chinese character data augmentation method DropStroke [17], this paper is applied to Yi character samples. Assuming that the number of strokes of Yi characters is n, m of them can be randomly dropped, so the number of new samples generated can be calculated as 2^{m-1} according to the Combination Number Formula. It must be noted that in order to ensure that the samples retain the overall skeleton after the DropStroke, we restrict the number of strokes for the operation to follow $m \le n-2$, implying the samples after DropStroke must retain at least two strokes. The results of some samples after the complete DropStroke process are shown in Fig. 3(a).

We propose MergeStroke method to mimic cursive handwriting in Yi, where each stroke in the original sample has the probability of being concatenated with its subsequent stroke to produce a new sample. The specific process is shown in Algorithm 1, and Fig. 3(b) shows some of the samples after the MergeStroke process.

Algorithm 1. MergeStroke

Input: Original Sample,S; Probability of Merge,m
Output: New Samples After Generated,$\grave{S}$
Symbol Definitions:
  $Stroke_i$: Current Strokes; $Stroke_k$: Previous Strokes
  $\leftarrow$: Assignment
  $<=$: Add the Right Element to the Left Collection
1: $\grave{S} \leftarrow \{\}$
2: **for** each *stroke* in S **do**
3:   **if** $random() < m$ **then**
4:     $Stroke_k <= Stroke_i$
5:   **else**
6:     $\grave{S} <= Stroke_i$
7:   **end if**
8: **end for**
9: **return** $\grave{S}$

4.2.2 Stroke Structure Information The structural information can represent the detail of the location, size, writing direction and length of the strokes in the samples, and the samples of the same character formed after pre-processing possess similar sequences among them. In this paper, we combine traditional structure-based and neural network-based methods, used the structural information and the potential spatio-temporal correlation between strokes as data augmentation to assist the model in making the final decision.

The stroke summary method used is as follows: (x_i, y_i) denotes the i-th point in the stroke, Δx_i and Δy_i denote the displacement of point i relative to point

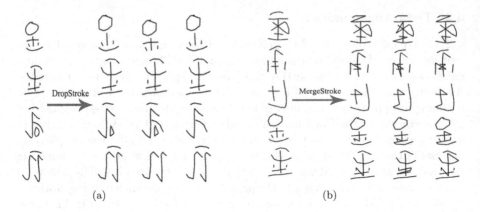

(a)   (b)

Fig. 3. Process of DropStroke and MergeStroke.

$i - 1$ in the x-axis and y-axis direction, x_{max}, x_{min}, y_{max} and y_{min} denote the maximum and minimum values of the points in the sample in the x-axis and y-axis direction, and let m denote the number of points in the stroke, then the stroke structure information shown in Eq. (3) can be obtained.

$$(x_{min}, y_{min}, x_{max} - x_{min}, y_{max} - y_{min}, x_m - x_1, y_m - y_1) \tag{3}$$

In the equation, $x_m - x_1$ and $y_m - y_1$ represent a simplification of the cumulative movements of strokes in the x-axis and y-axis directions ($\sum_{i=1}^{m} x_i - x_{i-1}$ and $\sum_{i=1}^{m} y_i - y_{i-1}$), which can be combined with the horizontal and vertical information in $x_{max} - x_{min}$ and $y_{max} - y_{min}$ to better summarize the stroke's information and draw a distinction between different strokes.

We apply the data augmentation methods such as DropStroke, MergeStroke and stroke structure information to the database, so that the number of sample points is reduced to no more than 50 points (see Fig. 4). At the same time, the number of sample strokes is at most 13, and the order of writing strokes is sorted by decreasing the number of strokes. These measures help the subsequent construction of our model.

5   Experiments

In this paper, we conduct experiments on the YI-OLHWDB database and perform several comparative experiments based on RnnNet-Yi and ParallelRnnNet-Yi models.

5.1   Implementation Details

The optimization algorithm of the network is Adam, with a training epoch of 4. The mini-batch size of each batch is 256 and the learning rate is 0.001. The

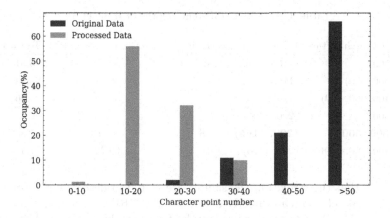

Fig. 4. Changes in the number of database characters.

Glorot uniform distribution is used to initialize the weight matrix in the neural network, the bias vector is initialized to 0, and the random orthogonal matrix is used to initialize the recurrent layer states of the GRU. 2/3 of the samples in the database are used for training and 1/3 of the samples are used for validation. All the models were implemented under the Keras platform on our workstation with 3.6 GHz CPU, RTX 2080 Super 8G GPU and Ubuntu 64-bit OS.

5.2 Experiments Result

5.2.1 Depths of RnnNet-Yi RnnNet-Yi is built as a deep recurrent network, we experiment with GRU of different depths, including 1–3 GRU layers and 1–2 Fully Connected layers, and the recognition performance of neural network is shown in Table 1. It can be seen that two Fully Connected layers $(ID_4 \sim ID_6)$ converge faster and with higher accuracy than one Fully Connected layer $(ID_1 \sim ID_3)$. At the same time, the model's training time increases with the network's depth, while the recognition accuracy on the test set is essentially the same for 2-layers GRU (ID_5) and 3-layers GRU (ID_6). Therefore, the final RnnNet-Yi uses 2 layers of GRU and 2 layers of Fully Connected layer (ID_5).

Before feeding into the model for training and recognition, the samples were pre-processed and the same samples written by different writers had similar normalized sequences, so RnnNet-Yi chooses deep recurrent networks instead of using bidirectional recurrent neural networks.

5.2.2 Dimensions of ParallelRnnNet-Yi The number of strokes in all samples didn't exceed 13, so there are 13 neural networks from M_1 to M_{13} for recognizing strokes and one network M_s for recognizing structure information in ParallelRnnNet-Yi. Since ParallelRnnNet-Yi is a series of parallel GRU networks working independently, we need to determine all networks' sequence lengths and output dimensions. The sequence length corresponds to the length of individual strokes of Yi characters. It is worth bearing in mind that since normalization

Table 1. Results of different Depths of RnnNet-Yi.

ID	GRU architecture	FC architecture	Train time (min)	Train acc. (%)	Test acc. (%)
1	(500)	(1165)	42	99.13	96.51
2	(300, 500)	(1165)	44	98.71	96.94
3	(100, 300, 500)	(1165)	52	98.88	96.96
4	(500)	(200, 1165)	35	99.65	96.85
5	**(300, 500)**	**(200, 1165)**	**38**	**99.43**	**97.41**
6	(100, 300, 500)	(200, 1165)	41	99.72	97.48

is used so that the strokes in the sample are sorted in descending order by the number of points, we can directly determine the GRU network that recognizes the strokes based on the maximum length of each stroke sequence. The detailed specification of the length of each sequence is shown in Table 2.

Table 2. The length sequence of ParallelRnnNet-Yi.

	M1	M2	M3	M4	M5	M6	M7	M8	M9	M10	M11	M12	M13	Ms
Length	22	12	9	6	5	4	3	2	2	1	1	1	1	13

When the recurrent neural network is applied to the classification task, the output dimension represents the number of categories that the network can recognize. The more categories that need to be classified, the larger the output dimension of the network is, and the more parameters need to be set. Therefore, the output dimension of sub-networks M_1 to M_{13} in ParallelRnnNet-Yi will directly influence the final recognition results. We experiment with several different specifications of dimension, and several representative specifications are listed in Table 3. These models have the same network skeleton, and the different scales are reflected in the output dimension of $M_1 \sim M_{13}$.

Table 3. The output dimension of ParallelRnnNet-Yi.

ID	M1	M2	M3	M4	M5	M6	M7	M8	M9	M10	M11	M12	M13
1	10	10	10	10	10	10	10	10	10	10	10	10	10
2	32	32	32	32	32	32	32	32	32	32	32	32	32
3	64	64	64	64	64	64	64	64	64	64	64	64	64
4	64	32	32	16	16	8	8	4	4	2	2	2	2
5	128	128	128	128	128	128	128	128	128	128	128	128	128

The experimental results obtained after training the models with different output dimensions are shown in Table 4, and we find that the price of the model accuracy improvement is the increase in training time when the output dimension increases. It should be noted that ParallelRnnNet-Yi does not have a high test accuary compared to RnnNet-Yi, so we list the Top5 accuary in the table to reflect the usability of the model, considering that there are usually multiple candidate characters for users to choose from in realistic input scenarios. Comparing the models from ID_3 to ID_5, we find that setting the output dimension according to the sequence length of GRU can appropriately reduce the output dimension of the model, which has almost no effect on the final recognition accuracy and also reduces the training and prediction time. Based on these results, the model finally selects ID_4 as the final structure of the network.

Table 4. Results of different dimensions of ParallelRnnNet-Yi.

ID	Train time (min)	Train acc. (%)	Test acc. (%)	Test top5 acc. (%)
1	2	95.14	82.59	89.56
2	8	97.22	85.64	95.85
3	13	98.83	88.60	97.34
4	5	**98.37**	**89.59**	**97.37**
5	150	98.65	89.66	97.35

5.3 Impact of Data Augmentation

Since the stroke structure information is already used in ParallelRnnNet-Yi and not necessary in RnnNet-Yi, we only discuss the effect of MergeStroke and DropStroke on the model training results. According to Table 5, both methods improve the performance of RnnNet-Yi and ParallelRnnNet-Yi models, and results are better when they are used together. It can be seen that data augmentation plays a role in reducing model overfitting.

Table 5. Effects of MergeStroke and DropStroke on models.

MODEL	Neither used (%)	MergeStroke (%)	DropStroke (%)	Both used (%)
RnnNet-Yi	97.41	98.48	98.46	98.63
ParallelRnnNet-Yi	89.59	90.83	90.95	91.68

5.4 Comparison with Other Approaches

To show the performance of the models in this paper, Table 6 lists the training results of some popular online Chinese online handwriting models on YI-OLHWDB, these models are all built based on convolutional neural networks and have won prizes in competitions such as ICDAR or published in high-level journals. These results indicate that RnnNet-Yi surpassed previous advanced methods and help us improve the model accuracy by exploiting the information of spatial order in sequence data and outperforming other models in terms of train time. While ParallelRnnNet-Yi has the most excellent training time, although it loses some performance, it is a model suitable for Yi characters and other languages with a small number of strokes by introducing feature extraction into deep learning and training each stroke individually.

The RnnNet-Yi model pays a longer time cost but has higher recognition accuracy than the ParallelRnnNet-Yi model, which is more suitable for use in scenarios requiring higher accuracy such as industry or teaching. While ParallelRnnNet-Yi, as a lightweight model, although the accuracy on the test set is less satisfactory, the Top5 accuracy is sufficient for use in realistic scenarios with rapid model training and lower computational resource requirements, which is effective for applying the model to phones and embedded devices.

Table 6. Performance results for all models on the YI-OLHWDB.

Method	Ref	Accuracy (%)	Train time (min)
Traditional	[22, 28]	96.01	45
Multilayer perceptron	[28]	96.55	276
Path Signature feature + CNN	[10, 28]	96.83	5865
DropDistortion + Domain knowledge + CNN	[15]	97.74	2254
DropWeight+GlobalPooling + CNN	[27]	97.28	413
RnnNet-Yi	Ours	97.41	38
ParallelRnnNet-Yi	Ours	89.59	5
RnnNet-Yi + Data augmentation	Ours	98.63	40
ParallelRnnNet-Yi + Data augmentation	Ours	91.68	6

6   Conclusion

In this paper, we have constructed a canonical database named YI-OLHWDB for the problem of online handwriting recognition of Yi characters, and have proposed two RNN-based models for different application scenarios. The two models achieve 98.63% and 91.68% accuracy on the constructed database. This work fills the gap in online handwriting recognition in Yi scripts and accelerates the progress of digital informatization of the Yi language.

Nevertheless, the two proposed models still have significant room for improvement in terms of model size and recognition accuracy compared to handwriting recognition in other languages. The models can only perform recognition of a single Yi character, and the problem of recognizing Yi handwriting in consecutive multiple lines has not been explored. In future work, we will try to use Transformer and Attention Mechanism to solve the above problems.

References

1. Alwajih, F., Badr, E., Abdou, S., Fahmy, A.: DeepOnKHATT: an end-to-end Arabic online handwriting recognition system. Int. J. Pattern Recogn. Artif. Intell. **35**(11), 2153006 (2021)
2. Bharath, A., Madhvanath, S.: HMM-based lexicon-driven and lexicon-free word recognition for online handwritten Indic scripts. IEEE Trans. Pattern Anal. Mach. Intell. **34**(4), 670–682 (2011)
3. Bhattacharya, S., Maitra, D.S., Bhattacharya, U., Parui, S.K.: An end-to-end system for Bangla online handwriting recognition. In: 2016 15th International Conference on Frontiers in Handwriting Recognition (ICFHR), pp. 373–378. IEEE (2016)
4. Chen, S., Wang, X., Han, X., Liu, Y., Wang, M.: A recognition method of ancient Yi character based on deep learning. J. Zhejiang Univ. (Sci. Ed.) **46**(3), 261–269 (2019)
5. Chowdhury, S., Garain, U., Chattopadhyay, T.: A weighted finite-state transducer (WFST)-based language model for online Indic script handwriting recognition. In: 2011 International Conference on Document Analysis and Recognition, pp. 599–602. IEEE (2011)
6. Chung, J., Gulcehre, C., Cho, K., Bengio, Y.: Empirical evaluation of gated recurrent neural networks on sequence modeling. arXiv preprint arXiv:1412.3555 (2014)
7. Cui, S., Su, Y., Ji, Y., et al.: An end-to-end network for irregular printed Mongolian recognition. Int. J. Doc. Anal. Recogn. (IJDAR) **25**(1), 41–50 (2022). https://doi.org/10.1007/s10032-021-00388-y
8. Foroozandeh, A., Askari Hemmat, A., Rabbani, H.: Use of the Shearlet transform and transfer learning in offline handwritten signature verification and recognition. Sahand Commun. Math. Anal. **17**(3), 1–31 (2020)
9. Ghods, V., Kabir, E.: A study on Farsi handwriting styles for online recognition. Malays. J. Comput. Sci. **26**(1), 44–59 (2013)
10. Graham, B.: Sparse arrays of signatures for online character recognition. arXiv preprint arXiv:1308.0371 (2013)
11. Han, X.: Research and implementation of characters detection and recognition in ancient Yi books master candidate of software engineering (2020)
12. Ibrayim, M., Simayi, W., Hamdulla, A.: Unconstrained online handwritten Uyghur word recognition based on recurrent neural networks and connectionist temporal classification. Int. J. Biom. **13**(1), 51–63 (2021)
13. Jia, X.: Research on the application of handwritten Yi character recognition technology based on deep learning (2017)
14. Kim, B.H., Zhang, B.T.: Hangul handwriting recognition using recurrent neural networks. KIISE Trans. Comput. Pract. **23**(5), 316–321 (2017)
15. Lai, S., Jin, L., Yang, W.: Toward high-performance online HCCR: a CNN approach with DropDistortion, path signature and spatial stochastic max-pooling. Pattern Recogn. Lett. **89**, 60–66 (2017)
16. Liang, J., Nguyen, C.T., Zhu, B., Nakagawa, M.: An online overlaid handwritten Japanese text recognition system for small tablet. Pattern Anal. Appl. **22**(1), 233–241 (2019). https://doi.org/10.1007/s10044-018-0746-8
17. Liu, M.: Online handwritten Chinese characters analysis and recognition based on deep learning (2018)
18. Maalej, R., Kherallah, M.: Improving the DBLSTM for on-line Arabic handwriting recognition. Multimed. Tools Appl. **79**(25), 17969–17990 (2020). https://doi.org/10.1007/s11042-020-08740-w

19. Nguyen, H.T., Nguyen, C.T., Bao, P.T., Nakagawa, M.: A database of unconstrained Vietnamese online handwriting and recognition experiments by recurrent neural networks. Pattern Recogn. **78**, 291–306 (2018)
20. Simayi, W., Ibrahim, M., Zhang, X.Y., Liu, C.L., Hamdulla, A.: A benchmark for unconstrained online handwritten Uyghur word recognition. Int. J. Doc. Anal. Recogn. (IJDAR) **23**(3), 205–218 (2020). https://doi.org/10.1007/s10032-020-00354-0
21. Singh, S., Sharma, A., Chhabra, I.: A dominant points-based feature extraction approach to recognize online handwritten strokes. Int. J. Doc. Anal. Recogn. (IJDAR) **20**(1), 37–58 (2017). https://doi.org/10.1007/s10032-016-0279-x
22. Su, T.H., Liu, C.L., Zhang, X.Y.: Perceptron learning of modified quadratic discriminant function. In: 2011 International Conference on Document Analysis and Recognition, pp. 1007–1011. IEEE (2011)
23. Suwanwiwat, H., Das, A., Saqib, M., Pal, U.: Benchmarked multi-script Thai scene text dataset and its multi-class detection solution. Multimed. Tools Appl. **80**(8), 11843–11863 (2021). https://doi.org/10.1007/s11042-020-10143-w
24. Tolosana, R., Vera-Rodriguez, R., Fierrez, J., Ortega-Garcia, J.: Reducing the template ageing effect in on-line signature biometrics. IET Biom. **8**(6), 422–430 (2019)
25. Wang, W., Li, Z., Cai, Z., Lv, X., Zhaxi, C., Han, Y.: Online Tibetan handwriting recognition for large character set on new databases. Int. J. Pattern Recogn. Artif. Intell. **33**(10), 1953003 (2019)
26. Wei, W., Guanglai, G.: Online handwriting Mongolia words recognition based on HMM classifier. In: 2009 Chinese Control and Decision Conference, pp. 3912–3914. IEEE (2009)
27. Xiao, X., Yang, Y., Ahmad, T., Jin, L., Chang, T.: Design of a very compact CNN classifier for online handwritten Chinese character recognition using dropweight and global pooling. In: 2017 14th IAPR International Conference on Document Analysis and Recognition (ICDAR), vol. 1, pp. 891–895. IEEE (2017)
28. Yin, F., Wang, Q.F., Zhang, X.Y., Liu, C.L.: ICDAR 2013 Chinese handwriting recognition competition. In: 2013 12th International Conference on Document Analysis and Recognition, pp. 1464–1470. IEEE (2013)
29. Yong-Hua, W., Yan-Qing, L., Ya-Li, G.: The recognition system of old-Yi character based on the image segmentation. J. Yunnan Natl. Univ. (Nat. Sci. Ed.) **17**(1), 76–79 (2008)
30. Yun, X.L., Zhang, Y.M., Yin, F., Liu, C.L.: Instance GNN: a learning framework for joint symbol segmentation and recognition in online handwritten diagrams. IEEE Trans. Multimed. **24**, 2580–2594 (2021)
31. Zhang, X.Y., Yin, F., Zhang, Y.M., Liu, C.L., Bengio, Y.: Drawing and recognizing Chinese characters with recurrent neural network. IEEE Trans. Pattern Anal. Mach. Intell. **40**(4), 849–862 (2017)
32. Zhu, L., Wang, J.: Off-line handwritten Yi character recognition based on the multi-classifier ensemble with combination features. J. Yunnan Natl. Univ. Nat. Sci. Ed. **19**, 329–333 (2010)
33. Zhu, Z., Wu, X.: Principles and implementation of an off-line printed Yi character recognition system. Comput. Technol. Dev. **22**(2), 85–88 (2012)

Self-attention Networks for Non-recurrent Handwritten Text Recognition

Rafael d'Arce[1]([✉])(iD), Terence Norton[1], Sion Hannuna[2], and Nello Cristianini[2]

[1] Adarga, Bristol, UK
rafael.darce@adarga.ai
[2] University of Bristol, Bristol, UK
{sh1670,nello.cristianini}@bristol.ac.uk

Abstract. Handwritten text recognition is still an unsolved problem in the field of machine learning. Nevertheless, this technology has improved considerably in the last decade in part thanks to advancements in recurrent neural networks. Unfortunately, due to their sequential nature, recurrent models cannot be effectively parallelised during training. Meanwhile, in natural language processing research, the transformer has recently become the dominant architecture; replacing the recurrent networks that were once popular. These new models are far more efficient to train than their predecessors because their primary building block, the self-attention network, can process sequences entirely non-recurrently. This work demonstrates that self-attention networks can replace the recurrent networks of state-of-the-art handwriting recognition models and achieve competitive error rates, while reducing the time required to train and the number of parameters significantly.

Keywords: Handwritten text recognition · Self-attention networks · Recurrent neural networks

1 Introduction

1.1 Handwritten Text Recognition

Handwritten text recognition (HTR) is the task of automatically transcribing a representation of handwriting into a digital text format. With its origins in the mid 20th Century [3,16], HTR has been a long-standing machine learning problem, closely tied with some of the most significant advancements in the field. For example, some of the first practical applications of artificial neural networks were for recognising handwritten digits [23,24]. HTR technology has, so far, been applied mostly in commercial settings; however, in recent years, the recognition of historical documents for academic and archival purposes has become increasingly accessible due to the development of the Transkribus platform [35].

In this work, we are concerned specifically with the line-level, offline HTR task: that is, transcribing from images of handwritten text lines, which are comparable to sentences. It is a common practice for an offline HTR system to

© The Author(s), under exclusive license to Springer Nature Switzerland AG 2022
U. Porwal et al. (Eds.): ICFHR 2022, LNCS 13639, pp. 389–403, 2022.
https://doi.org/10.1007/978-3-031-21648-0_27

make use of both an *optical model* and a *language model* [10,29]. Given an input image, the optical model predicts a sequence of probability distributions over some alphabet which, once decoded into a specific transcription, is passed to the language model for post-processing spelling correction. We assumed that the impact of using a language model would be independent of the optical model; this work only explores the role of the latter.

At its core, offline HTR is both an image processing and a sequence transduction problem: an optical model is required to extract visual features from its inputs and generate output sequences based on them. Current state-of-the-art methods use *convolutional neural networks* (CNN) and *recurrent neural networks* (RNN) which are trained end-to-end [29]. The *long short term memory* (LSTM) recurrent unit, in particular, saw adoption for many sequence transduction tasks, including HTR [25]. In fact, one of the first successful applications of the modern LSTM was for handwritten text recognition [13]. Meanwhile, the use of CNNs for HTR was enabled by breakthroughs achieved in the field of computer vision [22].

Training a sequence transduction model can be challenging because, usually, the input and output sequences are of different lengths and the alignment between their elements is unknown. The *connectionist temporal classification* (CTC) algorithm is a popular framework for alignment-free sequence transduction, such as for HTR, that can be used both as a loss function during training and as a means to decode output probabilities. CTC is used to train state-of-the-art CNN-RNN optical models, such as those available on the Transkribus platform (Sect. 2.2).

RNNs process sequence elements sequentially; at each time step t, generating a hidden state h_t using the current input element x_t and the previous hidden state h_{t-1}. These hidden states maintain a running context of the features extracted from the sequence, allowing the network to learn long-term dependencies between its elements. Unfortunately this sequential nature precludes parallelisation, making recurrent models generally inefficient to train; especially as the size of the sequences increase [44]. A report written during the development of Transkribus states that *"the time it takes to train and use a model is also important when judging the performance"* of an optical model and suggested searching for *"a faster, and maybe better, alternative to the LSTM"* as a possible improvement for its current architectures [28]. While there exists simplifications of the LSTM, such as the *gated recurrent unit* [4], that reduce the overall computational complexity; recurrency still remains the major bottleneck in training.

1.2   Attention

All neural networks have some form of implicit attention [15]. That is, they learn to respond more to the most relevant features of a noisy input, loosely mimicking cognitive attention found in humans. However, the use of an explicit attention mechanism has proved to be a powerful technique in sequence transduction.

An attention mechanism is traditionally defined as a mapping between a query vector and a set of key-value vector pairs to an output vector, where the output is computed as the weighted sum of the values [1]. The weight assigned to each value is determined by a similarity function of the query and the corresponding key. Ideally, these weights sum to 1 so they form a discrete probability distribution over the values sequence.

There are several ways to interpret the attention mechanism. One way is as a fuzzy feature look-up method: for a given query feature, it searches a set of candidates and produces an amalgamation of the most similar features. Or, given that the attention weights form a discrete probability distribution, it is akin to taking the expectation of the values given the similarity between the query and the keys. Attention can also be seen as an infinitely-wide convolutional filter with weights that change dynamically based on context.

Attention mechanisms were first popularised by the so-called *sequence-to-sequence* architectures, which relied heavily on recurrent layers [1,37]. However, in 2017, A. Vaswani et al. [44] introduced the *transformer*: an entirely non-recurrent architecture for sequence transduction that takes attention to its logical extremes. The transformer design is based on that of sequence-to-sequence, consisting of an encoder and autoregressive decoder; however, it eliminates all recurrent layers. Sequences are instead repeatedly transformed using layers of self-attention mechanisms, which capture relationships between arbitrarily distant elements within the same sequence.

These *self-attention networks* (SAN) are more parallelisable during training than RNNs, as a self-attention layer requires a constant number of sequential operations to connect all the elements of a sequence, while the number required by a recurrent layer scales linearly with the length of the sequence [44]. Self-attention layers are also less computationally complex than recurrent layers when the length of a sequence is smaller than the dimensionality of its vector elements, which is usually the case [44]. SAN-based models, therefore, can be significantly faster to train than their RNN-based counterparts.

1.3 Motivation and Contributions

Since its introduction, the transformer has become the state-of-the-art architecture for NLP; setting new records in performance for tasks such as language understanding, text classification, text summarisation, sentiment analysis and question answering [5,26,32,48]. The efficiency of self-attention has enabled the pretraining of large transformer models on datasets that were previously intractable for RNNs. These large language models, such as BERT [5] and GPT [32], can be adapted to downstream tasks with state-of-the-art performance using little-to-no fine-tuning. Transformer-based architectures have also started to replace recurrent networks outside of NLP, such as for speech recognition [6] and image generation [33].

There have already been several attempts to build a transformer-based model for offline HTR [2,20,46]. These are all claimed to match or surpass the state-of-the-art error rates of recurrent methods. However, despite the more efficient training afforded to them by the use of SANs, transformer models are often still expensive to train due to their large size and autoregressive decoders. Meanwhile, in the field of automatic speech recognition, J. Salazar et al. have recently proposed an entirely SAN-based sequence transduction model and demonstrated that training it with CTC is tractable [34].

Inspired by this, we propose a non-recurrent and non-autoregressive architecture for offline HTR based on SANs. We replaced the recurrent layers of a leading architecture with self-attention layers and showed that, under the same experimental setup, such a model could closely approximate the state-of-the-art recognition accuracy, while requiring significantly less time to train and fewer parameters. In order to achieve these results, the models were first pretrained on a large synthetic dataset, before being fine-tuned on human-generated data.

2   Technical Background

2.1   Connectionist Temporal Classification

Here we define CTC as per the works of Graves et al. [14] and J. Salazar et al. [34], in the context of an optical model for offline HTR. Consider a T-length sequence of D-dimensional features $\boldsymbol{X} \in \mathbb{R}^{T \times D}$, extracted from an input image using a CNN. Let $\boldsymbol{y} = (y_1, y_2, ..., y_U) \in \mathcal{A}^U$ represent its corresponding ground-truth transcription, a U-length output sequence of characters from some alphabet $\mathcal{A}$. Under the CTC framework, we assume that $U \leq T$ and define an intermediate alphabet $\mathcal{A}' = \mathcal{A} \cap \{\text{-}\}$ with an extra symbol "$-$" called *blank* for separating characters. A *path* $\boldsymbol{\pi} = (\pi_1, \pi_2, ..., \pi_T) \in \mathcal{A}'^T$ is a sequence of intermediate characters that can be associated to an output sequence via a many-to-one mapping $\mathcal{B} : \mathcal{A}'^T \to \mathcal{A}^{U'}$, where $U' \leq T$, by collapsing repeated characters and removing blanks. A path, therefore, can be regarded as a one of many possible alignments between an input and output sequence, for example:

$$(h, e, -, l, -, l, l, o, -,\ \ , -, w, o, o, -, r, l, d) \mapsto (h, e, l, l, o,\ \ , w, o, r, l, d)$$

CTC is able to model the probability of an output $\boldsymbol{y}$ given an input $\boldsymbol{X}$ by marginalising over all paths (Eq. 1) and by assuming that the elements of a path π_t are conditionally-independent (Eq. 2).

$$P(\boldsymbol{y}|\boldsymbol{X}) = \sum_{\boldsymbol{\pi} \in \mathcal{B}^{-1}(\boldsymbol{y})} P(\boldsymbol{\pi}|\boldsymbol{X}) \tag{1}$$

$$P(\boldsymbol{\pi}|\boldsymbol{X}) = \prod_{t=1}^{T} P(\pi_t, t|\boldsymbol{X}) \tag{2}$$

One approximates $P(\pi_t, t|\boldsymbol{X})$, the probability distribution over intermediate characters for a given output position and input sequence, using an optical model. This model can be trained end-to-end to minimize the CTC loss function $\mathcal{L}_{CTC}(\boldsymbol{y}, \boldsymbol{X}) = -\log P(\boldsymbol{y}|\boldsymbol{X})$ [14] or have its predictions decoded from the probabilities using a method such as beam search [18].

2.2   State-of-the-Art Architecture

We believe that the current state-of-the-art optical model for offline HTR is best represented by the so-called *Puigcerver* architecture, first proposed by J. Puigcerver [29] in 2017. It consists of a block of convolutional layers followed by a block of one-dimensional and bi-directional LSTM layers, which are trained end-to-end using CTC. There are 9.59 million total trainable parameters. We reserve the formal definition of this architecture to the work of J. Puigcerver [29].

In HTR literature, Puigcerver is a popular architecture to benchmark new methods against [9,10,20,39]. It is also used in industry by the Transkribus platform [45]. Transkribus provides users with two engines for training optical models: *PyLaia* and *HTR+* [42,43]. The former engine directly uses the Puigcerver architecture [30], while the latter uses a convolutional-recurrent architecture inspired by that of Puigcerver [28,45].

2.3   Self-attention Networks

We define a SAN as a series of self-attention layers (Fig. 1), where a self-attention layer is identical to an encoder layer of the canonical transformer architecture [44]. A self-attention layer consists of two sub-layers: a multi-head attention component, followed by a feed-forward network. Each sub-layer has a residual connection around it, followed by a layer normalisation.

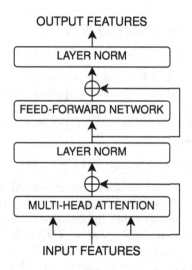

Fig. 1. A self-attention layer.

The particular attention mechanism used by in a self-attention layer is known as the *scaled dot-product attention* and it takes queries $Q \in \mathbb{R}^{T \times D}$, keys $K \in \mathbb{R}^{T \times D}$ and values $V \in \mathbb{R}^{T \times D}$ as input (SDPA; Eq. 3). The dot-product similarity function is used to calculate the attention weights, the results of which are scaled by a factor of $\frac{1}{\sqrt{D}}$ and transformed into a discrete probabilty distribution using the softmax function.

$$\text{SDPA}(\boldsymbol{Q}, \boldsymbol{K}, \boldsymbol{V}) = \text{softmax}\left(\frac{\boldsymbol{Q}\boldsymbol{K}^\mathsf{T}}{\sqrt{D}}\right)\boldsymbol{V} \tag{3}$$

Note that SDPA computes attention for multiple queries simultaneously by packing them into a matrix, which allows it to compare whole sequences with each other in a single step. SANs are self-attentional because they use the same vector sequence for all of the inputs of their attention mechanisms (i.e. $\boldsymbol{Q} = \boldsymbol{K} = \boldsymbol{V}$). SDPA, therefore, transforms a sequence of features based on how its elements relate to one another. This enables a SAN to model dependencies non-recurrently.

Rather than only using a single D-dimensional SDPA mechanism, a SAN applies self-attention H-many times in parallel using multiple, $\frac{D}{H}$-dimensional heads. For each head$_h$, the queries, key and values are first projected to unique subspaces using the learned linear projections $\boldsymbol{W}_h^Q \in \mathbb{R}^{D \times \frac{D}{H}}$, $\boldsymbol{W}_h^K \in \mathbb{R}^{D \times \frac{D}{H}}$ and $\boldsymbol{W}_h^V \in \mathbb{R}^{D \times \frac{D}{H}}$ respectively. The outputs of the different heads are concatenated to produce a single output sequence, which is finally projected back to D-dimensions using $\boldsymbol{W}^O \in \mathbb{R}^{\frac{D}{H} \times D}$. This *multi-head attention* (MHA; Eq. 4) enables a self-attention layer to jointly attend to information in different representational subspaces and to model different types of sequence dependency at the same time [44].

$$\text{MHA}(\boldsymbol{Q}, \boldsymbol{K}, \boldsymbol{V}) = \text{concat}(\text{head}_1(\boldsymbol{Q}, \boldsymbol{K}, \boldsymbol{V}), ..., \text{head}_H(\boldsymbol{Q}, \boldsymbol{K}, \boldsymbol{V}))\boldsymbol{W}_o$$
$$\text{where head}_h(\boldsymbol{Q}, \boldsymbol{K}, \boldsymbol{V}) = \text{SDPA}(\boldsymbol{Q}\boldsymbol{W}_h^Q, \boldsymbol{K}\boldsymbol{W}_h^K, \boldsymbol{V}\boldsymbol{W}_h^V) \tag{4}$$

A *feed-forward network* (FFN; Eq. 5), consisting of two linear transformations with a ReLU activation in between, is applied position-wise to each output element $\boldsymbol{x} \in \mathbb{R}^D$ of the MHA sublayer. The FFN sublayer contains all of the learnable parameters of a SAN, and it is responsible for transforming its intermediate features non-linearly.

$$\text{FFN}(\boldsymbol{x}) = \max(0, \boldsymbol{x}\boldsymbol{W_1} + \boldsymbol{b_1})\boldsymbol{W_2} + \boldsymbol{b_2} \tag{5}$$

Attention mechanisms are inherently content-based: they only considers the values of sequence elements and is not aware of their absolute or relative positions. The transformer embeds this information directly in the elements by adding an element-wise *positional encoding* $\boldsymbol{PE} \in \mathbb{R}^{T \times D}$ to input sequences before they are passed into its encoder or decoder. While $\boldsymbol{PE}$ can be learnt during training as an additional set of parameters, the original implementation of the transformer [44] defined it statically based on the sine and cosine functions (Eq. 6). The sinusoidal method was chosen because it can be more easily extended for larger inputs.

$$\boldsymbol{PE}_{t,2d} = \sin(t^{-1} \times 10000^{\frac{2d}{D}})$$
$$\boldsymbol{PE}_{t,2d+1} = \cos(t^{-1} \times 10000^{\frac{2d}{D}}) \tag{6}$$

2.4   Proposed Architecture

We propose a non-recurrent and non-autoregressive architecture for offline HTR based on the self-attention network, as defined in Sect. 2.3. Our architecture uses the same convolutional block as the Puigcerver architecture but replaces the recurrent block with a self-attention block (Fig. 2).

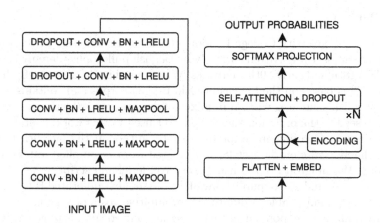

Fig. 2. Proposed CNN-SAN architecture with N-many self-attention layers.

The first block consists of five sets of convolutional (Conv), batch normalisation (BN) and leaky ReLU activation (LReLU) layers. The first three sets are each followed by a max pooling layer, while dropout layers precede the last two. The convolutional layers have kernels of size 3×3 and strides of 1×1. The number of convolutional filters in each set, in order, are 16, 32, 48, 64 and 80. The LReLUs use $\alpha = 0.001$, the dropouts use a rate of 0.2 and the max poolings have 2×2 kernels. The output of the convolutional block is passed to the self-attention block.

This output of the convolutional block is flattened into a sequence of vectors before it passed to the self-attention block. It is then embedded using a learned, D-dimensional embedding and has the sinusoidal positional encoding added to it. Next, the sequence is repeatedly transformed by N-many self-attention layers. Dropout is applied to the output of each self-attention layer. Finally, a position-wise linear transformation layer, with a softmax activation, projects the sequence elements to probability distributions over the intermediate CTC alphabet.

The number of self-attention layers N, the number of attention heads per SAN layer H, the dimensionality of the vector embeddings D, the size of the FFN hidden layers F and the dropout rate R are hyperparameters. They were optimised using grid search over a combination of the non-synthetic dataset used in our experiments (Sect. 3.1). Ultimately, the configuration $(N, H, D, F, R) = (4, 4, 192, 512, 0.1)$ was used in our experiments as it achieved the lowest CER on the combined test set. This configuration imposes 1.74 million total trainable parameters.

3   Experimental Evaluation

Our experiments were executed on a single Nvidia Tesla P100 GPU and results were averaged over 3 attempts. The codebase used was initially forked from an open-source framework for training offline HTR models with CTC [31].

3.1   Datasets

Compiled by University College London, the *Bentham* dataset [11,38] was created from manuscripts written cursively by English philosopher Jeremy Bentham in the late 18th and early 19th Centuries. Widely used in HTR research, the *IAM* dataset [27] is based on the Lancaster-Oslo-Bergen corpus and contains contemporary English sentences written in mixed styles by over 500 writers. The *Saint Gall* dataset [7], the only non-English dataset used in this work, contains Latin text written in Carolingian script by a single author in the late 9th Century. Lastly, the *Washington* dataset [8] was created from recompiled letters written by George Washington in 1755 using a cursive, longhand script.

The number and size of publicly-available, manually-annotated HTR datasets are relatively small (Table 1) due to the prohibitive costs of tagging such data. A cheaper alternative, popular in text recognition literature, is to generate additional training data synthetically [19–21]. Inspired by the work of L. Kang et al. [20], we created a large *synthetic* dataset by generating grayscale images of text using handwritten-style electronic fonts and then distorting them to simulate noise (Fig. 3). Sentences were randomly sampled from the Brown corpus [41] and were rendered as images using one of over 400 freely-available handwriting-style fonts collected from the Internet and an open-source data generation tool [40].

Table 1. The number of text line images per partition of each dataset used.

Dataset	Number of images		
	Train	Test	Valid
Bentham	8808	1372	820
IAM	6161	1861	1840
Saint Gall	468	707	235
Washington	325	168	163
Synthetic	40000	5000	5000

(a) Mrs. Kahler joined them.

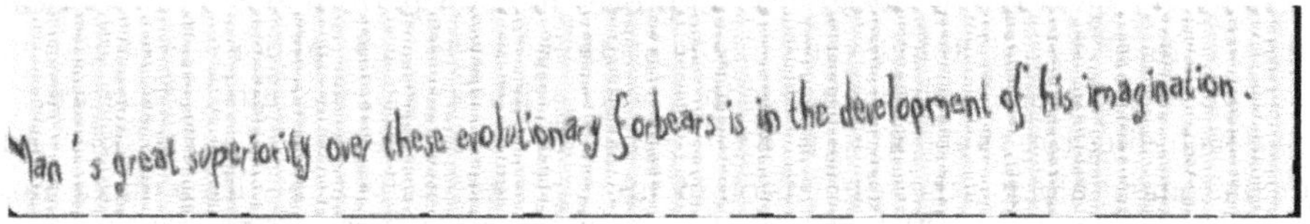

(b)   Man's greater superiority over these evolutionary forbears is in the development of his imagination.

(c) She arranged the letters carefully, one on top of the other.

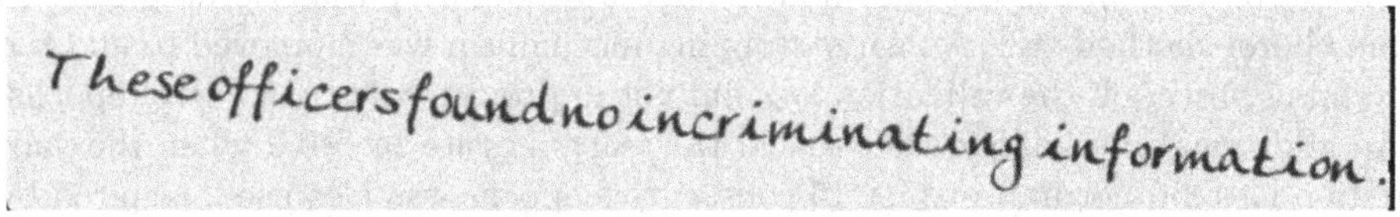

(d) These officers found no incriminating information.

Fig. 3. Example images of synthetically generated text lines and their corresponding transcriptions.

For each image generated, the height was set to 128 and the following properties were randomly configured: the font, the number of pixels between characters, and the number of pixels in each margin. The tool also provides some basic text distortions, the following of which were used and also randomly configured: the skewing of the texts horizontal direction, the vertical displacement of each character based on a sine or cosine signal and Gaussian background noise.

We also employed the elastic deformation method proposed by P. Simard et al., with $\sigma = 4$ and $\alpha = 34$, which is said to simulate *"uncontrolled oscillations of the hand muscles, dampened by inertia"* [36].

3.2   Preprocessing and Dynamic Augmentations

All training images are first preprocessed by normalising their pixel values and by resizing them into grayscale images to a width of 1024 and height of 128. The original aspect ratio is maintained through zero-padding. Meanwhile, for all ground-truth transcription strings, accents are removed from characters (for example, é ↦ e). For training and validation, the ground-truths are encoded from strings into fixed-length sequences of one-hot vectors using a character-level tokenizer. The tokenizer used a base alphabet of the first 95 printable

ASCII characters, extended with two special tokens for padding ground-truth transcriptions to the fixed length of 128 and to represent characters not in the base alphabet. The special tokens are removed from the final predictions.

It is well established in the HTR literature that using dynamic image augmentations during training can greatly improve the final performance of an optical model [28,29,47]. The following augmentations were each randomly configured and applied to each training image in a batch with 50% probability: *affine warping*, including rotations of up to 3° in either direction, resizing of up to 5%, and displacement along either dimension by up to 5% each; *dilation* with a maximum kernel size of 3 × 3; and erosion with a maximum kernel size of 5 × 5.

3.3   Training Setup

The training setup used in this work closely approximated that used by J. Puigcerver [29]. Specifically, models were trained to minimise the CTC loss function using the RMSProp gradient descent algorithm [17] with a learning rate of 0.0003, batches of size 16 and model weights randomly initialised according to the Glorot method [12]. An early stopping mechanism was employed to end the training process if the validation loss did not improve for 20 consecutive epochs. An additional mechanism that scaled the learning rate by ×0.2 when the validation loss did not improve for 15 consecutive epochs was also used, inspired by the work of A.F.S. Neto et al. [10].

A maximum of 1000 epochs were performed in each experiment. After each epoch, the average loss across the training and validation sets were logged, and the training batches randomly shuffled. When training finished, whether by the early stopping mechanism or epoch limit, the weight configuration that achieved the lowest validation loss was used for evaluation to the test set. Character probabilities were CTC decoded into concrete predictions using the *vanilla beam search* algorithm [18] with a beam width of 10. The standard HTR metrics of *character error rate* (CER) and *word error rate* (WER) were used to measure model performance, which are equivalent to the character- and word-level Levenshtein distances between ground-truth and predicted transcriptions. For each model, we also measured the time and number of epochs required to train it.

Unless stated otherwise, assume that the setup used for pretraining and fine-tuning was the same as that described above. For pretraining on the synthetic dataset, a fixed 100 epochs were performed and the early stopping mechanism was disabled. For fine-tuning on a specific dataset, the model weights were initialised to the configuration that achieved the lowest validation loss during pretraining. This experimental setup was used for both the recurrent and non-recurrent architectures.

3.4   Results

Our primary results are provided in Table 2. Across all datasets, both when first pretrained and when trained from random initial weights, our proposed architecture required 0.35× the time that Puigcerver required to be trained. After fine-tuning, our models achieved 1.08× and 1.07× the character and word error rates of Puigcerver models, respectively. Synthetic pretraining provided notable performance gains for both architectures; especially on the smaller Washington dataset, where training previously diverged without it. Given enough pretraining data, our CNN-SAN architecture achieved comparable results to a leading RNN-based model while requiring significantly less time to train. Despite the slight degradation in terms of CER and WER, we argue that is a reasonable trade-off given the substantial reduction in training time.

Table 2. Average model performance metrics for both our and Puigcerver's architectures on each public dataset, both when trained from randomly initialised weights and when fine-tuned from synthetically pretrained weights.

Dataset	Architecture	Test error (%)		Time to train	Total epochs
		CER	WER	(hours)	
Bentham	Puigcerver	7.55	31.99	2.09	86
	+ Synth	6.51	30.03	1.83	78
	Ours	9.19	35.24	1.22	116
	+ Synth	7.97	34.36	0.96	84
IAM	Puigcerver	7.68	23.48	1.53	88
	+ Synth	5.61	17.78	0.90	52
	Ours	10.26	32.62	1.05	134
	+ Synth	7.50	25.46	0.43	61
Saint Gall	Puigcerver	9.40	40.70	0.22	141
	+ Synth	8.46	39.58	0.05	28
	Ours	10.76	49.73	0.08	111
	+ Synth	7.12	35.68	0.03	39
Washington	Puigcerver	97.34	100.00	0.03	22
	+ Synth	6.65	21.33	0.07	56
	Ours	97.11	100.00	0.02	23
	+ Synth	7.66	27.00	0.03	66

It is worth noting that use of sinusoidal positional encoding had no impact on the model error rates, as noted in Table 3. We believe that monotonic alignment enforced by CTC during training acted as an inductive bias for sequence position, removing the need for positional encoding [34]. Regardless, its use may result in fewer total epochs so we continued its use.

Table 3. The average performance metrics of our proposed model architecture, with and without sinusoidal positional encoding, when trained from random initial weights on the combination of the Bentham, IAM, Saint Gall and Washington datasets.

Positional Encoding	Test error (%)		Time to train	Total epochs
	CER	WER	(hours)	
Yes	9.66	36.44	1.70	107
No	9.67	36.30	2.19	129

4 Conclusion

This work demonstrates that self-attention layers can be efficient replacements for the recurrent layers of offline HTR models. We hope that these results inspire the adoption of SANs more broadly in HTR, aligning the field with recent advancements in natural language processing and computer vision. Future work should focus on training larger SAN architectures on a greater amount of synthetic data, with the aim of creating generic pretrained models that can be employed with little-to-no fine-tuning. Other potential areas of improvement include the configuration of the convolutional layers, how the learning rate is scheduled during training and the quality of the synthetic data.

References

1. Bahdanau, D., Cho, K., Bengio, Y.: Neural machine translation by jointly learning to align and translate. CoRR abs/1409.0473 (2015)
2. Barrere, K., Soullard, Y., Lemaitre, A., Coüasnon, B.: Transformers for historical handwritten text recognition. In: Doctoral Consortium - ICDAR 2021. Nibal Nayef and Jean-Christophe Burie, Lausanne, Switzerland (2021). https://hal.archives-ouvertes.fr/hal-03485262
3. Bledsoe, W.W., Browning, I.: Pattern recognition and reading by machine. In: Papers Presented at the 1–3 December 1959, Eastern Joint IRE-AIEE-ACM Computer Conference, pp. 225–232. IRE-AIEE-ACM 1959 (Eastern), Association for Computing Machinery (1959)
4. Cho, K., van Merrienboer, B., Gülçehre, Ç., Bougares, F., Schwenk, H., Bengio, Y.: Learning phrase representations using RNN encoder-decoder for statistical machine translation. CoRR abs/1406.1078 (2014)
5. Devlin, J., Chang, M., Lee, K., Toutanova, K.: BERT: pre-training of deep bidirectional transformers for language understanding. CoRR abs/1810.04805 (2018)
6. Dong, L., Xu, S., Xu, B.: Speech-transformer: a no-recurrence sequence-to-sequence model for speech recognition. In: 2018 IEEE International Conference on Acoustics, Speech and Signal Processing (ICASSP), pp. 5884–5888 (2018)
7. Fischer, A., Frinken, V., Fornés, A., Bunke, H.: Transcription alignment of Latin manuscripts using hidden Markov models. In: Proceedings of the 2011 Workshop on Historical Document Imaging and Processing, pp. 29–36. Association for Computing Machinery (2011)

8. Fischer, A., Keller, A., Frinken, V., Bunke, H.: Lexicon-free handwritten word spotting using character HMMs. Pattern Recogn. Lett. **33**(7), 934–942 (2012)
9. Neto, A.F.S., Bezerra, B.L.D., Toselli, A.H.: Towards the natural language processing as spelling correction for offline handwritten text recognition systems. Appl. Sci. **10**(21), 1–29 (2020). https://doi.org/10.3390/app10217711
10. Neto, A.F.S., Bezerra, B.L.D., Toselli, A.H., Lima, E.B.: HTR-Flor: a deep learning System for Offline Handwritten Text Recognition. In: 2020 33rd SIBGRAPI Conference on Graphics, Patterns and Images (SIBGRAPI), pp. 54–61 (2020)
11. Gatos, B., et al.: Ground-truth production in the transcriptorium project. In: 2014 11th IAPR International Workshop on Document Analysis Systems, pp. 237–241 (2014)
12. Glorot, X., Bengio, Y.: Understanding the difficulty of training deep feedforward neural networks. In: Proceedings of the Thirteenth International Conference on Artificial Intelligence and Statistics, pp. 249–256. JMLR Workshop and Conference Proceedings (2010)
13. Graves, A.: Supervised sequence labelling. In: Supervised Sequence Labelling with Recurrent Neural Networks. Studies in Computational Intelligence, vol. 385, pp. 5–13. Springer, Berlin (2012). https://doi.org/10.1007/978-3-642-24797-2_2
14. Graves, A., Fernández, S., Gomez, F., Schmidhuber, J.: Connectionist temporal classification: labelling unsegmented sequence data with recurrent neural networks. In: Proceedings of the 23rd International Conference on Machine Learning, pp. 369–376. ICML 2006, Association for Computing Machinery, New York, NY, USA (2006)
15. Graves, A., Fernández, S., Liwicki, M., Bunke, H., Schmidhuber, J.: Unconstrained online handwriting recognition with recurrent neural networks. In: Proceedings of the 20th International Conference on Neural Information Processing Systems, pp. 577–584. NIPS 2007 (2007)
16. Grimsdale, R.L., Bullingham, J.M.: Character recognition by digital computer using a special flying-spot scanner. Comput. J. **4**(2), 129–136 (1961)
17. Hinton, G., Srivastava, N., Swersky, K.: Neural networks for machine learning lecture 6a overview of mini-batch gradient descent (2012)
18. Hwang, K., Sung, W.: Character-level incremental speech recognition with recurrent neural networks. CoRR abs/1601.06581 (2016)
19. Jaderberg, M., Simonyan, K., Vedaldi, A., Zisserman, A.: Synthetic data and artificial neural networks for natural scene text recognition. CoRR abs/1406.2227 (2014)
20. Kang, L., Riba, P., Rusiñol, M., Fornés, A., Villegas, M.: Pay attention to what you read: non-recurrent handwritten text-line recognition (2020)
21. Krishnan, P., Jawahar, C.V.: HWNet v2: an efficient word image representation for handwritten documents. CoRR abs/1802.06194 (2018)
22. Krizhevsky, A., Sutskever, I., Hinton, G.E.: ImageNet classification with deep convolutional neural networks. Commun. ACM **60**(6), 84–90 (2017)
23. LeCun, Y. et al.: Handwritten digit recognition with a back-propagation network. In: Touretzky, D. (ed.) Advances in Neural Information Processing Systems. vol. 2. Morgan-Kaufmann (1990)
24. LeCun, Y., et al.: Backpropagation applied to handwritten zip code recognition. Neural Comput. **1**(4), 541–551 (1989)
25. Lipton, Z.C.: A critical review of recurrent neural networks for sequence learning. CoRR abs/1506.00019 (2015)
26. Liu, Y., et al.: RoBERTa: a robustly optimized BERT pretraining approach. CoRR abs/1907.11692 (2019)

27. Marti, U., Bunke, H.: The IAM-database: an English sentence database for offline handwriting recognition. Int. J. Doc. Anal. Recogn. **5**, 39–46 (2002)
28. Michael, J., Weidemann, M., Labahn, R.: D7.9 HTR engine based on NNs P3(2022). https://readcoop.eu/wp-content/uploads/2018/12/Del_D7_9.pdf
29. Puigcerver, J.: Are multidimensional recurrent layers really necessary for handwritten text recognition? In: 2017 14th IAPR International Conference on Document Analysis and Recognition (ICDAR), vol. 01, pp. 67–72 (2017)
30. PyLaia (2022). https://github.com/jpuigcerver/PyLaia. Accessed 10 July 2022
31. Handwritten text recognition (HTR) using TensorFlow 2.x (2022). https://github.com/arthurflor23/handwritten-text-recognition. Accessed 10 July 2022
32. Radford, A., Wu, J., Child, R., Luan, D., Amodei, D., Sutskever, I.: Language models are unsupervised multitask learners. OpenAI Blog **1**(8), 9 (2019)
33. Ramesh, A., et al.: Zero-shot text-to-image generation. CoRR abs/2102.12092 (2021)
34. Salazar, J., Kirchhoff, K., Huang, Z.: Self-attention networks for connectionist temporal classification in speech recognition. In: 2019 IEEE International Conference on Acoustics, Speech and Signal Processing (ICASSP) (2019)
35. Seaward, L., et al.: Transforming scholarship in the archives through handwritten text recognition: transkribus as a case study. J. Documentation **75**(5), 954–976 (2019)
36. Simard, P., Steinkraus, D., Platt, J.: Best practices for convolutional neural networks applied to visual document analysis. In: Seventh International Conference on Document Analysis and Recognition, 2003. Proceedings, pp. 958–963 (2003)
37. Sutskever, I., Vinyals, O., Le, Q.V.: Sequence to sequence learning with neural networks. CoRR abs/1409.3215 (2014)
38. Sánchez, J.A., Romero, V., Toselli, A.H., Vidal, E.: ICFHR2014 competition on handwritten text recognition on transcriptorium datasets (HTRtS). In: 2014 14th International Conference on Frontiers in Handwriting Recognition, pp. 785–790 (2014)
39. Sánchez, J.A., Romero, V., Toselli, A.H., Villegas, M., Vidal, E.: A set of benchmarks for handwritten text recognition on historical documents. Pattern Recogn. **94**, 122–134 (2019)
40. Text Recognition Data Generator (2022). https://github.com/Belval/TextRecognitionDataGenerator. Accessed 10 July 2022
41. The Brown Corpus (2022). https://www.nltk.org/book/ch02.html#brown-corpus. Accessed 10 July 2022
42. Transkribus Glossary HTR+ (2022). https://readcoop.eu/glossary/htr-plus/. Accessed 10 July 2022
43. Transkribus Glossary PyLaia (2022). https://readcoop.eu/glossary/pylaia/. Accessed 10 July 2022
44. Vaswani, A., et al.: Attention is all you need. CoRR abs/1706.03762 (2017)
45. Weidemann, M., Michael, J., Grüning, T., Labahn, R.: D7.9 HTR engine based on NNs P3 (2022). https://readcoop.eu/wp-content/uploads/2017/12/Del_D7_8.pdf
46. Wick, C., Zöllner, J., Grüning, T.: Transformer for handwritten text recognition using bidirectional post-decoding. In: Lladós, J., Lopresti, D., Uchida, S. (eds.) ICDAR 2021. LNCS, vol. 12823, pp. 112–126. Springer, Cham (2021). https://doi.org/10.1007/978-3-030-86334-0_8
47. Wigington, C., Stewart, S., Davis, B., Barrett, B., Price, B., Cohen, S.: Data augmentation for recognition of handwritten words and lines using a CNN-LSTM network. In: 2017 14th IAPR International Conference on Document Analysis and Recognition (ICDAR), vol. 01, pp. 639–645 (2017)

48. Yang, Z., Dai, Z., Yang, Y., Carbonell, J.G., Salakhutdinov, R., Le, Q.V.: XLNet: generalized autoregressive pretraining for language understanding. CoRR abs/1906.08237 (2019)

An Efficient Prototype-Based Model for Handwritten Text Recognition with Multi-loss Fusion

Ming-Ming Yu[1,2(✉)], Heng Zhang[1], Fei Yin[1], and Cheng-Lin Liu[1,2]

[1] National Laboratory of Pattern Recognition (NLPR), Institution of Automation, Chinese Academy of Sciences, Beijing 100190, China
{fyin,liucl}@nlpr.ia.ac.cn
[2] School of Artificial Intelligence, University of Chinese Academy of Sciences, Beijing 100049, China
{yumingming2020,heng.zhang}@ia.ac.cn

Abstract. Prototype learning has achieved good performance in many fields, showing higher flexibility and generalization. In this paper, we propose an efficient text line recognition method based on prototype learning with feature-level sliding windows for classification. In this framework, we combine weakly supervised discrimination and generation loss for learning feature representations with intra-class compactness and inter-class separability. Then, dynamic weighting and pseudo-label filtering are also adopted to reduce the influence of unreliable pseudo-labels and improve training stability significantly. Furthermore, we introduce consistency regularization to obtain more reliable confidence distributions and pseudo-labels. Experimental results on digital and Chinese handwritten text datasets demonstrate the superiority of our method and justify advantages in transfer learning on small-size datasets.

Keywords: Text recognition · Prototype learning · Consistency regularization · Connectionist temporal classification · Pseudo-label

1 Introduction

Text recognition has drawn much research interest in the computer vision community due to its wide applications. The text recognition problem is to map the input image to the corresponding sequence of characters. With the development of deep learning, excellent performance is achieved in many scenarios, such as scene text recognition and handwritten text recognition. Based on the encoder-decoder framework, visual features are usually extracted by the CNN (Convolutional Neural Networks) encoder, followed by the RNN (Recurrent Neural Networks) or Transformer network to extract context features. Finally, the Connectionist Temporal Classification (CTC) [5] or attention mechanism is used to align feature sequences and labels. Some other works [9,34] remove the sequence modeling stage and only use the CNN encoder, improving the parallelization

© The Author(s), under exclusive license to Springer Nature Switzerland AG 2022
U. Porwal et al. (Eds.): ICFHR 2022, LNCS 13639, pp. 404–418, 2022.
https://doi.org/10.1007/978-3-031-21648-0_28

and reducing the computation cost. Text recognition methods have achieved high performance, but the interpretability and generalization are still inadequate [3,31]. Different from discriminative models, the Convolution Prototype Network (CPN) [31,32] is a discriminative and generative hybrid model with better generalization and robustness. CPN has been successfully applied to open set recognition and few-shot learning in character classification tasks. However, it is challenging to apply CPN on weakly-supervised text line recognition because a text line is composed of different characters with variable-length and unknown character positions.

To address the abovementioned issues, we propose a prototype classifier for text recognition to improve the generalization and robustness based on weakly-supervised discrimination and generation learning. Our model adopts the convolution-CTC framework with sliding windows at the feature level. The blank class can be considered as all the non-character samples dynamically generated in the window sliding process, so we also set a prototype for the blank class to fully use these non-character samples and enhance the discriminant of the model. The blank prototype is only used in discriminative learning rather than generative learning. Dynamic weighting and pseudo-label filtering are also adopted to weaken unreliable pseudo-labels and improve training stability. Moreover, consistency regularization [13,25] is introduced to obtain more reliable confidence distributions and pseudo-labels for improving the performance. The experimental results show that the AR and CR of our method can reach 93.66% and 93.90%, respectively, on the ICDAR-2013 dataset without a language model. Moreover, the string accuracy is 95.53% and 95.49% on CAR-A and CAR-B test sets, respectively. We also demonstrate that our model can better transfer the knowledge learned from the isolated character data to the text line, with obvious performance gain.

Our main contributions are summarized as follows: (1) We design a prototype-based text line recognition method with feature-level sliding windows for classification. Our feature-level sliding window method shows less memory footprint and higher inference speed than sliding windows on the input image. (2) In order to obtain robust pseudo-labels, we propose two methods to weaken unreliable pseudo-labels: pseudo-label filtering and dynamic weighting. Moreover, we introduce consistency regularization to enhance model robustness and further improve recognition performance. (3) Our method has achieved state-of-the-art performance on three handwritten text datasets and justifies the advantages in transfer learning from character recognition to text recognition with small-size datasets.

2 Related Work

2.1 Text Recognition

The existing text line recognition methods can be divided into two methods based on explicit and implicit segmentation. In the explicit segmentation approach, the most representative one is recognition based on the over-segmentation strategy [21,24,27]. Specifically, these methods first generate candidate characters by merging continuous primitives and then search for the optimal segmentation-recognition path in the candidate lattice. Finally, the score of the segmentation-recognition path is given by integration of the character classifier, language model, and geometric models. With only string-level annotations, Wang et al. [24] proposed a weakly supervised training strategy for character classifier training under the over-segmentation framework. Moreover, Peng et al. [11] formulate a new segmentation-based text recognition framework for segmenting and recognizing characters end-to-end. The models based on implicit segmentation do not need to perform explicit character segmentation and only use string-level annotations, including Hidden Markov Model (HMM) [2], CTC, Attention. With the development of deep learning, CTC and attention-based methods have been widely used in text recognition. Grave et al. [5] first applied CTC to handwritten text line recognition with RNN for contextual feature representation. Shi et al. [14] proposed the CRNN (Convolution Recurrent Neural Networks) model combining the feature extraction capability of CNN with the sequence modeling capability of RNN. Yin et al. [34] proposed a fully convolutional model with multi-scale sliding window classification. Liu et al. [9] proposed context beam search to combine the Transformer-based language model with a visual model. On the other hand, attention-based text recognition uses a 1D soft-attention model to select relevant local features during character decoding. In this manner, the model can learn a character-level language model from the training data. Shi et al. [15] proposed an end-to-end framework with RNN and attention for scene text recognition. SAR [8] used the 2D attention to recognize irregular texts. Wang et al. [19] employed the Transformer to replace RNN structure to capture long distance context.

2.2   Prototype Learning

Prototype learning is a method to classify samples based on template matching. The prototype (template) refers to a representative point in the sample or feature space. K-Nearest-Neighbor (KNN) and Learning Vector Quantization (LVQ) [6] are both classical prototype classifiers. Traditional prototype models cannot be optimized end-to-end, because feature extraction and prototype learning are performed in separate stages. With the development of deep learning, some works combined DNN with prototype learning to improve model performance. Snell et.al [16] applied the prototype concept in CNN for few-shot learning. Yang et al. [31,32] proposed the Convolutional Prototype Network (CPN) by training the CNN feature extractor and prototypes end-to-end. With the supervision of discriminative and generative learning, the CPN shows better robustness and generalization in multiple scenarios such as open set recognition and few-shot learning. Gao et al. [3] first proposed a prototype-based handwritten text recognition method; however, it still suffers from unreliable pseudo-labels and the high computation cost caused by sliding windows on the input image.

2.3   Consistency Regularization and Pseudo-labels

Consistency regularization and pseudo labels are usually used in semi-supervised learning. Based on the assumption that the model should output consistent probability distribution for the same input with slight disturbances, consistency regularization [13,18,29] can minimize the difference between prediction distributions from different disturbances of the same data. Random data augmentation, dropout [17], and exponential moving average (EMA) [13,18] are commonly used to add disturbances. Kullback-Leibler Divergence, Jensen-Shannon Divergence, cross-entropy, and mean square error (MSE) are frequently-used methods to measure the difference between two probability distributions. After training an initial model on a few labeled data, pseudo-labels [7] are given by the prediction on unlabeled data. For handwritten recognition, Gao et al. [4] use pseudo-labels in the CTC loss and get state-of-art results.

3   Method

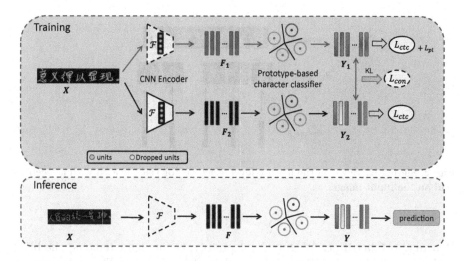

Fig. 1. An illustration of the proposed text recognition method based on prototype learning.

The overall framework of the proposed method is illustrated in Fig. 1. As the figure shows, we use the CTC-based convolutional prototype learning framework. The text image is first normalized by scaling to a fixed height with the aspect ratio preserved. Then the normalized text image is encoded by CNN for feature representation of candidate characters. After that, the candidate character features are classified by the prototype-based character classifier. In prototype learning (PL), the input text image will go through the training model twice with different dropout for consistency regularization computation. The two different sub-models are shown in Fig. 1. Besides, the CTC loss L_{ctc} and PL loss L_{pl} are also computed and combined with consistency regularization loss L_{con}. In prediction, we use the CTC decoding algorithm to obtain the final recognition results.

3.1   CNN Encoder and Prototype Classifier

CNN Encoder. Given a normalized text line image $\mathbf{X} \in \mathbb{R}^{H \times W \times C}$, the encoder $\mathcal{F}$ first produces the feature map $\mathbf{F} \in \mathbb{R}^{H' \times W' \times D'}$, where H', W', and D' denote the height, width, and channel number of the feature map, respectively. Then the feature sequence $\mathbf{F}$ with L elements is formed by sliding windows on the feature map. In our experiments, the window width and height are both set to H', and the feature dimensionality is $H' \times H' \times D'$. As illustrated in Fig. 2, each feature vector in the feature sequence corresponds to a local region in the original image through the receptive field. The whole process of feature extraction can be formalized as:

$$\mathbf{F} = \mathcal{F}(\mathbf{X}) = \{\mathbf{f}^1, \mathbf{f}^2, ... \mathbf{f}^L\}. \tag{1}$$

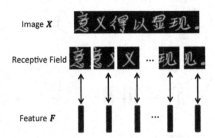

Fig. 2. Sliding windows on the feature level. Each vector is associated with a receptive field on the input image.

Prototype Classifier. We set $K + 1$ prototypes for characters and the blank in this paper, where K is the number of character categories. The prototypes are denoted as $\mathbf{C} = \{\mathbf{c}_1, \mathbf{c}_2, ..., \mathbf{c}_{K+1}\}$, $\mathbf{c}_i \in \mathbb{R}^d$. The blank in CTC can be denoted as the non-characters in the sliding window-based recognition framework. In order to make full use of non-character samples in the text lines and enhance the discriminant of the model, we set a prototype for the blank class. Following [22, 32], under the assumption of Gaussian distribution with equal identity covariance matrix, the negative euclidean distance can be used to measure the similarity between the feature $\mathbf{f}^t$ and the prototype $\mathbf{c}_k$. And the posterior probability that the t-th feature $\mathbf{f}^t$ belongs to category k can be defined as:

$$y_k^t = \frac{e^{-\gamma \|\mathbf{f}^t - \mathbf{c}_k\|_2^2}}{\sum_{k=1}^{K+1} e^{-\gamma \|\mathbf{f}^t - \mathbf{c}_k\|_2^2}}, \tag{2}$$

where γ is a hyper-parameter that controls the hardness of probability assignment. Therefore, the confidence distribution of the model output can be expressed as $\mathbf{Y} = \{\mathbf{y}^1, \mathbf{y}^2, ..., \mathbf{y}^L\}$, where $\mathbf{y}^t \in \mathbb{R}^{K+1}$ represents the probability distribution of the t-th feature over all classes.

3.2 Weakly Supervised Discrimination Loss

Unlike character classification, text line recognition is a weakly supervised learning task, i.e., the ground truth of each feature is unknown. Therefore, we use CTC loss as our discrimination loss, which can be directly learned from sequence labels, avoiding labeling the position of each character. In the CTC recognition framework, the input is a sequence $\mathbf{Y} = \{\mathbf{y}^1, \mathbf{y}^2, ...\mathbf{y}^L\}$. The corresponding label sequence is denoted as $\mathbf{l} = \{l_1, l_2, ..., l_T\}$, where $l_i (i \in \{1, ..., T\})$ denotes the ith character and T is the total character number. The conditional probability of a path π is defined as:

$$p(\pi|\mathbf{Y}) = \prod_{t=1}^{L} y_{\pi_t}^t, \tag{3}$$

where $y_{\pi_t}^t$ is the probability of generating the character π_t at timestep t. After that, the sequence-to-sequence mapping function $\mathcal{M}$ is performed by removing the repeated characters and blanks from the given path π. The conditional probability $p(\mathbf{l}|\mathbf{Y})$ is defined as the sum of probabilities of all π which are mapped by $\mathcal{M}$ onto $\mathbf{l}$.

$$p(\mathbf{l}|\mathbf{Y}) = \sum_{\pi \in \mathcal{M}^{-1}(\mathbf{l})} p(\pi|\mathbf{Y}). \tag{4}$$

Finally, the CTC loss function L_{ctc} is defined as the negative log-likelihood of the ground-truth conditional probability:

$$L_{ctc} = -\ln p(\mathbf{l}|\mathbf{Y}). \tag{5}$$

3.3 Weakly Supervised Generation Loss

Generation loss in our work is a supervised loss and can regarded as the maximum likelihood (ML) regularization under the standard Gaussian density assumption for class-specific features. Since there are no annotations for candidate characters in each text line, we can only calculate the generation loss through pseudo-labels. This subsection describes the generation of reliable pseudo-labels and robust L_{pl} computation based on dynamic weighting and pseudo-label filtering.

Pseudo-label Generation. We adopt the soft pseudo-label distribution z_k^t of feature $\mathbf{f}^t$ as in [3,4]:

$$z_k^t = \frac{\sum_{\{\pi|\pi \in \mathcal{M}^{-1}(\mathbf{l}), \pi_t=k\}} p(\pi|\mathbf{Y})}{p(\mathbf{l}|\mathbf{Y})}, k = 0, ..., K, \tag{6}$$

where the numerator is the probabilities of feasible alignment paths through character k at time t, and the denominator is the sum of the probabilities of all feasible alignment paths. Then the soft pseudo-label distribution matrix for a text line can be represented by $\mathbf{Z}$. Compared with semi-supervised learning using predicted probabilities as pseudo-labels, our method uses the alignment rule of CTC to integrate text line label l and so more reliable.

Pseudo-label Filtering. We use the best path decoding to decode the pseudo-label $\mathbf{Z}$ and use the decoding results to measure the reliability of pseudo-labels. The decoding result $\mathbf{s}$ is calculated by $\mathbf{s} = \mathcal{M}(\arg\max_\pi p(\pi|\mathbf{Z}))$, i.e., taking π_t with the maximum probability at each time step and mapping the π onto $\mathbf{s}$ by $\mathcal{M}$. If the decoding result $\mathbf{s}$ is the same as the real label l, the pseudo-label is considered to be reliable. Then, only the filtered reliable pseudo-labels are used to compute generation loss L_{pl}:

$$
L_{pl} = \begin{cases} 0 & \text{if } \mathbf{s} \neq \mathbf{l} \\ \sum_t \sum_{k \neq blank} z_k^t \|\mathbf{f}^t - \mathbf{c}_k\|_2^2 & \text{others} \end{cases}. \tag{7}
$$

Dynamic Weighting. Due to the inaccurate confidence of the model outputs in the early stage, a large number of unreliable pseudo-labels are generated. However, as the number of training epochs increases, generated pseudo-labels are relatively reliable. So we dynamically set the L_{pl} weight $\lambda(m)$, increasing with the number of training epochs:

$$
\lambda(m) = \begin{cases} 0 & m \leq m_s \\ \alpha \times e^{-5(1 - \frac{m - m_s}{m_{max}})} & m > m_s \end{cases}, \tag{8}
$$

where m is the current epoch, α and m_{max} are two hyperparameters that control the maximum value and increment of $\lambda(m)$. When $m < m_s$, the weight is set to 0. The network ensures the accuracy of classification by discrimination loss and improves the reliability of the network confidence. When $m > m_s$, the generation loss weighting function $\lambda(m)$ ramps up, starting from zero, along a Gaussian curve. The generation loss can be regarded as a further regularization to improve the model performance.

3.4 Consistency Regularization

Consistency regularization is introduced to improve the robustness and generate more reliable pseudo-labels. Consistent regularization assumes that the consistency probability distribution should be output for the same input, although slightly disturbed. Following [25], we regard dropout as the perturbation in model learning. Specifically, the input image $\mathbf{X}$ is fed to go through the forward pass of the model twice. Thus we can get two confidence distributions, $\mathbf{Y}_1$ and $\mathbf{Y}_2$. As shown in the top part of Fig. 1, at the training phase, the dropped units of the first path for the confidence distribution $\mathbf{Y}_1$ are different from that of the second path for distribution $\mathbf{Y}_2$, so the confidence distributions $\mathbf{Y}_1$ and $\mathbf{Y}_2$ are different for the same input. Therefore consistency regularization constrains the consistency of the predicted probabilities by minimizing the KL divergence between two output confidence distributions for the same sample, which is defined as:

$$L_{con} = 0.5(\mathcal{D}_{KL}(\mathbf{Y}_1 \| \mathbf{Y}_2) + \mathcal{D}_{KL}(\mathbf{Y}_2 \| \mathbf{Y}_1)). \tag{9}$$

Furthermore, we also use the dynamic weighting strategy to fuse consistency regularization loss L_{con}.

Then the discrimination loss is correspondingly defined as the combination of two forward propagations:

$$L_{ctc} = -0.5(\ln p(1|\mathbf{Y_1}) + \ln p(1|\mathbf{Y_2})). \tag{10}$$

3.5 Total Loss

We sum the loss functions defined above. The overall objective function for training our proposed model includes three parts: weakly supervised discrimination loss L_{ctc}, weakly supervised generation loss L_{pl}, and consistency regularization loss L_{con}. The final hybrid loss is defined as :

$$L = L_{ctc} + \lambda(m)L_{pl} + \lambda(m)L_{con}, \tag{11}$$

where $\lambda(m)$ is the dynamic weight of the generation loss and consistency regularization loss for multi-loss fusion.

4   Experiments

4.1   Datasets

We conduct both Chinese and digital handwritten text recognition experiments. For handwritten Chinese recognition, we evaluate the proposed approach on ICDAR-2013 [33]. We compare our method with the state-of-the-art methods and conduct a series of ablation studies to explore the effect of each part of our models. In addition, we validate the generalization of our approach by transferring character recognition to text line recognition with small sample size. For handwritten digital text recognition, we evaluate the performance of our method on the ORAND-CAR [1].

CASIA-HWDB [10] is a large offline Chinese handwriting database, which is divided into six sub-dataset. CASIA-HWDB1.0-1.2 contain 3,118,447 isolated character samples of 7,356 classes. HWDB2.0-2.2 have 52,230 text lines, which are segmented from 5,091 handwritten pages. **ICDAR2013** competition dataset [33] includes 3,432 text lines, which are segmented from 300 handwritten pages. For Chinese datasets, we use the character samples from CASIA-HWDB 1.0-1.2 to synthesize 700,000 synthetic text images following the method proposed by Wu et al. [26]. The synthetic and real text images from HWDB2.0-2.2 are used to train our model. **ORAND-CAR** [1] is a digital handwritten text line database. It contains 11,719 images in total, divided into CAR-A and CAR-B sub-datasets. The CAR-A database consists of 2,009 images for training and 3,784 images for testing. The CAR-B database contains 3,000 training images and 2,926 testing images.

4.2 Implementation Details

We implement experiments based on the framework of Pytorch with 4 NVIDIA RTX 24G GPUs. The architecture of CNN encoder is derived from the SeRes-Net1111 [9] with the residual and squeeze-and-excitation structures. We only set one prototype for each class, including the blank. All prototypes are initialized as zero vectors.

As for handwritten Chinese text recognition, the images are normalized to the height of 128 pixels and maintain their aspect ratios. The height and width of the feature map are $\frac{H}{128}$ and $\frac{W}{32}$, respectively. The Adam optimizer is applied to train our model with a learning rate initialized to 1×10^{-3}, and the learning rate will be decayed by timing 0.1 after 30 epochs. The training stops at 90 epochs. The weight decay is set to 1×10^{-4}. For the dynamic weighting in Sect. 3.3, we set m_s, m_{end} and α to 0, 60, and 0.001, respectively.

For the digital text recognition task, images are resized and padded to 32×256. Furthermore, the last three pooling layers in SeResNet1111 adopt 1×2 sized pooling windows instead of 2×2 to reduce feature dimension along the height axis only. Therefore, the shape of the feature map is $\frac{H}{32} \times \frac{W}{4} \times 512$. The model is also trained from scratch using an Adam optimizer with the base learning 1×10^{-3}, and the learning rate will be decayed by timing 0.1 after 3,000 iterations. The whole training process contains 10,000 iterations. We set m_s, m_{end} and α to 3,000, 6,000, and 0.001, respectively.

4.3 Ablation Experiments

Table 1. Recognition results of models with different training strategies on the ICDAR 2013 competition set without synthesized data and language model. (CR: correct rate; AR: accurate rate [26])

Methods	Without LM (%)	
	AR	CR
Linear Classifier + L_{ctc}	90.41	90.81
Proto + L_{ctc}	90.54	90.87
Proto + L_{ctc} + L_{pl}	90.63	90.92
Proto + L_{ctc} + L_{pl} + L_{con}	**90.96**	**91.26**

In this part, we design several variants of our model to validate the contributions of different components. We take real samples from CASIA-HWDB2.0-2.2

to train and evaluate our model on the ICDAR2013 data. In our model, K is set to 7357, including 7356 character classes in HWDB1.0-1.2 and one "unknown" token for characters not in the character dictionary. In the following subsection, we keep the same category setting.

The experimental results are shown in Table 1. Proto represents the prototype classifier. We can see that our model can achieve comparable performance with the traditional linear classifier. Furthermore, our model can perform better by introducing generation loss with dynamic weighting and pseudo-labels filtering. The weakly supervised generation loss can be used as the regularization to improve the generalization of the model. Meanwhile, dynamic weighting and pseudo-labels filtering make the weak-supervision training more stable. Consistency regularization aims to generate similar confidence distributions when the input is disturbed, thus improving the robustness of the model. Thanks to more reliable confidence distribution and pseudo-labels, performance can be further enhanced when we adopt both generation loss and consistency regularization.

4.4   Comparison with State-of-the-art Methods

Table 2. Comparison with existing methods on the ICDAR 2013 competition set. The results marked by "*" denotes using the powerful Transformer-based language model rather than the traditional n-gram language model, and the results marked by "†" denotes using contextual regularization to integrate contextual information. "†*" means using both the above two strategies.

Methods	Without LM (%)		With LM (%)	
	AR	CR	AR	CR
Wu et al. [26]	86.64	87.43	90.38	–
Wang et al. [23]	88.79	90.67	94.02	95.53
Wang et al. [24]	87.00	89.12	95.11	95.73
Gao et al. [3]	90.30	90.92	96.23	96.64
Peng et al. [12]	89.61	90.52	94.88	95.51
Xie et al. [30]	91.25	91.68	96.22	96.70
Xie et al. [28]	91.55	92.13	96.72	96.99
Liu et al. [9]	93.62	–	97.51*	–
Peng et al. [11]	93.05	93.30	–	–
Peng et al. [11]	**94.50**†	**94.76**†	**97.70**†*	**97.91**†*
Ours	93.66	93.90	97.04	97.23

In order to further improve the performance, we use synthetic and real text images for model training and experimental comparison. The comparison results with the existing methods are shown in Table 2, where we use a 5-gram statistical language model for context fusion. Without the language model, our method achieves comparable performance with AR 93.66% and CR 93.90%. The method proposed by Peng et al. [11] performs slightly better, owing to the contextual regularization with BLSTM layers. With the language model, our approach achieves comparable performance to Liu et al. [9] and Peng et al. [11]. However, we only use the 5-gram statistical language model instead of the transformer-based language model. In addition, we also compared the parameter sizes of different methods on the ICDAR-2013 dataset, and the results are shown in Table 3. It is worth noting that we only use SeResNet1111 as the feature extractor, making a good trade-off between performance and parameter size. Especially compared with the method in [9], our method achieves similar performance with about half the size of parameters.

Table 3. The parameter size comparison with different methods.

Methods	Params (MB)	AR
Wu et al. [26]	71 MB	86.64
Liu et al. [9]	203 MB	93.62
Peng et al. [11]	119 MB	94.50
Ours	115MB	93.66

For handwritten digital recognition, the comparison of our approach with state-of-the-art methods on ORAND-CAR is shown in Table 4. Our model achieves higher accuracy and reduces the error rate by 12% compared with the previous best result, demonstrating the effectiveness of our model.

Table 4. String accuracies of different models on the digital handwritten datasets.

Methods	CAR-A	CAR-B	Average
BeiJing [1]	80.73	70.13	75.43
FSPP [20]	82.61	83.32	82.97
CRNN [14]	88.01	89.79	88.90
ResNet-RNN [35]	89.75	91.14	90.45
Gao et al. [3]	94.83	94.70	94.77
Gao et al. [4]	95.01	94.74	94.88
Our	**95.53**	**95.49**	**95.51**

Figure 3 shows some recognition results of complex samples without the language model. Our method is robust to different writing styles and slanted texts. Meanwhile, good recognition results can be achieved for punctuation marks. Failure cases are mainly due to similar glyph characters. We visualize text location in Fig. 4. In the sliding-window based text recognition framework, the character classifier can recognize characters with high scores when these characters are

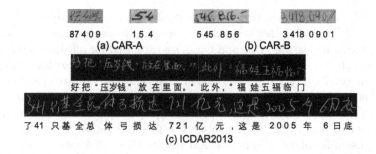

87 4 0 9   1 5 4   5 45 8 5 6   3 418 0 9 0 1

(a) CAR-A   (b) CAR-B

好 把 " 压 岁 钱 " 放 在 里 面 。 " 此 外 , " 福 娃 五 福 临 门

了 41 只 基 全 总 体 弓 损 达 7 2 1 亿 元 , 这 是 2 0 0 5 年 6 日 底

(c) ICDAR2013

Fig. 3. Visualization of recognition results for our proposed method.

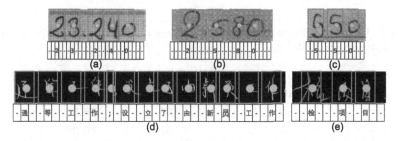

Fig. 4. Visualization of character locations, where yellow circles represent character centers. (Color figure online)

located in the center of the sliding windows. On the contrary, when the centers of sliding windows and characters are misaligned, the text recognizer will output blank labels or low character scores. Inspired by Non-Maximum Suppression, we can get the best candidate character center position by selecting the one with the highest confidence among adjacent character frames. The width of the bounding box can be obtained according to a prior of the character width. For Chinese characters in ICDAR-2013, we assume that the Chinese character width is $\frac{5}{8}$ of the image height, and for punctuation, it is $\frac{1}{4}$ of the height. For digital handwritten recognition, we assume that the digital character width is $\frac{1}{2}$ of the image height. So we can get the position of the characters by the center and width of the character bounding box.

4.5 Generalization Experiment

To demonstrate the generalization of our method, we construct a transfer learning experiment. We regard character images as short text lines to unify the character recognition and text line recognition into one framework. By sliding windows on character images, multi-frame features are generated. We can assume that the middle frame is the most aligned frame of each character and the edge frames are blank for prototype classifier training. CTC loss is uniformly used for linear classifiers trained on character and text line data. After ten epochs on character data pre-training, the linear classifier and our model can achieve comparable performance, i.e., AR being 96.15% and 96.04%, respectively. Then,

1%, 2%, 5%, and 10% real text lines are used for finetuning. Moreover, we also compare the linear classifier trained from scratch. In order to make a fair comparison, the same learning rate and batch size are used for all three models. The results are shown in Table 5. With only 1% real text lines for finetuning, our model can get 75.42% AR, 7.12% higher than finetuned linear classifier. However, when more text lines are used to finetune the model, the performance gap between the three models becomes smaller. This shows that with the increase in the number of text lines, the advantages of pre-training on isolated characters become insignificant. The experimental results show that our model has better generalization performance in transfer learning with fewer text lines. It can better apply the knowledge learned from character data to the text lines.

Table 5. AR under different percentages of training samples on the Chinese handwritten text dataset ICDAR-2013.

Sample rates (%)	LC (from scratch)	LC (finetuning)	Our (finetuning)
1	0	68.30	**75.42**
2	12.77	73.64	**78.36**
5	57.47	77.65	**80.30**
10	73.22	78.75	**80.97**

5   Conclusions

In this paper, we propose an efficient handwritten text recognition method based on the prototype classifier. By sliding windows at the feature level, our method is more efficient and can get character positions. Moreover, to improve the stability of the training and the performance of the recognition model, we propose dynamic weighting and pseudo-label filtering to weaken unreliable pseudo-labels. Furthermore, consistency regularization is used to give more reliable confidence distributions. Experimental results can demonstrate the effectiveness of our method. Furthermore, the transfer experiment from characters to text lines with small data size for fine-tuning proves that our proposed method has higher generalization.

Acknowledgements. This work has been supported by the National Key Research and Development Program Grant 2020AAA0109702, the National Natural Science Foundation of China (NSFC) grant 61936003.

References

1. Diem, M., et al.: ICFHR 2014 competition on handwritten digit string recognition in challenging datasets (HDSRC 2014). In: Proceedings of the 14th International Conference on Frontiers in Handwriting Recognition, pp. 779–784. IEEE (2014)
2. Du, J., Wang, Z.R., Zhai, J.F., Hu, J.S.: Deep neural network based hidden Markov model for offline handwritten Chinese text recognition. In: 2016 23rd International Conference on Pattern Recognition (ICPR), pp. 3428–3433. IEEE (2016)

3. Gao, L., Zhang, H., Liu, C.-L.: Handwritten text recognition with convolutional prototype network and most aligned frame based CTC training. In: Lladós, J., Lopresti, D., Uchida, S. (eds.) ICDAR 2021. LNCS, vol. 12821, pp. 205–220. Springer, Cham (2021). https://doi.org/10.1007/978-3-030-86549-8_14
4. Gao, L., Zhang, H., Liu, C.L.: Regularizing CTC in expectation-maximization framework with application to handwritten text recognition. In: 2021 International Joint Conference on Neural Networks (IJCNN), pp. 1–7. IEEE (2021)
5. Graves, A., Liwicki, M., Fernández, S., Bertolami, R., Bunke, H., Schmidhuber, J.: A novel connectionist system for unconstrained handwriting recognition. IEEE Trans. Pattern Anal. Mach. Intell. **31**(5), 855–868 (2008)
6. Kohonen, T.: The self-organizing map. Proc. IEEE **78**(9), 1464–1480 (1990)
7. Lee, D.H., et al.: Pseudo-label: the simple and efficient semi-supervised learning method for deep neural networks. In: Workshop on Challenges in Representation Learning, ICML, vol. 3, p. 896 (2013)
8. Li, H., Wang, P., Shen, C., Zhang, G.: Show, attend and read: a simple and strong baseline for irregular text recognition. In: Proceedings of the AAAI Conference on Artificial Intelligence, vol. 33, pp. 8610–8617 (2019)
9. Liu, B., Sun, W., Kang, W., Xu, X.: Searching from the prediction of visual and language model for handwritten Chinese text recognition. In: Lladós, J., Lopresti, D., Uchida, S. (eds.) ICDAR 2021. LNCS, vol. 12823, pp. 274–288. Springer, Cham (2021). https://doi.org/10.1007/978-3-030-86334-0_18
10. Liu, C.L., Yin, F., Wang, D.H., Wang, Q.F.: Casia online and offline Chinese handwriting databases. In: Proceedings of the International Conference on Document Analysis and Recognition, pp. 37–41. IEEE (2011)
11. Peng, D., et al.: Recognition of handwritten Chinese text by segmentation: a segment-annotation-free approach. IEEE Trans. Multimedia (2022)
12. Peng, D., Jin, L., Wu, Y., Wang, Z., Cai, M.: A fast and accurate fully convolutional network for end-to-end handwritten Chinese text segmentation and recognition. In: 2019 International Conference on Document Analysis and Recognition (ICDAR), pp. 25–30. IEEE (2019)
13. Samuli, L., Timo, A.: Temporal ensembling for semi-supervised learning. In: International Conference on Learning Representations (ICLR), vol. 4, p. 6 (2017)
14. Shi, B., Bai, X., Yao, C.: An end-to-end trainable neural network for image-based sequence recognition and its application to scene text recognition. IEEE Trans. Pattern Anal. Mach. Intell. **39**(11), 2298–2304 (2016)
15. Shi, B., Wang, X., Lyu, P., Yao, C., Bai, X.: Robust scene text recognition with automatic rectification. In: Proceedings of the IEEE Conference on Computer Vision and Pattern Recognition, pp. 4168–4176 (2016)
16. Snell, J., Swersky, K., Zemel, R.: Prototypical networks for few-shot learning. Adv. Neural Inf. Process. Syst. **30** (2017)
17. Srivastava, N., Hinton, G., Krizhevsky, A., Sutskever, I., Salakhutdinov, R.: Dropout: a simple way to prevent neural networks from overfitting. J. Mach. Learn. Res. **15**(1), 1929–1958 (2014)
18. Tarvainen, A., Valpola, H.: Mean teachers are better role models: weight-averaged consistency targets improve semi-supervised deep learning results. Adv. Neural Inf. Process. Syst. **30** (2017)
19. Wang, P., Yang, L., Li, H., Deng, Y., Shen, C., Zhang, Y.: A simple and robust convolutional-attention network for irregular text recognition. arXiv preprint arXiv:1904.01375 6(2), 1 (2019)

20. Wang, Q., Lu, Y.: A sequence labeling convolutional network and its application to handwritten string recognition. In: Proceedings of the International Joint Conference on Artificial Intelligence, pp. 2950–2956 (2017)
21. Wang, Q.F., Yin, F., Liu, C.L.: Handwritten Chinese text recognition by integrating multiple contexts. IEEE Trans. Pattern Anal. Mach. Intell. **34**(8), 1469–1481 (2011)
22. Wang, Q.F., Yin, F., Liu, C.L.: Improving handwritten Chinese text recognition by confidence transformation. In: 2011 International Conference on Document Analysis and Recognition, pp. 518–522. IEEE (2011)
23. Wang, S., Chen, L., Xu, L., Fan, W., Sun, J., Naoi, S.: Deep knowledge training and heterogeneous CNN for handwritten Chinese text recognition. In: Proceedings of the 15th International Conference on Frontiers in Handwriting Recognition, pp. 84–89. IEEE (2016)
24. Wang, Z.X., Wang, Q.F., Yin, F., Liu, C.L.: Weakly supervised learning for over-segmentation based handwritten Chinese text recognition. In: Proceedings of the 17th International Conference on Frontiers in Handwriting Recognition, pp. 157–162. IEEE (2020)
25. Wu, L., et al.: R-drop: regularized dropout for neural networks. Adv. Neural Inf. Process. Syst. **34** (2021)
26. Wu, Y.C., Yin, F., Chen, Z., Liu, C.L.: Handwritten Chinese text recognition using separable multi-dimensional recurrent neural network. In: Proceedings of the 14th IAPR International Conference on Document Analysis and Recognition, vol. 1, pp. 79–84. IEEE (2017)
27. Wu, Y.C., Yin, F., Liu, C.L.: Improving handwritten Chinese text recognition using neural network language models and convolutional neural network shape models. Pattern Recogn. **65**, 251–264 (2017)
28. Xie, C., Lai, S., Liao, Q., Jin, L.: High Performance offline handwritten Chinese text recognition with a new data preprocessing and augmentation pipeline. In: Bai, X., Karatzas, D., Lopresti, D. (eds.) DAS 2020. LNCS, vol. 12116, pp. 45–59. Springer, Cham (2020). https://doi.org/10.1007/978-3-030-57058-3_4
29. Xie, Q., Dai, Z., Hovy, E., Luong, T., Le, Q.: Unsupervised data augmentation for consistency training. Adv. Neural. Inf. Process. Syst. **33**, 6256–6268 (2020)
30. Xie, Z., Huang, Y., Zhu, Y., Jin, L., Liu, Y., Xie, L.: Aggregation cross-entropy for sequence recognition. In: Proceedings of the IEEE/CVF Conference on Computer Vision and Pattern Recognition, pp. 6538–6547 (2019)
31. Yang, H.M., Zhang, X.Y., Yin, F., Liu, C.L.: Robust classification with convolutional prototype learning. In: Proceedings of the IEEE Conference on Computer Vision and Pattern Recognition, pp. 3474–3482 (2018)
32. Yang, H.M., Zhang, X.Y., Yin, F., Yang, Q., Liu, C.L.: Convolutional prototype network for open set recognition. IEEE Trans. Pattern Anal. Mach. Intell. (2020)
33. Yin, F., Wang, Q.F., Zhang, X.Y., Liu, C.L.: ICDAR 2013 Chinese handwriting recognition competition. In: Proceedings of the 12th International Conference on Document Analysis and Recognition, pp. 1464–1470. IEEE (2013)
34. Yin, F., Wu, Y.C., Zhang, X.Y., Liu, C.L.: Scene text recognition with sliding convolutional character models. arXiv preprint arXiv:1709.01727 (2017)
35. Zhan, H., Wang, Q., Lu, Y.: Handwritten digit string recognition by combination of residual network and RNN-CTC. In: Liu, D., Xie, S., Li, Y., Zhao, D., El-Alfy, E.S. (eds.) ICONIP 2017. Lecture Notes in Computer Science, vol. 10639, pp. 583–591. Springer, Cham (2017). https://doi.org/10.1007/978-3-319-70136-3_62

Handwriting Datasets and Synthetic Handwriting Generation

Urdu Handwritten Ligature Generation Using Generative Adversarial Networks (GANs)

Marium Sharif[1][(✉)] [ID], Adnan Ul-Hasan[2] [ID], and Faisal Shafait[1,2] [ID]

[1] School of Electrical Engineering and Computer Sciences, National University of Sciences and Technology, Islamabad, Pakistan
{msharif.msee19,faisal.shafait}@seecs.edu.pk
[2] Deep Learning Lab, National Center of Artificial Intelligence, National University of Sciences and Technology, Islamabad, Pakistan
adnan.ulhassan@seecs.edu.pk

Abstract. Deep learning has significantly improved handwriting text recognition, esp. for Latin scripts. Arabic scripts including Urdu is a family of complex scripts and they pose difficult challenges for deep learning architectures. Data availability is a significant obstacle in developing Urdu handwriting recognition systems. Since gathering data is a costly and challenging task, there is a need to increase training data using novel approaches. One possible solution is to make a model that can generate similar yet different samples from the existing data samples. In this paper, we propose such models based on Generative Adversarial Networks (GANs) that have the ability to synthesize realistic samples similar to the original dataset. Our generator is class conditioned to produce Urdu samples of varying characters that differ in style. Visual and quantitative analysis convey that generated samples are of realistic nature and can be used to increase datasets. Synthesized samples integrated with the existing training set is shown to increase the performance of a handwriting recognition system.

Keywords: Generative adversarial networks · Handwriting generation

1 Introduction

Handwritten documentation of knowledge is known to be a great achievement of mankind: the transition into history from prehistory is marked by the first ever written record. Mostly historic events have been documented as handwritten markings and scripts. Handwritten data has numerous applications, especially in healthcare and financial sectors whereby all data is still being predominantly saved in handwritten form. All of this written data has to be processed one way or the other. Although, modern Optical Character Recognition Systems (OCRs) have exhibited good performance against printed text [1,2], handwritten text recognition is still lacking and could do with better performance results. This

© The Author(s), under exclusive license to Springer Nature Switzerland AG 2022
U. Porwal et al. (Eds.): ICFHR 2022, LNCS 13639, pp. 421–435, 2022.
https://doi.org/10.1007/978-3-031-21648-0_29

might be because of the lack of available data containing versatile handwritten text. This is more so in the case of Arabic script as it comprises of cursive characters [8]. The shape of the characters also varies with respect to its placement in a given word. The many diacritics and dots are used to define the correct grammar and pronunciation of the word. There is also a difference in the inter-word and intra-word spacing in the Arabic script. This holds true for all languages that make use of the Arabic Script including Urdu, Farsi, Sindhi, Punjabi and many others. Most of the words in the Arabic script have one or more ligatures that are made up of a combination of two or more characters. This behaviour of the Arabic script further create complications in addition to the fact that every writer has their own unique handwriting style.

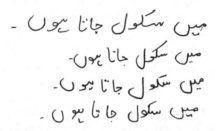

Fig. 1. Urdu phrase written by different writers. Some letters can get confused with others due to very similar shapes.

As in Fig. 1, major changes in shape, geometry, orientation and size of words can be seen as a result of difference in handwriting styles. Due to this property of the Arabic script, mostly work is done on Arabic ligatures, which are the connected components of the Arabic alphabet. This work is focused on generation of these ligatures to increase training data samples having versatility.

Our work is focused on networks designed for generative modeling known as Generative Adversarial Networks (GANs) [11]. These adversarial networks consists of two components; the discriminator D and the generator G. G synthesizes new image samples and tries to replicate the original data images. D then tries to distinguish whether the image being scrutinized is real or fake. The primary objective of this study is thus to design and build a system that is capable of generating handwritten text images for the Urdu language thereby extending already available datasets with realistic samples.

This paper is further organized as follows. Section 2 outlines the relevant work done previously. Section 3 provides the detail description of the proposed approach. Section 4 describes the experimental details and Sect. 5 discusses the results, compares them with other similar works and also evaluates performance enhancement of a text recognizer by extending dataset with generated images. Section 6 concludes the paper with a summary of our contributions and future directions.

2   Previous Work

Available training data is often limited as data gathering is a costly and challenging task. This constricts the ability of a learning classifier to be successfully trained. One possible solution for increasing training samples is data augmentation [18]. Different data augmentation methods, mostly affine transformations such as scaling and rotation are used in handwritten images. Another novel solution is generative models wherein images are synthesized from an arbitrary input. This method is relatively new, gaining popularity in the deep learning world and rapidly improving.

Generative Adversarial Networks (GANs) and auto encoders are two examples employing generative modeling. Both have certain drawbacks; Auto encoders output blurry images [6] while traditional GAN outputs incomprehensible and noisy images. Further variations in the architectures and loss function of the original GAN network were introduced, each resulting in different improvements causing generation of higher resolution images such as that required for medical applications.

Alonso et al. [4] presents a GAN architecture that is able to produce images that contain string of characters. The network comprises of discriminator D and generator G. Two other networks are also introduced; first is a bidirectional LSTM network and second is a convolutional neural network (CNN) that has convolutional layers with LSTM layers at the end. Their work focused on generation of French and Arabic handwritten strings of fixed length and width. The generated images were incorporated with the existing dataset and improved accuracy was observed.

Fogel et al. [10] introduces a GAN network where along with the discriminator D, the generated image is also passed through a text recognizer R for evaluation. G and D used in this work are fully convolutional. G is proposed as a concatenation of generators that are class conditioned where each class is a single individual alphabet. Each generator outputs a patch having its respective input alphabet. All these patches are then upsampled causing them to overlap and ultimately output a string of characters. Evaluation results by R incorporated with the discriminator loss is used for weight update.

Farooqui et al. [9] focuses on the task of improvement of hand written Urdu word spotting using generation of data samples. GAN variants were used to generate sample images of handwritten Urdu Ligatures for increasing the training data. Seven GAN variants are implemented. Each GAN has been trained to produce only a single class of Urdu ligature at a time or at most 10 Urdu ligatures for class conditioned variants of GANs.

Chang et al. [7] proposes an architecture that uses samples of a font to construct samples of another font. Architecture compromises of an encoder network that produces low dimensional representation for input image. Feature representation of output font is generated by a transfer module. The decoder network generates the target font character and discriminator network is used to classify this generated font character. The produced images are also evaluated by HCCRGoogleNet [19] classifier.

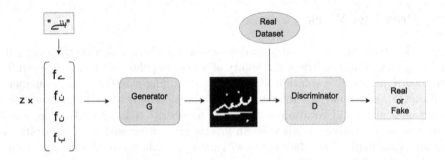

Fig. 2. Flowchart of proposed approach. For a four character ligature, four filters are concatenated. Note that filter for Urdu alphabet "ں" is used twice. Concatenated vector is multiplied with noise vector z and fed to G. Mix of resulting image from G and real data is given to D for evaluation.

3 Proposed Approach

The concept of GAN was first introduced by Goodfellow et al. in [11], whereby the network made use of two separate neural networks i.e. the Generator G and the Discriminator D. G approximates the training dataset by mapping random noise vector z and produces realistic data samples $G(z)$. We can express it as $G : G(z) \rightarrow R^{|x|}$ where $z \in R^{|z|}$ is latent vector noise, $x \in R^{|x|}$ is image generated from latent space and $|\cdot|$ is number of dimensions. The task of D, on the other hand, is to make an estimation such that $D : D(x) \rightarrow (0, 1)$, meaning it scores input image as either coming from the dataset (real$\simeq$1) or from the generator G (fake$\simeq$0). This approach permits G to learn about the underlying data distribution of the training dataset. In this way, the two network join in on a zero-sum min-max two player game. Figure 2 clearly demonstrates this concept being carried out for an Urdu ligature. Our approach focuses on reconstructing architectures for producing handwritten data samples for the Urdu language including all of its perplexing intricacies using an offline database. Analyzing the process in depth and the resultant samples also gives insight as to how the Urdu language poses challenges not present in case of the languages using the Latin script.

3.1 Fully Convolutional Generator

Handwriting is considered to be a local process when for this proposed network. Each handwritten character is only influenced by the letter before and after it respectively. Evidence supporting this theory can be found in previous works like [10] where they have successfully trained a generative architecture to produce strings of characters to form complete words for the Latin script. The proposed approach closely follows the concept of using a fully convolutional generator architecture as used in [10]. G can be thought of as generating each individual Urdu alphabet, instead of the whole ligature at once. Then due to the overlapping feature of receptive fields in CNN [16], influence of neighbouring letters will be taken into account and effect the overall output of the architecture. Consequently, the generator is seen to be a concatenation of multiple generators that are identical and class conditioned. The class of each generator is a single alphabetical character in the Urdu lexicon. A single patch containing the required character is produced by each individual generator.

Convolution layers with upsampling layers are used for each layer that widens the overlap between the neighbouring characters thereby widening the receptive field. This works in a similar fashion as convolutional-transpose layer, which is the opposite of a convolutional layer. This allows these neighbouring characters to interact and create smooth transition within a ligature. For every character in a given ligature, a filter f^* is chosen from the filter bank that is as wide as the Urdu alphabet lexicon. For a ligature containing four characters, four filters from the filter bank will be concatenated and then multiplied by latent vector z of comparable dimensions. Region generated by filter of each character f^* is of the same dimension and receptive field of adjacent filters end up overlapping. This allows for flexible size and versatile cursive style for the output character. The overlap is responsible for different alphabets combining together to form connected ligatures which is a necessity. Moreover, learning dependencies between neighbouring characters allows the generator network to create different shapes and variations of the same character depending on the adjacent characters. This behavior is specially desired in Arabic script which assumes different character shape depending on its position in a ligature.

3.2 Fully Convolutional Discriminator

In the traditional GAN, role of D is to accurately distinguish between the original data samples and samples synthesized by G. Similarly, in our proposed model, D is used to score images as real or fake. The dicriminator is fully convolutional, just like the generator with its architecture almost the opposite to that of G. Actual handwritten samples mixed in with the synthesized samples are both given as input to the discriminator that evaluates these images and gives output. This output is then used in the loss function to update the weights of both the generator and discriminator respectively.

3.3 Objective Function

Training of GAN is a delicate and unstable process that may result in blurry images due to the diversification of dataset. Previously, researchers have tried different customization and optimizations to achieve different GAN variations thus attaining better learning stability [17].

For generation of diverse and distinct image samples, different loss functions have been implemented. The first is DCGAN [16] that uses the same loss function as that of the Standard GAN but differs in the architectures of the G and D. Along with that, the other two implementations carried out are Wasserstein GAN (WGAN) [5] and Wasserstein GAN with Gradient Penalty (WGAN-GP) [12]. Using improved variations of standard GAN, provides better learning stability and helps in evading the potential mode collapse and balancing problems that are usually encountered with the training process of the traditional GAN.

DCGAN: Similar to standard GAN, a Deep Convolutional GAN (DCGAN) [16] employs the same training process and the same objective function. For DCGAN, the distinguishing factor from standard GAN is the change in architectures of G and D whereby convolutional layers replace the fully connected layers. Evidently, these layers prove to be more suitable for learning intrinsic properties of images. For G, transpose convolution (upsampling) is used. Objective function remains the same as that of standard GAN as is shown in Table 1.

Table 1. Objective functions for respective D and G networks

Model	Discriminator loss function	Generator loss function
DCGAN	$\max_D L_D = E_{x \sim p_{data}}[log(D(x))] + E_{z \sim p_z}[log(1 - D(G(z)))]$	$\min_G L_G = E_{z \sim p_z}[log(1 - D(G(z)))]$
WGAN	$\max_D L_D = E_{x \sim p_{data}}[D(x)] - E_{z \sim p_z}[D(G(z))]$	$\min_G L_G = -E_{z \sim p_z}[D(G(z))]$
WGAN-GP	$\max_D L_D = L_D - \lambda E_{z \sim p_{data}}[\|\|\nabla D(\alpha x + (1 - \alpha G(z)))\|\| - 1)^2]$	$\min_G L_G = -E_{z \sim p_z}[D(G(z))]$

WGAN: Conventionally in the training phase, G is pushed to produce samples whose distribution $p_g(x)$ matches real sample distribution $p_d(x)$. Ideally, this should work, but that is not always the case and gradient disappearance problem can occur making the training process unstable. To overcome the instability, Wasserstein distance is incorporated that quantifies the minimum cost that is utilized in transporting mass for converting data distribution q to data distribution p.

Wasserstein GAN (WGAN) [5] provides a much better gradient update for generator than the conventional cost function. Cost function is dependant upon D, also termed as the critic, satisfying strong conditional lipschitz continuity. For implementation purposes, D parameters are clipped to a certain range for lipschitz continuity. Respective loss functions of G and D are mentioned in Table 1.

WGAN-GP: Wasserstein GAN does not introduce any change in the architecture but rather improves performance by improvising the imposed constraint in WGAN. Limiting the D weights to comply to the conditional Lipschitz continuity causes the gradients to either explode or vanish. This problem was easily solved by applying a gradient penalty method as proposed by Gulrajani et al. in [12] and was named as Wasserstein GAN with Gradient Penalty (WGAN-GP). Weight trimming was replaced by calculation of weight gradient in accordance with the D network inputs which then penalizes the gradient norm so that it satisfies the Lipschitz constraint [5]. Modified objective function of discriminator is mentioned in Table 1.

4   Experiment

4.1   Dataset

To test the proposed network paradigm, we use UCOM database [3]. The dataset contains only 48 unique lines of Urdu text, written by 100 different authors. Adopting the scheme explained in [9], the ligatures are segmented out from images of Urdu sentences through binarization, segmentation and then resizing to get ligatures of fixed dimension. The Urdu language comprises of a total of 40 unique alphabets. Standalone Urdu alphabets are also considered to be ligatures of a single character, most of which have between 200 to 300 repetitions. This holds true for some of the most common used 2 and 3 character ligatures as well. All of this pre-processing yields a total of 317 unique handwritten ligatures with varying number of samples in each class. A total of approximately 32k sample ligatures were obtained from the dataset.

4.2   Implementation Details

The network architecture is set to generate images at a fixed size of 64×64 pixels. G comprises of a filter bank as large as the Urdu alphabet. Size of each filter has been set to 32×8192. As described in Fig. 2, for generation of a n-character ligature, n filters of the filter bank are selected in accordance to the characters. These filters are then concatenated and multiplied by a latent vector z to yield a vector of size 8×8192. This tensor is reshaped and then passed onto the convolutional layers followed by upsampling layer. LReLU and batch normalization [14] is used between these layers and a sigmoid activation function is used to produce the final output of size 64×64.

 D network is almost the opposite of the G network without the spatial embeddings layer, that is, the filter bank. An image of 64×64 is given as input to the discriminator, which is passed through a series of layers i.e., the convolutional layer, LReLU layer, batch normalization, and max pool layer. Last layer is a linear layer that outputs a single output representing the score or probability of image being *real* or *fake*.

 Same D and G networks are used for all three variants of GANs with only the varying loss functions being the determinant factor. Table 2 and Table 3 show implemented architectures of D and G respectively. For each GAN variant, different hyper parameter settings were explored and the ones with the lowest FID score were then used for the generation of samples. For each mini-batch, the weights of D are updated 5 times as compared to a single weight update for G. Weight clipping was employed for both networks and D loss was given a gradient penalty of 10 in case of WGAN-GP implementation Table 1.

Table 2. Generator architecture

Generator	Activation	Output shape
Embedding Layer × Latent Vector z	–	8192×8
Conv	LReLU	$32 \times 4 \times 256$
Batch Normalization	–	$32 \times 4 \times 256$
Conv	LReLU	$32 \times 8 \times 128$
Batch Normalization	–	$32 \times 8 \times 128$
Conv	LReLU	$32 \times 16 \times 128$
Batch Normalization	–	$32 \times 16 \times 128$
Conv	LReLU	$32 \times 32 \times 64$
Conv	LReLU	$64 \times 64 \times 64$
Conv	LReLU	$32 \times 64 \times 64$
Conv	LReLU	$16 \times 64 \times 64$
Conv	Sigmoid	$1 \times 64 \times 64$

Table 3. Discriminator architecture

Discriminator	Activation	Output shape
Input Vector	–	$64 \times 64 \times 1$
Conv	LReLU	$64 \times 64 \times 32$
Conv	LReLU	$32 \times 32 \times 64$
Conv	LReLU	$16 \times 16 \times 128$
Conv	LReLU	$16 \times 8 \times 256$
Conv	LReLU	$16 \times 8 \times 256$
Batch Normalization	–	$16 \times 8 \times 256$
Conv	LReLU	$16 \times 4 \times 256$
Batch Normalization	–	$16 \times 4 \times 256$
Conv	LReLU	$16 \times 4 \times 256$
Batch Normalization	–	$16 \times 4 \times 256$
Conv	LReLU	$16 \times 2 \times 256$
Linear	Sigmoid	1×1

5   Results

5.1   Qualitative Analysis

Application of GANs is the production of samples similar to the dataset whereby the performance of the generative models are analyzed using the quality of samples generated. Figure 3 shows the ligatures generated by different GAN variants. Ligatures generated by DCGAN show more diversity and have refined quality

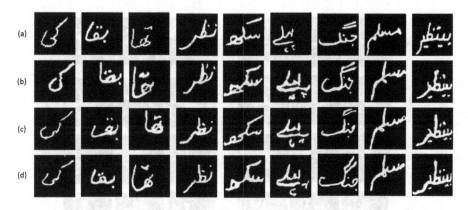

Fig. 3. Comparison of ligatures generated by GAN variants. Samples arranged in subfigures are (a) Original ligature samples (b) DCGAN (c) WGAN (d) WGAN-GP.

because of convolutional nature of the network architectures. This combined with the usage of techniques of batch normalization and Leaky ReLUs, increases the performance and stability of both the networks.

WGAN and WGAN-GP introduce further stability in the training process with imposed constraint on D to comply to the conditional Lipschitz continuity. Quality of sample ligatures produced are more diverse and finer as WGAN exercises reduction in distance between the generated $G(z)$ and real samples x. WGAN accurately enhances the model's capability to learn the probability distribution of the diversified ligatures belonging to the same class. This includes minuscule details such as a one or more dots or the tiny slash that is drawn diagonally over some of the ligatures. In comparison to WGAN, further quality improvement is seen in the samples produced by WGAN-GP. It practices further constraint for optimization of Wasserstein loss function. No hyperparameter tuning is required and successful training can be acheived for a number of image synthesizing tasks but, the convergence rate is the slowest in case of WGAN-GP as observed in the training process.

Comparison to Farooqui et al. [9] Main focus of their work was on the task of Urdu work spotting. They had explored various techniques for increasing dataset including augmentation and generative modeling. Seven GANs variants were implemented for dataset expansion to achieve better results for the aforementioned task. UNHD [3] dataset was used for this work. A single GAN was implemented to learn to generate only a single ligature class or at most 10 ligatures for class conditioned variants. Our model however is a sinlge architecture that is capable of producing all ligatures present in the dataset.

Since their main focus was not on the generation of images, their GAN variants had not been evaluated quantitatively and were subjected to qualitative evaluation only. Figure 4, subfigrure (a)–(g) displays the results of their work while ours is displayed in Fig. 4 subfigure (h). Observing these samples side by

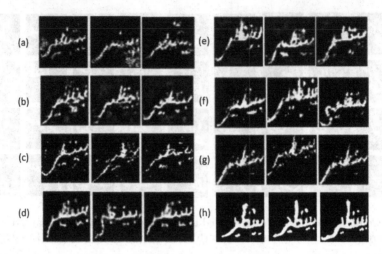

Fig. 4. Comparison of ligatures generated by different GANs for target ligature "بینظیر" ". Samples arranged in subfigures (a)–(g) are GAN variant outputs of Farooqui et al. [8] namely, (a) Standard GAN, (b) DCGAN, (c) CGAN, (d) CycleGAN, (e) ACGAN, (f) WGAN, (g) WGAN-GP, subfigures (h) are outputs from our models

side, it can be seen that for GAN variants (a)–(c), the results are a bit blurred and have additional pepper noise present in each sample. For GAN samples (e)–(g), the results are a bit better but alot of details that are crucial for the identification of ligatures are lost. As they have claimed in their work, visual inspection agrees that the samples from CycleGAN [20] gave the best results. This variant best learns the details enclosed in a ligature and finer samples than the rest are also achieved. Our samples however are still better in quality with lesser noisy pixels and are visually more resolute. They are more accurately detailed with cleaner and sharper edges around each character and dots are drawn with better pixel intensities.

Comparison to Alonso et al. [4]. Work in [4] was done on two databases namely the RIMES (French) and OpenHaRT (Arabic) database. Their model produced whole Arabic words rather than ligatures. All their evaluations were carried out on the RIMES dataset, no evaluation scores are available for the OpenHART database. Thus, qualitative comparison is carried out. As seen in Fig. 5, the results given by their model do pick up details but the edges of words are rather bleeding out and a little murky whereas our samples have more clearer and crisper edges while encapsulating all the finer details.

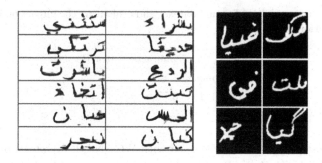

Fig. 5. Comparing results presented in Alonso et al. [6] on the left with our results on the right

Table 4. FID and GS comparison for GANs Variants

GANS	FID Score	Geometric score
DCGAN	21.45	7.82×10^4
WGAN	17.97	7.46×10^4
WGAN-GP	15.74	7.14×10^4

5.2 Quantitative Analysis

For evaluating the performance of proposed method quantitatively, Fréchet Inception Distance (FID) [13] and Geometric Score [15] were used. FID is used in the measurement of the feature distance between the generated and the real samples i.e. it measures the similarity between two sets of images. It is obtained by fitting two Gaussians on the feature representations of Inception Network and then calculating the Fréchet distance between them. GS compares the geometrical properties of the fundamental real and fake data manifold and provides a means to quantify mode collapse.

For every experiment conducted, FID was computed on the whole dataset vs equivalent number of generated samples i.e. approximately 32k samples, and GS was calculated on 5k real vs 5k generated samples with default parameter settings. Experiments run with different hyper parameters had FID calculated after every 10 iterations. Best FID was chosen from all experiments carried out and GS was also computed for this model setting. Visual inspection was relied upon for the verification of textual content. FID had shown to be in correlation with human judgement for visual quality of generated image samples and GS scores were also in favour of these findings.

Table 4 shows FID Scores and Geometric Scores computed for the different GAN implementations to better compare their performance. Lower score is better for both the indicators. As such, no quantitative results are available for comparison in the Urdu or Arabic language and hence Table 4 only shows results of our implementations for analysis. These are validated by observing the performance of a handwriting recognition system in the next section.

Table 5. Extending UCOM ligature dataset and evaluating the impact on handwriting recogniser performance

Data	Character Error Rate
UCOM only	7.12
UCOM + 15k	6.77

The lowest FID score was recorded with WGAN-GP architecture which was marked to be 15.74 at it's lowest. This score is comparable to state of the art scores available for other languages. DCGAN, with all its various settings, scored higher than both WGAN and its gradient penalty variant. This is also in accordance with the qualitative analysis where it was made evident that the other two architectures were better in picking up and thus producing more detailed samples as compared to DCGAN. Lowest possible FID score achieved by DCGAN was 21.45 while it was 17.97 with architecture of WGAN. GS scores were also in correlation with FID scores with WGAN-GP giving the lowest score, implying it to be the best model out of all three. Followed by WGAN and then DCGAN giving the highest score, GS scores proved true to the findings based on FID.

5.3   Data Generation for Handwritten Text Recognition

Main purpose for data generation using GANs is to increase performance of any model making use of respective dataset. For this purpose, an experiment is carried out to evaluate performance of a handwriting recognition system with and without generated samples. The recogniser consists of six convolutional layers, two Bidirectional Gated Recurrent Units (BiGRU) layers and a Connectionist Temporal Classification (CTC) output layer. Samples are created using WGAN-GP variant. An additional 50 samples are created for each ligature class resulting in approximately 15k synthesized samples. All samples of size 64×64 are scaled down and padded to have a size of 64×128 to be used as a training set for the recognition system. Table 5 evaluates recogniser performance in terms of Character Error Rate (CER). It is observed that recogniser performance is slightly increased by incorporating synthetic samples.

5.4   Discussion

Urdu alphabets, similar to the English alphabets, have varying widths and so does the resulting ligatures when these are joined together. Some ligatures, formed mostly from 6 or more alphabets, might result in cramped ligatures as a result of the initial pre-processing step. The proposed method fails to fully learn to distinguish between the small details in such ligatures and thus, squeezed variations of these ligatures are produced which do not clearly encapsulates all the details. An output width of 64 can thus be ruled as insufficient to cater to all characters required to be produced for ligatures with larger number of alphabets.

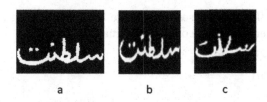

a   b   c

Fig. 6. Ligature "ﺳﻠﻄﻨﺖ". Subfigure (a) is the actual ligature sample without any resizing, subfigure (b) is resized sample after pre-processing steps, subfigure (c) is output of proposed model

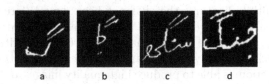

a   b   c   d

Fig. 7. Alphabet "ک". Subfigure (a) is single alphabet that is used as standalone ligature, subfigure (b)–(d) shows the various shape the alphabet acquires when used at the start, middle and end of ligature respectively.

Figure 6, subfigure (a) shows that the actual width of the ligature is almost twice that of the resized ligature in subfigure (b). Output from our architecture in subfigure (c) shows three alphabets jumbled together and are not easily distinguishable. There cases are rare for Urdu ligatures whereby ligatures largely are made of less than 6 alphabets and hence the proposed approach is sufficient for production of ligatures. For production of complete Urdu words however, generators with variable length outputs should be implemented.

Furthermore, the proposed model is capable of only producing 317 ligatures accurately for which it was trained. It fails to generate sequence of alphabets that are not present in the training set. Most of the Urdu alphabets change their shape depending on whether it is at the start, middle or end of a ligature and have a different shape when they are being used as standalone ligatures i.e. being used as a single character ligature. The generator is unable to learn these characteristics of the Urdu language fully. Giving character sequences for ligatures not available in the training set may result in production of arbitrary shapes that do not resemble any ligature from the Urdu language. Figure 7 shows the varying shapes the alphabet "ک" takes when used in different positions of a ligature.

In view of the above, we can conclude that the diversity in the Urdu alphabets differs from the English alphabets wherein there is some sort of a definite correlation between widths of the lowercase and uppercase forms of an alphabet. They are also considered to be different characters in the lexicon. Urdu alphabets however do not follow any such pattern, a single character has varying width and shape depending on its neighbouring characters that needs to be learned by the architecture. This also poses a challenge for implementing a generator with variable output length in case of Urdu language.

6 Conclusion

In this study, different architectures were implemented for generation of Urdu handwritten samples using a small dataset. Convolutional nature of architectures allows for cursive handwriting and synthesis of connected components which is a requirement for the generation of Urdu ligatures. A single generator is capable of producing multiple ligatures that has not been done before solely for Urdu language and evaluations were carried out that can be used later for future studies. Keeping in mind the complex nature of Urdu handwritten samples, the proposed approach, especially network architecture in conjunction with WGAN-GP objective function, is able to produce high quality images and give FID scores comparable to those present for state of the art in other languages. Lastly, the synthesized ligature samples were shown to improve performance of a handwriting recognition system validating the fact that even with a small dataset, performance for Urdu handwriting recognition can still be further improved with data generation.

The proposed adversarial model is a modified architecture used for the task of generation of Urdu handwriting samples which does not require any auxiliary networks and is not a massive network either as compared to others used for this task previously. Other GAN improvements and optimizations can be incorporated to the fully convolutional networks to further increase the quality of images. For generation of more diverse ligatures, a larger dataset, consisting of even more unique and complex ligatures, can be synthesized and used for training of GANs. The proposed model is sufficient for production of ligatures but variability in output sequence needs to be incorporated for production of Urdu words.

References

1. Amazon textract: intelligently extract text and data with OCR (2019)
2. Cloud vision API: detect text in images (2019)
3. Ahmed, S.B., Naz, S., Swati, S., Razzak, I., Umar, A.I., Khan, A.A.: UCOM offline dataset-an Urdu handwritten dataset generation. Int. Arab J. Inf. Technol. (IAJIT) 14(2) (2017)
4. Alonso, E., Moysset, B., Messina, R.: Adversarial generation of handwritten text images conditioned on sequences. In: 2019 International Conference on Document Analysis and Recognition (ICDAR), pp. 481–486. IEEE (2019)
5. Arjovsky, M., Chintala, S., Bottou, L.: Wasserstein generative adversarial networks. In: International Conference on Machine Learning, pp. 214–223. PMLR (2017)
6. Pierre, B.: Autoencoders, unsupervised learning, and deep architectures. In: Proceedings of ICML Workshop on Unsupervised and Transfer Learning, pp. 37–49. JMLR Workshop and Conference Proceedings (2012)

7. Chang, B., Zhang, Q., Pan, S., Meng, L.: Generating handwritten Chinese characters using cyclegan. In: 2018 IEEE Winter Conference on Applications of Computer Vision (WACV), pp. 199–207. IEEE (2018)
8. Dehghan, M., Faez, K., Ahmadi, M., Shridhar, M.: Handwritten Farsi (Arabic) word recognition: a holistic approach using discrete hmm. Pattern Recogn. 34(5), 1057–1065 (2001)
9. Farooqui, F.F., Hassan, M., Younis, M.S., Siddhu, M.K.: Offline hand written Urdu word spotting using random data generation. IEEE Access 8, 131119–131136 (2020)
10. Fogel, S., Averbuch-Elor, H., Cohen, S., Mazor, S., Litman, R.: Scrabblegan: semi-supervised varying length handwritten text generation. In: Proceedings of the IEEE/CVF Conference on Computer Vision and Pattern Recognition, pp. 4324–4333 (2020)
11. Goodfellow, I., et al.: Generative adversarial nets. Adv. Neural Inf. Process. Syst. 27 (2014)
12. Gulrajani, I., Ahmed, F., Arjovsky, M., Dumoulin, V., Courville, A.C.: Improved training of wasserstein gans. Adv. Neural Inf. Process. Syst. 30 (2017)
13. Heusel, M., Ramsauer, H., Unterthiner, T., Nessler, B., Hochreiter, S.: Gans trained by a two time-scale update rule converge to a local nash equilibrium. Adv. Neural Inf. Process. Syst. 30 (2017)
14. Ioffe, S., Szegedy, C.: Batch normalization: accelerating deep network training by reducing internal covariate shift. In: International Conference on Machine Learning, pp. 448–456. PMLR (2015)
15. Khrulkov, V., Oseledets, I.: Geometry score: a method for comparing generative adversarial networks. arXiv preprint arXiv:1802.02664 (2018)
16. Radford, A., Metz, L., Chintala, S.: Unsupervised representation learning with deep convolutional generative adversarial networks. arXiv preprint arXiv:1511.06434 (2015)
17. Salimans, T., Goodfellow, I., Zaremba, W., Cheung, V., Radford, A., Chen, X.: Improved techniques for training gans. Adv. Neural Inf. Process. Syst. 29 (2016)
18. Shorten, C., Khoshgoftaar, T.M.: A survey on image data augmentation for deep learning. J. Big Data 6(1), 1–48 (2019)
19. Zhong, Z., Jin, L., Xie, Z.: High performance offline handwritten Chinese character recognition using googlenet and directional feature maps. In: 2015 13th International Conference on Document Analysis and Recognition (ICDAR), pp. 846–850. IEEE (2015)
20. Zhu, J.Y., Park, T., Isola, P., Efros, A.A.: Unpaired image-to-image translation using cycle-consistent adversarial networks. In: Proceedings of the IEEE International Conference on Computer Vision, pp. 2223–2232 (2017)

SCUT-CAB: A New Benchmark Dataset of Ancient Chinese Books with Complex Layouts for Document Layout Analysis

Hiuyi Cheng[1], Cheng Jian[1], Sihang Wu[1], and Lianwen Jin[1,2(✉)]

[1] South China University of Technology, Guangzhou, China
{eechenghiuyi,eechengjian,eesihangwu,eelwjin}@mail.scut.edu.cn
[2] SCUT-Zhuhai Institute of Modern Industrial Innovation, Zhuhai, China

Abstract. Ancient books are the cultural heritage of human civilization, among which there are quite a few precious collections in China. However, compared to modern documents, the absence of large-scale historical document layout datasets makes the digitalization of ancient books still in its infancy and awaiting excavation and decryption. To this end, this paper proposes a large-scale dataset named SCUT-CAB for layout analysis of ancient Chinese books with complex layouts. The dataset is established by manually annotating 4000 images of ancient books, including 31,925 layout element annotations, which contains different binding forms, fonts, and preservation conditions. To facilitate the multiple tasks involved in document layout analysis, the dataset is segregated into two subsets: SCUT-CAB-Physical for physical layout analysis and SCUT-CAB-Logical for logical layout analysis. SCUT-CAB-Physical contains four categories, whereas SCUT-CAB-Logical contains 27 categories. Furthermore, the SCUT-CAB dataset comprises the labeling of the reading order. We compare various layout analysis methods for SCUT-CAB, i.e., methods based on object detection, instance segmentation, Transformer, and multi-modality. Extensive experiments reveal the challenges of layout analysis for ancient Chinese books. To the best of our knowledge, SCUT-CAB may be the first large-scale public available dataset for ancient Chinese document layout analysis. The dataset will be made publicly at https://github.com/HCIILAB/SCUT-CAB_Dataset_Release.

Keywords: Dataset · Ancient books · Document layout analysis

1 Introduction

Ancient books are the crystallization of human wisdom and non-renewable cultural resources. In addition to having primary historical research value, it is also a precious rare cultural relic and artwork. Resolving the contradiction between the protection and utilization of ancient books has become an urgent task, and

---

H. Cheng and C. Jian—Equal contribution.

© The Author(s), under exclusive license to Springer Nature Switzerland AG 2022
U. Porwal et al. (Eds.): ICFHR 2022, LNCS 13639, pp. 436–451, 2022.
https://doi.org/10.1007/978-3-031-21648-0_30

the digitization of ancient books could be a potential and valuable solution. The digitization of ancient books allows them to be more efficiently preserved, utilized, and transmitted without causing any damage to the original books. It can maximize ancient books' function as knowledge carriers and facilitate cultures' inheritance and dissemination.

Layout analysis is crucial in the digitization of ancient books. It is to identify objects in document images and determine the categories to which each region belongs, such as text, table, and figure. In recent years, layout analysis [24] has actively performed for automatic document retrieval, office automation, and other tasks. However, the layout analysis of ancient Chinese books was rarely investigated. The primary challenges encountered in this respect include (1) the lack of ancient Chinese book layout datasets. (2) the diversity and complexity of ancient book layouts. (3) the preservation conditions for ancient books, as ancient books inevitably degrade after centuries of usage.

Therefore, a layout dataset dedicated to the complex layout of ancient Chinese books is valuable to promote the research in ancient document processing. To this end, we propose a new dataset, SCUT-CAB, containing 4000 images of ancient Chinese books with various layouts. SCUT-CAB consists of two subsets: the SCUT-CAB-Logical dataset for logical analysis, which contains 31,925 layout element annotations in 27 categories, and the SCUT-CAB-Physical dataset for physical analysis, which contains 31,925 layout element annotations in four categories. Furthermore, the dataset includes the reading order of all the pages.

The contributions of this paper are summarized as follows:

- We propose the SCUT-CAB dataset, which, to the best of our knowledge is the first complex layout analysis dataset for ancient Chinese books.
- We propose two subsets: SCUT-CAB-Logical and SCUT-CAB-Physical, to facilitate the multiple tasks involved in layout analysis, we also provide the reading order of SCUT-CAB.
- Our dataset embodies both layout diversity and complexity. It derives from ancient books of different periods and layouts, which are in numerous different binding forms such as warp-fold binding, butterfly binding, and so on. In addition, the preservation quality of the ancient books varies, which further increases the variation of layouts.
- We provide comprehensive and comparative baseline experiments using different methods on the SCUT-CAB, including object detection-based, instance segmentation-based, Transformer-based and multi-modality-based methods, which demonstrate the challenge of our dataset and further enlighten future research.

2 Related Work

2.1 Existing Datasets

Existing datasets for document layout analysis can be categorized in modern and historical documents. We organized and listed the most frequently used datasets in Table 1.

Table 1. Document layout analysis datasets

Dataset	Year	No. images	Language	Physical	Logical	No.categories	Document type
Modern documents							
PRImA [1]	2009	305	English	✓	✓	10	Magazines, technical articles
PubLayNet [39]	2019	360K	English	✓	×	5	PDF articles on PubMed Central
DocBank [18]	2020	500K	English	×	✓	12	PDF files on arXiv.com
DSSE200 [37]	2017	200	English	×	✓	6	Pictures, brochure documents, old newspapers, PPT, scanned documents
Historical documents							
DIVA-HisDB [31]	2016	150	Italian, Latin	✓	×	3	Three medieval manuscripts from 11th and 14th century
Saint Gall [9]	2011	60	Latin	✓	×	3	9th century manuscript
Parzival [10]	2012	47	German	✓	×	2	13th century manuscript
George Washington [10]	2012	20	English	✓	×	2	18th century manuscript
IMPACT [26]	2013	Over 600k	18 language	×	✓	9	Newspapers, books, pamphlets typewritten notes
European Newspapers [7]	2015	528	13 language	✓	×	5	newspapers
HJDataset [30]	2020	2271	Japanese	✓	×	7	Biography scans
MTHv2 [22]	2020	3199	Chinese	✓	×	1	Tripitaka Koreana in Han (TKH), Multiple Tripitaka in Han(MTH)
							Include the annotation of text lines, no complete layout annotation information
SCUT-CAB	2022	4000	Chinese	✓	✓	27	Chinese ancient books containing different binding forms, fonts, and preservation conditions

The datasets of modern documents are abundant and well-developed. They contain various type of document such as PDF files and images for magazines, newspapers, articles, and books. For example, the PRImA Layout Analysis dataset [1] contains 305 pages of magazines and technical articles; PubLayNet [39] contains 360k PDF articles from modern research publications; DocBank [18] contains 500k PDF files from arXiv.com; DSSE [37] contains 200 brochures, documents and old newspapers.

However, the layout of ancient books is more complex than that of modern documents, particularly in terms of the arrangement of the text. The manifestation is that modern document text alignment is in the top-to-bottom, left-to-right, row-based manner, whereas ancient documents alignment in a top-to-bottom, right-to-left, column-based manner. Therefore, the existing layout analysis techniques and digitization tools for modern documents are not adaptive enough to identify the structure of the ancient books and recover the digitized results correctly.

There are still deficiencies in the development of ancient book datasets. (1)The layout categories of ancient books are not comprehensive. The MTHv2 [22] Chinese ancient book dataset only annotates the page box category, and the HJDataset [30] contains seven categories: Page Frame, Row, Title Region, Text Region, Title, Subtitle, and Other. (2) The scale of datasets is generally small. There are only 20, 47, 60, and 150 image pages in the datasets George Washington [10], Parzival [10], Saint Gall [9], and DIVA-HisDB [31], respectively. (3) Most of them are in Western languages, which hinders the exploration of layout analysis of ancient Chinese books. IMPACT [26]: large-scale datasets from European libraries; MTHv2 [22] is a dataset of ancient Chinese books. The DIVA-HisDB [31] is a Latin and Italian dataset; the European Newspaper Dataset [7] is a 13-language dataset. Parzival [10] is a German dataset, and George Washington [10] is an English dataset.

Currently, to the best of our knowledge, SCUT-CAB may be the first large-scale publicly available dataset for ancient Chinese document layout analysis, which contains a variety of complex layouts, professional and detailed definitions of layout analysis categories, and supports analysis in logical layout and physical layout.

2.2 Layout Analysis Methods

Document layout analysis methods are developed mainly from traditional image processing methods to data-driven deep learning-based methods.

Traditional layout analysis methods are generally based on complex heuristic rules. They can be grouped into top-down and bottom-up approaches. For top-down approaches, such as X-Y Cut [23], RLSA [34], and blank area analysis [2], they generally starts with a large document region and then divides the larger region into smaller regions, such as lines of text, according to some homogeneity rules. The analysis stops when no more regions are segmented or when certain stopping conditions are met. For Bottom-up approaches. It generally begin with a finer level of imagery, such as pixels, components, or text, progressively aggregated regions, and stop when a predefined analysis goal is reached, as in the Docstrum algorithm [25], Voronoi diagram-based algorithm [14] and run-length smearing algorithm [29]. However, traditional methods are limited to certain simple types of documents as they largely rely on heuristic rule design.

Modern layout analysis methods mainly involve the deep learning methods, which has been greatly developed in recent years. It can be divided into two categories: Detection-based and Segmentation-based approaches. For Detection-based approaches, Zhong et al. [39] directly used Faster R-CNN [28], and Mask-RCNN [11] for paper document element detection and classification. Li et al. [17] studied cross-domain document target detection based on the combination of Faster R-CNN and a domain adaptation module, where the page layout analysis is considered as an object detection task. For Segmentation-based approaches, Ma et al. [22] proposed a framework for ancient book layout analysis. They considered page layout analysis as a pixel-level classification problem and their framework can handle multiple tasks such as page segmentation, baseline extraction, and layout analysis.

3   SCUT-CAB Dataset

With the development of thousands of years, ancient books show diversity in content, layout, and image quality. Therefore, we construct a comprehensive dataset for ancient Chinese book layout analysis based on these three diversities.

3.1   Multiple Contents

We have collected ancient books of different content, including (1) Buddhist scriptures, mainly the Tripitaka, which is broad in content. They contain not only Buddhist teachings, but also discussions on politics, ethics, philosophy, literature, art, and customs. (2) Reproductions of Chinese Rare Editions Series, which covers the essence of Chinese cultural classics to the greatest extent. It has a high documentary value, academic research value and preservation value. All the books in this series are in the form of photocopying, According to the original book layout, they are photographed, printed, and collected, and they are carefully crafted. (3) Local chronicles, one of the essential sources for studying Chinese history over the past thousand years. They contain copious materials on local administration, local economies, local cultures, local dialects, local officials, and local dignitaries.

We selected 1000 images with multiple layouts from the MTHv2 [22] dataset, Then, we collected 962 images of Reproductions of Chinese Rare Editions Series and 2038 images of other ancient books with various contents and complex layouts, such as local chronicles, from the Internet. Hence, SCUT-CAB contains a total of 4000 images of ancient books.

3.2   Multiple Layouts

With the development of Chinese history and changes in book materials and production methods, various binding styles have been developed, including scroll binding, whirlwind binding, warp-fold binding, butterfly binding, wrapped-back binding, stitched binding, and photocopies. Different binding forms result in different layout styles and forms of illustration (see Fig. 1).

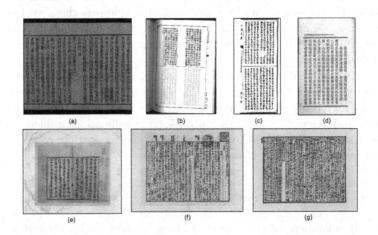

Fig. 1. Multiple layouts of SCUT-CAB. (a) Warp-fold binding. (b) Wrapped-back binding. (c) Photocopies. (d) Layout with the header. (e) Butterfly binding. (f) Layout with the marginal note. (g) Layout with the ear note.

Layouts included in SCUT-CAB are as follows:

- Warp-fold binding: It began in the late Tang Dynasty. It is a kind of improvement of the scroll packaging by the ancient Chinese Buddhist followers who learned the advantages of the traditional Indian binding method, they folded the Buddhist scriptures evenly left and right according to a certain number of rows and widths, as shown in Fig. 1(a).
- Wrapped-back binding: It began in the mid-to-late Southern Song Dynasty, with the pages folded back-to-back so that the side with the text is facing outward, as shown in Fig. 1(b).
- Photocopies: It refers to books printed with modern photographic technology. Most of them are wrapped-back binding and stitched binding. The structure of stitched binding and wrapped-back binding is roughly the same, as shown in Fig. 1(c).
- Butterfly binding: It is a product of engraving and printing technology in the Northern Song Dynasty. It folds the pages in half according to the center seam. The side with the printed text faces inward, as shown in Fig. 1(e).
- Other layouts: Due to differences between ancient books in different periods, there are some special layouts. (1) The Buddhist Sutras that contains a layout element "header", as shown in Fig. 1(d). (2) The butterfly bindings that contain a layout element "marginal note" (notes taken by ancient people while reading books), as shown in Fig. 1(f). (3) The butterfly binding that contains the category of "book ears" (which are at the top of the left column of the page box, and sometimes a tiny square appears, mainly found in the butterfly binding layout of the Song Dynasty), as shown in Fig. 1(g).

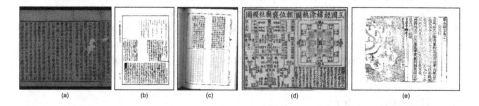

Fig. 2. Multiple image quality of SCUT-CAB. (a) Pages damaged. (b) Ink fading. (c) Back-through. (d) Exquisite. (e) Damages and unclear.

3.3 Multiple Image Quality

Ancient books have faded and corroded due to degradation from centuries of use. Therefore, in addition to the contents and layout diversity, we have taken into account multiple image qualities of the ancient books when constructing the dataset, including pages damaged (Fig. 2, a), ink fading (Fig. 2, b), and back-through (Fig. 2, c). Due to the rapid improvement printing techniques, modern photocopying technology has already surpassed the original lithography technology in terms of version identification and reproduction effect. This also makes the Reproductions of Chinese Rare Editions Series have more exquisite quality assurance, as shown in Fig. 2(d). The local chronicles datasets are mostly scanned black and white pictures, and there are many damages and unclear handwriting, as shown in Fig. 2(e).

3.4 SCUT-CAB-Logical and SCUT-CAB-Physical Subsets

Layout analysis can be classified into two categories: appearance-based and semantics-based layout analysis [16]. Appearance-based layout analysis is known as physical layout analysis, and its primary task is to separate different areas, such as text and figure. Semantic-based layout analysis is also known as logical layout analysis. It is to perform a more detailed logical analysis to understand the meaning of the different areas and then distinguish the semantic categories of the regions according to their meanings, such as heading, section heading, chapter, and paragraph.

Therefore, our dataset categorizes two subsets: SCUT-CAB-Logical and SCUT-CAB-Physical.

- **SCUT-CAB-Logical**: To better understand the meaning of different regions of ancient books, we define the categories of logical analysis data based on the basic knowledge of ancient books and "Introduction to Collation of Ancient Books" [13]. In this work, the following logical categories are annotated in SCUT-CAB: {EOV (end of the volume), author, bibliography, book number, caption, centerfold strip, chapter title, collation table, colophon, compiler, ear note, endnote, engraver, figure, foliation, header, interlinear note, marginal annotation, page box, part, section title, sub section title, subtitle, sutra number, text, title, volume number}. It contains 31,925 layout elements annotated in 27 categories. A sample of all categories is given in Fig. 3.
- **SCUT-CAB-Physical**: Physical analysis tasks do not need to classify as in detail as logical analysis tasks, so we modify the categories in SCUT-CAB-Logical. There are 4 categories: {centerfold strip, figure, page box, text}.

3.5 Reading Order

SCUT-CAB includes reading-order annotations. As shown in Fig. 4, the correct reading order is from top to bottom and left to right. We focused only on the reading order of the body text, and the labels outside were ignored, such as the volume number of sutra and centerfold strips in pages, as shown in Fig. 4(a) and 4(b). For photocopies, we filtered out labels that were extremely small, which could disrupt the reading order, as shown in Fig. 4(c).

3.6 Data Annotation

We used a quadrilateral bounding box for annotation for each label instance in each image. For skewed objects, we used multi-point annotation. The annotation was saved in COCO [21] format.

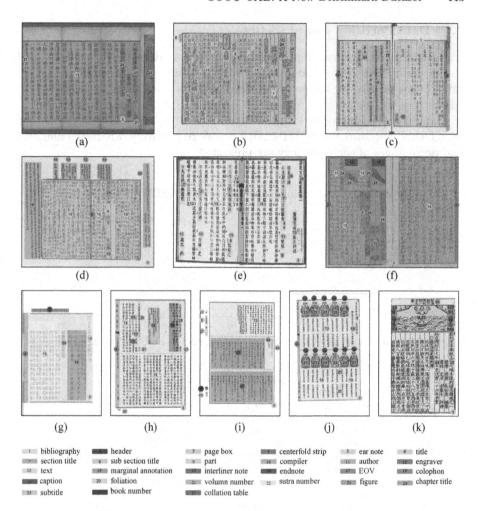

	bibliography		header		page box		centerfold strip		ear note		title
	section title		sub section title		part		compiler		author		engraver
	text		marginal annotation		interliner note		endnote		EOV		colophon
	caption		foliation		volumn number		sutra number		figure		chapter title
	subtitle		book number		collation table						

Fig. 3. Examples of SCUT-CAB images and annotations. 27 categories of layout elements are highlighted in different colors.

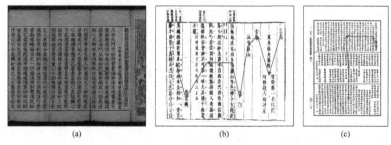

Fig. 4. The reading order of the SCUT-CAB dataset. (a) For the sutra dataset, we ignored the volume number and page number categories at the folds. (b) For the butterfly binding dataset, we ignored the centerfold strips categories and the marginal note categories outside the page box. (c) For the photocopies, we filtered out tiny labels, which could disrupt the reading order of the article.

3.7   Data Partition and Statistics

The Buddhist scriptures, Reproductions of Chinese Rare Editions Series, and the local chronicles were randomly splited into training and testing sets in a 4:1 ratio. Subsequently, we merged these three subsets' training and testing sets to ensure that the final segmented dataset contained different layout types.

Table 2 and Table 3 provide the statistics of the training and testing sets in the SCUT-CAB-Physical and SCUT-CAB-Logical, respectively, where the numbers and percentages of each category are presented.

Table 2. Statistics of training and testing sets in SCUT-CAB-Physical.

Category	Training		Testing	
	Number	Percentage (%)	Number	Percentage (%)
Centerfold strip	3036	11.84	757	12.05
Figure	361	1.41	64	1.02
Page box	3260	12.71	813	12.94
Text	18985	74.04	4649	73.99
Total	**25642**	**100**	**6283**	**100**

Table 3. Statistics of training and testing sets in SCUT-CAB-Logical.

Category	Training		Testing	
	Number	Percentage (%)	Number	Percentage (%)
EOV	17	0.07	4	0.06
Author	116	0.45	32	0.51
Bibliography	180	0.70	67	1.07
Book number	105	0.41	25	0.40
Caption	319	1.24	47	0.75
Centerfold strip	3036	11.84	757	12.05
Chapter title	2536	9.89	640	10.19
Collation table	25	0.10	8	0.13
Colophon	107	0.42	14	0.22
Compiler	1225	4.78	327	5.20
Ear note	205	0.80	42	0.67
Endnote	95	0.37	20	0.32
Engraver	53	0.21	10	0.16
Figure	361	1.41	64	1.02
Foliation	1035	4.04	250	3.98
Header	59	0.23	17	0.27
Interliner note	3	0.01	4	0.06
Marginal annotation	767	2.99	147	2.34
Page box	3260	12.71	813	12.94
Part	109	0.43	26	0.41
Section title	1081	4.22	241	3.84
Sub section title	151	0.59	41	0.65
Subtitle	1958	7.64	515	8.20
Sutra number	50	0.19	8	0.13
Text	7543	29.42	1875	29.84
Title	370	1.44	85	1.35
Volumn number	876	3.42	204	3.25
Total	**25642**	**100**	**6283**	**100**

4   Experiments

4.1   Baseline Models

Object Detection-Based Models: Page layout analysis can be considered an object detection task, identifying and locating layout objects within each page. We choose anchor-based one-stage methods, anchor-free one-stage methods, and anchor-based multistage methods as the baselines. Anchor-based one-stage methods adopt RetinaNet [20], YOLOv3 [27], and GFL [19]. Anchor-free one-stage methods adopt FCOS [32], FoveaBox [15]. Anchor-based multistage methods adopt Faster R-CNN [28], Cascade R-CNN [3], Mask R-CNN [11], Cascade Mask R-CNN [3], HTC [5], and SCNet [33].

Instance Segmentation-Based Models: Elements of the same class in the same image are located in different regions, and their font sizes may differ. Therefore, we regarded the layout object as a special instance and the page layout as an instance segmentation task. SOLO [35] and SOLOv2 [36] models are used to perform experiments.

Transformer-Based Models: Recently, query-based object detection frameworks have achieved promising performance. The main representatives are Deformable DETR [40] and QueryInst [8]. The former introduces deformable convolution into the DETR [4] framework, and the latter solves instance segmentation from a new perspective using parallel dynamic mask heads. This paper used Deformable DETR with iterative bounding box refinement and two-stage as the baselines.

Multi-modality Based Models: Recently, the layout analysis method based on multi-modality has also effectively developed. We adopt the VSR model as one of the baselines, VSR [38] model, which incorporates visual and two granularities of semantic information.

4.2   Experimental Settings

We evaluated the models mentioned above on the challenging SCUT-CAB benchmark. We report results with two standard backbones: a ResNet-50 [12] and a ResNet-101 [12].

Note that the aspect ratios of different categories of SCUT-CAB have significant differences, we adjust anchor ratios to ([0.0625, 0.125, 0.25, 0.5, 1.0, 2.0, 4.0, 8.0, 16.0]) accordingly instead of the original 3 anchor ratios ([0.5, 1.0, 2.0]) for anchor-based models.

Considering the different convergence of different models on the challenging SCUT-CAB benchmark, models are trained longer for 100 epochs to achieve higher performance. The learning rate is decayed at the 75th and 92nd epoch by a factor of 0.1, which is the same training scheme proportion as previous common settings [6]. We use multi-scale augmentation, resizing the input images so that the shortest side is at most 800 pixels and the longest one is at most 1333. In the inference stage, we keep the maximum image scale to 1333×800.

We adopt a standard COCO metric [21] consistent with Mask RCNN [11] to evaluate the performance of object detection and segmentation, including AP50, AP75, and average precision (AP). For the pure bounding box methods, the segmentation metrics were calculated using the detected bounding box as the segmentation mask. For the pure instances segmentation methods, we used the minimum bounding rectangle to calculate the metrics for the bounding box.

Table 4. AP50, AP75 and AP of each model on our testing sets.

Method	Backbone	Physical						Logical					
		Object Detection			Instance Segmentation			Object Detection			Instance Segmentation		
		AP_{50}	AP_{75}	AP	AP_{50}	AP_{75}	AP	AP_{50}	AP_{75}	AP	AP_{50}	AP_{75}	AP
Anchor-based one-stage													
RetinaNet [20]	R-50	91.6	80.9	73.9	91.5	80.4	73.1	76.7	57.4	52.4	76.6	56.5	52.0
RetinaNet [20]	R-101	91.5	82.9	74.7	91.5	81.6	73.8	78.3	61.2	55.1	78.3	61.7	55.0
YOLOv3 [27]	DarkNet-53	87.6	82.5	75.9	87.1	79.1	73.1	71.4	59.3	52.7	71.4	59.3	52.7
GFL [19]	R-50	92.4	74.8	72.9	92.5	72.6	71.9	77.0	60.1	53.5	74.6	58.5	52.9
GFL [19]	R-101	92.6	74.8	73.7	92.6	73.2	72.4	78.1	57.8	54.1	78.1	58.8	53.5
Anchor-free one-stage													
FCOS [32]	R-50	82.3	73.9	67.8	82.1	72.1	67.0	73.2	54.2	49.4	72.5	53.6	49.1
FCOS [32]	R-101	83.2	76.0	68.9	83.1	74.7	68.1	74.1	54.4	50.2	74.0	53.4	49.1
FoveaBox [15]	R-50	91.8	82.5	74.0	91.6	80.4	72.8	80.6	61.8	55.5	80.3	60.8	55.0
FoveaBox [15]	R-101	91.3	82.5	74.6	91.3	80.0	73.1	80.4	60.2	54.9	80.3	60.3	54.3
Anchor-based multi-stage													
Faster R-CNN [28]	R-50	92.2	86.5	77.9	91.8	83.5	75.7	77.6	59.8	53.3	77.5	58.8	52.8
Faster R-CNN [28]	R-101	91.3	86.1	77.5	91.0	83.4	75.3	77.4	61.3	54.9	77.3	60.6	54.2
Cascade R-CNN [3]	R-50	92.2	87.5	80.0	92.2	84.7	77.8	77.6	64.5	56.3	77.6	63.1	55.5
Cascade R-CNN [3]	R-101	91.4	87.8	79.9	91.4	84.8	77.4	77.5	62.3	55.9	77.5	60.9	55.4
Mask R-CNN [11]	R-50	91.6	84.6	77.1	91.3	87.0	77.9	78.4	61.2	54.8	77.9	62.0	54.6
Mask R-CNN [11]	R-101	92.1	87.7	79.1	91.7	87.2	79.5	78.5	61.9	55.1	77.7	63.1	55.3
Cascade Mask R-CNN [3]	R-50	91.7	86.5	79.3	91.2	85.9	79.3	78.0	63.4	56.7	77.9	62.3	56.2
Cascade Mask R-CNN [3]	R-101	92.1	88.6	80.9	92.1	88.4	81.0	78.0	62.7	56.8	77.9	61.8	56.3
HTC [5]	R-50	91.8	88.0	80.2	91.8	87.5	80.3	78.2	62.4	56.1	78.2	61.5	55.8
HTC [5]	R-101	92.8	**89.4**	**81.4**	92.8	<u>88.8</u>	<u>81.0</u>	80.1	65.2	58.3	80.0	63.1	58.0
SCNet [33]	R-50	92.6	87.3	79.9	92.6	87.3	80.3	83.0	69.1	60.1	82.6	68.7	60.3
SCNet [33]	R-101	**94.1**	<u>89.0</u>	<u>81.3</u>	**94.1**	**89.1**	**82.0**	83.6	67.3	60.2	83.6	68.0	60.3
Pure Instance Segmentation													
SOLO [35]	R-50	91.6	81.7	75.2	92.0	85.7	76.8	68.4	51.9	47.0	69.3	53.8	48.5
SOLO [35]	R-101	90.7	81.6	75.2	91.2	84.3	76.7	73.8	57.7	51.6	73.2	57.8	51.5
SOLOv2 [36]	R-50	<u>93.0</u>	80.3	74.9	<u>93.2</u>	86.2	79.3	72.8	49.7	47.7	73.9	56.8	50.8
SOLOv2 [36]	R-101	91.5	81.6	75.1	92.2	85.1	78.7	76.4	53.2	50.5	77.0	59.7	53.9
Query-based													
Deformable DETR [40]	R-50	92.3	87.1	79.9	92.1	84.3	77.9	**85.2**	**71.7**	**62.7**	**85.1**	**70.3**	**62.0**
Deformable DETR [40]	R-101	92.7	87.9	81.0	92.5	85.1	78.8	<u>84.6</u>	<u>69.8</u>	<u>61.6</u>	<u>84.6</u>	<u>69.9</u>	<u>61.1</u>
QueryInst [8]	R-50	92.1	87.5	79.9	92.1	87.3	79.8	80.9	65.6	58.1	80.6	63.7	57.6
QueryInst [8]	R-101	91.7	87.1	79.3	91.2	86.7	79.2	80.4	65.7	58.5	80.4	65.3	58.1
Multi-modality based													
VSR [38]	R-50	91.9	86.0	78.7	91.9	85.2	78.7	77.2	63.1	55.4	77.6	62.3	54.8
VSR [38]	R-101	90.4	85.5	78.5	90.4	84.5	78.2	78.3	61.6	55.7	78.2	61.1	55.1

Table 5. AP for each category in SCUT-CAB-Physical

	Centerfold-strip	Figure	Page-box	Text
RetinaNet [20]	66.0	59.0	97.7	72.7
YOLOv3 [27]	46.7	**72.6**	96.7	76.4
GFL [19]	49.7	65.5	98.6	75.9
FCOS [32]	39.6	66.8	95.2	70.6
FoveaBox [15]	61.9	62.8	94.5	73.3
Faster R-CNN [28]	70.4	58.0	97.9	75.0
Cascade R-CNN [3]	73.1	61.6	98.7	76.0
Mask R-CNN [11]	79.7	64.2	98.5	75.5
Cascade Mask R-CNN [3]	80.7	66.9	<u>98.8</u>	77.7
HTC [5]	<u>80.9</u>	66.9	<u>98.8</u>	77.6
SCNet [33]	**81.4**	<u>69.5</u>	98.7	**78.5**
SOLO [35]	75.9	60.4	97.3	73.3
SOLOv2 [36]	78.4	62.9	98.6	74.8
Deformable DETR [40]	71.9	66.3	<u>98.8</u>	<u>78.3</u>
QueryInst [8]	79.1	61.7	**98.9**	77.3
VSR [38]	78.8	59.8	98.6	75.6

4.3   Result and Analysis

Table 4 shows the performance of physical and logical analysis tasks under different models. Table 5 shows the results of AP@[0.5:0.95] for each category in SCUT-CAB-Physical on ResNet101 for different models. The experimental results show that one-stage method can achieve an effect comparable to other models at the AP50 in the Physical layout analysis task. However, the accuracy of AP75 and AP decreased significantly. This is because of the lower quality of the bounding box of the one-stage model at higher IoU levels than multi-stage models. In the physical analysis task, the effect of the SCNet [33] model is better. This is because SCNet involves self-calibration convolution, which enhances the feature conversion ability of CNN and enriches the output features, thereby allowing it to improve the performance without increasing model's size and complexity.

Table 6. AP for each category in SCUT-CAB-Logical.

	ID1	ID2	ID3	ID4	ID5	ID6	ID7	ID8	ID9	ID10	ID11	ID12	ID13	ID14
RetinaNet [20]	23.2	50.0	33.5	48.6	56.5	68.1	59.3	89.2	32.8	71.2	64.5	55.1	**47.5**	66.2
YOLOv3 [27]	6.0	52.8	45.1	_52.9_	35.2	54.5	63.8	**92.4**	37.8	**77.5**	61.9	49.3	40.5	**71.8**
GFL [19]	3.4	54.2	37.0	47.5	60.7	49.6	63.0	84.7	34.0	70.3	63.4	54.7	_44.2_	_66.4_
FCOS [32]	3.3	37.4	24.0	47.8	57.1	39.0	56.9	85.1	31.5	62.9	65.1	_59.6_	26.4	65.3
FoveaBox [15]	20.6	46.8	36.3	52.1	59.0	61.2	59.0	81.7	37.9	66.8	62.1	52.5	37.5	60.5
Faster R-CNN [28]	5.8	_57.6_	36.3	51.0	61.6	70.4	57.9	91.0	_40.6_	71.5	64.4	47.7	34.6	60.0
Cascade R-CNN [3]	11.6	53.0	41.2	50.6	61.8	72.3	59.2	90.2	38.5	72.1	65.4	53.0	40.1	62.7
Mask R-CNN [11]	11.6	55.1	38.0	52.3	60.5	78.8	58.1	85.9	40.6	71.5	65.1	57.7	40.8	59.5
Cascade Mask R-CNN [3]	20.6	**63.0**	48.7	48.5	62.1	_79.6_	58.2	85.5	35.0	75.3	64.6	55.5	40.7	66.4
HTC [5]	8.5	54.7	**50.5**	50.9	62.1	**80.7**	61.1	90.3	37.9	_75.8_	**65.6**	58.1	44.0	65.0
SCNet [33]	24.8	48.5	_49.6_	52.6	_66.0_	**80.7**	_65.0_	90.2	35.8	_75.8_	_65.5_	**64.1**	35.6	65.6
SOLO [35]	6.5	46.7	33.1	18.5	60.3	76.0	61.7	81.1	39.4	69.3	60.0	58.3	10.4	66.4
SOLOv2 [36]	2.1	49.0	29.3	41.9	58.0	71.6	58.4	85.7	39.2	64.8	57.8	43.9	16.7	60.9
Deformable DETR [40]	**42.2**	54.5	45.5	**55.1**	**67.8**	71.4	**65.3**	82.4	**40.9**	74.6	62.2	54.7	31.1	62.6
QueryInst [8]	_29.0_	36.5	36.5	46.4	58.7	77.0	61.9	89.1	39.1	70.3	60.7	49.0	27.2	61.4
VSR [38]	25.6	47.1	36.3	48.3	57.3	78.6	59.2	_91.9_	38.9	73.0	60.7	56.9	32.5	58.7

	ID15	ID16	ID17	ID18	ID19	ID20	ID21	ID22	ID23	ID24	ID25	ID26	ID27
RetinaNet [20]	46.2	55.4	17.5	67.9	98.2	52.3	44.3	29.2	73.2	41.4	89.9	54.3	50.9
YOLOv3 [27]	**50.2**	0.0	2.5	**73.8**	98.1	52.8	46.9	26.9	**75.2**	**47.8**	**93.1**	_56.4_	**56.6**
GFL [19]	46.9	53.8	8.6	69.0	_98.7_	_53.3_	45.6	23.8	72.8	43.2	90.4	53.1	53.6
FCOS [32]	45.5	58.1	13.7	66.8	95.8	49.7	37.7	13.9	68.4	44.3	86.6	50.7	49.9
FoveaBox [15]	43.5	54.9	46.1	66.9	94.5	52.9	46.6	34.0	69.6	39.8	87.8	47.4	49.4
Faster R-CNN [28]	45.7	57.5	33.0	65.1	97.8	50.0	40.7	15.2	71.3	43.2	89.6	51.7	51.6
Cascade R-CNN [3]	46.3	62.9	43.0	66.7	_98.7_	46.4	42.4	12.5	71.5	42.8	90.4	50.8	50.6
Mask R-CNN [11]	_47.4_	58.3	20.6	66.5	98.0	50.2	42.0	20.5	71.3	41.4	89.5	54.8	52.2
Cascade Mask R-CNN [3]	45.9	59.7	**71.1**	69.3	_98.7_	50.8	42.3	21.4	72.4	43.6	91.7	50.7	52.4
HTC [5]	46.8	_64.3_	51.1	67.4	**98.8**	50.4	47.1	19.5	_74.0_	43.2	92.2	54.6	51.6
SCNet [33]	47.0	62.9	61.5	70.9	98.7	50.0	_56.0_	_38.8_	73.8	_47.4_	_92.3_	55.2	_55.0_
SOLO [35]	35.2	62.7	30.4	66.4	97.0	49.4	46.8	36.0	70.2	29.9	89.1	50.5	42.4
SOLOv2 [36]	39.0	59.2	19.4	66.5	98.3	47.3	46.1	29.3	67.4	32.0	87.8	49.1	43.8
Deformable DETR [40]	44.3	**67.7**	_69.0_	_72.3_	**98.8**	**54.5**	**56.6**	**53.7**	73.0	46.7	92.3	**58.6**	52.6
QueryInst [8]	45.8	57.5	37.6	64.1	**98.8**	47.8	49.4	29.2	71.3	41.0	90.1	52.6	48.7
VSR [38]	46.2	58.4	37.9	64.7	98.7	42.7	42.7	29.8	71.6	35.8	91.0	52.8	50.7

*ID1:EOV,   ID2:author,   ID3:bibliography,   ID4:book   number,   ID5:caption, ID6:centerfold   strip,   ID7:chapter   title,   ID8:collation   table,   ID9:colophon, ID10:compiler,   ID11:ear   note,   ID12:endnote,   ID13:engraver,   ID14:figure, ID15:foliation,   ID16:header,   ID17:interlinear   note,   ID18:marginal   annotation, ID19:page box, ID20:part, ID21:section title, ID22:sub section title, ID23:subtitle, ID24:sutra number, ID25:text, ID26:title, ID27:volumn number.

Table 6 show the results of AP@[0.5:0.95] for each category in SCUT-CAB-Logical on ResNet101. In the logical analysis task, the experimental results show that the effect of the Deformable DETR [40] yields most of the best performances in the logical layout analysis task. Because Deformable DETR introduces a multi-level attention mechanism to better capture virtual features, Deformable DETR is still able to identify fewer sample categories, such as EOV, under conditions of extreme sample imbalance.

5 Conclusion

To facilitate the digitization of ancient Chinese books, we proposed the SCUT-CAB dataset, a new large-scale dataset of ancient Chinese books with complex layouts. This dataset contains two subsets: SCUT-CAB-Logical and SCUT-CAB-Physical. SCUT-CAB-Physical provides four appearance-based categories for physical layout analysis tasks that separate the different regions. SCUT-CAB-Logical provides 27 logical categories for logical layout analysis tasks to understands the meaning of different regions. Furthermore, we conducted comprehensive experiments on our dataset, including object detection, instance segmentation, Transformer-based object detection, and multi-modality methods on SCUT-CAB dataset. Experimental results show that the results of the existing methods for the task of analyzing the layout of Chinese ancient books are unsatisfactory. The main difficulty is the extreme imbalance of samples and the distinct logical categories. Therefore, better methods are urgently needed for Chinese ancient book layout analysis.

Acknowledgement. This research is supported in part by NSFC (Grant No.: 61936003), GD-NSF (no. 2017A030312006, No.2021A1515011870), Zhuhai Industry Core and Key Technology Research Project (no. 2220004002350), and the Science and Technology Foundation of Guangzhou Huangpu Development District (Grant 2020GH17)

References

1. Antonacopoulos, A., Bridson, D., Papadopoulos, C., et al.: A realistic dataset for performance evaluation of document layout analysis. In: Proceedings of the International Conference on Document Analysis and Recognition, pp. 296–300 (2009)
2. Breuel, T.M.: Two geometric algorithms for layout analysis. In: Lopresti, D., Hu, J., Kashi, R. (eds.) DAS 2002. LNCS, vol. 2423, pp. 188–199. Springer, Heidelberg (2002). https://doi.org/10.1007/3-540-45869-7_23
3. Cai, Z., Vasconcelos, N.: Cascade R-CNN: delving into high quality object detection. In: Proceedings of the IEEE Conference on Computer Vision and Pattern Recognition (2018)
4. Carion, N., Massa, F., Synnaeve, G., Usunier, N., Kirillov, A., Zagoruyko, S.: End-to-end object detection with transformers. In: Vedaldi, A., Bischof, H., Brox, T., Frahm, J.-M. (eds.) ECCV 2020. LNCS, vol. 12346, pp. 213–229. Springer, Cham (2020). https://doi.org/10.1007/978-3-030-58452-8_13
5. Chen, K., Pang, J., Wang, J., et al.: Hybrid task cascade for instance segmentation. In: Proceedings of the IEEE Conference on Computer Vision and Pattern Recognition (2019)
6. Chen, K., Wang, J., Pang, J., et al.: MMDetection: open MMlab detection toolbox and benchmark. arXiv preprint arXiv:1906.07155 (2019)
7. Clausner, C., Papadopoulos, C., Pletschacher, S., et al.: The ENP image and ground truth dataset of historical newspapers. In: Proceedings of the International Conference on Document Analysis and Recognition, pp. 931–935 (2015)
8. Fang, Y., Yang, S., Wang, X., et al.: Instances as queries. In: Proceedings of the IEEE International Conference on Computer Vision, pp. 6910–6919 (2021)

9. Fischer, A., Frinken, V., Fornés, A., et al.: Transcription alignment of latin manuscripts using hidden Markov models. In: Proceedings of the 2011 Workshop on Historical Document Imaging and Processing, pp. 29–36 (2011)
10. Fischer, A., Keller, A., Frinken, V., et al.: Lexicon-free handwritten word spotting using character HMMs. Pattern Recogn. Lett. **33**(7), 934–942 (2012)
11. He, K., Gkioxari, G., Dollar, P., et al.: Mask R-CNN. In: Proceedings of the IEEE International Conference on Computer Vision (2017)
12. He, K., Zhang, X., Ren, S., et al.: Deep residual learning for image recognition. In: Proceedings of the IEEE Conference on Computer Vision and Pattern Recognition (2016)
13. Huang, Y.: Introduction to Collation of Ancient Books (in Chinese). Shaanxi People's Publishing House (1985)
14. Kise, K., Sato, A., Iwata, M.: Segmentation of page images using the area Voronoi diagram. Comput. Vis. Image Underst. **70**(3), 370–382 (1998)
15. Kong, T., Sun, F., Liu, H., et al.: Foveabox: beyound anchor-based object detection. IEEE Trans. Image Process. **29**, 7389–7398 (2020)
16. Lee, J., Hayashi, H., Ohyama, W., et al.: Page segmentation using a convolutional neural network with trainable co-occurrence features. In: Proceedings of the International Conference on Document Analysis and Recognition, pp. 1023–1028 (2019)
17. Li, K., Wigington, C., Tensmeyer, C., et al.: Cross-domain document object detection: benchmark suite and method. In: Proceedings of the IEEE Conference on Computer Vision and Pattern Recognition (2020)
18. Li, M., Xu, Y., Cui, L., et al.: DocBank: a benchmark dataset for document layout analysis. In: Proceedings of the International Conference on Computational Linguistics, pp. 949–960 (2020)
19. Li, X., Wang, W., Wu, L., et al.: Generalized focal loss: learning qualified and distributed bounding boxes for dense object detection. In: Advances in Neural Information Processing Systems, vol. 33, pp. 21002–21012 (2020)
20. Lin, T.Y., Goyal, P., Girshick, R., et al.: Focal loss for dense object detection. In: Proceedings of the IEEE International Conference on Computer Vision (2017)
21. Lin, T.-Y., et al.: Microsoft COCO: common objects in context. In: Fleet, D., Pajdla, T., Schiele, B., Tuytelaars, T. (eds.) ECCV 2014. LNCS, vol. 8693, pp. 740–755. Springer, Cham (2014). https://doi.org/10.1007/978-3-319-10602-1_48
22. Ma, W., Zhang, H., Jin, L., et al.: Joint layout analysis, character detection and recognition for historical document digitization. In: Proceedings of the International Conference on Frontiers in Handwriting Recognition, pp. 31–36 (2020)
23. Nagy, G., Seth, S., Viswanathan, M.: A prototype document image analysis system for technical journals. Computer **25**(7), 10–22 (1992)
24. Namboodiri, A.M., Jain, A.K.: Document structure and layout analysis. In: Digital Document Processing: Major Directions and Recent Advances, pp. 29–48. Springer, London (2007). https://doi.org/10.1007/978-1-84628-726-8_2
25. O'Gorman, L.: The document spectrum for page layout analysis. IEEE Trans. Pattern Anal. Mach. Intell. **15**(11), 1162–1173 (1993)
26. Papadopoulos, C., Pletschacher, S., Clausner, C., et al.: The IMPACT dataset of historical document images. In: Proceedings of the International Workshop on Historical Document Imaging and Processing, pp. 123–130 (2013)
27. Redmon, J., Farhadi, A.: YOLOv3: an incremental improvement. arXiv preprint arXiv:1804.02767 (2018)
28. Ren, S., He, K., Girshick, R., et al.: Faster R-CNN: towards real-time object detection with region proposal networks. In: Advances in Neural Information Processing Systems, vol. 28 (2015)

29. Saini, R., Dobson, D., Morrey, J., et al.: ICDAR 2019 historical document reading challenge on large structured chinese family records. In: Proceedings of the International Conference on Document Analysis and Recognition, pp. 1499–1504 (2019)

30. Shen, Z., Zhang, K., Dell, M.: A Large dataset of historical Japanese documents with complex layouts. In: Proceedings of the IEEE Conference on Computer Vision and Pattern Recognition Workshops (2020)

31. Simistira, F., Seuret, M., Eichenberger, N., et al.: DIVA-HisDB: a precisely annotated large dataset of challenging medieval manuscripts. In: International Conference on Frontiers in Handwriting Recognition, pp. 471–476 (2016)

32. Tian, Z., Shen, C., Chen, H., et al.: FCOS: fully convolutional one-stage object detection. In: Proceedings of the IEEE International Conference on Computer Vision (2019)

33. Vu, T., Kang, H., Yoo, C.D.: SCNet: training inference sample consistency for instance segmentation. In: Proceedings of the AAAI Conference on Artificial Intelligence, pp. 2701–2709 (2021)

34. Wahl, F.M., Wong, K.Y., Casey, R.G.: Block segmentation and text extraction in mixed text/image documents. Comput. Graph. Image Process. **20**(4), 375–390 (1982)

35. Wang, X., Kong, T., Shen, C., Jiang, Y., Li, L.: SOLO: segmenting objects by locations. In: Vedaldi, A., Bischof, H., Brox, T., Frahm, J.-M. (eds.) ECCV 2020. LNCS, vol. 12363, pp. 649–665. Springer, Cham (2020). https://doi.org/10.1007/978-3-030-58523-5_38

36. Wang, X., Zhang, R., Kong, T., et al.: SOLOv2: dynamic and fast instance segmentation. In: Advances in Neural Information Processing Systems, vol. 33, pp. 17721–17732 (2020)

37. Yang, X., Yumer, E., Asente, P., et al.: Learning to extract semantic structure from documents using multimodal fully convolutional neural networks. In: Proceedings of the IEEE Conference on Computer Vision and Pattern Recognition (2017)

38. Zhang, P., Li, C., Qiao, L., et al.: VSR: a unified framework for document layout analysis combining vision, semantics and relations. In: Proceedings of the International Conference on Document Analysis and Recognition, pp. 115–130 (2021)

39. Zhong, X., Tang, J., Jimeno Yepes, A.: PubLayNet: largest dataset ever for document layout analysis. In: Proceedings of the International Conference on Document Analysis and Recognition, pp. 1015–1022 (2019)

40. Zhu, X., Su, W., Lu, L., et al.: Deformable DETR: deformable transformers for end-to-end object detection. In: Proceedings of the Conference on Learning Representations (2021)

A Benchmark Gurmukhi Handwritten Character Dataset: Acquisition, Compilation, and Recognition

Kanwaljit Kaur[1(✉)], Bidyut Baran Chaudhuri[2], and Gurpreet Singh Lehal[1]

[1] Punjabi University, Patiala, Punjab, India
kanwalvirk54@gmail.com
[2] Techno India University, Kolkata, India

Abstract. Gurmukhi script is used to write the official 'Punjabi' language of the people of the western part of Indian Punjab. The script is having approximately 160 million native speakers. Recognition of handwritten characters in the Gurmukhi script is still in its embryonic stage due to intricate character shapes and the scarcity of standard datasets. This paper introduces a new large-scale benchmark dataset "Gurmukhi_HWdb1.0" which is an important development in the handwritten character recognition of this script. This dataset has a total of 137,700 handwritten samples of 41 basic Gurmukhi characters and 10 numeral classes. Out of these, 110,160 images are used for training,13,770 images are set aside for validation, and 13,770 images are used for testing. Here, 265 individuals have contributed to the development of the dataset. Recognition of the script is carried out using a CNN architecture based on transfer learning on the VGG16 network. We fine-tuned the model and added our own fully connected layers needed for Gurmukhi characters. The proposed model is executed on this collected "Gurmukhi_HWdb1.0" dataset for evaluation. A detailed comparison with different batch sizes is performed to understand the functionality of the model. Experimental results show that the proposed model can be benchmarked against the concerned dataset with a test accuracy of 98.42% for Gurmukhi characters and 97.51% for Gurmukhi numerals.

Keywords: Transfer learning · VGG16 · Handwritten character recognition · "Gurmukhi_HWdb1.0"

1 Introduction

Identification and recognition of handwritten characters and digits are some of the prominent tasks of computer vision society [1]. It has achieved an abundance of interest for its diverse applications stretching from writer identification to recognizing numerals and alphabets in the number plate of traffic to bank check processing [2]. However, handwritten character recognition poses various challenges due to the prevailing problems in data acquisition and the variable

© The Author(s), under exclusive license to Springer Nature Switzerland AG 2022
U. Porwal et al. (Eds.): ICFHR 2022, LNCS 13639, pp. 452–467, 2022.
https://doi.org/10.1007/978-3-031-21648-0_31

writing styles of people. The Gurmukhi script includes a noteworthy character set with more curves, loops, and other details of the characters. The characters may be written differently by different writers. So, the script demands attention and a solution with the help of an efficient recognition system. The primary requirement of doing such work is a reasonable dataset written by a good number of people. The unavailability of a reasonable dataset poses a hindrance to the research efforts for this script. There is a lack of an exhaustive and extensive publicly available dataset for the Gurmukhi script. The importance of such handwritten datasets in developing the OCR to identify the history, culture, literature, and past events of people using the script is explained in [3]. There are some publicly available benchmark datasets for scripts like Chinese [4,5], Latin [6,7], Arabic [8,9], and Korean [10]. Among the Indic scripts, benchmark datasets have been developed for Bangla, Devanagari, Gujarati, Telugu, Tamil, Kannada, and Malayalam. Indian Statistical Institute, Kolkata, has done the pioneering work of developing datasets of some Indic scripts like Bangla [11], Devanagari [12], and Oriya [11]. CMATER Kolkata has developed datasets for numerals and characters of Devanagari [13,14] , Telugu [13], Bangla [15], and Urdu. Datasets for isolated characters of Tamil, Telugu, and Devanagari were developed by HP Lab India [16,17]. Efforts are also being done for the development of the Kannada handwritten character dataset for numerals and sentences [18]. As far as the Gurmukhi script is concerned, we could only find seven small size publicly available character datasets introduced in [19] (See Table 1). The problems with the above Gurmukhi datasets are small size, incomplete character and numeral sets, a small number of writers, and less diversity in age, gender, educational, and professional background, etc. In the present paper, attempts are made to overcome these shortcomings of the datasets. We introduce a new dataset that is larger and more significant than previous datasets. It considers all the basic characters and numerals of the script and the dataset is collected from students, professors, farmers, housewives, and retired persons covering a large range of Punjabi-speaking populations of different ages, gender, and professions, making it much more unconstrained. After the presentation in ICFHR, the developed

Table 1. Existing Gurmukhi datasets.

Dataset	Number of writers	Number of samples per class	Number of character classes	Training samples	Test samples	Total samples
HWR-Gurmukhi_1.1	1	100	35	2,450	1,050	3,500
HWR-Gurmukhi_1.2	10	10	35	2,450	1,050	3,500
HWR-Gurmukhi_1.3	100	1	35	2,450	1,050	3,500
HWR-Gurmukhi_2.1	1	100	56	3,920	1,680	5,600
HWR-Gurmukhi_2.2	10	10	56	3,920	1,680	5,600
HWR-Gurmukhi_2.3	100	1	56	3,920	1,680	5,600
HWR-Gurmukhi_3.1	200	1	35	4,900	2,100	7,000

dataset will be made publicly available to the research community. The rest of the paper is organized as follows: Sect. 2 introduces the script history. Section 3 provides the details of dataset creation and organization. Section 4 explains the proposed recognition methodology in detail. Section 5 shows the experimental results and performance analysis of the proposed model with the existing works. Finally, the Conclusion and future scope of work are discussed in Sect. 6.

2 Gurmukhi Script

Gurmukhi is the script primarily used to write Punjabi, an Indo-Aryan language spoken by people in the Punjab region of India and Pakistan. Punjabi is the first language of about 160 million people and is the 9th most spoken language in the world [20]. The origin of the Gurmukhi script [21] is attributed to Guru Angad dev Ji, the second Sikh guru who did modification and rearrangement of earlier characters and shaped them into a proper script. The Sikh gurus adopted 'proto-Gurmukhi' to write their holy scripture 'Sri Guru Granth Sahib'. Gurmukhi follows the Abugida type of writing system and is written from left to right direction. Modern Gurmukhi has thirty-five original letters, hence commonly termed as 'painti' plus six additional consonants, nine vowel diacritics, two diacritics for nasal sounds, and one diacritic that geminates the consonants, and three subscript characters, as given in Fig. 1. Six additional consonants are created by placing a dot at the foot of the consonant and are referred to as 'Navin Varg or 'Navin toli'. These are most often used for loanwords and pronunciation is different from the parent consonants giving subtle inflections to the tongue or throat. To express vowels, Gurmukhi uses diacritics called 'Laga-Matra'. One of the vowels has no symbol called 'Mukta' means 'liberated'. It is pronounced between every consonant where no other diacritic is present. This inherent vowel sound can be changed using the other nine dependent vowels attached to the bearing consonant. The first three letters out of 35 basic letters are the vowel carriers and are used to construct independent vowels where 'oora' and 'eeri' cannot be used independently without diacritics. 'Aira' when used independently produces the sound 'a' as in 'about'. The diacritics for gemination and nasalization are together called 'lagakkhar'. Gemination or consonant lengthening is an articulation of a consonant for a longer period than the utterance of a singleton consonant. It is often perceived as a doubling of the consonant. Three subscript letters called 'Pairin akhar' are utilized to make consonant clusters. Other than the comma, exclamation, and other western punctuation symbols, Gurmukhi also has its own three punctuation symbols called 'visarga' (represents abbreviation), 'dandi' (to mark the end of a sentence), and 'dodandi' (to mark the end of a verse). Gurmukhi also has its own set of numerals used extensively in the older text. Gurmukhi script was added to the Unicode Standard in October 1991 with the release of version 1.0. During the data collection process, the samples for 41 basic characters, 10 numerals, and three pairin akhar were included. As 'Pairin akhar' is always used with other consonants and cannot be used independently, we excluded these and only the samples of 41 basic characters and 10 numerals are considered for the experiment in this paper.

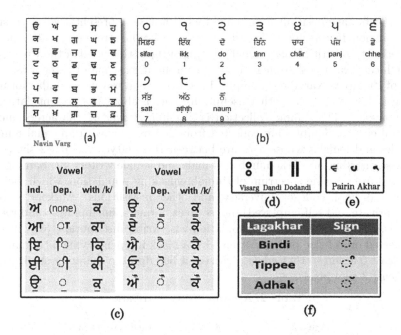

Fig. 1. (a) Gurmukhi alphabets (b) Gurmukhi numerals (c) Vowels (d) Punctuations (e) Pairin akhar (f) Lagakhar

Research efforts in HCR of Gurmukhi script gained momentum very late compared to other world and Indic scripts. Most of the works in the literature have examined conventional handcrafted techniques for detecting handwritten Gurmukhi units such as SVM [22–24], SVM with linear kernel [25] , SVM with RBF kernel [26], KNN [23,27], Random Forest [28] and Wavelet-based recognizer [29]. We have come across very few recognition efforts using deep learning as in [30,31]. The major reasons for this are that deep neural networks need a large amount of training data and when trained from scratch, require a substantial amount of time. In this article, a novel deep CNN architecture based on Transfer Learning (CNN-TL) on VGG16 is employed. CNN is more powerful in learning coarse to fine features compared to conventional classifiers, thus delivering excellent recognition results. We evaluated our proposed model on the developed dataset and showed that higher accuracies can be obtained compared to the state-of-the-art methods.

3   Dataset Development

3.1   Collection of Dataset Samples

One significant contribution of our work is the pioneering development of a big Gurmukhi handwritten isolated character and numeral dataset. Before collecting data, efforts have been made to make the dataset as representative as possible. The common factors for handwriting style variations are age, gender, profession, education, writing instrument, writing surface, the mood of the writer, etc. Enough care was taken to include samples from at least major categories under each of the above variables. Samples have been collected in a tabular form, as shown in Fig. 2(a). A few other useful information about the present dataset is as follows: (1) The sample collection is distributed over 41 characters and 10 numeral classes. Samples are collected from 265 persons with an equal number of males and females whose ages are between 15 to 60 years. Table 2 shows the age-wise distribution of the number of male and female writers. (2) Samples are collected from a wide spectrum of Punjabi-speaking population, which includes students (school, college, university, and other institutions), workers (government, private and self-employed), housewives, farmers, unemployed, and retired persons. Figure 2(b) shows the population-wise writer distribution. (3) The data is collected on A4 size paper written with a black ink pen having a tip diameter of at least 0.6 mm so that when digitized at 300 dpi, the character stroke width is at least 3 to 4 pixels.

Table 2. Age-wise writer distribution

Age (Low-High)	No. of males	No. of females	Total
15–26	25	41	66
27–38	39	40	79
39–50	37	40	77
51–60	32	11	43
Total	133	132	265

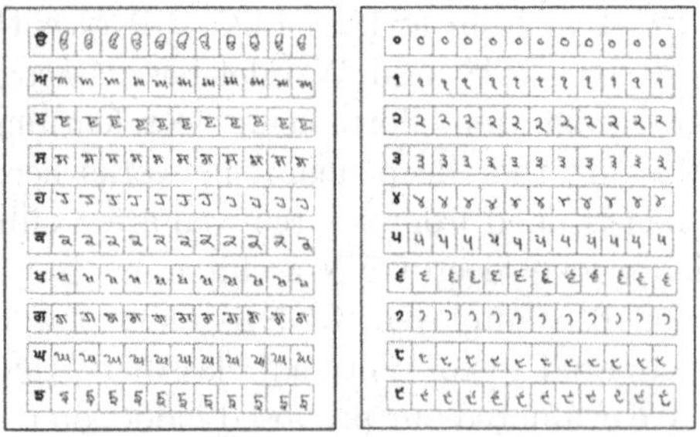

Category	No. of Writers
Research Scholars	28
Students	58
Academicians(College, University Professors)	24
Non-Academicians(Clerical and other Non-Teaching Persons)	28
Workers(other employees or Self-employed)	24
Unemployed(House-wives and other unemployed)	51
Farmers	32
Retired Persons	20
Total	265

(a)   (b)

Fig. 2. (a) Some samples of the collected Gurmukhi dataset (b) Population-wise writer distribution

(4) On A4 size paper, a matrix of horizontally arranged cells is drawn. In the first column of the matrix, one type of character is printed. There are 11 cells horizontally in subsequent columns where the writer writes that character repeatedly. In this form, the characters are written sequentially along each horizontal string of boxes. Some of the filled samples are illustrated in Fig. 2(a). (5)

The demographic information of the writer such as Name, age, gender, occupation, etc., is also included. This information can be used for other studies such as handwriting individuality analysis. (6) In addition to the character dataset, we have generated a Gurmukhi paragraph-level dataset, the details of which will be communicated in a separate manuscript.

3.2 Dataset Creation

All the collected samples are scanned using Canon CanoScan LiDE 120 scanner at 300dpi and saved as grayscale images in TIFF file format. For the dataset creation, scanned images are initially binarized using an experimented threshold. To crop the rows from the tabular form, a histogram of the image is generated in which peaks and valleys show the presence of text and space. After extracting useful rows, we detect the boundary area around these lines using the Hough transformation.

Table 3. Gurmukhi_HWdb1.0 Details

Dataset	Type	Number of writers	Number of classes	Number of samples per class	Total samples	Training dataset (80%)	Validation dataset (10%)	Test dataset (10%)
Gurmukhi_HWdb1.0	Character	265	41	2,700	110,700	88,560	11,070	11,070
Gurmukhi_HWdb1.0	Numeric	265	10	2,700	27,000	21,600	2,700	2,700

We have applied eight-connectivity on outer boundaries obtained through Hough transform in the previous stage to delete all boundary lines. Finally, vertical and horizontal projection profiles are applied to extract each character from the whole image. The cropped characters are not size-normalized. These are stored in original sizes ranging from 45 × 38 to 83 × 97 pixels. Initially, each character class has 2,915 images. We manually excluded some of the completely unrecognizable samples. Thus, there exist 2,700 images of each handwritten Gurmukhi character and numeral class (See Table 3).

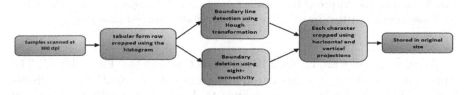

Fig. 3. Steps in dataset creation

Among the total 137,700-character images, 110,700 images are of 41 Gurmukhi character classes and 27,000 are of 0–9 Gurmukhi numerals. The above steps of dataset creation have been carried out using MATLAB R2010a version 7.10, as shown in Fig. 3. Figure 4 shows the steps in which the characters are cropped out of the document page. Some random samples of the first three consonants and first three numerals are shown in Fig. 6(a). The developed dataset is henceforth named Gurmukhi_HWdb1.0.

3.3 Shape Characteristics of the Samples

A few observations from the detailed study of the shape of handwritten samples of the dataset are as follows: (1) The character samples have either no headline (Fig. 5a(ii)), the headline is significantly oblique (Fig. 5a(i)), or extended headlines (Fig. 5a(iii)) introducing the additional variability among the samples of the same class. (2) Handwritten samples are found where all parts of the characters are not properly joined, thus giving it a different shape. A few examples are shown in Fig. 5(b). (3) There exist several groups of shape similar characters in which samples of one class look like the samples of another class.

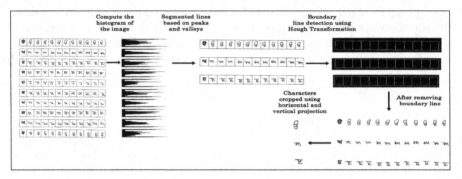

Fig. 4. Steps for cropping out characters from tabular forms.

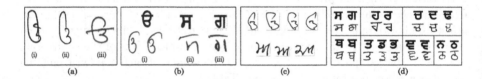

Fig. 5. (a) Variable nature of the presence of headline (b) Disjoint parts of characters (c) Some samples written by elderly persons (d) Similar shaped characters

All these samples are retained in the present dataset (see Fig. 5(d)). (4) The writing style of elderly persons introduces greater shape variations and distortions as shown in Fig. 5(c), hence helping to develop a generalized recognition model.

4   Recognition Scheme

In this paper, we have implemented VGG16 based deep CNN model to identify the Gurmukhi characters. This CNN architecture uses the concept of transfer learning where a pretrained VGG16 net is used with our training dataset. The VGG16 is originally trained on a subset of the ImageNet dataset, a collection of over 14 million images belonging to 22,000 categories. This network has achieved a 92.7% classification accuracy in the 2014 ImageNet Classification Challenge. The VGG16 architecture mainly consists of 13 convolutional layers, 5 max-pooling layers, and 3 dense layers, as shown in Fig. 6(b). We performed several experiments with the original VGG16 but the results were not so good. In an experiment with the original pretrained architecture, although we achieved 92% training accuracy, the test accuracy was only 49.58%. This lower accuracy rate is mainly due to training set overfitting. So, we detached its fully connected layer (FCL) and inserted a new FCL to compensate for the variations in the source and target tasks. Fine-tuning is also performed as shown in Fig. 7. To fine-tune our model following three steps are performed: i) Instantiate the VGG16 convolutional base and load its weights, ii) Freeze the convolutional layers of the original pre-trained model up to block 3 and Fully connected layers are also not included. iii) Then added our own fully connected network.

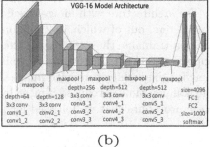

(a) (b)

Fig. 6. (a) Some samples of cropped characters (b) VGG16 architecture

We initialized the network with previously learned weights to prevent large gradient updates and then fine-tuning is performed on the convolutional weights. We only fine-tuned the convolutional block 4 and 5 instead of the whole network to prevent overfitting which can occur due to the high entropic ability of the entire network. The rest of the blocks are kept unchanged. Our fully connected network consists of two dense layers of 512 and 256 units with ReLU activation function and the last output dense layer of 41 units (for 41 Gurmukhi character classes) and 10 units (for 10 Gurmukhi numeric classes) with softmax activation function. CNN architectures may be biased towards the data encountered during training and may not generalize well for other samples. That's why, the network parameters are regularized using L1 and L2 regularization methods to avoid overfitting. We have used Kernel regularizers and activity regularizers to apply penalty on the layer's kernel and layer's output respectively.

The chosen value for the L1 and L2 regularization factors is 0.01. We have also used data augmentation to increase variations in training data to generalize the model better. The receptive field of VGG16 is 224 × 224 × 3. Initially, we experimented with the image size 224 × 224 × 3. It led to a higher number

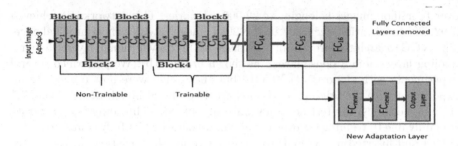

Fig. 7. Proposed Gurmukhi handwritten character recognition architecture.

of trainable parameters (25.9 million approx.), increased execution time, and inefficient usage of memory and C.P.U resources. We experimented with smaller image sizes viz. $32 \times 32 \times 3$ and $64 \times 64 \times 3$. Among these, the experiment with the image size $32 \times 32 \times 3$ was not giving us good results compared to $64 \times 64 \times 3$. The architecture with image size $64 \times 64 \times 3$ has approximately 15.9 million parameters, out of which 14.2 million parameters are trainable and 1.7 million are non-trainable thus, saving the execution time and C.P.U. resources. It takes approximately 6 h, and 20 min to train this architecture while the total training time for the architecture with image size $224 \times 224 \times 3$ was 31 h, 15 min. Thus, the training time is reduced almost by 80%. Algorithm 1 in Table 4 outlines the basic procedure that we followed for performing Transfer Learning.

Table 4. Transfer learning procedure

Algorithm 1: Transfer Learning Procedure
Input: A pre-trained architecture VGG16 with convolutional and fully connected layers as C_1 $C_{13} + FC_{14} + FC_{15} + Output\_Layer$.
Output: Trained VGG16 on target task with convolutional and fully connected layers as $C1$ $C_{13} + FC_{New1} + FC_{New2} + Output\_Layer$.
1. train data of Gurmukhi_HWdb1.0;
2. validate validation data of Gurmukhi_HWdb1.0;
3. Temp_model Extract_Layer(C1 C13 , VGG16) & Freeze(C1 C7 , VGG16)
4. Add_Layer(new_model, temp_model);
5. Add_Layer(new_model, FC_{New1});
6. Add_Layer(new_model, FC_{New2});
7. Add_Layer(new_model, Output_Layer);
8. Set Learning_rate, Optimizer, and batch_size.
9. while desired_epoch_not_reached do
10. Backpropagate;
11. Train_Model(train,validate,learning_rate, optimizer,batch_size);
12. end

5   Experimental Results

5.1   Dataset and System Used

The model is evaluated and validated on the proposed dataset "Gurmukhi_Hwdb1.0". Among the total 137,700 images, 110,700 images are of 41 Gurmukhi character classes and 27,000 are of 0–9 Gurmukhi numerals. 80% of data is used for training, 10% is set aside for validation and 10% is used for testing [see Table 3]. The whole experiment is executed in Keras and TensorFlow framework with GPU runtime on Google Colab pro deep learning cloud service. Google Colab pro provides a single 25 GB NVIDIA Tesla P100 GPU.

5.2   Design of Parameters and Classification Accuracy

All images are resized to 64× 64 × 3 for the experiment. We experimented with varied batch sizes for better recognition results. The model is trained by a Stochastic Gradient Descent optimizer with a low learning rate of 0.001 and momentum = 0.9. For cost function optimization, we used categorical cross-entropy, which is more often used for classification purposes. We used the Softmax function at the output layer. Real-time data augmentation has been performed to increase the generalizability of the proposed model. Augmentor [32], an image augmentation library in Python is used to apply rotation, shearing, and random distortions.

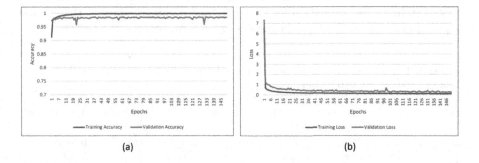

(a)   (b)

Fig. 8. (a) Training and validation accuracy for character dataset, batch size = 64 (b) Training and validation loss

These three operations are performed on 1,000 training images in each class to generate 3,000 augmented images [See Fig. 10(a)]. So, a total of 162,000 augmented images were generated for the training of 51 classes. Image normalization is also performed during training. As we previously set aside 80% (110,160) images for training, the size of the dataset for training now increases to 272,160 images. The model is trained for 150 epochs and the results are obtained for

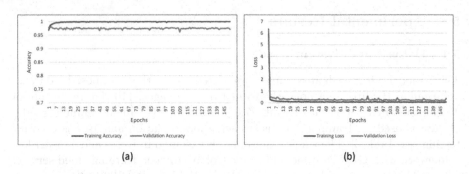

Fig. 9. (a) Training and validation accuracy for numeric dataset, batch size = 64 (b) Training and validation loss

Fig. 10. (a) Some samples of augmented images (b) Some samples misclassifying with each other.

different batch sizes as shown in Tables 5 and 6. Figures 8 and 9 show the accuracy of training and validation for character and a numeric dataset that almost approaches 100%, and the predicted loss value converges to zero fast.

Table 5. Classification performance of the proposed model for Gurmukhi character dataset for different batch sizes

Sr. No	Batch size	Optimizer	Training accuracy	Training loss	Validation accuracy	Validation loss	Test accuracy
1	64	**SGD**	**0.9994**	**0.1**	**0.9853**	**0.1**	**0.9842**
2	128	SGD	0.9989	0.1	0.9844	0.1	0.9803
3	256	SGD	0.9993	0.1	0.9830	0.1	0.9788

Table 6. Classification performance of the proposed model for Gurmukhi numeric dataset for different batch sizes

Sr. No	Batch Size	Optimizer	Training Accuracy	Training Loss	Validation Accuracy	Validation Loss	Test Accuracy
1	8	SGD	1.0000	0.04	0.9785	0.1	0.9744
2	16	SGD	0.9997	0.04	0.9711	0.3	0.9670
3	32	SGD	0.9997	0.05	0.9774	0.1	0.9725
4	64	**SGD**	**0.9996**	**0.07**	**0.9774**	**0.1**	**0.9751**

For the Gurmukhi character dataset, we have achieved the highest accuracy of 98.42% and for the numeric dataset, the test accuracy achieved is 97.51% for a batch size of 64. The proposed method delivers impressive recognition results. Still, there exists space for improvement. For that purpose, we analyzed the misclassified cases and found that the main reason for misclassification was the sheer similarity in character shapes, which made the convolution layers extract similar feature values. A few examples of such characters are provided in Fig. 10(b).

5.3 Comparative Analysis

The performance of the model is compared with the conventional classification techniques (See Table 7). From the results, it is evident that the proposed method achieves higher accuracies for the large-scale dataset in comparison to the classical methods implemented on small-scale datasets. The proposed model also provides promising results on the "Gurmukhi_HWdb1.0" dataset when compared to the existing work done using deep learning methods (See Table 8). Although the methods adopted in [31] and [33] yield higher accuracies for very small training and test dataset, our model performed very well on a relatively much larger dataset. The datasets developed in [31] and [33] are considering only 35 basic characters but the modern Gurmukhi script has 41 characters. So, our dataset of basic 41 Gurmukhi characters is more complete. Secondly, the dataset in [31] and [33] is not taken from a wide variety of populations. Also, they have taken one sample of each character from each person while our dataset is collected from a wide variety of populations in which each person is writing each character 11 times.

Table 7. Comparison with conventional techniques for "Gurmukhi_HWdb1.0" Character and Numeral dataset

Dataset name (Reference)	Dataset size	Dataset type	Classification technique	Accuracy
Kumar et al. [27]	3,500	Character	K-Nearest neighbour	98.10%
Kumar et al. [25]	10,500	Character	SVM with linear kernel	95.80%
Sinha et al. [23]	7,000	Character	SVM, KNN	95.11%, 90.64%
Siddharth et al. [24]	7,000	Character	SVM	95.04%
Singh et al. [26]	7,000	Character	SVM with RBF Kernel	94.29%
Jain and Sharma [34]	15,000	Character	Neocognitron N	92.78%
Garg et al. [35]	8,960	Character	PCA Linear, SVM+polySVM+kNN	92.30%
Proposed method	**110,700**	**Character**	**CNN**	**98.42%**
Singh and Budhiraja [29]	1,000	Numeric	Wavelet transforms	88.83%
Kumar et al. [24,28]	6,000	Numeric	Random forest classifier	87.9%
Kumar et al. [28]	6,000	Numeric	SVM	96.3%
Proposed method	**27,000**	**Numeric**	**CNN**	**97.51%**

Thus, it is much larger and more unconstrained having more variations in self-handwriting. The models in [31] and [33] are trained and tested on a very small number of samples. So, the likelihood of getting very high accuracy in a lab environment is also very high. But when this system will be tested in a different locality, the accuracy may be drastically reduced. But since our bigger dataset has been collected from a larger population, the system is likely to learn more variations and though the accuracy obtained at the lab is slightly lower, it is less likely to degrade drastically in a new environment. The results are also compared with available benchmark Gurmukhi character datasets (See Table 9). The proposed model when executed on the self-developed Gurmukhi_HWdb1.0 dataset outclasses the performance of other classifiers executed on existing benchmark smaller datasets.

Table 8. Comparison with Deep learning methods for "Gurmukhi_HWdb1.0" character and Numeral dataset

Dataset name (Reference)	Dataset size	Dataset type	Classification technique	Accuracy
Kumar and Gupta [33]	2,700	Character	Deep neural network	99.30%
Mahto et al. [31]	7,000	Character	Convolutional neural network	98.5%
Kaur and Rani [30]	2,450	Character	Convolutional neural network	92.08%
Proposed method	**110,700**	**Character**	**Convolutional neural network**	**98.42%**
Mahto et al. [31]	2,000	Numeric	Convolutional neural network	98.6%
Sarangi et al. [36]	1,500	Numeric	Artificial neural network	93.66%
Sarangi et al. [37]	1,500	Numeric	Hopfield neural network	95.4%
Proposed method	**27,000**	**Numeric**	**Convolutional neural network**	**97.51%**

Table 9. Comparison with existing benchmark Gurmukhi character datasets

Dataset name	Dataset size	Classification technique	Accuracy
HWR-Gurmukhi_1.1	3,500	knn, RBF-SVM, MLP	95.6%, 94.8%, 96.8%
HWR-Gurmukhi_1.2	3,500	knn, RBF-SVM, MLP	90.4%, 83.3%, 92.2%
HWR-Gurmukhi_1.3	3,500	knn, RBF-SVM, MLP	81.6%, 74.8%, 91.7%
HWR-Gurmukhi_3.1	7,000	knn, RBF-SVM, MLP	85.2%, 81.3%, 85.3%
Gurmukhi_HWdb1.0	**110,700**	**CNN**	**98.42%**

6   Conclusion

This article has presented a standard benchmark Gurmukhi handwritten character and a numeric dataset which will be made available for academic research. We have conferred various aspects of the collected samples of the present dataset.

All the samples have been taken from a reasonably wide variety of Punjabi-speaking populations. Our Dataset consists of a total of 137,700 samples out of which 110,700 samples are of 41 Gurmukhi characters and 27,000 samples are of 10 Gurmukhi numerals. Images of these samples are not exposed to any kind of operations to allow the researchers to engage in newer ideas on them at all stages including preprocessing. Further, a CNN architecture based on the VGG16 model using transfer learning is proposed to augment the success of computer vision in the character recognition domain. We fine-tuned its fourth and fifth blocks of convolutional layers and inserted our fully connected layer. Regularization techniques are adopted to modify the computation cost as well as to minimize the overfitting enigma. The results obtained outperform many state-of-the-art methods and the new benchmark performance with 98.42% accuracy for characters and 97.51% accuracy for the numeric dataset is obtained. We are still trying to enhance the performance of the model by increasing the number of epochs as well as working on larger test datasets. On training the model for more epochs , we have already obtained a hike in accuracy and the performance surpasses the results present in [31]. In near future, we shall extend our work to recognition of more intricate patterns such as compound characters, distinct words and sentence recognition in the Gurmukhi script.

Acknowledgments. The authors would like to acknowledge the time and efforts made by all the writers who have filled the samples towards the development of the dataset described in the present article.

References

1. Sharma, R., Kaushik, B.: Offline recognition of handwritten indic scripts: a state-of-the-art survey and future perspectives. Comput. Sci. Rev. **38**, 100302 (2020)
2. Memon, J., Sami, M., Khan, R.A., Uddin, M.: Handwritten optical character recognition (OCR): A comprehensive systematic literature review (SLR). IEEE Access **8**, 142642–142668 (2020)
3. Pal, U., Jayadevan, R., Sharma, N.: Handwriting recognition in indian regional scripts: a survey of offline techniques. ACM Trans. Asian Lang. Inf. Process. (TALIP) **11**(1), 1–35 (2012)
4. Liu, C.-L., Yin, F., Wang, D.-H., Wang, Q.-F.: Casia online and offline chinese handwriting databases. In: 2011 International Conference on Document Analysis and Recognition, pp. 37–41 (2011). IEEE
5. Su, T., Zhang, T., Guan, D.: Corpus-based hit-mw database for offline recognition of general-purpose Chinese handwritten text. IJDAR **10**(1), 27–38 (2007). https://doi.org/10.1007/s10032-006-0037-6
6. Marti, U.-V., Bunke, H.: The iam-database: an english sentence database for offline handwriting recognition. Int. J. Doc. Anal. Recogn. **5**(1), 39–46 (2002). https://doi.org/10.1007/s100320200071
7. Grother, P.J.: NIST special database 19. NIST handprinted forms and characters database (2017)
8. Lawgali, A., Angelova, M., Bouridane, A.: HACDB: handwritten arabic characters database for automatic character recognition. In: European Workshop on Visual Information Processing (EUVIP), pp. 255–259 (2013). IEEE

9. Mozaffari, S., Faez, K., Faradji, F., Ziaratban, M., Golzan, S.M.: A comprehensive isolated Farsi/Arabic character database for handwritten OCR research. In: Tenth International Workshop on Frontiers in Handwriting Recognition. Suvisoft (2006)

10. KIM, D.-H., Hwang, Y.-S., Park, S.-T., Kim, E.-J., Paek, S.-H., BANG, S.-Y.: Handwritten korean character image database pe92. IEICE Trans. Inf. Syst. **79**(7), 943–950 (1996)

11. Bhattacharya, U., Chaudhuri, B.: Databases for research on recognition of handwritten characters of Indian scripts. In: Eighth International Conference on Document Analysis and Recognition (ICDAR 2005), pp. 789–793. IEEE (2005)

12. Bhattacharya, U., Chaudhuri, B.B.: Handwritten numeral databases of Indian scripts and multistage recognition of mixed numerals. IEEE Trans. Pattern Anal. Mach. Intell. **31**(3), 444–457 (2008)

13. Das, N., et al.: A statistical-topological feature combination for recognition of handwritten numerals. Appl. Soft Comput. **12**(8), 2486–2495 (2012)

14. Basu, S., Chaudhuri, C., Kundu, M., Nasipuri, M., Basu, D.K.: Text line extraction from multi-skewed handwritten documents. Pattern Recogn. **40**(6), 1825–1839 (2007)

15. Das, N., Basu, S., Sarkar, R., Kundu, M., Nasipuri, M., et al.: An improved feature descriptor for recognition of handwritten bangla alphabet. arXiv preprint arXiv:1501.05497 (2015)

16. Agrawal, M., Bhaskarabhatla, A.S., Madhvanath, S.: Data collection for handwriting corpus creation in indic scripts. In: International Conference on Speech and Language Technology and Oriental COCOSDA (ICSLT-COCOSDA 2004), New Delhi, India November 2004 (2004). Citeseer

17. Agnihotri, V.P.: Offline handwritten devanagari script recognition. IJ Inf. Technol. Comput. Sci. **8**(1), 37–42 (2012)

18. Alaei, A., Nagabhushan, P., Pal, U.: A benchmark Kannada handwritten document dataset and its segmentation. In: 2011 International Conference on Document Analysis and Recognition, pp. 141–145 (2011). IEEE

19. Kumar, M., Sharma, R.K., Jindal, M.K., Jindal, S.R., Singh, H.: Benchmark datasets for offline handwritten Gurmukhi script recognition. In: Sundaram, S., Harit, G. (eds.) DAR 2018. CCIS, vol. 1020, pp. 143–151. Springer, Singapore (2019). https://doi.org/10.1007/978-981-13-9361-7_13

20. Punjabi Language. https://simple.wikipedia.org/wiki/Punjabi_language Accessed 17 May 2022

21. Gurmukhi. https://en.wikipedia.org/wiki/Gurmukhi. Accessed 31-05-2022

22. Aggarwal, A., Singh, K.: Handwritten Gurmukhi character recognition. In: 2015 International Conference on Computer, Communication and Control (IC4), pp. 1–5. IEEE (2015)

23. Sinha, G., Rani, R., Dhir, R.: Handwritten Gurmukhi character recognition using K-NN and SVM classifier. Int. J. Adv. Res. Comput. Sci. Soft. Eng. **2**(6), 288–293 (2012)

24. Siddharth, K.S., Jangid, M., Dhir, R., Rani, R.: Handwritten Gurmukhi character recognition using statistical and background directional distribution. Int. J. Comput. Sci. Eng. (IJCSE) **3**(06), 2332–2345 (2011)

25. Kumar, M., Jindal, M., Sharma, R.: Offline handwritten Gurmukhi character recognition: analytical study of different transformations. Proc. Natl. Acad. Sci. India Sect. A **87**(1), 137–143 (2017). https://doi.org/10.1007/s40010-016-0284-y

26. Singh, S., Aggarwal, A., Dhir, R.: Use of gabor filters for recognition of handwritten gurmukhi character. Int. J. Adv. Res. Comput. Sci. Soft. Eng. **2**(5) (2012)

27. Kumar, M., Sharma, R., Jindal, M.: Efficient feature extraction techniques for offline handwritten Gurmukhi character recognition. Natl. Acad. Sci. Lett. **37**(4), 381–391 (2014). https://doi.org/10.1007/s40009-014-0253-4

28. Kumar, M., Jindal, M., Sharma, R., Jindal, S.R.: Offline handwritten numeral recognition using combination of different feature extraction techniques. Natl. Acad. Sci. Lett. **41**(1), 29–33 (2018). https://doi.org/10.1007/s40009-017-0606-x

29. Singh, P., Budhiraja, S.: Offline handwritten Gurmukhi numeral recognition using wavelet transforms. Int. J. Mod. Educ. Comput. Sci. **4**(8), 34 (2012)

30. Kaur, H., Rani, S.: Handwritten Gurumukhi character recognition using convolution neural network. Int. J. Comput. Intell. Res. **13**(5), 933–943 (2017)

31. Mahto, M.K., Bhatia, K., Sharma, R.K.: Deep learning based models for offline gurmukhi handwritten character and numeral recognition. ELCVIA Electron. Lett. Comput. Vis. Image Anal. **20**(2) (2021)

32. Bloice, M.D.: Augmentor. https://augmentor.readthedocs.io/en/master/userguide/mainfeatures.html. Accessed 20 May 2022

33. Kumar, N., Gupta, S., Pradesh, H.: A novel handwritten gurmukhi character recognition system based on deep neural networks. Int. J. Pure Appl. Math. **117**(21), 663–678 (2017)

34. Jain, U., Sharma, D.: Recognition of isolated handwritten characters of Gurumukhi script using neocognitron. Int. J. Comput. Appl. **10**(8) (2010)

35. Garg, A., Jindal, M.K., Singh, A.: Offline handwritten Gurmukhi character recognition: K-NN vs. SVM classifier. Int. J. Inf. Technol. **13**(6), 2389–2396 (2021). https://doi.org/10.1007/s41870-019-00398-4

36. Sarangi, P.K., Sahoo, A.K., Kaur, G., Nayak, S.R., Bhoi, A.K.: Gurmukhi numerals recognition using ann. In: Cognitive Informatics and Soft Computing, pp. 377–386. Springer (2022). https://doi.org/10.1007/978-981-16-8763-1_30

37. Sarangi, P.K., Sahoo, A.K., Nayak, S.R., Agarwal, A., Sethy, A.: Recognition of isolated handwritten Gurumukhi numerals using hopfield neural network. In: Das, A.K., Nayak, J., Naik, B., Dutta, S., Pelusi, D. (eds.) Computational Intelligence in Pattern Recognition. AISC, vol. 1349, pp. 597–605. Springer, Singapore (2022). https://doi.org/10.1007/978-981-16-2543-5_51

Synthetic Data Generation for Semantic Segmentation of Lecture Videos

Kenny Davila[1]([✉]) [iD], Fei Xu[2] [iD], James Molina[1] [iD], Srirangaraj Setlur[2] [iD], and Venu Govindaraju[2] [iD]

[1] Universidad Tecnológica Centroamericana, Tegucigalpa, Honduras
kenny.davila@unitec.edu.hn,jjmolinaf12@unitec.edu
[2] University at Buffalo, New York, USA
{fxu3,setlur,govind}@buffalo.edu

Abstract. Lecture videos have become a great resource for students and teachers. These videos are a vast information source, but most search engines only index them by their audio. To make these videos searchable by handwritten content, it is important to develop accurate methods for analyzing such content at scale. However, training deep neural networks to their full potential requires large-scale lecture video datasets. In this paper, we use synthetic data generation to improve binarization of lecture videos. We also use it to semantically segment pixels into background, speaker, text, mathematical expressions, and graphics. Our method for synthetic data generation renders content from multiple handwritten and typeset datasets, and blends it into real images using random tight layouts and the location of the people. In addition, we also propose a mixed data approach that trains networks on two detection tasks at once: person and text. Both binarization and semantic segmentation are carried out using fully convolutional neural networks with a typical encoder-decoder architecture and residual connections. Our experiments show that pre-training on both synthetic and mixed data leads to better performance than training with real data alone. While final results are promising, more work will be needed to reduce the domain shift between synthetic and real data. Our code and data are publicly available.

Keywords: Semantic Segmentation · Lecture videos · Synthetic data

1 Introduction

Lecture videos represent a vast source of information useful for both students and teachers. There videos can provide additional explanations and examples to help students catch up with difficult topics. Searching these videos by content is highly desirable, but most search engines only consider the audio for indexing and ignore the handwritten content. Audio transcripts do not always describe elements such as handwritten mathematical expressions (ME) and drawings. Therefore, it is important to develop accurate methods for analyzing the handwritten content

U. Porwal et al. (Eds.): ICFHR 2022, LNCS 13639, pp. 468–483, 2022.
https://doi.org/10.1007/978-3-031-21648-0_32

at scale. However, training deep neural networks to their full potential on lecture video analysis tasks requires large-scale datasets.

Lecture videos include multiple elements that can be analyzed separately through semantic segmentation for a variety of applications such as navigation, summarization, indexing and search. One option is to broadly distinguish between content elements (text, ME and graphics) and other elements (speaker and background). Another option is to differentiate between all five classes.

In this paper, we propose a method for synthetic data generation for training models for semantic segmentation of lecture video elements. Here we only consider recordings of traditional in-person lectures and tutorials instead of those produced by Learning Management Systems (LMS). To approximate the characteristics of these videos, the generation of synthetic data needs to consider lecture video elements of each class. Our method renders content from different handwritten and typeset datasets, and blends it into real images using randomized tight layouts and the location of the people.

In addition, we also proposed a mixed data approach where deep neural networks are pretrained simultaneously on two tasks: person detection and text detection. The resulting networks cannot be used directly on the target lecture videos, but the learned weights are helpful when the network is modified and retrained on the final task.

The semantic segmentation of lecture video elements is carried out using fully convolutional neural networks (FCN) in this work. For binarization, we use the FCN-LectureNet model [4]. For semantic segmentation with five classes, we employ a typical encoder-decoder architecture with 5 output maps. This network uses modified basic ResNet blocks [15] with GeLU activations, and residual connections between compatible encoder-decoder scales are added.

Our experiments show that pretraining with our synthetic data leads to better performance than training with real data alone. Combining both types of pretraining leads to better performance for semantic segmentation. While final results are promising, more work will be needed to reduce the domain shift between synthetic and real data. Our code and data are publicly available[1].

2   Related Works

Our goal is to generate synthetic data to train networks for semantic segmentation of lecture video elements such as text and math. Our work falls at the intersection between synthetic data generation, detection or segmentation between text and MEs in documents, and lecture video analysis. Here, we provide a brief discussion about each one of these areas.

2.1   Synthetic Data Generation

Synthetic data is cheaper than manual annotations and it has become a popular way to improve the training of deep neural networks. Gupta et al. [11] use

---

[1] https://github.com/kdavila/ICFHR2022_Synth_Lectures.

pixel-wise depth map and segmentation methods to align text regions to the background image with its local texture and color cues to generate synthetic images for scene text recognition. SynthTIGER [29] is a synthesis engine to generate word images for scene text recognition, where each image includes spurious pieces of extra text. Villamizar et al. [26] generate artificial slides for text segmentation using open templates with various text content and drawings.

Other works have targeted the generation of synthetic handwritten data. Wolf et al. [28] create synthetic images for handwritten word-spotting. A fixed vocabulary is used to create multiple variations of each word by first rendering them using a small set of fonts that look like handwriting. Different style parameters such as skew, slant and gaps between characters are used to introduce variations to the synthetic images. Guan et al. [10] proposed a style-conditioned generative adversarial network (GAN) that can produce photo-realistic handwriting text line images using online samples for content and offline samples for styles. Springstein et al. [25] proposed an attention-based GAN with recurrent neural network to generate large amounts of synthetic handwritten formulas from rendered LaTex formulas for ME recognition.

2.2 Text and Math Detection

Many methods have been proposed for scene text detection, but we mostly care about those using instance segmentation. Most of these methods use semantic segmentation to identify candidate text pixels, which must be grouped into individual text regions. Multiple ideas have been proposed for this step such as predicting which neighboring pixels belong to the same region (PixelLink [7]), progressive scale expansion which differentiates between pixels at the center of text regions and borders (PSENet [27]), heatmaps which identify centers of regions as the peaks of a mountain (TextMountain [30]), and threshold maps that delimit the borders of the regions (Differentiable binarization [17]). Recent deep learning based methods for text detection are described by Long et al. [20].

Many methods for detecting MEs on scanned documents have been proposed. The types of noise on these images are very different from what is seen in scene text detection. Also, scanned documents contain a lot of text, and MEs can appear either as separated blocks or inline. Recent methods treat this problem as an instance of object detection. Hashmi et al. [12] used Cascade Mask-RCNN with deformable convolutions and dual-backbone with ResNeXt-101. Dey and Zanibbi [8] used a scanning single shot detection (ScanSSD) model that detects math regions on overlapping windows in high resolution document images. Phong et al. [23] propose a hybrid method that uses document layout analysis, follow by detection of isolated and inline MEs. There have been competitions to benchmark methods for detection of typeset formulas [21].

2.3 Lecture Video Analysis

Several applications can be facilitated by automatically processing lecture videos. In recent years, the main focus has been on audio and transcripts [16], to which many NLP models are applied to complete tasks such as temporal video segmentation and summarization. The focus has been mostly on text alone, and MEs are typically ignored.

Different methods have been proposed for the extraction of lecture video content. Earlier methods were based on heuristic features and traditional machine learning models [5]. Our recent work uses deep learning techniques for handwritten content [4]. Meanwhile, other works focus on typeset content on slide-based lecture videos. A few datasets and methods have been proposed for semantic segmentation of slide-based lecture video images including LectureVideoDB [9], WiSE [13], and synthetic slides [26].

Lecture video content can be indexed for search applications. For slide-based videos, many works rely on OCR systems [1,16] to automatically detect and extract the slide content. Extracted textual content can be further analyzed using NLP methods such as topic modeling [16] or named entity recognition [1]. However, these text-based systems typically have poor performance on MEs. The Tangent-V model can retrieve lecture videos using MEs as queries [6].

3 Mixed Data

Different classification problems need tailored features. These can be learned by deep neural networks as long as enough training data is available. However, the amount of annotated lecture video data that is publicly available is rather limited. Transfer learning can help neural networks learn better features when too little data is available for the target task. In this work, we transfer learning from two tasks related to lecture video analysis: text detection and person detection. Both tasks are treated as general semantic segmentation tasks.

It is easy to pretrain a network on either person or text detection. However, our target are lecture videos that include both people and text-like content in the same image. Training on a given task will make the network learn specialized features for that task. Further training on a different task can make the network forget features that were good for the first task. Here, we train the network on the two tasks simultaneously to help it learn good features for both.

We generate batches where each image is labeled for either person or text detection, and we only compute the Binary Cross Entropy (BCE) loss on the corresponding output. The final loss of the batch is the sum of the losses for both person detection and text detection images.

4 Synthetic Data

In this work, we propose a model for the generation of synthetic lecture video images. Creating effective images for pretraining a neural network on a given domain can be challenging. The more similar that these synthetic images are to the real ones, the more effective that they will be. The method is designed to create images on the fly, ideal for training of neural networks. Our method works through a series of steps namely content generation, layout generation, content blending and data augmentation as illustrated in Fig. 1.

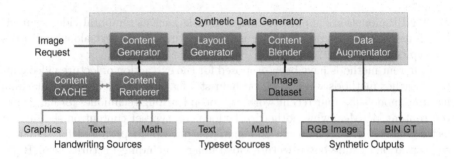

Fig. 1. Overview of the synthetic data generation process.

4.1   Content Generator

We target lecture videos where there is a speaker who writes by hand most of the lecture content on a whiteboard or chalkboard. Online lecture videos might also include typeset content. Our content generator distinguishes between three types of content: text, MEs and graphics. Each of these types might include multiple handwritten and typeset datasets. For handwritten content, we used the IAM Offline dataset [22] for text, CROHME 2019 [21] for MEs and QuickDraw for graphics. For typeset content, we used TotalText [2] for text and a random set of 30,000 MEs from Wikipedia [3].

Data Preprocesing. CROHME 2019 [21] and QuickDraw are online datasets (include trace information), while the rest are offline datasets. Each of these datasets comes in its own format. We transform these to a common format, where JSON files are used to create small groups of content elements (text lines, MEs, drawings). When writer information is available, all elements written by the same person are grouped in a single JSON file. Otherwise, these elements will be randomly grouped into small sets. For online sources, each JSON file keeps all trace information in a simplified format. For offline sources, each JSON file links to the relative path of each individual content element.

Size Normalization. Each content element can come at any arbitrary scale. Our goal is to control the size of the content regions included in the generated images. This requires resizing each element from its original scale to a given random size from a controlled distribution. Here, by original scale we mean the average height of the symbols that appear on a given content element. This can be hard specially because most sources do not provide symbol level ground truth. If these are available (as in CROHME 2019 and QuickDraw), one can directly use these to compute statistics of the mean and median symbol sizes.

When the symbol-level ground truth is not available, we consider two methods for estimating the average size of symbols: connected component (CC) analysis and baseline estimation. The method based on CCs, used for TotalText and Wikipedia datasets, assumes that most symbols will correspond to one CC on the image. Small CCs are filtered out first, and then the mean and the 75% quantile of the remaining CCs are computed. The second method that uses baseline estimation is required for datasets such as IAM Offline where each CC might correspond to several symbols. The method estimates the average height of the lower-case symbols by locating the baseline and corpus line. We first detect and correct the slant on the image by finding a rotation that minimizes the height between the 7.5% and 92.5% percentiles of the foreground pixels. Then, hierarchical clustering over the vertical pixel profile is used to identify the upper, middle and lower zones. To deal with text lines which do not include ascenders and/or descenders, we pick the densest cluster as middle zone instead of the middle cluster. In most cases, this procedure successfully estimates the average height of the lower-case characters, but it can easily fail if the text line is not straight. For this reason, we extract the height between the 2.5% and 97.5% percentiles as a secondary scaling value.

The estimated average height of the symbols and the secondary scaling values produced by each algorithm are then used to compute a scaling factor for the content region using Eq. 1. The goal is to make sure that the average symbol size on the re-scaled content can be controlled in most cases, and that failures will not be rendered at large resolutions. The render resolution is randomly sampled from a positively skewed distribution which controls the sizes of the content elements. Most elements will be relatively small but a few big elements will be occasionally allowed. In addition, hard minimum and maximum values are used to prevent issues caused by very small or very large render resolutions. Finally, this resolution is multiplied by a fixed scaling value that is set individually for each dataset.

$$s_f = min \left(\frac{Render\ Resolution}{AVG.\ Scaling\ value}, \frac{1.5 * Render\ Resolution}{Secondary\ Scaling\ value} \right) \qquad (1)$$

Content Rendering. Blending of content into the target images requires binary content images. For offline sources, binary images are already available, they just need to be resized using the scaling factor computed before. The resulting handwriting thickness will be proportional to the one in the original image. For online sources, rendering of the tracing information is required to generate binary images. First, the trace points must be transformed from their original space to a given pixel space based on a target average symbol height. Then, lines of a given handwriting thickness are drawn between each pair of consecutive points in each trace.

Content Ground Truth. Different type of pixel-level ground truth are generated for each task. For binarization, the ground truth is basically a copy of each content region. For semantic segmentation, pixel-level masks of entire regions are created by dilating the original content. Ground truth of each region needs to be generated only once, and they can be resized as needed.

Content Cache. Our goal is to generate synthetic images quickly enough to be used during the training phase without storing them. This allows to create large amounts of images without needing large amounts of disk space. However, normalization of content size, ground truth generation and rendering can be really expensive in practice. For this reason, we precompute and store in memory the binary rendered images, their binary ground truths and their scaling values. Every time a content region is reused, a different rendering resolution can be selected to create more variations of the same content. To reduce the memory required by the cache, all images are stored in a PNG format in memory, noting that is faster to decompress them than to load them from the disk. For the datasets selected in this work, the final cache size is estimated to require just around 16 GB in RAM.

Content Selection. This process happens every time that a new synthetic image must be generated. A small set of content elements are selected from each source. Cached elements can be taken from there and decompressed in memory, while the rest might need to be loaded from disk or rendered on the fly. User-given parameters control the number of regions that can be taken from each source. These regions all extracted from a small set of randomly selected groups of content regions (JSON files) per source. This last condition ensures a degree of consistency for datasets where the writer is known, allowing multiple samples of the same writer to be used on the same synthetic image.

Sometimes lecture videos include empty whiteboards or chalkboards. For this reason, the content selector might return an empty list of content regions with a very small probability, allowing to generate empty synthetic images.

4.2   Layout Generator

Handwritten content in whiteboards and chalkboards is characterized by having a semi-structured layout which is not as regular as textual paragraphs, but it is not completely random either. Layouts depend directly on the specific set of content regions that will be used to generate a synthetic image, and therefore they must be generated on the fly using efficient algorithms.

In this work, we use a greedy optimization procedure to generate layouts of content. These layouts follow a paragraph-like pattern based on lines, where each line might contain multiple content regions. The procedure starts by shuffling the input content regions to form a layout where there is exactly one content region per line. Then, an iterative greedy process is applied, where for each step, the two shortest lines are combined. If this change leads to a layout that has an aspect ratio closer to the target, then the change is kept and the process continues. Otherwise, the process stops and the previous layout is used.

The resulting layout is further refined. Lines are removed from the bottom until the layout fits within the target height. Random elements are also removed from lines that are wider than the target width until these fit. Lines that become empty because of this process are also removed. User-given parameters control the ranges of the vertical gaps (between lines) and horizontal gaps (between elements in the same line) of the layout. For the regions that are finally included in the layout, the horizontal and vertical gaps are randomly selected from these ranges. Horizontal and vertical offsets are also computed, allowing content to be randomly placed horizontally within the range of the layout width, and content

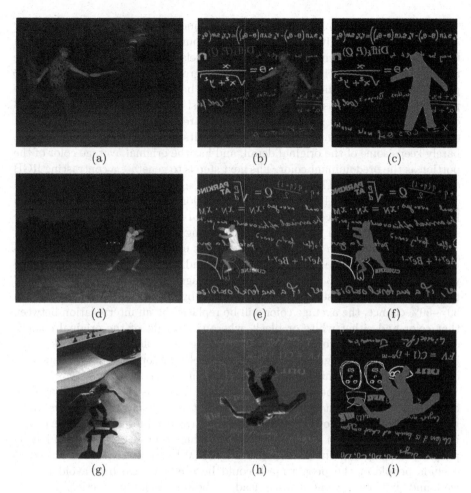

Fig. 2. Examples of synthetic images and their corresponding ground truth. (a), (d) and (g) are the input images. (b), (e) and (h) are variations generated by our algorithm. (c), (f) and (i) are the corresponding pixel-level ground truth images for these variations, where black = background, red = people, blue = text, green = ME, and yellow = graphics. (Color figure online)

regions within the same line can also be sightly miss-aligned vertically based on the range of the tallest region on each line.

4.3 Content Blender

The next step is to blend the selected content on the target image using the generated layout. The proposed framework requires an external dataset of images that has pixel-level ground truth for people. In this case, we used COCO [19].

We used two types of backgrounds that define two blending models: regular and simplified. For a given image, the background type is selected at random based on some user-given probability. Both models start by creating a copy of the target image. The model based on regular backgrounds only modifies the specific portions of the image where content regions will be blended. For each one of these portions, the contrast is locally reduced by first subtracting the average RGB color of the portion, then the resulting values are scaled by a factor of 0.05 and the average color is then added back. This results in a portion of the image that barely keeps some of the original detail, and has the original average color of the portion as the predominant color. The next step is to generate a contrasting RGB color for the content. The absolute channel-wise distance between the selected color and the average RGB color of the portion must be above a user-given threshold. Finally, the content region is blended into the background using the selected RGB color. An example of this blending is shown in Fig. 2b.

The model based on simplified backgrounds emulates whiteboards or chalk-boards by modifying the entire image background. A similar procedure is applied where the average RGB color for the entire image is computed and substracted from the image. The resulting values are scaled by a factor of 0.025. Then with a 50%-50% chance, the average color will be replaced by an interpolation between that color and either white or black, where the weight of the original color is chosen at random between 0 to 1. The resulting new color is added back to the image. One constrasting RGB color is finally selected for blending all regions of content. Examples of this blending are shown in Fig. 2e and 2h.

In lecture videos, people typically occlude portions of the whiteboard or chalkboards. Text might only occlude people if the video has text overlays. After blending all content onto the target image copy, the pixel-level mask of the people is used to reinsert them back into the image This generates the effect of having people occluding content regions (see Fig. 2). While text could be allowed to occlude people or the people mask could be used to have text avoid people, we found that our current strategy leads to better results in practice, probably because it is more consistent with lecture videos.

4.4   Data Augmentator

Previous steps can be used to generate a large variety of images from a fixed set of base images and content sources. To get even more from the synthetic data, traditional data augmentation techniques can still be applied. Some augmentations must be applied to both the RGB image and its binary ground truth. These include random croppping, zoom in/out, rotations, and horizontal/vertical flipping. Other variations like random changes of the hue, brightness, contrast, gamma, and saturation only affect the RGB image. Finally, gaussian and compression noise are added to the image. Lecture videos often suffer from artifacts caused by video compression. Based on some user-given probability, the images are compressed and decompressed in memory using the JPG codec with a randomly chosen quality factor. While this augmentation is expensive, it is also one of the most effective ones based on our preliminary tests.

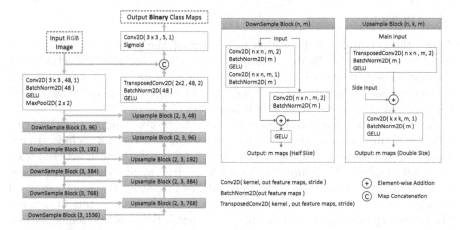

Fig. 3. Network architecture used for semantic segmentation of lecture video elements

5 Semantic Segmentation of Lecture Video Elements

In this work, the semantic segmentation of lecture video elements is carried out using fully convolutional neural networks. The simpler case is binarization, where the goal is to differentiate between content elements such as text, MEs and graphics from background and the speaker. Here, we use the FCN-LectureNet model [4] that utilizes two auxiliary branches, one for detection of text and another for estimation of background. These branches help to produce a simplified input that is then binarized by the main branch of the network. We refer the reader to the original FCN-LectureNet paper for additional details [4].

The more complex case, semantic segmentation, aims to distinguish between: background, people, text, MEs and graphics. For this problem, we also use a fully convolutional network with a typical encoder-decoder structure. This network uses modified basic ResNet blocks [15] with GeLU activations, and residual connections between compatible encoder-decoder scales are added. The architecture of this network is illustrated in Fig. 3.

6 Experiments

The effectiveness of the proposed synthetic data generation method for pretraining FCN models was measured using two experiments. The first one is for binarization of lecture video content. The second one is for semantic segmentation of lecture video elements. Due to the characteristics of lecture videos, some parameters of the data generation pipeline can be set up in a very intuitive way. Meanwhile, other parameters had to be fine-tuned based on multiple tests that took several days to complete. Due to space and time constrains, we only show results for the best combination of parameters that we found.

6.1   Dataset

We use the LectureMath dataset [4] to benchmark the performance of different networks pretrained with our proposed data generation model. LectureMath contains 34 lecture videos from 8 different online sources including both full lectures as well as short tutorials. It has both whiteboard and chalkboard based lectures, and most of the content is handwritten.

LectureMath includes ideal key-frames and their corresponding binarized versions for lecture video content binarization. However, it did not include ground truth for other elements such as people, and it did not distinguish between text, MEs and graphics. Therefore, in this work we had to further annotate these videos to separate content by type (text, ME or graphics). Polygons were used to indicate the location of each content region, and we also annotated the times when these regions were added, modified or deleted. For semantic segmentation, lecture video frames were sampled at one frame per second. Then, using the polygons and time stamps, we can easily generate pixel-level mask that indicate the position of each region at every frame of the sample where it appears. For people, creating such detailed annotations was unfeasible. Therefore, we used a trained person detector to automatically locate and annotate the speaker at every frame of the sample. The detector used is the Mask R-CNN model [14] which uses ResNet-50 [15] and FPN [18].

6.2   Binarization of Handwritten Content

This experiment follows the evaluation protocol for binarization of lecture video content used for FCN-LectureNet [4] based on H-DIBCO metrics [24]. These metrics measure the perceived quality of the resulting binary images. Following the original FCN-LectureNet work [4], here we considered multiple combinations of three types of pretraining: Reconstruction of the median filtered image (Med), text detection (TD) and synthetic data (Syn). During the fine-tuning stage, in addition to regular training (Reg), we also consider the foreground-focused training (Fg) and foreground-focused training with reduced learning rate (FR).

Table 1 shows the results of our ablation study. Rows with ID 1 to 7 are the same values reported by FCN-LectureNet [4]. Our experiments (IDs 8–23) reused the same weights of the pretrained networks from our previous work [4]. Our tests considered the effect of using synthetic data pretraining with and without Med and TD pretraining.

The network that has been trained only on synthetic data (ID 8) does not have a good performance. If we consider using only one type of pretraining followed by regular training without further fine-tuning (IDs 2, 4, 9), synthetic pretraining (ID 9) outperforms the other two options in all metrics. It is hard to pick an overall winner, but the highest recall, PSNR, and DRD metrics were achieved when only the Med and Syn pretrainings were used (IDs 14, 15). Meanwhile, the highest precision metrics are obtained when all three pretraining models are used (ID 21). In most cases, the fine-tuning seems to consistently trade-off pixel-wise precision for recall. If we compare using Med + TD (IDs 5, 6, and 7) with

Table 1. Ablation study for lecture video content binarization using different types of pretraining and fine-tuning. The evaluation uses H-DIBCO metrics on the LectureMath dataset and follows the same protocol used in FCN-LectureMath [4].

ID	Med	TD	Syn	Reg	Fg	FR	PSNR	DRD	Rec	Prec	F1	Rec	Prec	F1
	\multicolumn Pretraining			Finetuning					Standard			Pseudo		
1	−	−	−	✓	−	−	19.18	7.56	74.69	80.77	75.34	82.16	76.58	77.38
2	−	✓	−	✓	−	−	20.68	4.70	88.57	82.11	84.92	95.24	77.94	85.43
3	−	✓	−	✓	✓	−	21.19	4.56	**93.49**	81.25	86.59	97.59	77.22	85.82
4	✓	−	−	✓	−	−	21.13	4.74	84.05	87.47	84.97	93.23	85.20	88.49
5	✓	✓	−	✓	−	−	21.12	4.12	87.51	84.98	85.99	94.49	81.08	87.01
6	✓	✓	−	✓	✓	−	22.02	3.49	87.73	88.99	88.23	96.27	86.25	90.82
7	✓	✓	−	✓	✓	✓	21.99	3.57	91.47	86.07	88.55	97.10	82.57	89.05
8	−	−	✓	−	−	−	19.20	7.77	65.58	80.14	69.22	69.69	76.36	69.92
9	−	−	✓	✓	−	−	22.71	3.37	92.35	87.78	89.70	97.36	84.75	90.26
10	−	−	✓	✓	✓	−	22.78	3.25	93.27	87.38	90.05	98.24	84.06	90.67
11	−	−	✓	✓	✓	✓	22.79	3.23	93.28	87.47	90.13	98.29	84.73	90.79
12	✓	−	✓	−	−	−	18.81	8.38	63.58	77.15	66.88	66.76	73.13	67.00
13	✓	−	✓	✓	−	−	22.74	3.25	92.69	87.93	90.08	98.16	85.23	91.02
14	✓	−	✓	✓	✓	−	**22.99**	**2.97**	93.15	88.15	**90.43**	97.83	85.27	90.91
15	✓	−	✓	✓	✓	✓	22.89	3.11	**93.49**	87.64	90.33	**98.32**	84.83	90.88
16	−	✓	✓	−	−	−	19.04	8.14	65.05	78.72	68.20	69.06	74.72	68.55
17	−	✓	✓	✓	−	−	22.94	3.16	92.36	88.64	90.31	98.05	85.84	**91.35**
18	−	✓	✓	✓	✓	−	22.98	3.12	93.33	87.84	90.35	98.07	84.91	90.80
19	−	✓	✓	✓	✓	✓	22.82	3.35	93.28	87.26	89.97	98.25	84.48	90.59
20	✓	✓	✓	−	−	−	18.97	8.13	64.21	78.76	67.38	67.25	74.60	67.38
21	✓	✓	✓	✓	−	−	22.72	3.24	89.04	**90.34**	89.43	94.89	**87.81**	90.92
22	✓	✓	✓	✓	✓	−	22.96	3.16	92.39	88.47	90.23	97.70	85.74	91.11
23	✓	✓	✓	✓	✓	✓	22.92	3.17	92.73	88.03	90.16	97.84	85.29	90.92

their corresponding counter parts that also use Syn pretraining (IDs 21, 22, and 23), we can see that adding the synthetic pretraining leads to improvements for almost all metrics. Overall, using the proposed synthetic data pretraining seems to help and leads for either higher or comparable new peaks for all metrics.

6.3 Semantic Segmentation of Lecture Video Elements

Table 2. Ablation study for semantic segmentation of lecture video content using different pretraining types. The evaluation uses pixel-wise metrics on LectureMath.

M	S	R	Rec	Pre	F1	Rec	Pre	F1	Rec	Pre	F1	Rec	Pre	F1	Rec	Pre	F1	Rec	Pre	F1
\multicolumn Conds			Background			People			ME			Text			Graphic			Average		
−	✓	−	90.8	91.0	90.5	86.6	82.0	82.6	62.6	70.6	65.5	46.0	55.5	46.7	18.4	16.4	5.0	60.9	63.1	58.1
−	−	✓	93.6	91.1	92.0	92.3	87.6	89.6	64.1	70.0	64.8	35.2	55.1	40.1	20.8	23.9	8.5	61.2	65.5	59.0
✓	✓	−	93.6	94.1	93.6	90.8	88.2	88.4	76.5	75.2	74.5	58.8	65.8	58.5	41.3	31.0	23.7	72.2	70.9	67.8
✓	−	✓	95.8	95.0	95.3	94.6	94.0	94.2	75.3	76.6	74.8	55.1	55.3	53.5	46.8	34.1	28.4	73.5	71.0	69.3
−	✓	✓	95.0	95.4	95.2	95.7	93.2	94.4	79.1	81.4	79.6	73.5	64.4	67.9	50.7	33.9	30.9	78.8	73.6	73.6
✓	✓	✓	95.7	95.1	95.3	94.6	94.2	94.2	78.7	82.7	80.1	68.6	64.9	65.2	54.3	35.9	33.8	78.4	74.6	73.7

In this experiment, we use our synthetic and mixed pretraining models to train a network for semantic segmentation of lecture video elements. Pretaining on mixed data alone does not produce a network that can be used directly for semantic segmentation. Therefore, all conditions tested here finish after either training with synthetic data or real data. Here, we focus on the pixel-wise performance of the network measured by recall, precision and F1 metrics. We compute the macro-averages for each output map and for each metric.

Results for this experiment are shown in Table 2. When trained on synthetic data alone (S), the network achieves reasonable performance for the background class, and fair performance for the people class. Text and ME do not perform so well, and the graphics class is bad. When the network is trained on real data alone (R), the trends are similar, but the performance on people mask is better. The same speakers appear both in training and testing sets of LectureMath, but they do not appear on the pretraining datasets. Note that the ground truth for the people class was auto-generated using an existing person detector.

A considerable improvement is achieved when the mixed pretraining and synthetic pretraining are combined (M+S). All classes improve, but math and graphics are the ones with the highest performance boost. This is promising considering that these numbers represent the full generalization capabilities of the proposed pretraining models. As expected, these values further improve when the network is fine-tuned using real data from the target domain (M+S+R). A comparison between using either synthetic pretraining (S+R) or mixed pretraining (M+R) shows that synthetic pretraining alone (S+R) leads to better metrics. The difference between using synthetic data alone (S+R) and using both synthetic data and mixed data (M+S+R) for pretraining is rather small. The network that used both types of pretraining (M+S+R) gets better performance on graphics and math, but using synthetic data alone (S+R) achieved better performance on text. In all cases, the hardest class is graphics. This is not surprising considering that every piece of handwriting that is not text or math must be assigned to this class.

Looking further into the AVG F1 per video, we discovered that going from R to M+S+R leads to a bigger improvement for chalkboard-based sources (+21.49%) than for whiteboard-based sources (+2.64%). This makes sense since there is less training data for chalkboards (only 2 out of 8 sources). Mixed pretraining seems to help in most cases, but for a few sources with very thin handwriting it actually hurts a lot specially on the text class, leading to a drop in text recall for M+S+R. Overall, the benefit of using the proposed synthetic data generation model for pretraining a network for semantic segmentation is clear.

7   Conclusion

In this work, we have proposed a model for generating synthetic data that can be used for semantic segmentation tasks in the context of lecture video analysis. We also proposed a mixed pretraining model that combines training on two detection tasks simultaneously, thus helping to get a network that includes features that are good for both tasks. Our experiments using the LectureMath dataset showed that when synthetic data is used, the performance of the networks for binarization and semantic segmentation is boosted. Overall, the results achieved on the harder semantic segmentation task are promising, but a considerable amount of work is needed, specially to improve on the text, ME and graphics classes.

Acknowledgment. This material is based upon work partially supported by the National Science Foundation under Grant No. OAC/DMR 1640867.

References

1. Cagliero, L., Canale, L., Farinetti, L.: Visa: a supervised approach to indexing video lectures with semantic annotations. In: 2019 IEEE 43rd Annual Computer Software and Applications Conference (COMPSAC), vol. 1, pp. 226–235. IEEE (2019)
2. Ch'ng, C.K., Chan, C.S.: Total-text: a comprehensive dataset for scene text detection and recognition. In: 2017 14th IAPR International Conference on Document Analysis and Recognition (ICDAR), vol. 1, pp. 935–942. IEEE (2017)
3. Davila, K., Joshi, R., Setlur, S., Govindaraju, V., Zanibbi, R.: Tangent-V: math formula image search using line-of-sight graphs. In: Azzopardi, L., Stein, B., Fuhr, N., Mayr, P., Hauff, C., Hiemstra, D. (eds.) ECIR 2019. LNCS, vol. 11437, pp. 681–695. Springer, Cham (2019). https://doi.org/10.1007/978-3-030-15712-8_44
4. Davila, K., Xu, F., Setlur, S., Govindaraju, V.: Fcn-lecturenet: extractive summarization of whiteboard and chalkboard lecture videos. IEEE Access **9**, 104469–104484 (2021)
5. Davila, K., Zanibbi, R.: Whiteboard video summarization via spatio-temporal conflict minimization. In: 2017 14th IAPR International Conference on Document Analysis and Recognition (ICDAR), vol. 1, pp. 355–362. IEEE (2017)
6. Davila, K., Zanibbi, R.: Visual search engine for handwritten and typeset math in lecture videos and latex notes. In: 2018 16th International Conference on Frontiers in Handwriting Recognition (ICFHR), pp. 50–55. IEEE (2018)
7. Deng, D., Liu, H., Li, X., Cai, D.: Pixellink: detecting scene text via instance segmentation. In: Proceedings of the AAAI Conference on Artificial Intelligence, vol. 32 (2018)
8. Dey, A., Zanibbi, R.: ScanSSD-XYc: faster detection for math formulas. In: Barney Smith, E.H., Pal, U. (eds.) ICDAR 2021. LNCS, vol. 12916, pp. 91–96. Springer, Cham (2021). https://doi.org/10.1007/978-3-030-86198-8_7
9. Dutta, K., Mathew, M., Krishnan, P., Jawahar, C.: Localizing and recognizing text in lecture videos. In: 2018 16th International Conference on Frontiers in Handwriting Recognition (ICFHR), pp. 235–240. IEEE (2018)
10. Guan, M., Ding, H., Chen, K., Huo, Q.: Improving handwritten OCR with augmented text line images synthesized from online handwriting samples by style-conditioned GAN. In: 2020 17th International Conference on Frontiers in Handwriting Recognition (ICFHR), pp. 151–156. IEEE (2020)
11. Gupta, A., Vedaldi, A., Zisserman, A.: Synthetic data for text localisation in natural images. In: Proceedings of the IEEE Conference on Computer Vision and Pattern Recognition, pp. 2315–2324 (2016)

12. Hashmi, K.A., Pagani, A., Liwicki, M., Stricker, D., Afzal, M.Z.: Cascade network with deformable composite backbone for formula detection in scanned document images. Appl. Sci. **11**(16), 7610 (2021)
13. Haurilet, M., Roitberg, A., Martinez, M., Stiefelhagen, R.: Wise-slide segmentation in the wild. In: 2019 International Conference on Document Analysis and Recognition (ICDAR), pp. 343–348. IEEE (2019)
14. He, K., Gkioxari, G., Dollár, P., Girshick, R.: Mask r-cnn. In: Proceedings of the IEEE International Conference on Computer Vision, pp. 2961–2969 (2017)
15. He, K., Zhang, X., Ren, S., Sun, J.: Deep residual learning for image recognition. In: Proceedings of the IEEE Conference on Computer Vision and Pattern Recognition, pp. 770–778 (2016)
16. Husain, M., Meena, S.: Multimodal fusion of speech and text using semi-supervised LDA for indexing lecture videos. In: 2019 National Conference on Communications (NCC), pp. 1–6. IEEE (2019)
17. Liao, M., Wan, Z., Yao, C., Chen, K., Bai, X.: Real-time scene text detection with differentiable binarization. In: Proceedings of the AAAI Conference on Artificial Intelligence, vol. 34, pp. 11474–11481 (2020)
18. Lin, T.Y., Dollár, P., Girshick, R., He, K., Hariharan, B., Belongie, S.: Feature pyramid networks for object detection. In: Proceedings of the IEEE Conference on Computer Vision and Pattern Recognition, pp. 2117–2125 (2017)
19. Lin, T.-Y., et al.: Microsoft COCO: common objects in context. In: Fleet, D., Pajdla, T., Schiele, B., Tuytelaars, T. (eds.) ECCV 2014. LNCS, vol. 8693, pp. 740–755. Springer, Cham (2014). https://doi.org/10.1007/978-3-319-10602-1_48
20. Long, S., He, X., Yao, C.: Scene text detection and recognition: the deep learning era. Int. J. Comput. Vis. **129**(1), 161–184 (2021). https://doi.org/10.1007/s11263-020-01369-0
21. Mahdavi, M., Zanibbi, R., Mouchere, H., Viard-Gaudin, C., Garain, U.: ICDAR 2019 CROHME+ TFD: competition on recognition of handwritten mathematical expressions and typeset formula detection. In: 2019 International Conference on Document Analysis and Recognition (ICDAR), pp. 1533–1538. IEEE (2019)
22. Marti, U.V., Bunke, H.: The iam-database: an English sentence database for offline handwriting recognition. Int. J. Doc. Anal. Recogn. **5**(1), 39–46 (2002)
23. Phong, B.H., Hoang, T.M., Le, T.L.: A hybrid method for mathematical expression detection in scientific document images. IEEE Access **8**, 83663–83684 (2020)
24. Pratikakis, I., Zagoris, K., Barlas, G., Gatos, B.: ICFHR 2016 handwritten document image binarization contest (H-DIBCO 2016). In: 2016 15th International Conference on Frontiers in Handwriting Recognition (ICFHR), pp. 619–623. IEEE (2016)
25. Springstein, M., Müller-Budack, E., Ewerth, R.: Unsupervised training data generation of handwritten formulas using generative adversarial networks with self-attention. In: Proceedings of the 2021 Workshop on Multi-Modal Pre-Training for Multimedia Understanding, pp. 46–54 (2021)
26. Villamizar, M., Canévet, O., Odobez, J.M.: Multi-scale sequential network for semantic text segmentation and localization. Pattern Recogn. Lett. **129**, 63–69 (2020)
27. Wang, W., et al.: Shape robust text detection with progressive scale expansion network. In: Proceedings of the IEEE/CVF Conference on Computer Vision and Pattern Recognition, pp. 9336–9345 (2019)
28. Wolf, F., Brandenbusch, K., Fink, G.A.: Improving handwritten word synthesis for annotation-free word spotting. In: 2020 17th International Conference on Frontiers in Handwriting Recognition (ICFHR), pp. 61–66. IEEE (2020)

29. Yim, M., Kim, Y., Cho, H.-C., Park, S.: SynthTIGER: synthetic text image GEnerntoR towards better text recognition models. In: Lladós, J., Lopresti, D., Uchida, S. (eds.) ICDAR 2021. LNCS, vol. 12824, pp. 109–124. Springer, Cham (2021). https://doi.org/10.1007/978-3-030-86337-1_8

30. Zhu, Y., Du, J.: Textmountain: accurate scene text detection via instance segmentation. Pattern Recogn. **110**, 107336 (2021)

Generating Synthetic Styled Chu Nom Characters

Jonas Diesbach[1], Andreas Fischer[1,2], Marc Bui[3],
and Anna Scius-Bertrand[1,2,3(✉)]

[1] iCoSys, HES-SO, Fribourg, Switzerland
{jonas.diesbach,andreas.fischer,anna.scius-bertrand}@hefr.ch
[2] DIVA, University of Fribourg, Fribourg, Switzerland
[3] EPHE-PSL, Paris, France
marc.bui@ephe.sorbonne.fr

Abstract. Images of historical Vietnamese steles allow historians to discover invaluable information regarding the past of the country, especially about the life of people in rural villages. Due to the sheer amount of available stone engravings and their diverseness, manual examination is difficult and time-consuming. Therefore, automatic document analysis methods based on machine learning could immensely facilitate this laborious work. However, creating ground truth for machine learning is also complex and time-consuming for human experts, which is why synthetic training samples greatly support learning while reducing human effort. In particular, they can be used to train deep neural networks for character detection and recognition. In this paper, we present a method for creating synthetic engravings and use it to create a new database composed of 26,901 synthetic Chu Nom characters in 21 different styles. Using a machine learning model for unpaired image-to-image translation, our approach is annotation-free, i.e. there is no need for human experts to label character images. A user study demonstrates that the synthetic engravings look realistic to the human eye.

Keywords: Synthetic handwriting generation · Generative adversarial networks (gan) · Contrastive Unpaired Translation (CUT) · Historical vietnamese steles · Chu Nom characters

1 Introduction

Created mostly between the 17th to 19th century, Vietnamese steles contain very valuable information about the history of the country [17,18]. The reason for this is that the current knowledge of Vietnamese history is mainly based on annals from the royal court and clergy, and those official records only tell about the great national history, diplomatic conventions, nomination of mandarins and wars. In contrast, the steles are the sole historical source able to describe the village life. They not only speak about the large history of the country, but also

© The Author(s), under exclusive license to Springer Nature Switzerland AG 2022
U. Porwal et al. (Eds.): ICFHR 2022, LNCS 13639, pp. 484–497, 2022.
https://doi.org/10.1007/978-3-031-21648-0_33

about the social, economic and cultural life of the rural communities. In short, they contain a vast amount of important information for historians.

To preserve the ancient history written on the stone steles, whose average size exceeds a square meter, the French School of the Far East (EFEO) began creating a collection of stamps between 1910 to 1954. This work was then later continued by the Han-Nôm institute, starting in 1995. A stamp is created by first gluing paper onto the steles with banana-juice acting as the adhesive. Then, an ink-covered roll is used to paint over the whole sheet leaving the characters in white and coloring the background in black. Thus, creating a faithful representation of the engravings while keeping the original scale. Finally, the stamps are photographed to obtain digital images of the steles. Examples of stele images are shown in Fig. 1, illustrating the heterogeneous nature of both the stele backgrounds as well as the engravings.

Fig. 1. Historical Vietnamese stele images.

Many steles have now disappeared, but their stamps remain. Because there exist images of thousands of unique steles, automated reading of the Chu Nom characters would greatly support the historians, who are currently investigating the steles in the context of the Vietnamica[1] project. A significant step towards automated reading has been taken in [22], where an annotation-free character detection system has been introduced. Avoiding human labelling effort, the system is using printed Chu Nom characters to train deep convolutional detection networks and performs unsupervised self-calibration in order to adapt to to the stele images. The result of this study are hundreds of thousands of automatically detected character images.

---

[1] https://vietnamica.hypotheses.org/.

The next step would be not only to detect where the characters are located but also to read them. Due to the lack of annotated data, a promising line of research is to generate a large amount of synthetic engravings that are close to the real characters in order to train systems for keyword spotting and automatic transcription. In this paper, we leverage the automatically detected character images and use generative adversarial networks (GANs) to transfer different engraving styles to printed Chu Nom characters. The synthesis remains annotation-free for creating characters, that is no human labelling of character images is required. However, there are still a few human interactions necessary for the proposed method. First, we use a printed Chu Nom font that was created manually in the past (but not specifically for our task). Secondly, a brief human interaction was necessary to manually choose 21 different engraving styles among the stele images.

In the present paper, we rely not only on CycleGAN [29], but also on recent advancements in image-to-image translation, namely we use Contrastive Unpaired Translation (CUT) [19], a machine learning model which is based on the GAN framework and uses contrastive learning. The latter model is used to generate a total of 21 synthetic engraving styles of about 26,901 printed Chu Nom characters each. We made the database freely available[2] with the aim to support the development of automatic reading systems in the future. To evaluate the general quality of the synthesis, we have conducted a user study that will be discussed below.

2   Related Work

Various methods to generate styled handwritten word or character images have been studied. Many of which are based on GAN, a fairly recent and powerful generative model, which is based on a generator-discriminator architecture where the two networks compete against each other to improve the generated results. Due to the impressive results achieved by GANs as well as their wide applicability, they have also been largely adopted for synthetic image generation tasks.

In the context of Latin scripts, one of the most recent works focusing on the generation of handwritten words based on the Latin alphabet is GANwriting [11], which generates words conditioned on a given style. Even more recently, HiGAN [3] was proposed, which, in contrast to GANwriting, can generate variable-length handwritten words. Both of those approaches use a threepart objective: An adversarial loss to discriminate between real and fake samples, a word recognition loss to preserve the textual content and a writer identification loss to match the calligraphic style. Unfortunately, we cannot use a word (or more accurately in our case, a character) recogniser, because we do not have the necessary labelled data. Removing the recogniser part from those networks noticeably affects the results, which is why we decided to not further investigate in this direction.

---

[2] https://github.com/asciusb/21SyntheticStylesNom-Database/.

Other works include SC-GAN [5] and JointFontGAN [27], both of which would require skeletons of the extracted Chu Nom characters, which are not available. Because of the noise in the image, the skeletons risk to not be readable without a considerable work of denoising before. GlyphGAN [7] is a style-consistent font generation model, which takes a style and a character class vector as its input. The latter is a one-hot encoded vector associated with the character class of each sample image. In our case, this would require manual annotation of each and every individual stele character image with its character class label, which is not feasible.

In the context of Asian logograms, many different works to generate Chinese characters have already been proposed. One of them is zi2zi [23], which is an extension of the Pix2Pix framework [9] specifically built to model Chinese characters. Other works such as CalliGAN [26] and MTfontGAN [25] show that multiple calligraphy styles of a character can be generated using a single model.

Further proposed methods all serve different purposes: [24] handles the generation of thin strokes, SCFont [10] tries to generate characters with correct structure and without artefacts, MSMC-CGAN [13] generates realistic multi-scale and multi-class handwritten characters, TH-GAN [1] helps improve historical Chinese character recognition and [20] uses a DenseNet-Pix2Pix model to restore incomplete calligraphy fonts. Furthermore, LSCGAN [12] proposed a stroke-based font generation method where the styles of two existing font characters are fused together to create a new style and FontGAN [14] synthesises characters with a specified style using character stylisation and de-stylisation to improve content consistency. However, not all proposed works focus exclusively on Chinese characters. For example, [15] uses the Pix2Pix architecture to generate Bangla characters and DM-Font [2] decomposes each glyph into several components (sub-glyphs) before reassembling them to new Korean or Thai glyphs. Unfortunately, all of the aforementioned methods have one thing in common: They all use paired training data. Such a dataset would require a large amount of annotated steles, which do not yet exist.

Due to the lack of paired training data, a model which works with unpaired data is needed. And indeed, such a method has already been published in 2021 to generate handwritten Chinese characters. The model, named StrokeGAN [28], is based on CycleGAN and introduces a one-bit stroke encoding to alleviate the mode collapse issue. Unfortunately, this method cannot be easily applied to Chu Nom characters, since the required stroke encodings are not available.

In the present paper, we chose two annotation-free models for generating synthetic Chu Nom engravings that do not require paired data, namely a Cycle-GAN model built on Pix2Pix and a CUT model. Both are described in more detail in the next section.

3 Unpaired Generative Adversarial Networks

The concept of Generative Adversarial Networks was introduced in 2014 by Ian Goodfellow et al. [4]. They then became rapidly popular because of their successful application in various domains, such as image processing, computer vision, music generation, natural language processing and also in the medical field [6,8]. GANs proved to be especially useful for image-to-image translation tasks, e.g. CycleGAN [29].

The basic underlying structure of a GAN consists of a generator and a discriminator model. The goal of the generator is to create samples which look indistinguishable from real ones. The discriminator, which is usually a binary classifier, tries to accurately discriminate between real and generated samples. The general idea of this network is that the generator tries to deceive the discriminator and the discriminator tries to detect the generated samples from the generator. Thus, GANs introduced the concept of adversarial learning.

One of the main issues with GANs is non-convergence, of which the most common catastrophic problem is mode collapse [4]. This problem occurs when the generator learns to map several different input values to the same mode, even though samples of the missing modes existed in the training data. In the worst case of a complete collapse, the generator produces only a single output.

Another typical limitation of GANs is the requirement of paired samples between the source domain and the target domain. In the following, two unpaired GAN models are described, which are able to overcome this limitation, namely CycleGAN and CUT.

3.1   CycleGAN

CycleGAN [29], which is built upon the Pix2Pix framework, is able to translate images from a source domain X into a target domain Y when there is no paired data available. The goal is to learn a mapping $G : X \rightarrow Y$ such that the generated samples $G(x)$, where $x \in X$, are indistinguishable from the real images $y \in Y$. An adversarial loss $\mathcal{L}_{GAN}$ is used for this purpose. However, this translation does not suffice to produce compelling results, because the mapping G is highly under-constrained. Moreover, optimisation of the standard adversarial objective ($\mathcal{L}_{GAN}$) in practice often leads to mode collapse, which is why an inverse mapping $F : Y \rightarrow X$ is introduced to exploit the cycle consistency property of translations (see Fig. 2a).

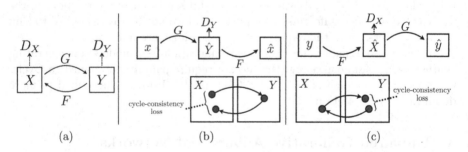

Fig. 2. Overview of the CycleGAN architecture and the two cycle-consistency losses. Images from [29].

The meaning behind cycle consistency is that a translation from source domain to target domain and back should return the original sample, i.e. the learned mapping functions F and G should be cycle-consistent and thus enforce $x \rightarrow G(x) \rightarrow F(G(x)) \approx x$ (forward cycle-consistency, see Fig. 2b) and $y \rightarrow F(y) \rightarrow G(F(y)) \approx y$ (backward cycle-consistency, see Fig. 2c). This behaviour is encouraged by the following cycle consistency loss:

$$\mathcal{L}_{cyc}(G, F) = \mathbb{E}_x[\|F(G(x)) - x\|_1] + \mathbb{E}_y[\|G(F(y)) - y\|_1]. \tag{1}$$

where $\mathbb{E}_x, \mathbb{E}_y$ are the expected values with respect to the source and target domain, respectively.

Additionally, adversarial losses are applied to both mapping functions using a discriminator D_X to distinguish between images x and generated images $F(y)$ and similarly, a discriminator D_Y to discriminate between y and $G(x)$. Combining the aforementioned losses yields the final objective of the CycleGAN model:

$$\mathcal{L}_{CycleGAN}(G, F, D_X, D_Y) = \mathcal{L}_{GAN}(G, D_Y) + \mathcal{L}_{GAN}(F, D_X) \\ + \lambda \mathcal{L}_{cyc}(G, F). \tag{2}$$

3.2   CUT

Recently, a new unpaired image-to-image translation model using contrastive learning named CUT [19] was proposed by the authors of CycleGAN. The idea behind this method is to maintain the content of the source image by maximising the mutual information between corresponding input and output patches, which is done via a noise contrastive estimation (NCE) framework [16]. This method only requires to learn a mapping in one direction, which simplifies the training. To do so, the generator G is split up into an encoder and a decoder. Combined sequentially, they generate an output image $G(x) = G_{dec}(G_{enc}(x))$, where x is an image from the source domain X. A sample image-to-image translation problem demonstrating this method is displayed in Fig. 3.

In contrastive learning, a *query* z should be strongly associated to its corresponding *positive* counterpart z^+ while being disassociated from all N *negatives* z_n^-. An $(N + 1)$-way classification problem is created, where the softmax cross entropy loss $\ell(z, z^+, z^-)$ is used to predict the probability of the query belonging to the same class as its positive.

The query, positive and negatives are sampled from different layers $l \in \{1, \ldots, L\}$ and spatial locations $s \in \{1, \ldots, S_l\}$ of the encoder and are passed through a small two-layer Multilayer Perceptron H_l to project both the input and output patches into a shared embedding space. With respect to the embedded query z_l^s sampled from the output image $G(x)$, the embedded positive $z_l^{s,+}$ at the same position in the input image x, and embedded negatives $z_l^{S\backslash s,-}$ at positions different to s in the input image, the PatchNCE loss is defined as

$$\mathcal{L}_{PatchNCE}^X(G, H) = \mathbb{E}_x[\sum_{l=1}^{L} \sum_{s=1}^{S_l} \ell(z_l^s, z_l^{s,+}, z_l^{S\backslash s,-})]. \tag{3}$$

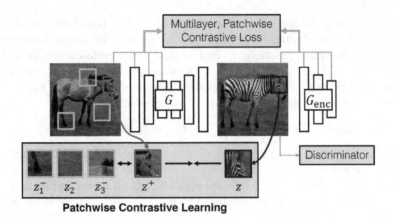

Fig. 3. Patchwise contrastive learning method overview. Figure from [19].

The same loss $\mathcal{L}^Y_{PatchNCE}(G,H)$ can also be computed for images of the target domain to prevent the generator from making changes to images of the target domain. Finally, by combining the GAN loss with the two PatchNCE losses, the objective function of the Contrastive Unpaired Translation (CUT) is defined as

$$\mathcal{L}_{CUT}(G,D,H) = \mathcal{L}_{GAN}(G,D) + \mathcal{L}^X_{PatchNCE}(G,H) + \mathcal{L}^Y_{PatchNCE}(G,H). \quad (4)$$

4 Data

The combined collection of stone engravings by the EFEO and the Han-Nôm institute comprises about 40,000 unique copies. From all those stele stamp images, we had a subset of 2,036 images at our disposal. A single image consists of around three million pixels on average. The only preprocessing step performed on these images was to invert their colour, turning the characters black. Note that all those images are of varying sizes, have different amounts of text columns and characters in total. Furthermore, certain steles have irregular text columns or even columns which split into two. Other steles show clear signs of deterioration, which were caused by the weather or are simply due to their age. Most notably, though, is the huge range of different styles, be that due to the size (stele itself, text area, characters, borders), the ornamentation or the handwriting [21].

In addition to the stele images, we also had the bounding box coordinates of automatically detected characters at our disposal for about 1,800 steles, which were obtained by the annotation-free character detection system [22]. For this subset, the detection system was able to locate at least 100 characters on each stele with sufficient confidence.

Important to note here is that the bounding boxes not necessarily encircle an actual character, because the coordinates only show where the detection system assumes a character is located. Note further that the characters were only detected automatically but not classified, i.e. we do not know to which characters they correspond to. The average size of the extracted character images is 19.1 × 20.0 pixels. The set of all character images constitutes the superset of style target domains used in future image-to-image translations. The source domain consists of 26,901 printed characters (black font on white background), which are based on a Chu Nom font, courtesy of the Vietnamese Nom Preservation Foundation[3].

5 Evaluation

To evaluate the performance of the proposed method for generating synthetic Chu Nom characters, we comment in the following sections on the training setup and behavior of CycleGAN and CUT, followed by a user study to assess whether or not the generated characters look realistic to the human eye.

5.1 Setup

Because neither a single CycleGAN nor CUT model can generate multiple different styles, we tried using K-Means clustering to find character images from various steles with similar engraving styles such that a large range of styles can be covered with few trained models. Unfortunately, this did not work well, since the difference between character images from the same cluster was too large.

Instead of retraining a model on a new style for some additional epochs, we decided to manually select a handful of styles, because we achieved better results with individual models. Therefore, we combed through a subset of 300 steles, where we chose 21 steles from which we extracted the characters in order to create the target domain set for the training of the models. The advantages of this approach are twofold. First, a noticeable difference between the various styles can be guaranteed, which would have not been the case with automated clustering. Second, a good quality of training data can be ensured, because steles with a lot of poorly detected characters can simply be discarded as long as there exists a stele of a very similar style with better detection results.

The datasets used to train the two models consist of a source and a target set. The extracted character images from a single engraving style constitute the target domain. Depending on the selected stele, there are as few as 202 images and as many as 544 images. If there were more than 600 detected characters of a given stele, the set was reduced to 400 images by randomly removing characters to improve the models training performance. On average, for the 21 engraving styles, there were 366 images in the target domain. The source domain consists of the same amount of binary printed font character images, which were created anew for each style.

---

[3] http://www.nomfoundation.org.

5.2   CycleGAN

To train the model on our data, we used the default settings of CycleGAN, the sole change being a smaller input image size. Concretely, we only scaled the extracted character images up to 128 × 128 pixels instead of 256 × 256. The networks are trained from scratch over 200 epochs with a learning rate of 0.0002 for the first 100 epochs. The learning rate linearly decays over the remaining 100 epochs to zero and the hyperparameter λ in Eq. 2 is set to 10.

Figure 4 displays a selection of sample results after training the CycleGAN model on one of the engraving style datasets. The generated images (middle row) should show the source characters (top row) in the style of target domain (bottom row), but they clearly display different occurrences of the mode collapse issue.

Fig. 4. Three separate instances of a CycleGAN mode collapse.

5.3   CUT

Similar to CycleGAN, we have also used the standard setup of CUT when training the model, i.e. no parameters were fine-tuned. The source and target domain images were scaled up to 256 × 256 pixels and the networks were trained for 400 epochs, which takes around 12 to 13 h on a GeForce GTX 1080 with a training dataset consisting of 400 images in both domains. Once a model has been trained, a single character image can be generated in around one eight of a second, which means it still takes almost a full hour to produce all 26,901 Chu Nom characters in one engraving style.

Figure 5 illustrates some exemplary results after training the CUT models until convergence. The generated images (middle row) combine the content of the printed source images (top row) with the style of the target domain (bottom row), thus allowing us to recreate the entire printed font with 26,901 Chu Nom characters in the 21 engraving styles.

We notice that the visual features of the generated characters are not perfect, e.g. some strokes are missing, making it difficult or even impossible to read some

Fig. 5. Generated CUT results of three different styles

of the characters. However, on the stele images similar problems are encountered with real characters, which are not well readable due to low resolution for small characters, damages in the stone, etc. Therefore, we would expect that using the generated characters for training downstream tasks, such as character detection, keyword spotting, and transcription, may still be helpful, especially when combined with real images.

5.4   User Study

In order to evaluate the quality of our generated character images, we conducted a user study with 14 participants. Due to the superior quality of the CUT results, we exclusively used those generated character images for this study. A survey was handed out to the participants either as a hard copy or in electronic form as a PDF. In total, there were 14, mostly male, participants of ages ranging from 23 to 59 years with various occupations. We also provided some example images of steles and real character images to at least give the participants a rough idea how the actual steles and characters look like, because none of them had prior experience with ancient Vietnamese stone engravings (nor with Asian logograms in general). We presented each participant with the same grid of 54 characters images, which consists of equal parts real and synthetic images but this information was withhold from the participants. The 27 real character images stem from the 21 steles used as basis for the generated styles. At least one character and at most two were selected from each stele. Similarly, 27 generated images were also chosen from a small random subset (32 images) of each generated style. Each participant was then asked to mark every single image they thought was synthetically generated.

To evaluate the results of the user study, we compute different metrics, which are shown in Table 1. When compared with a random selection process, i.e. an expected recall and precision of 0.5, the participants found significantly less synthetic images (average recall of 0.333). However, among the images marked as synthetic, the participants performed slightly better than random (average precision of 0.582) hinting at the possibility that there might be some visual artefacts in the synthetic samples that can be identified.

Table 1. Quantitative evaluation of the user study. Average metrics for the detection of synthetic characters.

Recall	Precision	F1 Score
0.333	0.582	0.413

We have also evaluated the qualitative feedback that was given by the participants. The task of distinguishing real from fake character images was perceived as difficult by every single one of them. Some of the most frequent reasons are as follows: The amount of different and diverse styles made it hard to identify a synthetic character of a particular style. Also, the participants were unable to detect any obvious patterns, i.e. they were unable to find characteristics of fake images which would have given them away. Individual character images are qualitatively consistent and do not contain visible artefacts or impeccable areas. Additionally, some characters are unrecognisable, incomplete or there does not exist a clear distinction between symbol and background, which adds to the difficulty as well. Furthermore, individual character images are rather small and not of very high-resolution quality. Finally, the lack of familiarity with logograms, not to mention Chu Nom symbols, also made the task more complicated. To illustrate the difficulty of the task, Fig. 6 shows a subset of 27 images from the survey.

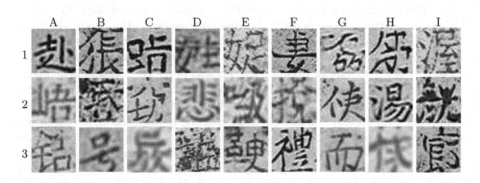

Fig. 6. Selection of real and artificially generated character images. Try to identify the synthetic samples (Synthetic: 1B, 1C, 1E, 1G, 1H, 1I, 2A, 2C, 2E, 2F, 2I, 3A, 3C, 3E, 3H and 3I).

The bottom line is that the images are indeed visually convincing and it is not at all evident which ones are real and which ones were generated. There is not a single style which shows obvious signs that those particular images were computer-generated. Clearly, not every single image of the 21 * 26, 901 generated ones was inspected and evaluated in detail. Thus, there is the possibility that certain character images of some styles might display clear indications that they are not real. Nonetheless, from the images we have looked at so far, we can confidently say that the CUT model was successful in imitating styled stele characters. This statement is supported by the fact that, on average, two-thirds of generated images were missed and not a single participant found more than half of the generated characters. This indicates that a very good approximation of the real images has been achieved. Also, the qualitative feedback from the participants clearly emphasises that the generated characters seem genuine to the human eye.

6   Conclusion and Future Work

Due to a lack of annotated data, we explored the possibility of generating synthetic Chu Nom engravings with different styles to make a significant step towards automated reading. A lot of generative models for character synthesis use paired data, i.e. character images that are annotated with their correct character label, which are not available in our case. Therefore, we have investigated two models that do not require paired data: CycleGAN and CUT. With Cycle-GAN, we rapidly encountered mode collapse problems that produced unreadable results. With CUT, however, we succeeded in the generation of readable Chu Nom engravings, although some of the generated characters were missing important strokes. In a user study, it was demonstrated that the generated characters seemed realistic to an inexperienced human observer.

Although our method does not require human experts to annotate individual character images, a brief manual interaction was still necessary to choose 21 well-distinguishable engraving styles. In future work, we aim to automate this step as well, to cover a larger number of different styles present in the entire stele dataset.

Furthermore, an interesting line of future research would be to synthesize whole stele images for training and improving the annotation-free character detection method. Another line of research would be to use the synthetic characters for training keyword spotting and transcription systems.

References

1. Cai, J., Peng, L., Tang, Y., Liu, C., Li, P.: TH-GAN: generative adversarial network based transfer learning for historical Chinese character recognition. In: International Conference on Document Analysis and Recognition (ICDAR), pp. 178–183 (2019)
2. Cha, J., Chun, S., Lee, G., Lee, B., Kim, S., Lee, H.: Few-shot compositional font generation with dual memory. In: Vedaldi, A., Bischof, H., Brox, T., Frahm, J.-M. (eds.) ECCV 2020. LNCS, vol. 12364, pp. 735–751. Springer, Cham (2020). https://doi.org/10.1007/978-3-030-58529-7_43
3. Gan, J., Wang, W.: HiGAN: handwriting imitation conditioned on arbitrary-length texts and disentangled styles. AAAI Conf. Artif. Intell. 35(9), 7484–7492 (2021)
4. Goodfellow, I., et al.: Generative adversarial nets. In: International Conference on Neural Information Processing Systems (NIPS), pp. 2672–2680 (2014)

5. Guan, M., Ding, H., Chen, K., Huo, Q.: Improving handwritten OCR with augmented text line images synthesized from online handwriting samples by style-conditioned GAN. In: International Conference on Frontiers in Handwriting Recognition (ICFHR), pp. 151–156 (2020)
6. Gui, J., Sun, Z., Wen, Y., Tao, D., Ye, J.: A review on generative adversarial networks: Algorithms, theory, and applications. IEEE Trans. Knowl. Data Eng. pp. 1–20 (2021)
7. Hayashi, H., Abe, K., Uchida, S.: GlyphGAN: style-consistent font generation based on generative adversarial networks. Knowl. Based Syst. **186**, 1–13 (2019)
8. Hong, Y., Hwang, U., Yoo, J., Yoon, S.: How generative adversarial networks and their variants work. ACM Comput. Surv. **52**(1), 1–43 (2019)
9. Isola, P., Zhu, J.Y., Zhou, T., Efros, A.A.: Image-to-image translation with conditional adversarial networks. In: IEEE Conference on Computer Vision and Pattern Recognition (CVPR), pp. 5967–5976 (2017)
10. Jiang, Y., Lian, Z., Tang, Y., Xiao, J.: SCFont: structure-guided Chinese font generation via deep stacked networks. AAAI Conf. Artif. Intell. **33**(01), 4015–4022 (2019)
11. Kang, L., Riba, P., Wang, Y., Rusiñol, M., Fornés, A., Villegas, M.: GANwriting: content-conditioned generation of styled handwritten word images. In: Vedaldi, A., Bischof, H., Brox, T., Frahm, J.-M. (eds.) ECCV 2020. LNCS, vol. 12368, pp. 273–289. Springer, Cham (2020). https://doi.org/10.1007/978-3-030-58592-1_17
12. Lin, X., Li, J., Zeng, H., Ji, R.: Font generation based on least squares conditional generative adversarial nets. Multimedia Tools Appl. **78**(1), 783–797 (2018). https://doi.org/10.1007/s11042-017-5457-4
13. Liu, J., Gu, C., Wang, J., Youn, G., Kim, J.-U.: Multi-scale multi-class conditional generative adversarial network for handwritten character generation. J. Supercomput. **75**(4), 1922–1940 (2017). https://doi.org/10.1007/s11227-017-2218-0
14. Liu, X., Meng, G., Xiang, S., Pan, C.: FontGAN: a unified generative framework for Chinese character stylization and de-stylization. CoRR abs/1910.12604 (2019)
15. Nishat, Z.K., Shopon, M.: Synthetic class specific Bangla handwritten character generation using conditional generative adversarial networks. In: International Conference on Bangla Speech and Language Processing (ICBSLP), pp. 1–5 (2019)
16. van den Oord, A., Li, Y., Vinyals, O.: Representation learning with contrastive predictive coding. CoRR abs/1807.03748 (2019)
17. Papin, P.: Aperçu sur le programme «?publication de l'inventaire et du corpus complet des inscriptions sur stèles du viêt-nam?». Bulletin de l'Ecole française d'Extrême-Orient **90**(1), 465–472 (2003)
18. Papin, P.: Les inscriptions anciennes du viêt-nam, source d'une nouvelle vision des xviie et xviiie siècles. Good Morning 105 (2010)
19. Park, T., Efros, A.A., Zhang, R., Zhu, J.-Y.: Contrastive learning for unpaired image-to-image translation. In: Vedaldi, A., Bischof, H., Brox, T., Frahm, J.-M. (eds.) ECCV 2020. LNCS, vol. 12354, pp. 319–345. Springer, Cham (2020). https://doi.org/10.1007/978-3-030-58545-7_19
20. Qin, M., Chen, X.: Restore the incomplete calligraphy based on style transfer. In: Chinese Control Conference (CCC), pp. 8812–8817 (2019)
21. Scius-Bertrand, A., Voegtlin, L., Alberti, M., Fischer, A., Bui, M.: Layout analysis and text column segmentation for historical Vietnamese steles. In: Proceedings of 5th International Workshop on Historical Document Imaging and Processing (HIP), pp. 84–89 (2019)

22. Scius-Bertrand, A., Jungo, M., Wolf, B., Fischer, A., Bui, M.: Annotation-free character detection in historical Vietnamese stele images. In: International Conference on Document Analysis and Recognition (ICDAR), pp. 432–447 (2021)
23. Tian, Y.: Master Chinese calligraphy with conditional adversarial networks (2017). https://github.com/kaonashi-tyc/zi2zi
24. Wen, C., et al.: Handwritten Chinese font generation with collaborative stroke refinement. In: IEEE/CVF Winter Conference on Applications of Computer Vision (WACV), pp. 3882–3891 (2021)
25. Wu, L., Chen, X., Meng, L., Meng, X.: Multitask adversarial learning for Chinese font style transfer. In: International Joint Conference on Neural Networks (IJCNN), pp. 1–8 (2020)
26. Wu, S.J., Yang, C.Y., Hsu, J.Y.J.: CalliGAN: style and structure-aware Chinese calligraphy character generator. CoRR abs/2005.12500 (2020)
27. Xi, Y., Yan, G., Hua, J., Zhong, Z.: JointFontGAN: joint geometry-content GAN for font generation via few-shot learning. ACM Int. Conf. Multimedia, pp. 4309–4317 (2020)
28. Zeng, J., Chen, Q., Liu, Y., Wang, M., Yao, Y.: StrokeGAN: reducing mode collapse in Chinese font generation via stroke encoding. CoRR abs/2012.08687 (2021)
29. Zhu, J.Y., Park, T., Isola, P., Efros, A.A.: Unpaired image-to-image translation using cycle-consistent adversarial networks. In: International Conference on Computer Vision (ICCV), pp. 2242–2251 (2017)

UOHTD: Urdu Offline Handwritten Text Dataset

Aftab Rafique[✉] and M. Ishtiaq

Foundation University Islamabad, School of Science and Technology, Islamabad,
Pakistan
armalik001f20@gmail.com, ishtiaq@fui.edu.pk

Abstract. We present our Offline Urdu Handwritten Text Dataset
(UOHTD) in this paper by collecting 800 Urdu handwritten samples
written by 800 native language writers. It consists of images in the form
of a dataset containing written text samples scanned with multiple spatial resolutions. 8000 text lines and 40000 words as patches have been
extracted from sample pages and checked manually and formally using
a ground truth database. Machine Learning Tools have been utilized to
extract sample pages and segment them into lines and words. Initial tri-
als on demographic (gender and age group) classification of Urdu writers
with samples of Offline Urdu Handwritten Text Dataset (UOHTD) has
produced promising results (85% for gender and 79% for age group classification) using CNNs. The database would be made available to the
researcher worldwide for study into various handwritten-related topics
including text recognition, identification of the writer's age, ethnicity,
demographics, gender, and handedness, as well as verification.

Keywords: UOHTD · CNNs · Segmentation · Machine learning ·
Demographic classification

1 Introduction

Classification of handwritten text proves an attractive issue owing to its multiple
uses, including converting texts to digital medium, automatically reading building numbers with address, recognizing postal addresses, application in forensic
document analysis, and robotics. [1,2,5] & [9]. The classification of handwritten letters and words is a tough as complexity relies on the limits imposed on
the development of the original handwriting [1,4]. Moreover, handwritten text
identification is difficult since writing styles differ from individual to individual. In addition, a felt tip pen might make an arbitrary handwritten word with
isolated, touching, or overlapping letters, cursive fragments, or entirely cursive
sentences and varied degrees of neatness attainable, ranging from exceedingly
messy to extremely neat [1]. The Urdu language is extremely important being
one of the world's major languages and the native language of Pakistan [5].
Urdu language stands upright being widely spoken in Southern Asian region.

© The Author(s), under exclusive license to Springer Nature Switzerland AG 2022
U. Porwal et al. (Eds.): ICFHR 2022, LNCS 13639, pp. 498–511, 2022.
https://doi.org/10.1007/978-3-031-21648-0_34

Urdu is one of the world's major languages, being the first tongue of over 100 million people worldwide. When coupled with Hindi, Pashto, Punjabi, Sindhi, and Balochi, it surpasses 330 million people because these languages are almost identical in spoken and written form, with the exception of Hindi [8]. Despite its popularity, it plainly lacks computational support and is written in Romanized Urdu script which has parallels to Arabic and Persian language [4,5] & [6]. There are languages with extensive research and many databases, however this facet is relatively restricted and limited in the case of Urdu [7,8]. Curious researchers face a paucity of publicly available and complete Urdu handwritten datasets, demonstrating the language's scarcity of study on text recognition and demographic categorization in compared to other languages [7].

Unfortunately, there appears to be little or no effort on Urdu language processing, owing to a lack of linguistic resources. Besides that, unlike English and other significant language groups, Urdu text classification as well as recognition is more difficult due to the presence of diacritics, that are also observed in Arabic and Persian, hence any study objective on Urdu text classification and recognition would eventually ease out significant improvement in research work on handwritten text classification of several languages of the same family. This research presents a framework for gender and age classification of Urdu writers, a relatively unexplored subject in the Urdu language based on handwritten samples. There has not been a publicly accessible dataset of handwritten Urdu text. To solve this issue, we present Urdu Offline Handwritten Text Dataset (UOHTD), a unique Urdu dataset. The idea came from the reality that there is no standard collection of Urdu handwritten text which could be used as a foundation for new research. In addition, for gender and age group prediction, we employed a convolutional neural network (CNN) that was trained with 85% of the information and assessed with the remaining 15%. Both have been tested and validated in TensorFlow and PyTorch libraries with promising results proving the reasonable quality of UOHTD.

This research has following major contributions that will prove helpful for other researchers:

- Creation of Urdu Handwritten Text Dataset (UOHTD) scanned as images in gender and different age groups.
- UOHTD will benefit the global scientific community and be considered as a valuable asset for scientific research.
- Novel venues of research on Urdu language can be explored in the field of computer vision and pattern recognition as well as natural language processing.

The organization of paper is Sect. 2 describes the previous work in literature on the creation of Urdu offline handwriting databases. Section 3 presents data collecting and statistics. Section 4 has a task definition. Section 5 discusses data verification. Section 6 presents the experimental achievements of Urdu text classification on UOHTD in terms of gender and age group. Section 7 contains the findings and conclusions.

2   Literature Review

Handwriting recognition has made significant improvements in various current life areas, including mail sorting, check reading, and the processing of historical documents during the last few years [2]. The usage of soft biometrics, such as gender, age, and ethnicity, has recently gained popularity in several applications where such information is needed. This kind of information is beneficial in demographic studies and forensics applications. Various research studies have been conducted on this subject in recent years. Writers' demographic classification by handwriting analysis started in 2001 by Cha et al. [10] who applied data mining technique to identify discriminating patterns in demographic sub-categories using a priori Algorithm to select only sub-categories with appropriate support and classification rate among all the available ones. This section elaborates on the databases accessible for scientific study in Urdu.

Traditional techniques of identifying writers rely heavily on complicated hand-crafted characteristics. The textural and allographic features for writer identification were designed by Bulacu and Schomaker [9]. To represent online signatures, the authors of [11] used interval-valued symbolic qualities. The authors of [12] identified 16 different types of histogram-based signature verification characteristics. In ongoing decade, Machine Learning Algorithms especially CNNs have been employed successfully for writer identification [13] because CNN is better suited to dealing with fixed-size pictures as revealed in [14], stroke and character segmentation were necessary as a preprocessing step, and hand-crafted features were still required to obtain high accuracy for CNN-based writer identification systems. Moreover, a huge dataset is required to feed neural networks in machine learning, so we highlight datasets created in the past for Urdu language before going into the details of UOHTD.

CENPARMI Urdu Database has been developed by the Center of Pattern Recognition and Machine Intelligence (CENPARMI) [15]. It consists of Urdu words, characters, numerals. The database consists of 318 date samples, 60,329 isolated digits, 12,914 strings of numerals, 1,705 occurrences of 4 special symbols, 14,890 samples of 37 basic characters and 19,432 examples of 57 words related to finance. 343 native Urdu speakers were accessed from around the world to collect the data.

Urdu Handwritten Sentence Database (UHSD) [9] contains 2051 handwritten sentences consisting of 23,833 different words. Forms with six different categories of News data were used for data collection. 200 writers participated in data collection process and each one provided 2 forms. The dataset could be used to evaluate the performance of segmentation and recognition of text and writer identification.

UCOM Offline Handwritten Dataset was developed in 2013, the COMSATS Institute of Information Technology (CIIT) Abbottabad proposed building an Urdu handwritten dataset. The preliminary samples were collected from bachelor students [16]. On A4 paper with specified baselines, there had been 48 Urdu text lines with a few Urdu number acquired from students. Later, a dataset of 600 forms was collected from 100 Urdu writers. Each writer replicated six specified pages (forms), each of which included eight text lines. The database has 6200 total words. The database includes both text transcriptions and writer information, allowing for both character recognition and writer identification. This dataset was started expanding to include UNHD data.

UNHD Offline Handwritten Dataset is recent dataset collected from 500 native Urdu writers with 6–8 lines per page contributed by each writer. It consists of 10,000 text lines, having only 700 unique text lines. It contains 31,2000 handwritten words. Only 700 unique text lines are not sufficient to solve demographic classification problem of real-world scenario [17].

IPC-Handwritten database was developed by Ahsen Raza to help researchers in the field of handwriting analysis, specifically writer identification/verification. It contributed towards experiment and quantifying their algorithms/techniques on this common benchmark to determine the cumulative and overall value. This dataset includes handwritten examples in English and Urdu. Each writer provided two handwriting for Urdu. The dataset comprises 176 scanned handwriting samples contributed by 44 writers having a total of 88 samples digitized in ".png" format at a resolution of 300 dpi [18]. Urdu Corpus of Handwritten Text Image is the collection of Urdu text, containing 1000 handwritten text images written by 500 unique writers. Each image has 4 to 6 lines of Urdu text, with 60–80 words each line, for a total of approximately .35 million words. The terms were chosen from six categories in order to contain as many diverse words as possible. The corpus would be annotated for line and word segmentation, where a word is a single character or component [11].

CENIP-UCCP [12] is a novel offline sentence dataset of Urdu handwritten texts that includes certain preprocessing and text line segmentation methods. The authors constructed an Urdu handwriting database, CENIP-UCCP, based on their own generated and maintained corpus (Center for Image Processing-Urdu Corpus Construction Project). The corpus consists of a collection of Urdu writings that have been used to improve forms. Native writers filled up these forms in their own handwriting. Six text categories have been used to create these forms, each with around 66 forms. There are presently 400 digital forms in the collection, produced by 200 different writers. The database is completely labelled for both content information and content detection, and it allows for the test of various algorithms such as Urdu.

Centre for Pattern Recognition and Machine Intelligence (CENPARMI) designed a dataset for Urdu named as "A large new Urdu handwriting database" [13]. It comprises isolated digits, numeral strings with & without decimal points, five special symbols, 44 isolated characters, 57 Urdu words, and Urdu dates in various patterns. 343 different writers from various professional backgrounds, qualifications, and ages contributed to the writing process. This database for offline Urdu handwriting recognition and classification problems. Furthermore, the database is available in three separate formats: true colour, grey level, and binary.

CALAM (Cursive and Language Adaptive Methodologies) is a Four-Tier Annotated Urdu Handwritten Text Image Dataset for Urdu Script Multidisciplinary Research [19]. The authors describe CALAM, a large handwritten text document picture corpus collection for Urdu language. The database comprises unconstrained handwritten phrases as well as structural annotations for offline handwritten text pictures in XML format. CALAM includes 1,200 photos of handwritten writing, 3,043 lines, 46,664 words, plus 101,181 ligatures. Data collection is divided into 6 categories and 14 subcategories to capture the most variation in words and handwriting styles. Handwritten forms were completed by 725 unique authors from various geographical locations, ages, and genders and varying educational backgrounds.

UHaT Dataset comprises handwritten Urdu characters and numerals [7]. Over 900 people contributed to the samples. Organization and description Image resolution: All photos are kept at a resolution of 28 by 28 pixels. It has been divided into three categories: training (678 photos), test (147 images), and validation (145 images). The dataset allows academics to research on Urdu text recognition and classification.

Table 1 given below, indicates available in literature datasets for Urdu handwritten text containing maximum possible information (characters, patches words, text lines, and digits in Urdu language).

Table 1. Major datasets in Urdu literature

Dataset	No of writers	Types of samples
UCOM	100	Continuous writing
UNHD	500	5 to 8 lines per page
Urdu corpus of handwritten text	500	.35 million words
CALAM	725	Unconstrained phrases and structural annotations for offline text
CENIP-UCCP	200	400 forms
CENPRAMI	343	Characters, words, sentences and numerals
UHaT dataset	900	970 text images in 3 categories

We are currently unaware of almost any large comprehensive open vocabulary Urdu handwritten text corpus that is large enough to represent the naturalness of Urdu writing. The UOHTD dataset has been produced to address this requirement. It will also be made freely available to interested researchers.

3   Data Collection and Statistics

UOHTD has been collected from people belonging to every sphere of field and in maximum possible age groups. Information (biodata) of each contributing writer is given in Table 2 while Table 3 shows collection of sample from different categories of people with different subjects. Different topics to encompass complete Urdu alphabets and different linguistic scenarios have also been covered. However, targeted audience for data collection belong to Pakistan including overseas Pakistanis only with a balance class strategy of males, females, right-handed and left-handed. As there is paucity of left-handed people, maximum possible samples have been obtained [26]. Tables 4 shows number of different and same text sample obtained from males and females while Table 5 shows age group and education details. In Table 6, we suggest that composition for dataset for training, testing, and validation.

Table 2. Information (biodata) collection of contributing writers

Name:_____	Father's name:_____	Education:_____	Profession:_____
Age:_____	Gender:_____	Right hand:_____	Left hand:_____
District:_____	Province:_____	Cell no/email:	Date:_____
		_____	

Table 3. Type of resources used to collect dataset with topics and paragraphs

Sample collection category (people)	Resource (sample donors)
Education	3 (teachers, students and clerical staff)
Serving (working in office (government and private sector)	6 (all categories)
Health	3 (doctors, nurses and patients)
House wives	3 (<25, >25 & 50 above)
Retired persons	2 (all categories 60 above)
Overseas Pakistanis	3 (people like teachers, students and literate laboures)

Table 4. Gender of participants

Type of text	Male	Female
Different	300	300
Same	100	100
Total	400	400

Table 5. Age and education of participants

Sr. no	Age group	Education/qualification
1	10–20	7th grade to bachelors and masters
2	21–30	
3	31–40	
4	41–50	
5	51–60	
6	61–70	
7	71–80	

Table 6. Dataset division

Ser no	Type of dataset	Pages	Lines	Words
1	Training	600	6000	30000
2	Test	100	1000	5000
3	Validation	100	1000	5000
Total		800	8000	40000

4   Data Verification

Each Urdu handwriting sample in the UOHTD dataset has underwent multiple stages of verification. The first phase checks information at the page level to confirm that it is appropriate for inclusion in the database. The second phase confirms the validity of the scanned page's ground truth by confirming that it resembles the handwritten text on the sample page. The third stage assures that the dataset is error-free following the scanning procedure and segmentation at the line, word, or patch level. To optimize data accuracy, the fourth and final stage is validation of all previous three phases.

4.1   Page Level Verification

Reaffirmation of scanned images at page level has been carried out to check any sort of information missing from the sample and cropping error to make it completely readable. Aim is to discard sample not proving the desired standard. Different samples are related with printed and same has been shown in Fig. 1 and 2 below.

4.2   Ground Truth Verification

The scanned sample pages were divided into different sets at this stage. Each set of sample pages viz a viz ground truth was assigned to a separate reviewer for verification purposes to check for writer deviance. While reviewing it has been

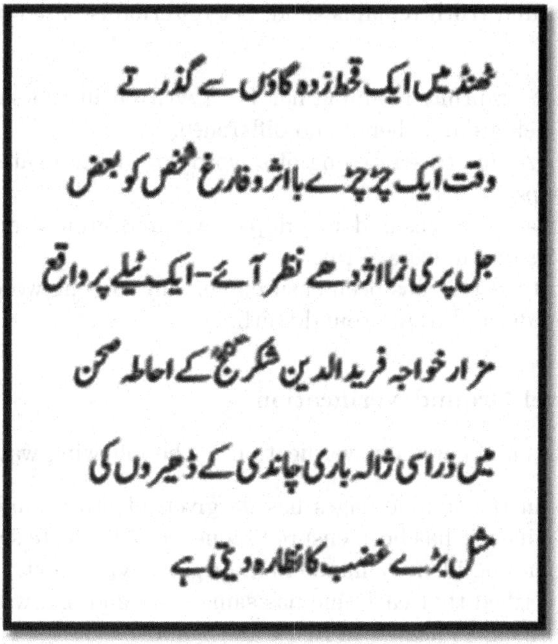

ٹھنڈ میں ایک قحط زدہ گاؤں سے گذرتے
وقت ایک چڑچڑے بااثر و فارغ شخص کو بعض
جل پری نما اژدھے نظر آئے۔ ایک ٹیلے پر واقع
مزار خواجہ فرید الدین شکر گنج کے احاطہ صحن
میں ذرا سی ژالہ باری چاندی کے ڈھیروں کی
مثل بڑے غضب کا نظارہ دیتی ہے

Fig. 1. Original text

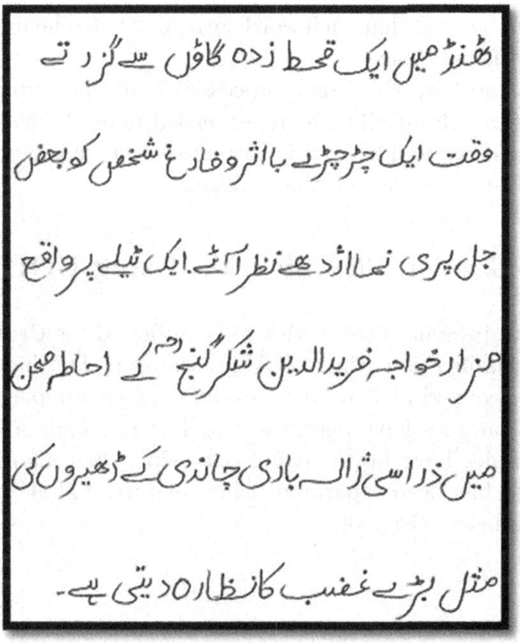

ٹھنڈ میں ایک قحط زدہ گاؤں سے گزرتے
وقت ایک چڑچڑے بااثر و فارغ شخص کو بعض
جل پری نما اژدھے نظر آئے۔ ایک ٹیلے پر واقع
مزار خواجہ فرید الدین شکر گنج رحمۃ کے احاطہ صحن
میں ذرا سی ژالہ باری چاندی کے ڈھیروں کی
مثل بڑے غضب کا نظارہ دیتی ہے۔

Fig. 2. Handwritten text

ensured that ground truth remains valid. Set criterion is based on the following principles:-

1. Ensuring that preprinted sample has been written in the same way as per language guidelines and there is no difference.
2. Variance in size due to scratching of a word writers and categorizing it into different groups.
3. Space equalization is ensured in sample text and ground according to the bulk of writers on an average rule.
4. 100% similarity of text has been ensured i.e., any missing word in the sample has also been deleted from ground truth.

4.3   Line Level Ground Verification

We generate lines and carry out verification in the following way:

1. As each line in the sample pages has its given identity number. While segmenting into lines, it has been ensured sequence of lines are segmented as per their corresponding identity numbers in original typed text.
2. It has been verified that each line has same start and end word independent of length of line as indicated in Fig. 1 and 2 above

4.4   Word Level Ground Verification

At word level, it is ensured that each word corresponds to the line, no information has been lost and is readable.

In the fourth and final step, each process is reaffirmed from the initial verification process through all different reviewer till final verification process used in above mentioned steps. The goal is to remove and repair any problems that may have been overlooked during the process.

5   Age and Gender Classification Using UOHTD

In this section, we present the experiments conducted for Urdu writer's gender and age group classification using UOHTD dataset. The handwritten sample pages have been scanned at 300 DPI. As different colour pens have been used by writers so all samples have been converted to grayscale and then Binarization of these samples have been carried out using Otsu algorithm. To remove noise, background has been separated using adaptive OTSU method. Detailed flowchart is given below (Fig. 3):-

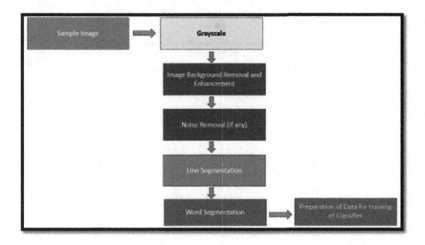

Fig. 3. Flowchart of method adopted for age and gender classification

5.1 Convolutional Neural Network

The Convolutional Neural Networks introduced in the early 1990s and failed to capture the interest of the research community owing to the scarceness of large databases, computational complexity, and ample training time. With recent advent in technology during the decade, it introduced incredibly effective Graphical Processing Units (GPUs) and the development of large datasets like ImageNet [20] [43] have improved performance on various learning tasks thereby lowering the training time. Convolutional Neural Networks gained prominence when Krizhevesky et al. [21] employed them in the ImageNet Large Scale Visual Recognition competition in 2012. The CNN-based approach beat previous methods and had a much lower error rate. CNNs have quickly become the most prevalent solution for classification tasks [22,23] and have been employed for a range of pattern classification applications such as character recognition [24,25]. The traditional networks need input in a single vector, with a high number of weights per neuron due to the fully linked topology. Such networks do not scale well for the image-based dataset, leading to overfitting. CNNs are designed to deal with images since they presume the input is fundamentally an image. The number of weights associated with a neuron is substantially lower than a fully linked architecture since each neuron is connected to a particular portion of the image. The neuron is grouped in three dimensions in CNN's: height, width, and depth. The scalability is improved by connecting a neuron to a subset of the preceding neurons rather than all of them. Convolutional and pooling layers are placed on top of each other in a CNN, which is then followed by the fully connected layer. The detailed flow chart in given in Fig. 4.

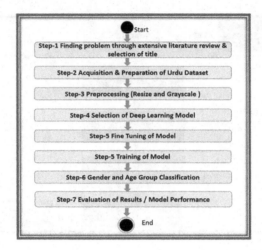

Fig. 4. Flowchart of method adopted for gender age and classification using UOHTD

6 Experiments for Age and Gender Classification on UOHTD

In this study, we analyze gender and group classification rate on newly developed UOHTD where training and test sets have different context as in real world scenarios. Experiments were conducted on TensorFlow and PyTorch environment. Initially data was labelled into male and female categories and then this labelled data was split into training, testing, and validation sets for gender classification. For age prediction, 4 × age groups were categorized. We performed experiments at line level images and then at word or patch level images in both cases. Details of the experiments performed and in each experiment has been shown in the Table 6 and 7 on each environment respectively (Table 8).

Table 7. Experiment-1 classification accuracy on the UOHTD using CNNs in Tensor-Flow environment on text lines and words level (patches)

Type of data	No. of classes	Classification accuracy
Gender prediction (text lines)	2 (Male & female)	66%
Gender prediction (text patches)	2 (Male & female)	78.14%
Age group prediction (text lines)	4	49.5%
Age group prediction (text patches)	4	56%

Table 8. Experiment-2 classification accuracy on the UOHTD using CNNs in PyTorch environment on text lines and word level (patches).

Type of data	No. of classes	Classification accuracy
Gender prediction (text lines)	2 (Male & female)	74%
Gender prediction (text patches)	2 (Male & female)	72%
Age group prediction (text lines)	4	85%
Age group prediction (text patches)	4	79%

7 Conclusions

In this study, we bring forth a complete Urdu Offline Handwritten Text Dataset (UOHTD) created by 800 native Urdu writers. It contains people of various ages, literacy levels, gender, and handedness. Each writer provided a sample page that had been scanned at various resolutions of 200, 300, and 600 dpi. The validity was done manually as well as formally with ground truth and was examined at page, line, and word levels. The pages have been divided into lines, and the lines have been divided into words. Machine Learning Tools for extracting lines and words are being created. The findings of preliminary experiments on Urdu Offline Handwritten Text categorization using UOHTD are provided. Existing datasets meet the needs of researchers interested in Urdu text classification and recognition, writer identity, signature verification, forensic analysis and bank cheques verification, segmentation, and many other topics. The database will be made accessible to any interested scholars for free.

References

1. Hull, J.J.: A database for handwritten text recognition research. IEEE Trans. Pattern Anal. Mach. Intell. **16**(5), 550–554 (1994)
2. He, S., Schomaker, L.: FragNet: writer identification using deep fragment networks. IEEE Trans. Inf. Forensics Secur. **15**, 3013–3022 (2020)
3. Lee, C., Leedham, C.: A new hybrid approach to handwritten address verification. Int. J. Comput. Vis. **57**, 107 (1994). https://doi.org/10.1023/B:VISI.0000013085.47268.e8
4. Plötz, T., Fink, G.A.: Markov models for offline handwriting recognition: a survey. Int. J. Doc. Anal. Recogn. **12**(269), 3013–3022 (2009). https://doi.org/10.1007/s10032-009-0098-4
5. Latif, A., Rasheed, A., Sajid, U., et al.: Content-based image retrieval and feature extraction: a comprehensive review. Math. Probl. Eng. (2019). Article ID: 9658350
6. Ratyal, N., Taj, I.A., Sajid, M., et al.: Deeply learned pose invariant image analysis with applications in 3D face recognition. Math. Probl. Eng. (2019). Article ID: 3547416
7. Ali, H., Ullah, A., Iqbal, T., et al.: UHat SN. Appl. Sci. **2**, 152 (2020). https://doi.org/10.1007/s42452-019-1914-1

8. Uddin, I., Javed, N., Siddiqi, I.A., Khalid, S., Khurshid, K.: Recognition of printed Urdu ligatures using convolutional neural networks. J. Electron. Imaging **28**(3), 033004 (2019)
9. Bulacu, M., Schomaker, L.: Text-independent writer identification and verification using textural and allographic features. IEEE Trans. Pattern Anal. Mach. Intell. **29**(4), 701–717 (2007)
10. Cha, S.-H., Srihari, S.N.: A priori algorithm for sub-category classification analysis of handwriting. In: Proceedings of Sixth International Conference on Document Analysis and Recognition. IEEE (2001)
11. Guru, D., Prakash, H.: Online signature verification and recognition: an approach based on symbolic representation. IEEE Trans. Pattern Anal. Mach. Intell. **31**(6), 1059–1073 (2009)
12. Sae-Bae, N., Memon, N.: Online signature verification on mobile devices. IEEE Trans. Inf. Forensics Secur. **9**(6), 933–947 (2014)
13. Yang, W., Jin, L., Liu, M.: Character-level Chinese writer identification using path signature feature, dropstroke and deep CNN. In: Proceedings of the International Conference on Document Analysis and Recognition, pp. 546–550 (2015)
14. Yang, W., Jin, L., Liu, M.: DeepWriterID: an end-to-end online text-independent writer identification system. IEEE Intell. Syst. **31**(2), 45–53 (2016)
15. Sagheer, M.W., He, C.L., Nobile, N., Suen, C.Y.: Holistic Urdu handwritten word recognition using support vector machine. In: 2010 20th International Conference on Pattern Recognition, pp. 1900–1903. IEEE (2010)
16. Ahmed, S.B., Naz, S., Swati, S., Razzak, I., Umar, A.I., Khan, A.A.: UCOM online dataset-an Urdu handwritten dataset generation. Int. Arab J. Inf. Technol. (IAJIT) **14**(2) (2017)
17. Ahmed, S.B., Naz, S., Swati, S., Razzak, M.I.: Handwritten Urdu character recognition using one-dimensional BLSTM classifier. Neural Comput. Appl. **31**(4), 1143–1151 (2019). https://doi.org/10.1007/s00521-017-3146-x
18. Ahsen Raza. https://sites.google.com/site/artificialtextdataset/home/atabipc-handwritten-dase-urdu-english
19. Papadatou-Pastou, M., Martin, M., Munafó, M.R., Jones, G.V.: Sex differences in left-handedness: a meta-analysis of 144 studies. Psychol. Bull. **134**(5), 677 (2008)
20. Russakovsky, O., et al.: ImageNet large scale visual recognition challenge. Int. J. Comput. Vis. **115**(3), 211–252 (2015). https://doi.org/10.1007/s11263-015-0816-y
21. Krizhevsky, A., Sutskever, I., Hinton, G.E.: ImageNet classification with deep convolutional neural networks. In: Advances in Neural Information Processing Systems, vol. 25, pp. 1097–1105 (2012)
22. Simonyan, K., Zisserman, A.: Very deep convolutional networks for large-scale image recognition. arXiv preprint arXiv: 1409.1556 (2014)
23. Szegedy, C., et al.: Going deeper with convolutions. In: Proceedings of the IEEE Conference on Computer Vision and Pattern Recognition, pp. 1–9 (2015)
24. Bouchain, D.: Character recognition using convolutional neural networks. Institute for Neural Information Processing (2006/2007)
25. Uijlings, J.R.R., Van De Sande, K.E.A., Gevers, T., Smeulders, A.W.M.: Selective search for object recognition. Int. J. Comput. Vis. **104**(2), 154–171 (2013). https://doi.org/10.1007/s11263-013-0620-5
26. Sagheer, M.W., He, C.L., Nobile, N., Suen, C.Y.: A new large Urdu database for off-line handwriting recognition. In: Foggia, P., Sansone, C., Vento, M. (eds.) ICIAP 2009. LNCS, vol. 5716, pp. 538–546. Springer, Heidelberg (2009). https://doi.org/10.1007/978-3-642-04146-4_58

27. Raza, A., et al.: An unconstrained benchmark Urdu handwritten sentence database with automatic line segmentation. In: 2012 International Conference on Frontiers in Handwriting Recognition. IEEE (2012)
28. Choudhary, P., Nain, N.: A four-tier annotated Urdu handwritten text image dataset for multidisciplinary research on Urdu script. ACM Trans. Asian Low-Resource Lang. Inf. Process. (TALLIP) 15(4), 1–23 (2016)
29. Morera, Á., Sánchez, Á., Vélez, J.F., Moreno, A.B.: Gender and handedness prediction from offline handwriting using convolutional neural networks. Complexity 2018 (2018)
30. Moetesum, M., Siddiqi, I., Djeddi, C., Hannad, Y., Al-Maadeed, S.: Data driven feature extraction for gender classification using multi-script handwritten texts. In: 2018 16th International Conference on Frontiers in Handwriting Recognition (ICFHR), pp. 564–569. IEEE (2018)

Document Analysis and Processing

DAZeTD: Deep Analysis of Zones in Torn Documents

Chandranath Adak[1]([✉])[iD], Priyanshi Sharma[2][iD], and Sukalpa Chanda[3][iD]

[1] Department of CSE, Indian Institute of Technology Patna, Patna 801106, India
chandranath@iitp.ac.in
[2] Indira Gandhi Delhi Technical University for Women, Delhi 110006, India
[3] Department of CSC, Østfold University College, 1757 Halden, Norway
sukalpa@ieee.org

Abstract. In a crime scene, document fragments with similar contents might lead to significant evidence. A criminalist when encounters such a scene with an enormous amount of torn document pieces, automated analysis becomes imperative in procuring potential evidence in a fast and reliable way. To analyze document fragments with similar contents, a processing module to segment the homogeneous zones based on the content type is a prerequisite. This paper proposes a deep learning-based module DAZeTD that can detect textual (printed/handwritten) and non-textual ragged zones. For classifying the content of the zones, we adopt the scheme of vision transformer; and to draw the zone boundaries, we employ outer isothetic cover. We created a dataset of 881 torn documents on which we performed rigorous experiments. We obtained an overall 87.71% mAP@0.5, which is quite promising.

Keywords: Deep learning · Object detection · Torn document · Vision transformer · Zone classification

1 Introduction

Forensics and crime scene analysis are extremely data-intensive. In some scenarios, crime analysts need to deal with truckloads of torn paper documents and data from heterogeneous sources [1]. To extract a piece of possible evidence from the massive pile of torn documents, the analyst needs to sort the similar document fragments first. Here, the notion of similarity can be primarily based on the *physical* appearance, *content*, and *context* of the document fragments. The physical appearance mostly depends on document color, surface textures, shape, ragged edges, etc. The content here usually refers to the substance type, e.g., text: handwritten, typed, printed; non-text: image, seal, logo; script, font, etc. The context refers to the topic and circumstance of a document. The physical attributes, content, and contextual information of torn documents may be very random at a crime scene. Manual effort to accomplish the torn document sorting/analysis task is quite unpragmatic and infeasible. Therefore, automated

© The Author(s), under exclusive license to Springer Nature Switzerland AG 2022
U. Porwal et al. (Eds.): ICFHR 2022, LNCS 13639, pp. 515–529, 2022.
https://doi.org/10.1007/978-3-031-21648-0_35

analysis of torn documents can aid crime analysts in such situations [2]. In the automation process, the physical and content-based similarity understanding can be achieved by the techniques of the computer vision domain. However, linguistic information from the document is required for context-level similarity, which can be adopted from natural language processing. In this paper, we focus only on the content type, where the torn documents can be sorted/analyzed with respect to the textual (handwritten/printed) and non-textual information. As a matter of fact, a processing module is required to categorize the text and non-text zones in the torn documents before they are analyzed further with appropriate segments. This paper mainly aims for this categorization.

In the domain of Document Analysis and Recognition (DAR), the last decade of the 20th century reported multiple research works on document *zone classification* [3], where the zones of a document image are classified into multiple categories, e.g., text, table, mathematical expression, pictures (binarized/halftone/color), graphics (flow-chart/plot/logo), for the logical understanding of a page structure [4]. In the first one and half decades of the 21st century, *layout analysis* of a document was explored extensively [5], where the content/structure of a document was annotated [6]. Document *image segmentation* is also being performed for the past two decades, where the target is to identify the homogeneous regions, e.g., paragraph, images [7,8]. Understanding the zone/layout of a document is "not dead yet" [9], and attracts researchers due to its interesting research challenges [5]. For the last one decade, due to the boom of deep learning, this problem is being tackled by various deep learning strategies instead of the past hand-crafted features [8,10]. Moreover, this problem has come up with great interest after being re-formulated as *object detection*, where the task is to detect various objects from a page, mostly tables, formulae, figures, etc. [11–13]. However, most of the past works dealt with printed documents, and very few were concerned with the handwritten/hybrid types [14]. This paper deals with printed, handwritten, and hybrid documents. Besides, the focus of this paper is torn documents, on which very few investigations were performed in the literature. For example, Chanda et al. [15] employed Gabor features and chain-code-based hand-crafted features with SVM; Klebber et al. [16] proposed a heuristic technique using the orientation of fragments, paper/ink colors. Most of the abovementioned methods demand a specific layout/format/type/orientation of the document that is quite unlikely to be present in the case of torn documents. Therefore, such methods are incompetent to deal with challenges encountered in automatic torn document analysis.

In this paper, we propose deep learning [17]-based object-zone detector for analyzing torn documents, where we focus on the zone content (text/non-text) irrespective of physical attributes, text scripts, and fonts. The proposed method explores a deep-learning-based approach for segmentation of a document piece based on its content (handwritten text, printed text, image, stamp, table, etc.), irrespective of physical attributes. We first divide a torn document into grids, and find the grids containing some objects, which are further classified into grids comprising the printed text, handwritten text, and non-textual image categories using a Vision Transformer (ViT) [18]. Finally, we draw the outer isothetic cover through the boundary of the grids containing the homogeneous object. Our major **contributions** to this paper are as follows:

(i) Torn document analysis has not been explored through an object detection framework in the literature. For the very first time, we propose a ViT [18]-based object detector in the context of zone segmentation in torn documents.

(ii) We aptly introduce the prowess of outer isothetic cover [19] to define the boundaries of a ragged zone, where the usual rectangular bounding boxes [20] did not work well.

(iii) We created a dataset containing 881 torn documents due to the unavailability of a public dataset. We also systematically analyzed the challenges associated with data, and performed extensive experiments, which may serve as a benchmark for future similar research.

The rest of the paper is structured as follows. The employed dataset details and the challenges related to the data are discussed in Sect. 2. The following Sect. 3 describes the proposed methodology. After that, Sect. 4 presents and analyzes the experimental results. Finally, Sect. 5 concludes this paper.

2 Dataset Details and Challenges

In this section, the employed dataset with the associated challenges is discussed.

The main aim of this research is to analyze torn documents, for which we required a dataset containing several such documents. Even though some researches [15, 16] exist on torn documents, but we could not find any publicly available datasets to the best of our searching capacity. Therefore, we had to create a dataset to fulfill our requirement. For this, we first collected different offline documents, e.g., handwritten notes, diary writings, newspapers, tabloids, and book pages. Then we tore up the documents into small pieces manually with the help of multiple volunteers of different age groups and educational/professional backgrounds to mimic the process of vandalizing a document arbitrarily. Next, the torn pieces were scanned using a flatbed scanner to generate the digital version of the torn documents used as input to our system.

Understanding zones from the torn documents often faces **challenges** due to the following instances occurred jointly/separately:

(i) *Degraded image*: Due to the ink smearing, poor page material/density/GSM, uneven illumination, contrast variation, improper reprography, and improper storage of documents, the quality of the document image may deteriorate. Often the image may contain a halftone [25] pattern that introduces more challenging scenarios.

(ii) *Complex background*: Sometimes, the background zones are pretty complex due to the presence of various undesirable patterns/objects, artifacts, ink bleed-through, etc.

(iii) *Partial view*: Due to tearing up a document into small pieces, an object zone may also be torn and present partially within a specific small piece. This also leads to the scattering of data into document fragments.

(iv) *Oriented view*: The small torn pieces of a document may be scanned arbitrarily in multiple orientations while digitizing in a large volume. This leads to having rotated object zones in the torn document.

 (v) *Damaged part/zone*: A specific portion/zone of a document may be damaged due to tearing into tiny pieces or may be missed during the scanning phase. Moreover, due to tiny paper fibers, tearing a document may produce an uneven ragged edge with undesired shears.

Moreover, the torn documents containing texts of various scripts and font-sizes/styles bring additional challenges.

Our database contains a total of 881 torn documents, among which 202 documents contain only images/non-textual objects, 412 documents comprise only textual objects, and 267 documents are hybrid that includes both textual and non-textual objects. This database contains texts from five different scripts, i.e., English, Bengali, Devanagari, Odia, and Arabic.

The task in this research is to detect the zones representing texts and non-texts from a torn document. We also find the categories of a text, i.e., printed or handwritten. For zone detection, a common choice is the bounding box (BB) strategy adopted from the literature on object detection [21,22], where the center, height, and width of the BB are predicted to localize the object [20,23]. However, the rectangular BB may include unwanted zones of torn documents. Therefore, to avoid predicting false-positive zones, we use outer isothetic covers [19,24] to define the zone boundary. For the ground-truth information, we semi-automatically mark a zone using free-form digital polygon [26] for its boundary and add its corresponding category/class information. In the leftmost column of Fig. 4, we present samples from our dataset with ground-truth, where some of the abovementioned challenging issues can also be found. More samples can be found at https://github.com/Priyanshi-Sharma-142/DAZeTD.

3   Proposed Methodology

In this section, we discuss our problem formulation followed by the proposed solution.

3.1   Problem Formulation

In this work, we are given a torn document image $\mathcal{I}$ as an input to our system, and our task is to detect the zones from $\mathcal{I}$. Zone detection consists of two sub-problems: (i) zone localization, (ii) zone classification. For zone localization, we usually draw a boundary over the zone. Here, we draw outer isothetic covers [19,24] for zone boundaries, as mentioned earlier in Sect. 2, and refrain from drawing rectangular bounding boxes [20] that may lead to erroneous boundary results due to covering undesirable regions by the rectangular shape. The outer isothetic cover is actually a digital polygon [26] and includes the pixel-level zone. In zone classification, we identify the category of the zone. As we mentioned earlier, we work with three categories, i.e., *printed* text, *handwritten* text, and non-textual *image*.

3.2 Solution Architecture (DAZeTD)

Our problem formulation for zone detection and the proposed solution architecture (DAZeTD) are different from the traditional object detection schemes [21,22]. In traditional object detection, the bounding boxes are predicted for localizing the object zones [20–23]. However, in DAZeTD, we use isothetic cover to localize the object zone, for which we need to divide the image into grids [19]. We present the workflow of DAZeTD in Fig. 1, and discuss its modules below. The input image $\mathcal{I}$ is first resized into $w_z \times w_z$ sized $\mathcal{I}_z$. During resizing, to maintain the aspect ratio, some columns and/or rows are filled with zeros. The resized image $\mathcal{I}_z$ is divided into $w_g \times w_g$ sized grids g_i (for $i = 1, 2, \ldots, n_g$). Here, we obtain n_g $(= \lfloor \frac{w_z}{w_g} \rfloor^2)$ number of grids. For our task, we choose $w_z = 1024$, and $w_g = 128$; therefore, $n_g = 64$.

Objectness Network (f_o). In DAZeTD, we only focus on the grids containing the zones/objects of our interest (ZoI). Therefore, in the preprocessing stage, we discard the background (BG) grids that contain background zones/unwanted objects [20].

Although we here work on text documents, it is not always the case that the ZoI (/BG) grid contains black-ink (/white) pixels due to the challenges discussed in Sect. 2. Moreover, we keep our architecture as a generic one to tackle complex backgrounds. Here, we use the *objectness network* (f_o) for separating ZoI and BG grids. In f_o, we employ upto the average_pool layer of ResNet50V2 [27], which takes a 128×128 sized grid as input and produce a 2048-dimensional deep feature vector. We used ResNet50V2 here, since it outperformed some contemporary

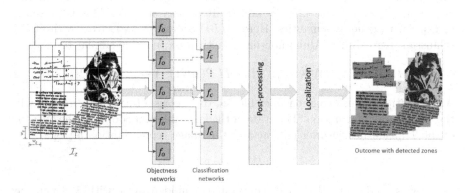

Fig. 1. Workflow of our proposed architecture DAZeTD.

deep architectures, e.g., VGG19, MobileNetV2, DenseNet121, etc. [17,28]. The extracted feature vector is then fed into three sequential fully-connected (FC) layers with 1024, 512, and 128 neurons, respectively. In the FC layers, we use Mish [29] activation function, since it performed better than some state-of-the-art activation functions, e.g., ReLU, PReLU, SELU, GELU, Swish [17,29], on our task. In the output layer of f_o, we keep two neurons with softmax activation function [17,30] to separate ZoI and BG grids. The ZoI grids are further fed into the classification network (f_c), which is discussed as follows.

Classification Network (f_c). A $w_g \times w_g$ sized ZoI grid is the input of the classification network f_c. The task of f_c is to classify whether the grid contains an object category of interest, i.e., printed text, handwritten text, and non-textual image.

In f_c, we employ the strategy of Vision Transformer (ViT) due to its lesser image-specific inductive bias than Convolutional Neural Network (CNN) [18]. The pictorial representation of f_c is shown in Fig. 2. Here, the input grid image is reshaped into a sequence of flattened patches. A $w_g \times w_g$ sized grid is divided into n_p ($= \lfloor \frac{w_g}{w_p} \rfloor^2$) number of patches x_p^i (for $i = 1, 2, \ldots, n_p$), each with size $w_p \times w_p \times C$, where C is the number of image channels of a ZoI grid. Here, $w_g = 128$, as mentioned earlier. In our task, $C = 1$, and we fix $w_p = 32$ that results $n_p = 16$. Each patch x_p^i is flattened and mapped to a D-dimensional latent vector, i.e., patch embedding z_0, using a trainable linear projection, as follows.

$$z_0 = [x_{class}; x_p^1 \mathbb{E}; x_p^2 \mathbb{E}; \ldots; x_p^{n_P} \mathbb{E}] + \mathbb{E}_{pos} \tag{1}$$

where, $\mathbb{E}$ is the patch embedding projection, $\mathbb{E} \in \mathbb{R}^{w_p \times w_p \times C \times D}$; $\mathbb{E}_{pos}$ is the position embeddings added to patch embeddings to preserve the positional information of patches, $\mathbb{E}_{pos} \in \mathbb{R}^{(n_p+1) \times D}$; $x_{class} = z_0^0$ is a learnable embedding.

After mapping patch images to the embedding space with positional information, we add a sequence of transformer encoders [18,31]. The internal view of a transformer encoder is shown in Fig. 3, which includes alternating layers of MSA (Multi-headed Self-Attention) [18] and MLP (Multi-Layer Perceptron) [30] modules. LN (Layer Normalization) [32] is employed before every module, and residual connections [27] are added after every module. This is presented in Eq. 2 with general semantics. Here, MLP comprises two layers with GELU (Gaussian Error Linear Unit) non-linear activation function similar to [18].

$$z_l' = MSA(LN(z_{l-1})) + z_{l-1}; \ z_l = MLP(LN(z_l')) + z_l'; \ l = 1, 2, \ldots, L \tag{2}$$

After multiple transformer encoder blocks, the <class> token [33] enriches with the contextual information. The state of the learnable embedding at the outcome of the Transformer encoder (z_L^0) acts as the image representation y [18].

$$y = LN(z_L^0) \tag{3}$$

Now, we add an MLP head containing a hidden layer with 128 neurons. To capture the non-linearity, we use Mish [29] here. In the output layer, we

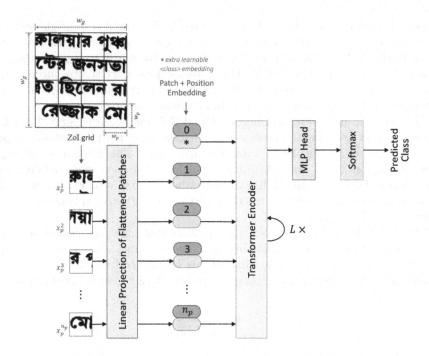

Fig. 2. Workflow of the classification network f_c.

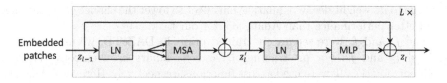

Fig. 3. Internal view of a transformer encoder.

keep three neurons with softmax activation function to classify a grid into the abovementioned three categories.

On every ZoI grid, we apply f_c and obtain its category. We now add a post-processing stage to localize the objects properly.

Post-processing and Localization. A ZoI grid of size $w_g \times w_g$ may contain a small object (covering area $< w_g \times w_g$), where the majority portion of the grid is of the background region. This is reasonably comprehended by the ink-pixel count of the binary grid image and the position of ink-pixel from the grid boundary. In post-processing, we squeeze the grid boundary towards the small object. Here, we put a rectangular bounding box over the small object inside a ZoI grid for precise localization. We call such a grid a *squeezed grid*.

A ZoI grid may also include partial objects of more than one categories. This is plausibly understood from the output layer of f_c, where the probabilities of multiple categories are pretty similar. Here, we look into Moore's eight neighborhoods of the concerned grid g_i. If n_M ($8 \geq n_M \geq T_M$) number of neighbors of g_i are from a category under consideration, then g_i is (re-)confirmed to that category. In this task, empirically, we decide $T_M = 6$. For $n_M < T_M$, g_i is marked as a *mixed* category (refer to Fig. 4). This additional category is introduced in the post-processing stage. Introducing the mixed category assists in reducing false positives, which simultaneously helps in increasing the precision.

Finally, the post-processed neighborhood ZoI and squeezed grids from the same category are joined to form a bigger region/zone. For localization, we put an outer isothetic cover [19,24] over this bigger zone. This helps in localizing a halftone region [25] as well. As a matter of fact, if a document contains multiple zones, DAZeTD can localize the zones in the same way. Finally, with the assistance from the previously traced aspect ratio, the detected zones on $\mathcal{I}_z$ are reverted to the original input $\mathcal{I}$.

In DAZeTD, the overall loss is the summation of zone classification loss and zone localization loss. For classifying the zone categories, we use cross-entropy [17,30] loss, whereas for zone localization, we employ IoU loss [23], which is the intersection over the union of areas of ground-truth ($\mathcal{Z}$) and predicted ($\hat{\mathcal{Z}}$) zones.

$$IoU = \frac{\mathcal{Z} \cap \hat{\mathcal{Z}}}{\mathcal{Z} \cup \hat{\mathcal{Z}}} \qquad (4)$$

The intersection and union areas of the zones are computed by counting the image pixels of digital polygons [26]. Here, we employed the AdamW optimizer [34] due to its weight decay regularization effect to lower the chance of overfitting, and better performance than Adam [35] for our task.

The f_o and f_c were separately trained prior. In DAZeTD, we used the pre-trained f_o and f_c. For pre-training f_o and f_c, binary cross-entropy [17] and cross-entropy [17,30] were used as loss functions, respectively. Here also, AdamW optimizer [34] was employed. The training details and hyper-parameter tuning is mentioned in Sect. 4.

4   Experiments and Discussion

This section presents the employed dataset followed by experimental results with discussions.

4.1   Database Employed

For performing the experimental analysis, we procured a total of 881 torn documents, as mentioned earlier in Sect. 2. In this database (DB), 412 samples contain only texts, 202 samples comprise only non-texts, and the rest of the 267 are hybrids containing both texts and non-texts.

The DB was split into training (DB_{tr}), validation (DB_v), and testing (DB_t) sets with a ratio of 7 : 1 : 2. As a matter of fact, DB_{tr}, DB_v, and DB_t sets were disjoint. In DB, the torn documents contained insufficient handwritten objects to train DAZeTD properly. Therefore, we took 110 English pages from IAM [36] and 110 Bengali pages from PBOK-Bangla [37], and semi-automatically cropped them in free-form to give the essence of torn documents, which were included in DB_{tr}. To reduce the overfitting [17] issue, we also augmented the documents of DB_{tr} by randomly changing the brightness and contrast of the document image.

For pre-training the f_o and f_c networks separately, we extracted $w_g \times w_g$ sized grids by sliding a window of the same size over the document images of DB_{tr}. We also validated the performances of f_o and f_c over the extracted grids from DB_v.

4.2 Experimental Results

In this subsection, we discuss our performed experimentation and outcome to analyze the efficacy of DAZeTD. All presented results were executed on DB_t. The performance of DAZeTD was evaluated with respect to mAP (mean Average Precision) [21, 22] that is a standard object detection metric.

The hyperparameters of the proposed architecture were tuned and fixed during model training with respect to the performance over the valida-tion/development set DB_v. The hyperparameters of AdamW [34] were selected as follows: initial_learning_rate $(\alpha) = 10^{-3}$; exponential decay rates for the 1st and 2nd moment estimates, i.e., $\beta 1 = 0.9$, $\beta 2 = 0.999$; zero-denominator removal parameter $(\varepsilon) = 10^{-8}$; and weight_decay $(\lambda) = 10^{-3}/4$. For training, the mini-batch size was chosen as 32. In f_c (ViT [18]), we empirically set the following hyperparameters: transformer_layers $(L) = 12$, hidden_size $(D) = 512$, mlp_size $= 1024$, and num_heads $= 12$.

In Table 1, we present the performance of our system DAZeTD in terms of average precision at IoU threshold $= 0.5$ (AP@0.5) on textual and non-textual object categories. We obtained 88.18%, 85.35%, and 89.62% AP@0.5 on printed text, handwritten text, and non-textual image categories, respectively. The mean AP (mAP) @0.5 on textual objects was 86.76%. The performance of detecting non-textual objects was better than the textual one. Overall, DAZeTD attained 87.71% mAP@0.5 on all object categories.

Table 1. Performance of DAZeTD on various object categories

Category	Text		Non-text	Text & Non-text
	Printed	Handwritten	Image	
AP@0.5 (%)	88.18	85.35	89.62	–
mAP@0.5 (%)	86.76		–	**87.71**

In Table 2, we show the performance of DAZeTD across different IoU thresh-olds, e.g., 0.5, and 0.75. We also observed the system performance over AP@μ_I

that is the average of ten APs at various IoU thresholds in a range of [0.5, 0.95] with a step of 0.05. Overall, DAZeTD performed the best for the IoU threshold, equal to 0.5.

Table 2. Performance of DAZeTD across different IoU thresholds

Threshold	AP (%)			mAP (%)
	Printed text	Handwritten text	Non-text	
@ 0.5	88.18	85.35	89.62	**87.71**
@ 0.75	84.45	79.72	85.47	83.21
@ μ_I	78.57	73.96	79.71	77.41

We checked the system performance over various grid sizes $w_g \times w_g$ (refer to Sect. 3.2), which can be seen in Table 3. DAZeTD achieved the best performance for 128×128 grid size. We observed that neither a very big-sized grid nor a tiny grid properly detected the zones.

Table 3. Performance of DAZeTD on various grid sizes

Grid size	AP@0.5 (%)			mAP@0.5 (%)
$(w_g \times w_g)$	Printed text	Handwritten text	Non-text	
32×32	63.83	60.31	65.02	63.05
64×64	81.72	79.36	82.95	81.34
128×128	88.18	85.35	89.62	**87.71**
256×256	43.86	41.64	46.80	44.10

We also examined the performance of DAZeTD over various classification networks f_c (refer to Sect. 3.2) and present in Table 4. The ViT [18]-based f_c performed the best when compared with various contemporary architectures [27,38–43].

As we mentioned in Sects. 3.2 and 4.1, we pre-trained the objectness network (f_o) and classification network (fc). The validation accuracies of f_o and f_c on extracted grids of DB$_v$ were 98.72% and 95.27%, respectively.

Table 4. Performance of DAZeTD using various classification networks (f_c)

Classification network (f_c)	AP@0.5 (%)			mAP@0.5 (%)
	Printed text	Handwritten text	Non-text	
VGG19 [38]	74.70	72.19	76.48	74.45
MobileNetV2 [39]	77.92	75.82	79.66	77.80
DenseNet201 [40]	80.64	77.06	82.19	79.96
InceptionV3 [41]	81.28	77.84	82.68	80.60
ResNet152V2 [27]	82.37	78.18	83.83	81.46
Xception [42]	83.26	79.16	84.29	82.23
Inception-ResNetV2 [43]	84.73	81.39	86.38	84.16
ViT [18]-based [Ours]	88.18	85.35	89.62	**87.71**

4.3 Comparison

In this subsection, we compare DAZeTD with some major related works of the literature. In Table 5, we present the quantitative comparison results on DB_t. Some qualitative comparisons are also shown in Fig. 4.

EAST [44] and CRAFT [45] are two popular deep learning-based scene text detectors that mostly generate bounding boxes (BBs) over word-level textual objects. To match with our task, here we took the area union of the word-level BBs. We compared with [44] and [45] for detecting the textual objects only due to their sole objective of text detection, and obtained 80.70% and 85.57% mAP@0.5 over textual objects (printed & handwritten), respectively. These methods [44,45] cannot separately detect printed and handwritten texts; however, we individually tested on printed and handwritten texts for comparative analysis. Here, although CRAFT [45] produced the best AP@0.5 for printed text, it performed poor for detecting handwritten zones (refer to Fig. 4). Our DAZeTD obtained 86.76% mAP@0.5 over textual objects, which is better than CRAFT [45].

Table 5. Comparison with some state-of-the-art methods

Method	AP@0.5 (%)			mAP@0.5 (%)
	Printed text	Handwritten text	Non-text	
EAST [44]	85.29	76.12	–	–
CRAFT [45]	**89.69**	81.46	–	–
SSD [46]	68.23	63.86	71.58	67.89
Faster R-CNN [47]	71.36	67.47	74.55	71.12
YOLO-v3 [20]	75.94	71.83	77.12	74.96
BigyaPAn [48]	79.46	80.24	82.91	80.87
DAZeTD [ours]	88.18	**85.35**	**89.62**	**87.71**

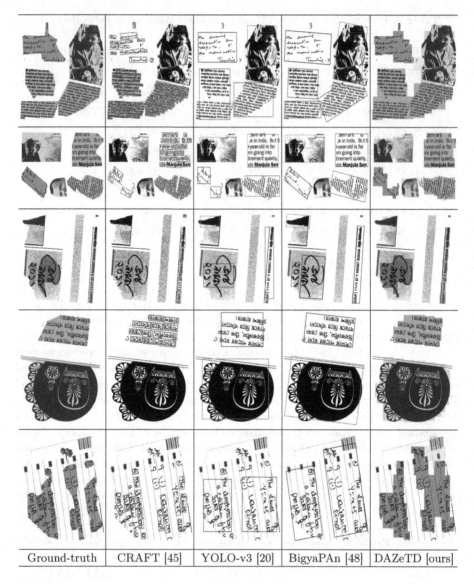

| Ground-truth | CRAFT [45] | YOLO-v3 [20] | BigyaPAn [48] | DAZeTD [ours] |

Fig. 4. Comparison of qualitative results: *printed* text, *handwritten* text, and non-textual *images* are enclosed by red, blue, and green covers/boxes, respectively. CRAFT [45]-detected texts are marked by red boxes. The *mixed* category's object is marked by a yellow cover in the outcome of DAZeTD. (Softcopy exhibits better display.) (Color figure online)

We also compared some contemporary methods, e.g., two-stage object detector: Faster R-CNN [47], one-stage object detectors: SSD [46], YOLO-v3 [20], and BigyaPAn [48] that produce BBs over the detected objects. We treated the area covered in a BB as the detected zone for comparison purposes. As a matter of fact, BBs include false-positive zones. SSD [46], Faster R-CNN [47], and YOLO-v3 [20] cannot detect oriented objects; here [20] outperformed [46] and [47], as shown in Table 5. BigyaPAn is a reinforcement learning-based orientation-aware object detector [48], and performed better than [46,47], and [20]. BigyaPAn achieved slightly better performance for handwritten text than printed text. In Fig. 4, we show some outcomes from YOLO-v3 and BigyaPAn. Overall, as evident from Table 5, our DAZeTD performed better than major state-of-the-art methods for detecting object zones of torn documents. We could not compare with [15] and [16], since we formulated the task in terms of object detection.

Some **limitations** of our system can be comprehended from the qualitative results as presented in Fig. 4. In DAZeTD, nearby zones of the similar category can be overlapped, e.g., printed text zones of the sample of Fig. 4's row-1. Our system hardly detects fine-level zones inside a grid that contains objects from multiple categories, e.g., yellow-colored zones in DAZeTD's output of Fig. 4. Sometimes a tiny character-level isolated text object may be ignored by DAZeTD, e.g., samples of Fig. 4's rows 3 and 5. Misclassification of categories can also occur, e.g., the printed text has been marked as handwritten in the sample of 4's rows 3.

5 Conclusion

In the forensic analysis of torn documents, automated detection of the zones from the ragged fragments is essential. In this paper, we propose an architecture DAZeTD to explore the possibilities of the object detection framework for zone segmentation in Torn documents. DAZeTD first divides a torn document into grids and finds the grids of interest. Then, it employs ViT to classify such grids into printed text/handwritten text/non-textual image categories. Finally, it merges the homogeneous grids based on the printed/handwritten/image categories, and draws the outer isothetic covers. We created a dataset comprising 881 torn documents and performed extensive experiments. On this dataset, DAZeTD attained 88.18%, 85.35%, and 89.62% AP@0.5 on printed text, handwritten text, and non-textual image categories, respectively, i.e., overall an 87.71% mAP@0.5. A comparative study with some major state-of-the-art object detectors was performed, where DAZeTD performed the best. Our future endeavors include improving the performance of DAZeTD, and analyzing the contextual information of torn fragments.

References

1. Brooks, E., et al.: Forensic physical fits in the trace evidence discipline: a review. Forensic Sci. Int. **313**, 110349 (2020)
2. Diem, M., et al.: Document analysis applied to fragments: feature set for the reconstruction of torn documents, DAS, pp. 393–400 (2010)

3. Okun, O., Doermann, D., Pietikainen, M.: Page segmentation and zone classification: the state of the art, Technical report, Acc. no. ADA458676, Defense Technical Information Center, VA, USA (1999)
4. Wang, Y., Phillips, I.T., Haralick, R.M.: Document zone content classification and its performance evaluation. Pattern Recogn. **39**(1), 57–73 (2006)
5. Binmakhashen, G.M., Mahmoud, S.A.: Document layout analysis: a comprehensive survey. ACM Comput. Surv. **52**(6), 109 (2020)
6. Yang, X., et al.: Learning to extract semantic structure from documents using multimodal fully convolutional neural networks. In: CVPR, pp. 4342–4351 (2017)
7. Lu, T., Dooms, A.: Probabilistic homogeneity for document image segmentation. Pattern Recogn. **109**, 107591 (2021)
8. Oliveira, S. A., Seguin, B., Kaplan, F.: dhSegment: a generic deep-learning approach for document segmentation, arXiv:1804.10371 (2018)
9. Markewich, L., et al.: Segmentation for document layout analysis: not dead yet. IJDAR **25**(2), 67–77 (2022)
10. Biswas, S., Banerjee, A., Lladós, J., Pal, U.: DocSegTr: an instance-level end-to-end document image segmentation transformer, arXiv:2201.11438 (2022)
11. Yi, X., Gao, L., Liao, Y., Zhang, X., Liu, R., Jiang, Z.: CNN based page object detection in document images. ICDAR **1**, 1417–1422 (2017)
12. Saha, R., Mondal, A., Jawahar, C.V.: Graphical object detection in document images. In: ICDAR, pp. 51–58 (2019)
13. Bhatt, J., et al.: A survey of graphical page object detection with deep neural networks. Appl. Sci. **11**(12), 5344 (2021)
14. Xu, X., et al.: Multi-task layout analysis for historical handwritten documents using fully convolutional networks. In: IJCAI, pp. 1057–1063 (2018)
15. Chanda, S., Franke, K., Pal, U.: Document-zone classification in torn documents. In: ICFHR, pp. 25–30 (2010)
16. Kleber, F., Diem, M., Sablatnig, R.: Torn document analysis as a prerequisite for reconstruction. In: International Conference on Virtual Systems and Multimedia, pp. 143–148 (2009)
17. Zhang, A., et al.: Dive into deep learning (2021). d2l.ai. Accessed 8 June 2022
18. Dosovitskiy, A., et al.: An image is Worth 16 × 16 words: transformers for image recognition at scale, arXiv:2010.11929, ICLR (2021)
19. Biswas, A., Bhowmick, P., Bhattacharya, B.B.: Construction of isothetic covers of a digital object: a combinatorial approach. J. Vis. Commun. Image Represent. **21**(4), 295–310 (2010)
20. Redmon, J., Farhadi, A.: YOLOv3: an incremental improvement, arXiv:1804.02767 (2018)
21. Zaidi, S.S.A., et al.: A survey of modern deep learning based object detection models. Digital Signal Process. **126**, 103514 (2022)
22. Jiao, L., et al.: A survey of deep learning-based object detection. IEEE Access **7**, 128837–128868 (2019)
23. Yu, J., Jiang, Y., Wang, Z., Cao, Z., Huang, T.: Unitbox: an advanced object detection network. In: ACM Multimedia, pp. 516–520 (2016)
24. Sarkar, A., Biswas, A., Bhowmick, P., Bhattacharya, B.B.: Word segmentation and baseline detection in handwritten documents using isothetic covers. In: ICFHR, pp. 445–450 (2010)
25. Adak, C., Maitra, P., Chaudhuri, B.B., Blumenstein, M.: Binarization of old halftone text documents. In: IEEE TENCON, pp. 1–5 (2015)
26. Gillies, S.: The Shapely User Manual, Ver. 1.7.0, (2021). shapely.readthedocs.io/en/latest/manual.html#geometric-objects. Accessed 8 June 2022

27. He, K., Zhang, X., Ren, S., Sun, J.: Identity mappings in deep residual networks. In: Leibe, B., Matas, J., Sebe, N., Welling, M. (eds.) ECCV 2016. LNCS, vol. 9908, pp. 630–645. Springer, Cham (2016). https://doi.org/10.1007/978-3-319-46493-0_38

28. Alom, M.Z., et al.: The history began from AlexNet: a comprehensive survey on deep learning approaches, arXiv:1803.01164 (2018)

29. Misra, D.: Mish: a self regularized non-monotonic activation function. Paper #928, BMVC (2020)

30. Goodfellow, I., Bengio, Y., Courville, A.: Deep Learning. MIT Press (2016). deeplearningbook.org. Accessed 8 June 2022

31. Vaswani, A., et al.: Attention is all you need. In: NIPS, pp. 5998–6008 (2017)

32. Ba, J.L., Kiros, J.R., Hinton, G.E.: Normalization, arXiv:1607.06450 (2016)

33. Devlin, J., et al.: BERT: pre-training of deep bidirectional transformers for language understanding. ACL 1, 4171–4186 (2019)

34. Loshchilov, I., Hutter, F.: Decoupled weight decay regularization. In: International Conference on Learning Representations (ICLR) (2019)

35. Kingma, D.P., Ba, J.: Adam: a method for stochastic optimization. Poster # 9, ICLR (2015)

36. Marti, U., Bunke, H.: The IAM-database: an English sentence database for offline handwriting recognition. IJDAR 5, 39–46 (2002)

37. Alaei, A., Pal, U., Nagabhushan, P.: Dataset and ground truth for handwritten text in four different scripts. IJPRAI 26(4), 1253001 (2012)

38. Simonyan, K., Zisserman, A.: Very deep convolutional networks for large-scale image recognition. In: ICLR (2015). arXiv:1409.1556

39. Sandler, M., et al.: MobileNetV2: inverted residuals and linear bottlenecks. In: CVPR, pp. 4510–4520 (2018)

40. Huang, G., et al.: Densely connected convolutional networks. In: CVPR, pp. 2261–2269 (2017)

41. Szegedy, C., et al.: Rethinking the inception architecture for computer vision. In: CVPR, pp. 2818–2826 (2016)

42. Chollet, F.: Xception: deep learning with depthwise separable convolutions. In: CVPR, pp. 1251–1258 (2017)

43. Szegedy, C., Ioffe, S., Vanhoucke, V., Alemi, A.A.: Inception-v4, inception-ResNet and the impact of residual connections on learning. In: AAAI, pp. 4278–4284 (2017)

44. Zhou, X., et al.: EAST: an efficient and accurate scene text detector. In: CVPR, pp. 2642–2651 (2017)

45. Baek, Y., Lee, B., Han, D., Yun, S., Lee, H.: Character region awareness for text detection. In: CVPR, pp. 9357–9366 (2019)

46. Liu, W., et al.: SSD: single shot multibox detector. In: Leibe, B., Matas, J., Sebe, N., Welling, M. (eds.) ECCV 2016. LNCS, vol. 9905, pp. 21–37. Springer, Cham (2016). https://doi.org/10.1007/978-3-319-46448-0_2

47. Ren, S., He, K., Girshick, R., Sun, J.: Faster R-CNN: towards real-time object detection with region proposal networks. IEEE Trans. on PAMI 39(6), 1137–1149 (2017)

48. Adak, C., Tao, X.: BigyaPAn: deep analysis of old paper advertisement. In: IJCNN, pp. 1–9 (2021)

CNN-Based Ruled Line Removal
in Handwritten Documents

Christian Gold[(✉)] and Torsten Zesch

CATALPA, Center of Advanced Technology for Assisted Learning and Predictive
Analysis, FernUniversität in Hagen, Hagen, Germany
{christian.gold,torsten.zesch}@fernuni-hagen.de

Abstract. Even state-of-the-art neural approaches to handwriting
recognition struggle when the handwriting is on ruled paper. We thus
explore CNN-based methods to remove ruled lines and at the same time
retain the parts of the writing overlapping with the ruled line. For that
purpose, we devise a method to create a large synthetic dataset for training and evaluation of our models. We show that our best model variants
are capable of reconstructing characters that are overlapping with the
line to be removed, which is a problem that simpler approaches often fail
to solve. On a dataset of children handwriting, we show that removing
the ruled lines improves character recognition. We made our synthetic
dataset and all experimental code available to foster further research in
this area.

Keywords: Ruled Line Removal · Handwriting Recognition · CNN

1 Introduction

A frequently addressed issue in handwriting recognition (HWR) is the recognition itself, although it is preceded by many steps such as word- or line-level
segmentation. In addition, contrast enhancement, noise removal or struck-out
word detection can further improve the recognition performance. The main goal
of these preprocessing steps is to separate the handwritten text referred to as
foreground from the background in the best possible way. Focusing on HWR in
the educational domain, handwritten texts challenge this separation as printed
forms (exams, school workbooks, writing pads, ...) where ruled lines help guiding
the orientation and size of writing are commonly used. Although in most cases
only a single ruled line is used, especially younger children use paper with several
lines (see Fig. 1). Furthermore, in textual answers of exams, ruled lines still exist
to indicate that a textual answer must be given.

Ruled lines pose a challenge for several levels of HWR. First, for word-
segmentation, where the ruled lines cause a problem as they connect multiple
words together making it harder to determine the beginning and end of a word.
Second, as character recognition has to interpret stroke-based structures into
characters it additionally have to deal with a ruled line, which is sometimes very

© The Author(s), under exclusive license to Springer Nature Switzerland AG 2022
U. Porwal et al. (Eds.): ICFHR 2022, LNCS 13639, pp. 530–544, 2022.
https://doi.org/10.1007/978-3-031-21648-0_36

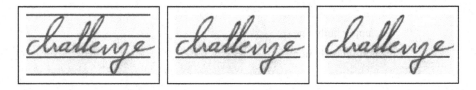

Fig. 1. The word 'challenge' from the IAM dataset with 4 ruled lines, 2 ruled lines and one ruled line.

dominant. The removal process of ruled lines faces the challenge of the retention of character-pixels, while pixels from the ruled line shall be removed. We address this challenge in our work using a neural network.

Although neural networks have recently been successfully used for a variety of image processing tasks, their use for ruled line removal has - to the best of our knowledge - not yet been explored. In this paper, we therefore propose a simple and effective convolutional neural network for line removal while retaining handwritten characters and brightness. Furthermore, the proposed model re-creates the shape of characters which align along the ruled line. For training and evaluation, we first create a synthetic dataset which can be adapted to any use case (e.g. number of ruled lines).

2   Related Work

Removing ruled lines with standard image processing algorithms usually follows three stages: *detection, removal,* and *text enhancement* [14]. In the *detection* stage, the ruled lines are extracted from the background and handwritten text. At the *removal* stage, the ruled lines are set to background color. In the *text enhancement* stage, the characters are usually re-connected and background noise is revised. Two stage solutions use connected component analysis during the *removal* stage to distinguish between ruled line pixels and text pixels and thus already perform the re-connection. An enhancement stage would then only remove noisy leftovers.

[12] use a canny filtering for the detection stage and a local intensity of black pixels with a morphological above-threshold filtering for the detection of connected components. The enhancement stage deals with the removal of dotted and broken lines and clearing of noise.

The method proposed by Refaey [13,14], use a windowed Hough-Transformation at the detection stage. The removal stage includes an intensity histogram (hue) with a local entropy to detect the connected components. A subsequent morphological operation enhances the text isolation from the background.

[6] make use of prior knowledge (e.g. average width of characters) and determined parameters such as the total number of characters and the total number

upper body line

bottom body line

top

body

bottom

Fig. 2. The word 'challenge' from the IAM dataset divided into a top, body and bottom part. The word is categorized as *full* as its characters span across all 3 parts.

of lines. With these determined, the detection and removal stage use structuring filters and merging (AND, OR and subtraction) of these results. At the text enhancement stage, the local entropy is calculated to re-connect disconnected characters. Finally, the noise is removed with a median filter.

In [3], the authors present a linear regression model which is capable of detecting ruled lines and especially dealing with broken lines. However, they do not propose a removal method after detection. Some classical approaches, e.g. [6] and [13], are dependant on predefined parameters (e.g. the average width of a character) or on the page layout. All classical approaches cannot deal with characters that overlay ruled lines (see example in Fig. 7). Thus, in this paper we leverage the power of neural networks to tackle the task of ruled line removal.

3   Synthetic Dataset Creation

To our knowledge, no freely available dataset containing ruled lines documents with and without the ruled line are published.[1] Thus, we create a synthetic dataset where ruled lines are generated and placed into sequences of concatenated handwritten word images, as shown in Fig. 3. The resulting dataset contains two version of text images with and without ruled lines.

3.1   Word Concatenation

The word images for the dataset were taken from IAM [10]. These word images come from various writers written without any constraints on size, alignment or pen type. Therefore, to create a realistic text line, we need to align these word images vertically along a virtual bottom body line. The virtual lines can be seen in Fig. 2. We categorize words into *bottom*, *top*, *body* or *full* (all) according to the characters' extension within each word. The characters which use the space below the bottom body line are: [f, g, j, p, q, y]. Characters using the top space are all upper case characters as well as [b, d, f, h, i, k, l, t]. The word images can then be vertically arranged according to their category (e.g. *bottom* categorized words will be aligned at the bottom, while *body* words will be aligned onto the bottom body line).

Some parameters, e.g. the number of text lines per page or the number of words per text line, are varied. However, for a consistent layout across the page,

---

[1] A potentially useful dataset [9] was not available under the published URL.

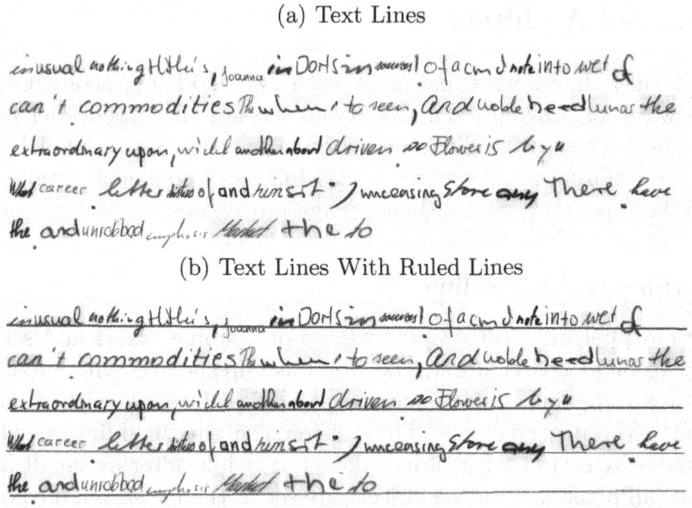

(a) Text Lines

(b) Text Lines With Ruled Lines

Fig. 3. An example page of the synthetic dataset displaying the text setting (a) and the ruled line placement (b).

we assign a random base value to some parameters (e.g. constant base gap between text lines). Each time such a parameter is used, the parameter will then be varied slightly from this base value.

3.2 Adding Ruled Lines

Simultaneously to the word concatenation, a copy of the image with inserted ruled lines will be created. While keeping the page layout consistent, we create pages with one, two or four ruled lines each text line (see Fig. 1). To further diversify our training data, we decided to add another type in which we place one ruled line somewhere over the handwritten text line. To make the inserted ruled lines even more realistic, we randomly add several defects. These include noise, differing grey level, small rotation as well as other techniques. An example of a final version with and without one ruled line each text line can be found in Fig. 3.

In summary, our proposed method is capable of generating unlimited training images while the specific configuration of lines (e.g. number of lines, line thickness, line length, or adding vertical lines) can be adapted to the requirements of the task. Despite the benefits of this synthetic dataset, we want to note that models solely trained on synthetic data always perform worse on real data. For our dataset, we did not specialize on one particular use case only, but rather aim at a widely applicable model by using a variety of line types.

4 Proposed Architectures

To remove ruled lines, we experiment with two modeling strategies: First, a straightforward all convolutional model and second, an autoencoder model. We explore several setups with different hyperparameters of each model to further increase the performance. As a simple baseline, we implement a naive morphological method based on conventional image processing.

4.1 Morphological Baseline

With our morphological filter approach we detect line pixels and set them to white (background color). First, the input is thresholded with Otsu [11] and followed by an opening and closing morphological filtering with a rectangular (horizontal) structuring element. Thus, we retrieve the ruled lines as white areas which can be extracted with object detection while filtering small areas out. Afterwards, all pixels within the objects are set to the background color, including pixels which intersect with character strokes. With this approach, we can analyze whether the recognition performance decreases if the character shape is not restored after removing the line.

4.2 ALL-CNN

As described in Sect. 2, earlier, ruled line removal has so far been done using regular image processing techniques, similar to the morphological approach. One basic technique is to apply rectangular filters on the image (e.g. opening and closing). In neural networks, convolutional layers (conv) can be seen as filters, although the filter values are trainable parameters. Instead of applying filters consecutively, several filters are trained in parallel, together with a bias value.

Our model consists of only three conv layers and thus we name our model ALL-CNN following [15]. We set the filter dimension of the first layer to 21×21 followed by two layers of dimension 3×3 each. By choosing a rather large filter dimension at the first layer, the network can gather more information around the centered line pixel. Therefore, the network is able to recognize dependencies of other character pixels around the ruled line (e.g. above and below) and thus to retain the shape of the characters. The number of filters is 32, which is later on varied (see Sect. 4.4). Additionally, we apply LeakyRelu [16] and Batch-Normalization [7] after each conv layer. An exemplary architecture of ALL-CNN can be seen in Fig. 4.

4.3 Autoencoder

Inspired by the various fields of application of autoencoders [5] such as cleaning [18] and denoising [17], we decided to test a CNN-based autoencoder architecture for the removal of ruled lines. The idea is to have an encoder at the beginning of the bottleneck and a decoding afterward, while filtering out the ruled lines in the taper.

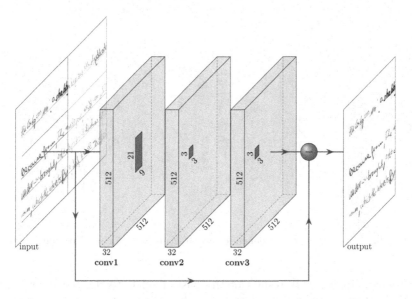

Fig. 4. Our proposed ALL-CNN (a variant with rectangle 21×9 filter) network with 3 conv layers, subtraction and with a Rectangular Filter (see Sect. 4.4). The Batch-Normalization and LeakyRelu after each conv layer are not visualized.

The basic architecture consists of 5 conv layers. A downsampling is applied after the second conv layer and an upsampling after another conv layer (see Fig. 5). The number of filters and their dimension of the first conv layer remain identical to the ALL-CNN model, as the purpose of the first layer remains the same. The remaining conv layers keep the number of filters, but the filter dimension is decreased to 5×5 (conv2) and 3×3 (conv3-5).

4.4 Hyperparameters

We experimented with several model parameters.

DNF - Decay of Number of Filters. Initially, we weight all conv layers alike and thus assign the same number of filters, a setup which we denote with *SNF* (Same-Number-of-Filters). We believe that the large filter (21×21) at the beginning will have the most influence for finding the ruled lines and the connection between characters intersecting the ruled line. We assume, that too many filters afterwards would only hold redundant information. Thus, we decay the number of filters after the first conv layer by cutting the number in half starting with 32 filters at conv layer 1, 16 at layer 2 and 8 for layer 3. This setup is called *DNF* (Decay-Number-of-Filters) and is only applied on the ALL-CNN model.

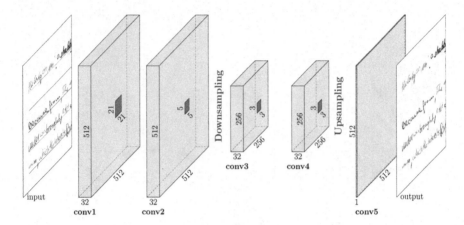

Fig. 5. Proposed CNN Autoencoder architecture with 5 convolutional layers.

Subtraction. In our first results of the basic models (Autoencoder and ALL-CNN), both predicted results were darker than the input images presented.[2] We assumed that this is due to the fact that the deeper the network was, the less information was still present from the input image. Furthermore, all pixels are influenced by the network although only the pixels of the ruled line should be affected. As only a few pixels must be changed, we referred to image processing techniques where only areas of attention (masks) are changed. We decided to use the output of the network to function as such a mask by subtracting the former output from the input layer and thereby present the input layer anew at the end of the network (see Fig. 4).

Rectangular Filter Shape. While we use the conv layers with a quadratic filter dimensions, the line to be reduced is of vertical shape. Thus, we adapt the filter dimension of 21 × 9 (width × height) which we denotes as *RectFilter*. While the ruled lines are horizontal, the shape of the characters crossing the ruled line [f, g, j, p, q, y] are rectangular with the longer side on the vertical axis. Therefore, we transpose the dimensions 9 × 21 as *RectFilter Transpose*.

5   Evaluation Setup

We describe how we evaluate results on synthetic data pixel-wise and on real data using CER as a derived extrinsic metric.

5.1   Synthetic Dataset

Although the synthetic data pages are created as full pages, they are divided into tiles of size 512 × 512 pixels. We choose this size as a tradeoff between

---

[2] Although the contrast in visualization might differ, we want to note that it can be seen in Fig. 6.

computation costs while covering several (~5) text lines. Furthermore, due to the varying number of text lines and varying text height, white tiles are common at the bottom of the page and thus are not ignored. We analyze the performance of our models on 100 synthetic images.

Evaluation Metric for Synthetic Data. With the synthetic dataset created, a pixel-wise evaluation can be performed. First, we calculate the root mean squared error (RMSE), which compares the delta pixel-wise and eliminates negative values by squaring.[3] Furthermore, we compute the delta between the predicted gray value (after the removal process) and the target gray value. Additionally, we assume that small deviations from the target value will not influence the handwriting recognition and thus set up a gray value threshold of $\pm$ 5. In this way, we can differentiate between three cases: no deviations, small deviations, and deviations of a greater extent.

5.2 Real Dataset

To test our methods on a use-case, children's handwritings are a particularly good example as they mainly include ruled lines. We decided to use the FD-LEX dataset [1] due to its clear structure (white pages with ruled lines) and availability. The dataset comprises freely-written texts with about 370,000 words from 938 learners in the 5th (age 9–11) and 9th (age 14–16) grade from the German school system written in German. For demonstration purposes, we took the first 50 pages of the set GYM 5.1 which are written from children of the 5th grade. In total, we consider 522 text lines. Unfortunately, the transcribed text was spellchecked. Thus, we had to manually transcribe the texts again.

Evaluation Metric for Real Data. As the FD-LEX dataset does not provide a ground truth for ruled line removal, the RMSE metric as used before is not applicable. Thus, we follow [2] and use an object-level metric, the recognition rate, as an extrinsic evaluation across all models, which makes it necessary to implement a recognizer. Although [2] use the word error rate, we use the character error rate (CER) as calculated with the edit distance. The CER can be seen as the inverse character accuracy, more precisely, the percentage of incorrectly predicted characters compared to the ground truth text. This way, we are able to analyze mispredicted characters in a more fine-grained way, especially those intersecting with the ruled lines.

Handwriting Recognizer. We use a straightforward text line handwriting recognizer with a CNN architecture and a Connectionist Temporal Classification (CTC [4]) at the end. The model is trained using the text line images from the IAM dataset while taking the split as defined in "Large Writer Independent Text Line Recognition Task"[4] for training, evaluation, and testing. It should be noted that we chose not to use a language model as it would correct spelling errors.

---

[3] It shall benoted that the RMSE is also used for the loss function during training.
[4] http://www.fki.inf.unibe.ch/databases/iam-handwriting-database.

(a)

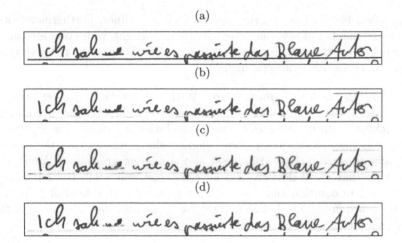

(b)

(c)

(d)

Fig. 6. Exemplary results of the first text line from the real dataset. (a) Original (b) Morphological Baseline, (c) Autoencoder and (d) ALL-CNN

The handwriting recognizer requires a text line segmentation as a preprocessing step. We use a straightforward segmentation with the l A^* path finding algorithm. Afterwards, we receive a pixel-wise mask of the individual text lines. As this segmentation process should not influence the recognition performance across all models, we apply the segmentation on one of the visually best methods (ALL-CNN with Subtraction) and transfer the segmentation masks to all other methods. In this way, we always segment the same part of the images, while having the individual results from the ruled line removal methods. After training, the handwriting recognizer has a CER of 11.52% on the IAM test set.[5]

6   Results and Discussion

Table 1 shows the results of our experiments. The character error rate (CER) on the real handwriting data with ruled lines is rather high (see the next subsection for an analysis of the reasons). However, as we hypothesized, removing ruled lines has a massive effect on recognition performance. Already the baseline removal method reduces CER from 66.3 to 35.4. Our neural models decrease CER even more. The best autoencoder version yields 31.1 while the best All-CNN models yields 28.5 which is a 6.9 percent point improvement over the baseline.

From those numbers alone it is hard to further analyze the individual performance of model variants. We thus now have a look at the synthetic test data where we know the true gold standard without the ruled lines and can perform a pixel-wise scoring.

---

[5] Note that CER values for the IAM dataset are typically computed per word, while we tackle the much harder task of recognizing whole lines. However, as we are only interested in the relative impact of ruled line removal, the performance of our recognizer is sufficient.

Table 1. Performance of all models on synthetic data with RMSE, amount of pixels which were changed !=0, amount of pixels with a greater difference then the threshold and on real data FD-LEX with CER.

		Synthetic			Real
		RMSE	Error		CER
			$\neq 0$	$> \pm 5$	
Original		–	–	–	66.3
Morphological baseline		6.6	2.3	1.8	35.4
Autoencoder	SNF	6.6	94.5	38.1	**31.1**
	SNF Subtraction	17.0	92.9	**7.5**	65.9
	SNF RectFilter	**6.3**	93.3	32.5	39.2
	SNF Subtraction RectFilter	11.1	92.8	19.4	55.4
	SNF RectFilter Transpose	6.7	93.8	36.7	31.3
	SNF Subtraction RectFilter Transpose	10.9	92.1	20.3	54.9
All-CNN	DNF 2L	4.7	85.6	5.6	32.3
	DNF 3L	4.5	84.9	6.0	31.8
	DNF 3L Subtraction	3.6	71.7	2.4	31.0
	SNF 2L	4.0	72.8	5.8	30.1
	SNF 3L	4.1	80.7	5.1	31.7
	SNF 3L Subtration	**2.6**	47.4	2.4	**28.5**
	SNF 3L RectFilter	3.6	68.0	4.6	33.8
	SNF 3L Subtraction RectFilter	2.9	63.5	3.4	29.4
	SNF 3L Subtraction RectFilter Transpose	3.0	**44.5**	**1.9**	29.5

Our Morphological method has an RMSE of 6.6 on the synthetic data. This denotes that on average, every pixel differs by 6.6 gray values from the ground truth. In addition, only 2.3% of all pixels were changed and 1.8% of these were above the threshold. This shows that those 1.8% have a bigger distance (e.g. a pixel that should be black was incorrectly changed to white) as they are able to increase the RMSE above the threshold of 5. The CER on the children's data is worse compared to our other methods but better than the original images. This proves our assumption that roughly removing the ruled lines increases the recognition performance. As this approach was naive - since only relevant pixels were changed - we use this as the baseline.

The best Autoencoder setup can reduce the RMSE value to 6.3. In general, however, the resulting images from the Autoencoder change nearly all pixels (above 92%). Furthermore, about one third of all pixels in the image have a difference above the threshold.

The best ALL-CNN architecture achieves the closest image in terms of the RMSE with only 2.6%. Although the amount of changed pixels is high, the amount of pixels exceeding the threshold is low with approx. 2–5%. This - together with the small RMSE - indicates that the delta is small contrary to the Baseline result where a few pixels increased the RMSE. In our results, the rather basic architecture with only 3 conv layers and the renunciation of down and up sampling resulted in our best recognition performance with a CER of

28.5% beating baseline by about 7 percent points. This is a strong result, as CER performance is also influenced by other factors (e.g. crossed-out words, tally marks) and our method focuses on ruled lines only.

6.1   CER Performance Level

While ruled line removal is able to improve recognition quality, the CER is still rather high. We now discuss some factors that additionally influence the results.

A major limitation is that training and application dataset do not match well (but there are not other datasets available). In particular, IAM texts were written by native English adults, while the FD-LEX texts were written by German children. Therefore, this is already an influence of two factors: the delta between adults and children, and between English and German. The latter might cause a bigger issue for the recognizer than is visible at first sight. One difference is the alphabet between German and English. The umlauts ä, ö, ü and the special character ß are not existing in English and thus are not present in the training data. To address this issue partly, we replaced those umlauts with a, o, u and s, respectively, in the ground truth text (including their upper-case variants).

Another, not so obvious gap arose from the different character bi-gram distributions which are learnt from the recognizer due to its temporal structure. For instance, 'th' is the most common bi-gram amongst English words whereas in German it is 'er' or 'en'.[6] In particular, frequencies of double consonants differ between both languages, too. As another difference, in German, three repeated consonants like 'Schifffahrt' (naut. 'shipping') exist. Therefore, the recognizer is biased towards an English dataset.

A further negative influence on the recognizer is, that tally marks are used in some of the FD-LEX images after every 10 words. Those are predicated as |, (or) most of the times. Unfortunately, the prediction of these vertical lines can sometimes influence neighboring characters (e.g. a| = d), too. This makes it difficult to filter these errors out or change the ground truth data to |.

Crossed-out words are another influencing factor as they are represented by a # sign in the transcription but can be interpreted by the recognizer as several characters leading to many errors. This further decreases the performance as children are presumably more likely to cross-out words. Ultimately, we decided to ignore all these issues as we focus on the ruled line removal itself. Therefore, we analyze the model performances with different metrics by looking at the improvement in relative performance rather than the overall performance.

Furthermore, some digitalization artefacts (e.g. gray border from the page during scanning) additionally decrease the performance.

6.2   Hyperparameters

SNF vs. DNF. Our hypothesis was that too many filters at the latter layers would only hold redundant information and thus do not influence the recognition

---

[6] https://www.staff.uni-mainz.de/pommeren/Cryptology/Classic/8_Transpos/ BigramsE1.html and *BigramsD1.html.

(a) ruled line   (b) original   (c) predicted

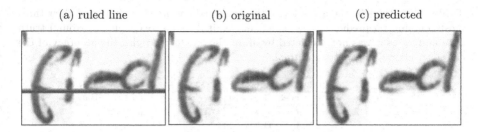

Fig. 7. Image segment as example of the model (best performing All-CNN variant) capability to remodel the letter 'e' as the bottom stroke overlaps with the ruled line.

performance. As the DNF version is always worse than the SNF version with all other parameters held equal, we can conclude that in the deeper layers, the amount of filters plays an important role.

Subtraction. For the ALL-CNN model, adding subtraction was a crucial step that decreased CER by 3 percent points. Subtraction is only beneficial for ALL-CNN, while it strongly decreases performance of the Autoencoder models.

Rectangular Filter. Focusing on the recognition results, neither the horizontal nor vertical rectangular filter shape resulted in an improvement compared to the quadratic shape. Only for one model using a rectangular shape came close to beat the best performance on the autoencoder models.

6.3   Remodeling of Overlapping Characters

A special problem for earlier approaches are characters overlapping with ruled lines as those cannot be reconstructed with filtering. In Fig. 7, the character 'e' of the word 'modified' is covered with a ruled line. Although the bottom stroke of the character 'e' overlaps with the ruled line, the CNN model was able to recreate the stroke partly (shorter in length). We assume, that this was possible due to the darkness information still being present in the ruled line as both, the character strokes' darkness and those of the ruled line interfere with each other. As these images were taken from the synthetic dataset and thus can't be tested on real data due to the loss of a real ground truth, we assume that our model should be capable to remodel characters likewise, if the darkness in the intersection area is superimposed.

6.4   Prediction of Characters with Restoration

We assumed that the Baseline approach would lead the recognizer to mispredict bottom characters as body characters as those characters cross the ruled line and are thus cut in half when the line is removed. E.g. 'g' would be predicted as 'a' or 'o' and a 'y' as 'u' or 'v'. Hence, we analyzed the misprediction of those bottom characters by using the Levenshtein Library[7] and return all edit

---

[7] https://pypi.org/project/python-Levenshtein/.

Table 2. On the left: the accuracy of the predicted *bottom* characters and their three most common mispredictions. 'y' is left out as it does not occur in the ground truth data and 'q' as it was not predicted by all methods. On the right: the accuracy of the character classes *upper, body, bottom*.

Model	'f' (247)		'g' (330)		'j' (330)		'p' (330)		Accuracy in %		
	%	pred	%	pred	%	pred	%	pred	up	body	bot
Original	33.2	f	51.2	g	57.1	j	30.9	p	42	39	46
	8.5	t	11.5	s	9.5	b	14.4	e			
	5.3	e	2.4	t	4.8	i	3.1	b			
	3.2	l	2.4	l	4.8	s	3.1	s			
Morphological baseline	61.9	f	84.2	g	76.2	j	66.0	p	73	75	72
	17.4	t	5.5	s	14.3	g	7.2	s			
	1.6	e	1.5	a	4.8	J	4.1	e			
	1.2	l	1.5	e	–	–	2.1	r			
Autoencoder	67.2	f	88.8	g	76.2	j	82.5	p	80	75	71
	13.4	t	2.7	s	9.5	g	3.1	e			
	2.8	l	1.2	y	4.8	o	2.1	r			
	2.0	e	0.9	a	4.8	F	1.0	f			
ALL-CNN	72.1	f	90.0	g	71.4	j	87.6	p	83	80	74
	11.7	t	1.8	s	14.3	g	2.1	o			
	1.6	l	0.9	y	4.8	f	1.0	k			
	1.6	e	0.9	q	4.8	o	1.0	j			

operations necessary to transform the predicted text line into the ground truth. The numbers in Table 2 correspond to the frequency of the operations 'equal' (correctly predicted characters) and 'replace' (incorrectly predicted characters). We do not consider 'insert' and 'delete' operations in this analysis.[8] However, using these numbers across all models makes them comparable again and at least indicate trends which can be seen in Table 2.

For instance, 'g' was incorrectly predicted as an 's', 'a' or 'e' in 8.5% for the Baseline method while for the Autoencoder and ALL-CNN the recognizer mispredicted them as bottom characters like 'y' or 'q', too. In general, the mispredicted characters are identical across all models except for the character 'g'.

6.5 Limitations

Both models were trained to detect and remove horizontal ruled lines. Unfortunately, some characters e.g. 'T' are build of likewise horizontal lines and thus are removed partly, too. Although these errors occur, we were unable to find upper characters (further examples: 'E', 't', 'F', 'H') being incorrectly predicted due to this result. We thus calculated the accuracy of correctly recognized upper characters (see Table 2).[9] We can see, that the recognition performance of upper

---

[8] For example, a 'y' could be recognized as 'ii' with two characters instead of one, which would lead to a 'replace' and 'delete' operation. However, the 'delete' operation could be assigned to the next character in the ground truth string and thus makes it invisible for us to analyze.

[9] We want to note, that again, only the operations 'replace' and 'equal' were considered from the Levensthein output and thus the true value might differ.

characters increased, in relation to the Baseline method, which leaves those characters untouched.

Furthermore, as all models are trained on an English dataset, we assume that the model will perform worse on scripts like Arabic which align strongly on ruled lines. For various reasons, it was not possible for us to set our results in relation to those procedures mentioned in Sect. 2. In many cases, the datasets and code used in the experiments were not freely available, whereas some links to the datasets were offline and the authors did not respond. For other approaches, the writing script (e.g. Arabic) differed too much. In addition, a large variety of different metrics were used, which often related to characteristics of the line (e.g. distance of lines, skew, position, ...), rather than the accuracy of the removal.

7 Summary

We evaluated a dataset of children handwriting and showed that ruled lines greatly affect character recognition. To this end, we trained different CNN models, namely an autoencoder and an ALL-CNN model, to remove the ruled lines. To do this, we created a flexible synthetic dataset in which we automatically place ruled lines onto handwritten text lines. This dataset is then used to train and evaluate the models. Finally, we tested the models on the children handwritings and showed that the error rate made by the recognizer was cut in half. In addition, we varied hyperparameters to further improve the model performance.

In conclusion, we found that a 3-layer CNN was already enough to solve the removal task. When presenting the input image anew, we were able to reduce the character error rate even further. In comparison, a morphological approach, where ruled lines where removed roughly, was outperformed by 7% by our best performing model (3 Layer ALL-CNN with subtraction) in terms of CER. Not only was the model capable of retaining the connections of characters intersecting with the ruled line, the model was also able to reconstruct strokes aligning with the ruled line. To foster further research, we make the model publicly available.[10]

In future work, we want to extend our training set by using the cvl-database which includes German words [8]. Being able to remove ruled lines is a prerequisite for training handwriting recognition models for children where almost all dataset have ruled lines. We plan to train a German children handwriting recognition model based on the cleaned FD-LEX data. However, children handwriting data remains challenging as it often contains other confounding factors like drawings, cross-outs, or tally marks.

References

1. Becker-Mrotzek, M., Grabowski, J.: FD-LEX (Forschungsdatenbank Lernertexte) (2018). textkorpus Scriptoria. Köln: Mercator-Institut für Sprachförderung und Deutsch als Zweitsprache. https://fd-lex.uni-koeln.de, https://doi.org/10.18716/FD-LEX/861

---

[10] https://github.com/catalpa-cl/ruled-line-removal-in-handwritten-documents.

2. Cao, H., Prasad, R., Natarajan, P.: A stroke regeneration method for cleaning rule-lines in handwritten document images. In: Proceedings of the International Workshop on Multilingual OCR, pp. 1–10 (2009)
3. Chen, J., Lopresti, D.: Model-based ruling line detection in noisy handwritten documents. Pattern Recogn. Lett. **35**, 34–45 (2014)
4. Graves, A., Fernández, S., Gomez, F., Schmidhuber, J.: Connectionist temporal classification: labelling unsegmented sequence data with recurrent neural networks. In: 23rd ICML, pp. 369–376 (2006)
5. Hinton, G.E., Salakhutdinov, R.R.: Reducing the dimensionality of data with neural networks. Science **313**(5786), 504–507 (2006)
6. Imtiaz, S., Nagabhushan, P., Gowda, S.D.: Rule line detection and removal in handwritten text images. In: 2014 ICSIP, pp. 310–315. IEEE (2014)
7. Ioffe, S., Szegedy, C.: Batch normalization: accelerating deep network training by reducing internal covariate shift. In: ICML, pp. 448–456. PMLR (2015)
8. Kleber, F., Fiel, S., Diem, M., Sablatnig, R.: Cvl-database: an off-line database for writer retrieval, writer identification and word spotting. In: ICDAR 2013, pp. 560–564. IEEE (2013)
9. Kumar, J., Doermann, D.: Fast rule-line removal using integral images and support vector machines. In: 2011th ICDAR, pp. 584–588. IEEE (2011)
10. Marti, U.V., Bunke, H.: The IAM-database: an English sentence database for offline handwriting recognition. IJDAR **5**(1), 39–46 (2002)
11. Otsu, N.: A threshold selection method from gray-level histograms. IEEE Trans. Syst. Man Cybern. **9**(1), 62–66 (1979)
12. Shobha Rani, N., Vasudev, T.: An efficient technique for detection and removal of lines with text stroke crossings in document images. In: Guru, D.S., Vasudev, T., Chethan, H.K., Sharath Kumar, Y.H. (eds.) Proceedings of International Conference on Cognition and Recognition. LNNS, vol. 14, pp. 83–97. Springer, Singapore (2018). https://doi.org/10.1007/978-981-10-5146-3_9
13. Refaey, M.: Fast detection and removal algorithms for ruled lines in full-color scanned handwritten documents. In: CSCESM, pp. 77–81. Citeseer (2014)
14. Refaey, M.A.: Ruled lines detection and removal in grey level handwritten image documents. In: ICICS, pp. 218–221. IEEE (2015)
15. Springenberg, J.T., Dosovitskiy, A., Brox, T., Riedmiller, M.: Striving for simplicity: the all convolutional net. arXiv preprint arXiv:1412.6806 (2014)
16. Xu, B., Wang, N., Chen, T., Li, M.: Empirical evaluation of rectified activations in convolutional network. arXiv preprint arXiv:1505.00853 (2015)
17. Yasenko, L., Klyatchenko, Y., Tarasenko-Klyatchenko, O.: Image noise reduction by denoising autoencoder. In: 2020 IEEE 11th International Conference on DESSERT, pp. 351–355. IEEE (2020)
18. Yin, K.: Cleaning up dirty scanned documents with deep learning (2019). https://medium.com/illuin/cleaning-up-dirty-scanned-documents-with-deep-learning-2e8e6de6cfa6. Accessed 30 July 2019

Complex Table Structure Recognition in the Wild Using Transformer and Identity Matrix-Based Augmentation

Bangdong Chen[1] , Dezhi Peng[1] , Jiaxin Zhang[1] , Yujin Ren[1] , and Lianwen Jin[1,2]([✉])

[1] South China University of Technology, Guangzhou, China
{eebdchen,eedzpeng,msjxzhang,eerenyj}@mail.scut.edu.cn,
eelwjin@scut.edu.cn
[2] SCUT-Zhuhai Institute of Modern Industrial Innovation, Zhuhai, China

Abstract. Tables are a widely used and efficient data structure. Although people can intuitively understand table contents, it remains challenging for machines, especially the tables taken in the wild. Previous methods mainly focus on scanned or PDF tables, but ignore investigating camera-based tables. This paper treats table structure recognition (TSR) as an image-to-sequence recognition task and adopts an end-to-end trainable model for complex TSR in the wild. Specifically, the model consists of a CNN-based encoder and two Transformer-based decoding branches, which can simultaneously predict the logical and physical structures of a table. Currently available camera-based table datasets are scarce, but deep learning methods heavily rely on large-scale datasets. To alleviate data insufficiency and boost model's performance, we propose a new and effective table data augmentation method, called *TabSplitter*. Due to the complex structure caused by cells spanning multiple rows or columns, directly cropping will lead to damage and change the properties of these cells. To solve this problem, we proposed a matrix representation, named *Identity Matrix (IM)*, to describe the table structure. Based on *IM*, we crop the tables and correct the cells whose attributes have changed, thus enhancing data diversity. Furthermore, the proposed *IM* facilitates the pre-processing of data and post-processing of predictions. Experimental results on several datasets demonstrate the effectiveness of the model and the *TabSplitter* for TSR, especially for complex tables in the wild.

Keywords: Table structure recognition · Transformer · Neural networks

1 Introduction

It is well known that tables serve as vital forms of information presentation, organizing data into a standard structure for easy information retrieval and comparison. With the popularity of smartphones and portable cameras, it is

U. Porwal et al. (Eds.): ICFHR 2022, LNCS 13639, pp. 545–561, 2022.
https://doi.org/10.1007/978-3-031-21648-0_37

becoming increasingly common for people to record and share tables as photographs. Therefore, the automatic extraction and parsing of table structures from camera-based images is a pressing and important problem.

A complete table recognition system should consider three subtasks: table detection (TD), table structure recognition (TSR), and table content recognition (TCR). The TD task aims to locate the table regions in an image, while the TCR task refers to the recognition of table content. Both parts can be treated as regular optical character recognition tasks. By contrast, the TSR task is more challenging and it is the core of the entire system. The TSR task aims to locate the position of cells and obtain the row-column information of them.

Most researchs have focused on TSR from digital images [1, 20, 22, 25, 32], a scenario where tables are usually converted directly from meta-data (e.g. PDF, LaTeX, etc.) to images. Generally, these tables are strictly aligned horizontally and vertically, and their backgrounds and appearances are free from interference. In this scenario, traditional methods [2, 7, 8, 10] usually adopt heuristic rules or handcrafted features to build TSR systems. However, heuristic-based algorithms generally have many hyperparameters that must be carefully adjusted according to different datasets, which may cause difficulties in application.

With the success of deep learning, recent methods for TSR tasks can be divided into three categories: 1) extracting cell bounding boxes using object detection or segmentation methods, and then using heuristic rules or clustering algorithms for table structure reconstruction [9, 16, 19, 20, 22, 25, 28, 32]; 2) extracting table elements by meta-files parsing or object detection methods, and then adopting graph models to predict the connection relationships between these elements [1, 12, 13, 21, 23, 31]; 3) designing image-to-sequence models to convert table images into markup sequences, such as LaTeX and HTML [3, 6, 11, 18, 33].

Although these methods can achieve promising results on digital-born image datasets, the performance suffers or even fails when switching to images taken in the wild. This is because of several challenges in interpreting camera-captured tables. First, tables in the wild often have various problems, such as paper deformation, lighting, and shadows. Furthermore, the annotation of tables in the wild requires more cost than the fully automated and semi-automated annotation methods used for digital-born image datasets, resulting in scarcity of table datasets in the wild. The above challenges limit the researchs and applications of TSR in the wild.

In this paper, we treat camera-based TSR as an image-to-sequence problem and adopt an end-to-end trainable model. Specifically, the model uses a CNN-based encoder and two Transformer-based decoding branches to obtain the table structure sequencee and the bounding boxes of cells. Additionally, to alleviate the problem of table data scarcity in the wild, we further propose a representation of table structure, called *Identity Matrix*. Based on this, a new data augmentation method is proposed, named *TabSplitter*. Experimental results on public datasets demonstrate the effectiveness of the adopted model and the *TabSplitter* method for TSR task.

In summary, the contributions of this paper are summarized as follows:

- We treat TSR in the wild as an image-to-sequence recognition problem and adopt an end-to-end trainable model that can obtain both the structural sequence of the table and the bounding box of each cell.
- We propose a matrix representation of the table structure, called *Identity Matrix (IM)*. Based on *IM*, we propose a new table data augmentation method named *TabSplitter*, which can mitigate the impact of data scarcity on model performance, especially tables in the wild. In addition, the proposed *IM* can also perform pre-processing and post-processing of table structure.
- Experimental results on public datasets demonstrate the superiority of the adopted model on the TSR problem for tables in the wild, and they also reveal the effectiveness of the proposed *TabSplitter* method.

2   Related Work

Traditional TSR methods [2,7,8,10] rely primarily on handcrafted features and heuristic rules. These methods are mainly applied to digital-born images with simple structure and no interference, or directly to PDF or LaTeX meta-files. In addition, these methods rely heavily on domain-specific heuristic rules, which limit their generalizability and applications. As deep learning flourishes, researchers have proposed many methods to solve the TSR problem in recent years, which can be divided into three types: cell extraction based methods [9,16,19,20,22,25,28,32], graph modeling based methods [1,12,13,21,23,31], and image to sequence based methods [6,11,18,33].

2.1   Cell Extraction Based Methods

Inspired by object detection and semantic segmentation, some researchers utilized detection or segmentation-based methods to extract table cells. The earliest approach DeepDeSRT [25] first used Faster RCNN [24] for TD task and then adopted FCN [15] for table row and column segmentation to obtain the bounding box of all cells. Similarly, TableNet [19] proposed an end-to-end and multitask semantic segmentation model, which could solve both TD and TSR tasks. Unlike most CNN-based approaches, Saqib Ali et al. [9] utilized bidirectional GRU to establish row and column boundaries in a context-driven manner. However, these methods encounter difficulties when identifying cells that span multiple rows and columns. SPLERGE [28] first splited the table into grids, and then merged the adjacent grid elements that need to be merged. GTE [32] performed cell detection and used a clustering algorithm for TSR. Cycle-CenterNet [16] used a cycle-pairing module to simultaneously detect table cells and combine them into structured tables, mainly addressing TSR of wired tables in the wild. LGPMA [22] applied a soft pyramidal mask learning mechanism to obtain aligned cell bboxes for subsequent heuristic reconstruction processes.

2.2   Graph Modeling Based Methods

Inspired by using graph neural network (GNN) to solve relational reasoning problems, many researchers have also tried to use graph models to model the relationships between table elements. GraphTSR [1] defined the TSR task as an edge prediction problem on a graph. It encoded the table by a stack of graph attention blocks, and then recognized the table structure by predicting the relationships between cells. Shah Rukh et al. [21] introduced DGCNN to fuse the words represented by appearance and geometric features, then used GNN to predict the relationship between nodes. It combined the benefits of CNN for visual feature extraction and GNN for dealing with the relational reasoning. Also based on DGCNN, GFTE [12] attempted to integrate image features, location features and text features together for accurate edge prediction with promising results. TabStruct-Net [23] combined cell detection and interaction modules to localize cells and predicted their row and column associations. TGRNet [31] was an end-to-end trainable GNN for TSR, which used a cell detection branch and a cell logical position branch to jointly predict the physical and logical positions of table cells. However, using GNN alone to model the relationships between table elements would introduce strong inductive bias, resulting in an inability to handle tables with complex structure. To address this problem, FLAG-Net [13] designed a FLAG module that adaptively aggregated Transformer-based and GNN-based contexts before performing relational reason.

2.3   Image-to-Sequence Based Methods

Recently, many researchers have treated the TSR problem as an image-to-sequence recognition task. TableBank [11] treated the TSR problem as a sequence prediction task, and used an encoder-decoder framework to directly predict the HTML sequence of the table structure. However, it only included HTML sequence of the table structure and ignored the physical location of table cells. EDD [33] also utilized the encode-decode framework, in which the decoder included the cell decoding branch and content decoding branch, which could obtain the table structure and content simultaneously. However, this approach was only applicable to English table datasets with a small number of categories. Because the model was trained end-to-end, it consumeed considerable computational resources and was unstable to train. To solve these problems, TableMAS-TER [6] decoupled table structure and table content and replaced the RNN-based decoder with a Transformer-based decoder on the framework of EDD. Unlike EDD, TableMASTER did not take the cell content decoding branch, but instead adopted a branch to regress the cell bounding boxes. Similarly, Table-Former [18] adopted a similar technical route to solve the TSR problem, and also achieved encouraging results. However, they both require large-scale data to obtain good performance and a stable training process due to the introduction of Transformer.

3   Method

3.1   Model Architecture

Figure 1 shows an overview of the adopted end-to-end trainable TSR model, which consists of a CNN-based encoder and a Transformer-based decoder. For a given table image, a high-dimensional feature vector of fixed size is first extracted by a CNN-based encoder and decoded by a Transformer-based decoder. As with other Transformer-based sequence prediction methods, the decoding branch stacks several Transformer decoding layers to predict the HTML structure sequence. The decoder contains two decoding branches, which can simultaneously obtain the structure category and the corresponding bbox of the category at each time step.

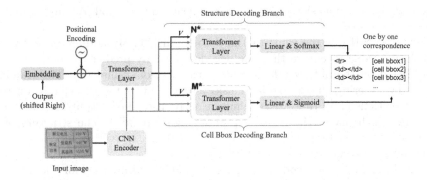

Fig. 1. Architecture of our proposed model. Given a table image, a high-dimensional feature vector of fixed size is first extracted by a CNN-based encoder, and then decoded by a Transformer-based decoder.

CNN-Based Encoder. Given a table image, we use the encoder proposed in MASTER [17] to obtain a high-dimensional feature vector. In order to avoid the original Global Context Attention (GCA) to perform the attention function in the global context, this encoder adopts the Multi-Aspect attention approach and introduces a Multi-Aspect GCA module on ResNet-50 [5], which facilitates the multiple attention function. Moreover, the input and output of this module have the same dimensionality and can be simply inserted into various CNN-based backbone networks as a plug-and-play module. The obtained high-dimensional features are then fed into the decoder.

Transformer-Based Decoder. As shown in Fig. 1, the decoder takes the high-dimensional features obtained from the encoder as input and passes first through a common Transformer layer, followed by two parallel decoding branches, each of which includes multiple stacked Transformer layers for global relationship modeling. By decoding the two branches in parallel, the structural sequence of the table and the bounding boxes of table cells can be predicted simultaneously.

Each Transformer layer includes three core components: a Multi-Head Attention (MHA), a Masked MHA, and a Feed-Forward Network (FFN).

MHA, first introduced in [29], is the main structure of Transformer. Specifically, MHA employs multiple heads to map the queries Q, keys K and values V transformations to different subspaces, and then apply the attention function in parallel to produce the output state $head_i$, which is computed as shown below:

$$Att(\mathbf{Q}, \mathbf{K}, \mathbf{V}) = softmax\left(\frac{\mathbf{Q}\mathbf{K}^{\mathbf{T}}}{\sqrt{d_k}}\right)\mathbf{V}, \tag{1}$$

$$head_i = Att(\mathbf{Q}\mathbf{W_i^q}, \mathbf{K}\mathbf{W_i^k}, \mathbf{V}\mathbf{W_i^v}), i \in \{1, 2, ..., h\}, \tag{2}$$

$$MHA(\mathbf{Q}, \mathbf{K}, \mathbf{V}) = Concat(head_1, ..., head_h)\mathbf{W^o}, \tag{3}$$

where d_k is the dimension of keys, h is the head number, and $\mathbf{W^o}$ are learnable linear projections.

Masked MHA is the same as the decoder in [29]. By adding Mask to the MHA, it is able to ensure that the decoder can only predict the current time step based on the output information of the previous time step. The training phase by lowering the triangular mask matrix enables the decoder to output the predictions of all time steps simultaneously instead of outputting them one by one, which makes the training process highly parallel. The structure of a standard FFN contains two FC layers and a ReLU activation function exists between the two layers.

At the end of both decoding branches, two FC layers are used for linear transformation. In structure decoding branch, a softmax function is added after the FC layers, which is used to calculate the predicted probabilities of all structure categories. In the training phase, our model predicts the structure category and cell bbox at current time step in parallel based on the GT at the previous time steps.

GT Generation. Since the original annotation of datasets are in HTML, XML or other formats, it cannot be directly used to train our model, so we need to build the GTs for training. According to the structural properties of tables, we define several structural categories, and all the detailed descriptions are shown in Table 1. Since there may be empty cells in the table whose bbox is not needed in subsequent TCR tasks, we classify the cells into: non-empty cells and empty cells, and represented by "$<td></td>$" and "$<td>\_</td>$", respectively.

For a cell spanning multiple rows, the original HTML tag "$<td\,rowspan = n></td>$" is broken into three categories: "$<td$", "$rowspan = N$" and "$></td>$". "$></td>$" and "$>\_</td>$" are used to distinguish whether a cell is empty. Because not all categories have corresponding cell bboxes, we must set a flag for each category with "1" and "0" referring to the existence of corresponding cell bboxes. Note that we need not obtain the bboxes for empty cells; therefore, only "$<td></td>$" and "$></td>$" have corresponding cell bboxes. In addition, four special symbol categories need to be added: "$<SOS>$", "$<EOS>$",

Table 1. Structure categories and detailed descriptions used for TSR.

Categories	Mask	Description
<tr>	0	Beginning of each row
</tr>	0	End of each row
<td></td>	1	None-empty cells
<td>_</td>	0	Empty cells
<td	0	Beginning of span cells
rowspan = '2'~'M'	0	Attributes of rowspan cells
colspan = '2'~'N'	0	Attributes of colspan cells
></td >	1	End of none-empty span cells
>_</td>	0	End of empty span cells

"$<PAD>$" and "$<UNK>$", which indicate the beginning of a sequence, the end of a sequence, padding symbols, and unknown characters, respectively. An example of building a training GT is shown in Fig. 2.

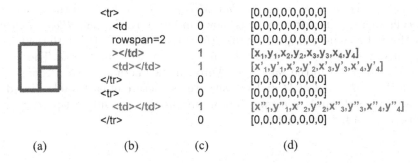

	<tr>	0	[0,0,0,0,0,0,0,0]
	<td	0	[0,0,0,0,0,0,0,0]
	rowspan=2	0	[0,0,0,0,0,0,0,0]
	></td>	1	$[x_1,y_1,x_2,y_2,x_3,y_3,x_4,y_4]$
	<td></td>	1	$[x'_1,y'_1,x'_2,y'_2,x'_3,y'_3,x'_4,y'_4]$
	</tr>	0	[0,0,0,0,0,0,0,0]
	<tr>	0	[0,0,0,0,0,0,0,0]
	<td></td>	1	$[x''_1,y''_1,x''_2,y''_2,x''_3,y''_3,x''_4,y''_4]$
	</tr>	0	[0,0,0,0,0,0,0,0]

(a)   (b)   (c)   (d)

Fig. 2. Example of building training GTs. (a) is the table image; (b) is the original GT in HTML format; (c) is the flag to indicate whether the structure category has a valid cell; (d) is the cell bboxes that the model needs to regress.

Loss Functions. For the structure decoding branch, at each decoding position, we use the standard cross-entropy to calculate the loss L_{struc} between the predicted probability and the ground truth. For the cell decoding branch, we use L_1 as the loss function. However, as in the above GT generation process, the decoding results of some time steps do not have corresponding cell bboxes; therefore, only the loss of valid cell bboxes needs to be calculated during the training, which can be calculated as follows:

$$L_{bbox} = \frac{1}{N} \sum_{i=0}^{maxL} \left| Cell_i - Cell'_i \right| \cdot Mask_i. \tag{4}$$

Finally, the global optimization can be defined as:

$$L = \lambda_1 L_{struc} + \lambda_2 L_{bbox}, \tag{5}$$

where λ_1 and λ_2 are the weight parameters. We set $\lambda_1 = 1$ and $\lambda_2 = 2$ in our experiments.

3.2 Matrix Representation of Table Structure

Current datasets usually adopt HTML, XML, LaTeX, and other formats to label table structure, but these formats are difficult to intuitively discover the characteristics of table structures. Therefore, we propose a new matrix representation to describe the table structure, called *Identity Matrix (IM)*. Given a table, we construct a *Row Identity Matrix (Row_M)* and *column Identity Matrix (Col_M)* to represent the row and column structure information, respectively.

Figure 3 shows an example of constructing *IM* of a table structure. First, we construct two $M \times N$ matrices denoting *Row_M* and *Col_M*, where M and N represent the number of rows and columns, respectively. Then, we need to fill in the *Row_M* and *Col_M* with values that represent the structure of the table. Taking *Row_M* as an example, if a cell does not span multiple rows is positioned in *i-th* row and *j-th* column, we then set $Row\_M[i][j] = 1$, as shown in the green box. For cells spanning multiple rows, as shown in blue box, the cell *"Blood Routine"* is in the first column and spans six rows (from the first row to the sixth row). Consequently, we set the $Row\_M[0][1] = 6$ and $Row\_M[1:6][1] = 0$. The *Col_M* is also constructed in the same way. Through this method, we can construct a unique *Row_M* and *Col_M* and visualize the structure of each table. Based on *IM*, we can perform data augmentation, pre-processing, and post-processing of the tables in a flexible manner.

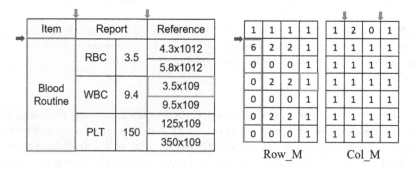

Fig. 3. Example of constructing *IM* of a table structure. (Color figure online)

Identity Matrix-based Augmentation *(TabSplitter)*. To increase the structural diversity of data and mitigate data insufficiency, we propose a new data augmentation method *TabSplitter* for tables in the wild, which is based on *IM*. Although there are many annotation formats for existing tabular datasets, the row and column structure information contained in all annotation formats can be converted into *IM*. For a table image labeled in HTML format, *TabSplitter* considers cropping the rows and columns of the table, expecting to obtain a new table image and corresponding HTML tags. For simple tables that do not have cells that span multiple rows or columns, we can directly obtain the cropped table image and the corresponding HTML tags. However, for complex tables with cells that span multiple rows or columns, direct cropping may break those cells and not yield corresponding HTML tags. The following two issues need to be considered: *"1. Where to cut from? 2. If the attributes of the cells are changed after cropping, how should the corresponding HTML tags be generated?"*

With the proposed *Row_M* and *Col_M*, we are able to solve these two problems in an elegant and simple way. For the first problem, given a table, we construct *Row_M* and *Col_M*. The rows that represent by *"0"* in *Row_M* indicates that they are cuttable and we can discover all cuttable positions, as shown by the red arrow in Fig. 3. Similarly, we can find all the positions where a column can be cut, as shown by the orange arrows.

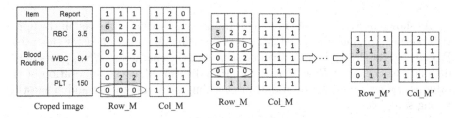

Fig. 4. Flowchart for correcting those cells whose span properties have been corrupted. (Color figure online)

For the second problem, as shown in Fig. 4, we cropped the table image, *Row_M* and *Col_M* according to the cropped position obtained in the previous step. From the cropped image, we observe that the spanning properties of some cells have changed. For example, the cell *"Blood Routine"* in the original image spans six rows, but after cropping it spans three rows. If the spanning properties of a cell have changed, the corresponding *Row_M* will appear in some rows that are all *"0"*. This is because in *Row_M*, *"0"* represents spanning rows, and when a row is all *"0"*, it indicates that the number of spanning rows in the current cell needs to be reduced. As shown in Fig. 4, by traversing the entire *Row_M* rows, we can get the property corrected *Row_M'*, and similarly we can get the corresponding *Col_M'*. Finally we can get the HTML sequence of table structure after cropping based on the newly obtained *Row_M'* and *Col_M'*.

Table Structure Pre-processing and Post-processing. Since the annotation of table structures is complicated, annotation errors may occur in the case of automatic or manual annotation. About this question, the proposed *IM* can be used as a filter tool. The elements of both *Row_M* and *Col_M* are initialized to "*N*" at the beginning. After constructing these two matrices, if there are still positions with elements of "*N*", then we can conclude that there is an error in the table structure. This conclusion can help us to pre-process the data and post-process the predicted results.

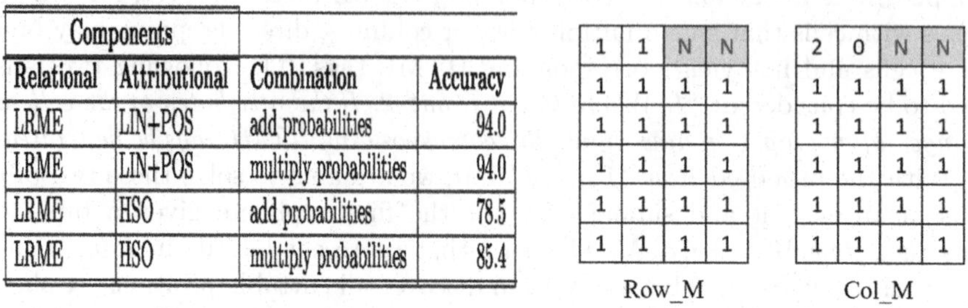

Components				Row_M				Col_M			
Relational	Attributional	Combination	Accuracy	1	1	N	N	2	0	N	N
LRME	LIN+POS	add probabilities	94.0	1	1	1	1	1	1	1	1
LRME	LIN+POS	multiply probabilities	94.0	1	1	1	1	1	1	1	1
LRME	HSO	add probabilities	78.5	1	1	1	1	1	1	1	1
LRME	HSO	multiply probabilities	85.4	1	1	1	1	1	1	1	1
				1	1	1	1	1	1	1	1

Fig. 5. The missing annotation of the structure leads to the presence of non-numeric items in *Row_M* and *Col_M*.

As shown in Fig. 5, an obvious error in the annnotation is the cell in the upper right corner are not annotated. We construct *Row_M* and *Col_M* for this table. From the right side of Fig. 5, it can be seen that the character "*N*" appears in the upper right corner of *Row_M* and *Col_M*, indicating that the annotation of the table is wrong. This can be useful in checking data annotation and suggesting wrong samples.

Furthermore, the proposed *IM* can be used as a post-processing method to filter the prediction results. After converting the prediction results into *Row_M* and *Col_M*, if we find character "*N*" in them, it means that there is an absolute error in the results, then we can set the confidence to 0. This can pick out the absolute wrong prediction results in practical applications, which is convenient for human intervention and can improve the robustness of the entire table recognition system.

4 Experiment

4.1 Datasets and Evaluation Metrics

Datasets. We evaluated the adopted model and *TabSplitter* method on three benchmark datasets: TAL_OCR_TABLE [27], WTW [16] and SciTSR [1]. Using the data pre-processing method mentioned in Sect. 3.2, we can filter the incorrectly labeled samples. The data statistics before and after filtering are shown in Table 2. Note that for a fair comparison with other methods, we use the filtered datasets for training and then the unfiltered test sets for testing.

Table 2. Statistics of existing datasets for TSR before and after data preprocessing.

Dataset	Split	Unfiltered	Filtered
TAL_OCR_TABLE [27]	Train	12800	12276
	Val	3200	3072
	Test	3000	3000
WTW [16]	Train	12567	11740
	Test	4048	3611
SciTSR [1]	Train	12000	11516
	Test	3000	2896
SciTSR-COMP [1]	Test	716	640

TAL_OCR_TABLE (TAL) [27] is obtained using a camera in an educational scenario. The training set consists of 16,000 images with cell bbox, cell content, and HTML sequence for tables; the test set (TAL_test) consists of 2,000 images with only the HTML sequence annotation. The official training set was randomly divided into a training set (TAL_train) and a validation set (TAL_val) in a 4:1 ratio.

WTW [16] is the first public dataset for TD and TSR tasks in the wild, which is constructed from photoing, scanning and web pages. It contains 14,581 real-scene images and the logical structure and the cell locations are well-annotated in this dataset.

SciTSR [1] is a large-scale table recognition dataset from scientific literature, which includes annotations of table structure and table content. It contains 12,000 training samples and 3,000 test samples. In addition, there are 716 complicated tables in the test set, which is named SciTSR-COMP [1].

Evaluation Metrics. We evaluate the performance of TSR in terms of both logical and physical structures. For the physical structure of the tables, we used precision, recall, and F-score under different IoUs for evaluation. For the logical structure of the tables, we used Tree-Edit-Distance-based Similarity (TEDS) [33] for evaluation, which can measure both the structural similarity and content similarity of two tables represented in HTML format. We also report the results that only consider the table structure represented as TEDS-Struc.

4.2 Implementation Details

In the experiments, the input images are uniformly resized to a fixed size and each mini-batch contained six images per GPU. According to different datasets, different maximum sequence lengths are set respectively. The maximum length of WTW and SciTSR datasets is 500, while TAL is 400. Both decoding branches stack 3 Transformer layers. We chose Ranger [30] as our optimizer, and used four GeForce RTX 2080 Ti GPUs for distributed training, with a total of 100k iterations. The learning rate is initialized as 1e−3 and divided by 10 at 40k and

80k iterations. During the testing phase, we set batchsize $= 1$ and stop prediction when the model predicts the end symbol.

4.3   Comparison with State-of-the-Arts

We compared our results on the TAL dataset with the PRCV2021 table recognition competition list [27] and two TSR methods. The competition used TEDS as an evaluation metric and required the assessment of the structural accuracy and content accuracy of tables. To compare with the top three results, we used the same TCR method as the first place. The BDN [14] detection model was first used as the text line detector, and then the CRNN model [26] with CTC decoding method [4] were used for text line recognition. As evident in Table 3, our method achieved the best results for both the TAL_val and TAL_test. Since SPLERGE [28] and CascadeTabNet [20] were designed for scanning tables, they could not applicable to tables taken in the wild. In addition, even though the best result in the competition was achieved using multi-model integration techniques to improve performance, we are able to surpass the first-place on TEDS and TEDS-Struc metrics using only a single model.

Table 3. Performance comparison of different methods on TAL dataset.

Methods	TAL_val		TAL_test	
	TEDS-Struc	TEDS	TEDS-Struc	TEDS
SPLERGE [28]	52.73	46.36	53.14	47.26
CascadeTabNet [20]	66.63	61.90	66.71	62.11
TAL_ Third_Place [27]	–	95.59	–	95.31
TAL_ Second_Place [27]	–	95.64	–	95.90
TAL_ First_Place [27]	99.40	95.92	99.39	95.91
Ours	**99.41**	**95.94**	**99.45**	**95.96**

The TEDS-Struc results on WTW is summarized in Table 4, the proposed method outperforms other SOTA methods, including the object detection method CenterNet [34], and three TSR methods: SPLERGE [28], CascadeTabNet [20], and Cycle-CenterNet [16]. Compared with Cycle-CenterNet, our method increases TEDS-Struc on all subclasses of the WTW dataset and obtains about 8% improvement on the overall test set. Notably, on the overlaid subclass, our method is able to obtain a significant boost of about 40%, which demonstrates its generalization ability.

Table 4. Performance comparison of different methods on WTW dataset.

Method	Simple	Inclined	Curved	Occluded and blurred	Extreme aspect ratio	Overlaid	Muti color and grid	**All**
CenterNet [34]	–	–	–	–	–	–	–	58.7
SPLERGE [28]	–	–	–	–	–	–	–	26.0
CascadeTabNet [20]	–	–	–	–	–	–	–	11.4
Cycle-centernet [16]	94.2	90.6	70.0	53.3	77.4	51.2	66.7	83.3
Ours	**100.0**	**98.1**	**92.4**	**72.2**	**92.6**	**90.9**	**87.3**	**91.9**

4.4   Ablation Study

Effectiveness of *TabSplitter*. We validate the effectiveness of *TabSplitter* in the proposed model on several datasets. First, these test sets are divided into simple and complex tables according to whether there are cells spanning multiple rows or columns, and then their TEDS-Struc metrics are calculated separately. The results are reported in Table 5, from which we can find that applying *TabSplitter* can indeed improve the performance of the model on all datasets, especially for complex tables.

Table 5. Effectiveness of *TabSplitter* on several datasets. "Simp" and "Comp" stand for simple and complex tables, respectively.

Train dataset	Test dataset	TabSplitter	TEDS-Struc		
			Simp	Comp	All
TAL_train	TAL_val	✗	99.51	97.56	99.20
		✓	**99.61**	**98.34**	**99.41**
TAL_train	TAL_test	✗	99.51	97.72	99.23
		✓	**99.58**	**98.73**	**99.45**
WTW	WTW	✗	92.74	89.50	90.36
		✓	**93.43**	**91.35**	**91.91**
SciTSR	SciTSR	✗	99.36	95.96	98.57
		✓	**99.67**	**97.55**	**99.19**

Robustness of *TabSplitter*. To explore the robustness of the proposed *TabSplitter* method, we used 20%, 40%, 60%, 80%, and 100% of the TAL_val dataset and applied *TabSplitter* for training. We then compared both the physical and logical structure of the tables on the TAL_val dataset to verify the effectiveness of *TabSplitter*. From Table 6 and 7, it can be seen that as the amount of training data increases from 20% to 100%, the performance of both the physical and logical structure of tables improves, indicating that the amount of data plays an critical role in model performance.

Table 6. Comparison of table logical structure metrics with baseline result for different percentage of TAL_train.

Training data used	TEDS-Struc (val)			TEDS-Struc (test)		
	Simp.	Comp.	All	Simp.	Comp.	All
Baseline	99.51	97.56	99.20	99.51	97.72	99.23
20%	99.19	94.42	98.44	99.07	95.04	98.44
40%	99.43	96.81	99.02	99.39	97.29	99.07
60%	99.50	97.56	99.19	99.37	98.07	99.17
80%	99.61	97.85	99.33	99.55	98.20	99.34
100%	**99.61**	**98.34**	**99.41**	**99.58**	**98.73**	**99.45**

Table 7. Comparison of table physical structure metrics with baseline result for different percentage of TAL_train.

Training data used	IoU = 0.5			IoU = 0.6			IoU = 0.7			IoU = 0.8			IoU = 0.9		
	P	R	F1	P	R	F1	P	R	F1	P	R	F1	P	R	F1
Baseline	97.5	98.0	97.8	96.1	96.6	96.4	92.8	93.3	93.0	83.8	84.2	84.0	52.8	53.1	52.9
20%	96.8	97.0	96.9	95.4	95.5	95.5	92.4	92.5	92.4	84.0	84.1	84.1	51.8	51.8	51.8
40%	97.7	98.1	97.9	96.8	97.2	97.0	93.1	93.5	93.3	83.9	84.3	84.1	54.9	55.1	55.0
60%	98.1	98.1	98.1	97.0	97.1	97.0	93.9	93.9	93.9	85.8	85.8	85.8	56.5	56.5	56.5
80%	98.2	98.3	98.3	97.2	97.3	97.3	94.7	94.8	94.7	86.4	86.3	86.4	58.8	58.8	58.8
100%	**98.7**	**98.8**	**98.7**	**97.7**	**97.8**	**97.8**	**94.8**	**94.9**	**94.8**	**87.9**	**87.9**	**87.9**	**59.6**	**59.6**	**59.6**

Table 6 compared the performance of the logical structure of tables, and the TEDS-Struc metric is reported. We found that if we use *TabSplitter* on top of the proposed model, we only need to use 60% of the training data and its performance is comparable to the baseline results. Table 7 compared the performance of the physical structure of tables, we reported the precision, recall, and F1-score at different IoUs. It could be found that using the *TabSplitter* can significantly improve the results compared to the baseline. And, with the increase of IoU, the obtained performance improvement is more obvious. For example, when IoU is equal to 0.9, applying *TabSplitter* can significantly increase the F1-score by about 7%. Furthermore, the performance is comparable to or better than the baseline using only 40% of the training data, which fully demonstrates that the proposed *TabSplitter* can improve the robustness of the model.

4.5 Effectiveness of Identity Matrix for Post-processing

In addition to being used for data argumentation, the proposed *IM* can be also used for post-processing to further improve the performance. This subsection demonstrates the effectiveness of the proposed *IM* in post-processing the prediction results of the model. In practical applications, we want the table recognition system to inform us of cases where the recognition results are wrong and return a confidence level for those cases which the system is not sure they are correct.

This will improve the robustness of the entire system and facilitate human intervention to correct samples that are incorrectly recognized. Our proposed *IM* can achieve this goal easily. We performed post-processing in the results to filter those that are absolutely wrong. As shown in Table 8, it can be found that on several benchmarks, the TEDS-Struc results can be significantly improved after using *IM*-based post-processing to filter out those absolutely wrong results, which can significantly improve the robustness of TSR systems in practical applications.

Table 8. Post-processing of the prediction results by *IM*. "Before" and "After" stand for the results before and after post-processing, respectively.

Test data	TEDS-Struc	
	Before	After
TAL_val	99.20	**99.52**
TAL_test	99.23	**99.45**
WTW	90.36	**94.76**
SciTSR	98.57	**99.01**

5 Conclusion

In this work, we treated the TSR problem in the wild as an image-to-sequence recognition task and adopted an end-to-end trainable model with a CNN-based encoder and two Transformer-based decoding branches that can simultaneously obtain the structural sequence of tables and bounding boxes of cells. To mitigate the impact of camera capture table data scarcity on model performance, we proposed a *IM* representation for table structure, based on which we develop a table data augmentation method called *TabSplitter*. In addition, the *IM* is also useful for data pre-processing and post-processing of model prediction. Experiments were conducted on several datasets to validate the effectiveness of the proposed methods. Experimental results showed that the adopted model achieves state-of-the-art performance, and the *TabSplitter* can mitigate the impact of data scarcity on models, especially in camera-captured tabular scenarios.

Acknowledgments. This research is supported in part by NSFC (Grant No. 61936003), GD-NSF (No. 2017A030312006, No. 2021A1515011870), Zhuhai Industry Core and Key Technology Research Project (No. 2220004002350), and the Science and Technology Foundation of Guangzhou Huangpu Development District (Grant 2020GH17).

References

1. Chi, Z., Huang, H., Xu, H.D., Yu, H., Yin, W., Mao, X.L.: Complicated table structure recognition. arXiv preprint arXiv:1908.04729 (2019)
2. Coüasnon, B., Lemaitre, A.: Recognition of tables and forms. In: Handbook of Document Image Processing and Recognition (2014)

3. Desai, H., Kayal, P., Singh, M.: TABLEX: a benchmark dataset for structure and content information extraction from scientific tables. In: Lladós, J., Lopresti, D., Uchida, S. (eds.) ICDAR 2021. LNCS, vol. 12822, pp. 554–569. Springer, Cham (2021). https://doi.org/10.1007/978-3-030-86331-9_36

4. Graves, A., Fernández, S., Gomez, F., Schmidhuber, J.: Connectionist temporal classification: labelling unsegmented sequence data with recurrent neural networks. In: International Conference on Machine Learning, pp. 369–376 (2006)

5. He, K., Zhang, X., Ren, S., Sun, J.: Deep residual learning for image recognition. In: IEEE Conference on Computer Vision and Pattern Recognition, pp. 770–778 (2016)

6. He, Y., et al.: PingAN-VCGroup's solution for ICDAR 2021 competition on scientific table image recognition to latex. arXiv preprint arXiv:2105.01846 (2021)

7. Hirayama, Y.: A method for table structure analysis using DP matching. In: Proceedings of 3rd International Conference on Document Analysis and Recognition, vol. 2, pp. 583–586 (1995)

8. Itonori, K.: Table structure recognition based on textblock arrangement and ruled line position. In: International Conference on Document Analysis and Recognition, pp. 765–768 (1993)

9. Khan, S.A., Khalid, S.M.D., Shahzad, M.A., Shafait, F.: Table structure extraction with bi-directional gated recurrent unit networks. In: International Conference on Document Analysis and Recognition, pp. 1366–1371 (2019)

10. Kieninger, T., Dengel, A.: The T-recs table recognition and analysis system. In: Lee, S.-W., Nakano, Y. (eds.) DAS 1998. LNCS, vol. 1655, pp. 255–270. Springer, Heidelberg (1999). https://doi.org/10.1007/3-540-48172-9_21

11. Li, M., Cui, L., Huang, S., Wei, F., Zhou, M., Li, Z.: TableBank: table benchmark for image-based table detection and recognition. In: Language Resources and Evaluation Conference, pp. 1918–1925 (2020)

12. Li, Y., Huang, Z., Yan, J., Zhou, Y., Ye, F., Liu, X.: GFTE: graph-based financial table extraction. In: International Conference on Pattern Recognition, pp. 644–658 (2021)

13. Liu, H., et al.: Show, read and reason: table structure recognition with flexible context aggregator. In: ACM International Conference on Multimedia, pp. 1084–1092 (2021)

14. Liu, Y., et al.: Exploring the capacity of an orderless box discretization network for multi-orientation scene text detection. Int. J. Comput. Vision 129(6), 1972–1992 (2021)

15. Long, J., Shelhamer, E., Darrell, T.: Fully convolutional networks for semantic segmentation. In: IEEE Conference on Computer Vision and Pattern Recognition, pp. 3431–3440 (2015)

16. Long, R., et al.: Parsing table structures in the wild. In: IEEE International Conference on Computer Vision, pp. 944–952 (2021)

17. Lu, N., et al.: MASTER: multi-aspect non-local network for scene text recognition. Pattern Recognit. 117, 107980 (2021)

18. Nassar, A., Livathinos, N., Lysak, M., Staar, P.: TableFormer: table structure understanding with transformers. In: Proceedings of the IEEE/CVF Conference on Computer Vision and Pattern Recognition, pp. 4614–4623 (2022)

19. Paliwal, S.S., D, V., Rahul, R., Sharma, M., Vig, L.: TableNet: Deep learning model for end-to-end table detection and tabular data extraction from scanned document images. In: International Conference on Document Analysis and Recognition, pp. 128–133 (2019)

20. Prasad, D., Gadpal, A., Kapadni, K., Visave, M., Sultanpure, K.: CascadeTab-Net: an approach for end to end table detection and structure recognition from image-based documents. In: IEEE Conference on Computer Vision and Pattern Recognition Workshops, pp. 572–573 (2020)
21. Qasim, S.R., Mahmood, H., Shafait, F.: Rethinking table recognition using graph neural networks. In: International Conference on Document Analysis and Recognition, pp. 142–147 (2019)
22. Qiao, L., et al.: LGPMA: complicated table structure recognition with local and global pyramid mask alignment. In: Lladós, J., Lopresti, D., Uchida, S. (eds.) ICDAR 2021. LNCS, vol. 12821, pp. 99–114. Springer, Cham (2021). https://doi.org/10.1007/978-3-030-86549-8_7
23. Raja, S., Mondal, A., Jawahar, C.V.: Table structure recognition using top-down and bottom-up cues. In: Vedaldi, A., Bischof, H., Brox, T., Frahm, J.-M. (eds.) ECCV 2020. LNCS, vol. 12373, pp. 70–86. Springer, Cham (2020). https://doi.org/10.1007/978-3-030-58604-1_5
24. Ren, S., He, K., Girshick, R., Sun, J.: Faster R-CNN: towards real-time object detection with region proposal networks. In: Advances in Neural Information Processing Systems, vol. 28 (2015)
25. Schreiber, S., Agne, S., Wolf, I., Dengel, A., Ahmed, S.: DeepDeSRT: deep learning for detection and structure recognition of tables in document images. In: International Conference on Document Analysis and Recognition, pp. 1162–1167 (2017)
26. Shi, B., Bai, X., Yao, C.: An end-to-end trainable neural network for image-based sequence recognition and its application to scene text recognition. IEEE Trans. Pattern Anal. Mach. Intell. **39**(11), 2298–2304 (2017)
27. TAL_OCR_TABLE: Tal table recognition technology challenge (2021). https://ai.100tal.com/dataset
28. Tensmeyer, C., Morariu, V.I., Price, B., Cohen, S., Martinez, T.: Deep splitting and merging for table structure decomposition. In: International Conference on Document Analysis and Recognition, pp. 114–121 (2019)
29. Vaswani, A., et al.: Attention is all you need. In: Advances in Neural Information Processing Systems, vol. 30 (2017)
30. Wright, L., Demeure, N.: Ranger21: a synergistic deep learning optimizer. arXiv preprint arXiv:2106.13731 (2021)
31. Xue, W., Yu, B., Wang, W., Tao, D., Li, Q.: TGRNet: a table graph reconstruction network for table structure recognition. In: IEEE International Conference on Computer Vision, pp. 1295–1304 (2021)
32. Zheng, X., Burdick, D., Popa, L., Zhong, X., Wang, N.X.R.: Global table extractor (GTE): a framework for joint table identification and cell structure recognition using visual context. In: IEEE Winter Conference on Applications of Computer Vision, pp. 697–706 (2021)
33. Zhong, X., ShafieiBavani, E., Jimeno Yepes, A.: Image-based table recognition: data, model, and evaluation. In: Vedaldi, A., Bischof, H., Brox, T., Frahm, J.-M. (eds.) ECCV 2020. LNCS, vol. 12366, pp. 564–580. Springer, Cham (2020). https://doi.org/10.1007/978-3-030-58589-1_34
34. Zhou, X., Wang, D., Krähenbühl, P.: Objects as points. In: arXiv preprint arXiv:1904.07850 (2019)

Author Index

Printed in the United States
by Baker & Taylor Publisher Services